Probability and Mathematical Statistics (Continued)

WILLIAMS • Diffusions, Markov Processes, and Martingales, Volume I: Foundations

ZACKS • Theory of Statistical Inferen

Applied Probability and Statistics

ANDERSON, AUQUIER, HAUC WEISBERG • Statistical Methods

ARTHANARI and DODGE • Matl

BAILEY • The Elements of Stochast Natural Sciences

BAILEY • Mathematics, Statistics and Systems for

BARNETT • Interpreting Multivariate Data

BARNETT and LEWIS • Outliers in Statistical Data

BARTHOLOMEW • Stochastic Models for Social Processes, *Third Edition*

BARTHOLOMEW and FORBES • Statistical Techniques for Manpower Planning

BECK and ARNOLD • Parameter Estimation in Engineering and Science

BELSLEY, KUH, and WELSCH • Regression Diagnostics: Identifying Influential Data and Sources of Collinearity

BENNETT and FRANKLIN • Statistical Analysis in Chemistry and the Chemical Industry

BHAT • Elements of Applied Stochastic Processes

BLOOMFIELD • Fourier Analysis of Time Series: An Introduction

BOX • R. A. Fisher, The Life of a Scientist

BOX and DRAPER • Evolutionary Operation: A Statistical Method for Process Improvement

BOX, HUNTER, and HUNTER • Statistics for Experimenters: An Introduction to Design, Data Analysis, and Model Building

BROWN and HOLLANDER • Statistics: A Biomedical Introduction

BROWNLEE • Statistical Theory and Methodology in Science and Engineering, *Second Edition*

BURY • Statistical Models in Applied Science

CHAMBERS • Computational Methods for Data Analysis

CHATTERJEE and PRICE • Regression Analysis by Example

CHERNOFF and MOSES • Elementary Decision Theory

CHOW • Analysis and Control of Dynamic Economic Systems

CHOW • Econometric Analysis by Control Methods

CLELLAND, BROWN, and deCANI • Basic Statistics with Business Applications, *Second Edition*

COCHRAN • Sampling Techniques, *Third Edition*

COCHRAN and COX • Experimental Designs, *Second Edition*

CONOVER • Practical Nonparametric Statistics, *Second Edition*

CORNELL • Experiments with Mixtures: Designs, Models and The Analysis of Mixture Data

COX • Planning of Experiments

DANIEL • Biostatistics: A Foundation for Analysis in the Health Sciences, *Second Edition*

DANIEL • Applications of Statistics to Industrial Experimentation

DANIEL and WOOD • Fitting Equations to Data: Computer Analysis of Multifactor Data, *Second Edition*

DAVID • Order Statistics, *Second Edition*

DEMING • Sample Design in Business Research

DODGE and ROMIG • Sampling Inspection Tables, *Second Edition*

DRAPER and SMITH • Applied Regression Analysis, *Second Edition*

DUNN • Basic Statistics: A Primer for the Biomedical Sciences, *Second Edition*

DUNN and CLARK • Applied Statistics: Analysis of Variance and Regression

ELANDT-JOHNSON • Probability Models and Statistical Methods in Genetics

ELANDT-JOHNSON and JOHNSON • Survival Models and Data Analysis

continued on back

SEQUENTIAL NONPARAMETRICS

Sequential Nonparametrics:

INVARIANCE PRINCIPLES AND STATISTICAL INFERENCE

Pranab Kumar Sen
University of North Carolina
Chapel Hill

JOHN WILEY & SONS
New York Chichester Brisbane Toronto Singapore

Library of Congress Cataloging in Publication Data:

Sen, Pranab Kumar, 1937–

Sequential nonparametrics.

(Wiley series in probability and mathematical statistics)

"A Wiley-Interscience publication."

Bibliography: p.

Includes index.

1. Sequential analysis. 2. Nonparametric Statistics. I. Title. II. Series.

QA279.7.S46 519.4 81-4432

ISBN 0-471-06013-5 AACR2

Printed in the United States of America

10 9 8 7 6 5 4 3 2 1

In memory of my father,
The late Nagendra Bhusan Sen

The music has entered the instrument,
and of that mode I have learnt a song.
And that music is always playing before me,
and concentration is the great teacher thereof.

Sadhaka Ram Prasad
Translation by Sister Nivedita

Preface

A vast amount of research on statistical procedures based on ranks and robust statistics has been carried out during the past few decades. In particular, the past decade has witnessed a major development on sequential procedures based on nonparametric statistics. This book deals with some basic aspects of *sequential nonparametrics* in a systematic and logically integrated form. For a variety of *nonparametric statistics*, some basic *invariance principles* are studied in detail and these are then incorporated in the foundation of the theory of *nonparametric sequential inference*. Special emphasis has been placed on the development of the *asymptotic theory*, which plays a vital and unifying role in this context and has taken place in the subject only in recent years.

This book is intended for research workers, teachers, and advanced graduate students with a background knowledge of *parametric sequential analysis*, (*nonsequential*) *nonparametric methods*, and *probability theory* at a graduate level. The classical books *Sequential Analysis* by Wald (1947) and *Theory of Rank Tests* by Hájek and Šidák (1967) should serve as excellent prerequisites for this one. Apart from these, some familiarity with the basic concepts of convergence of probability measures on function spaces (Billingsley, 1968) and embedding of Wiener processes should be useful too.

Composed of two parts (invariance principles and statistical inference) and an appendix, the book is divided into 12 chapters. The first part, consisting of Chapters 2 through 8, is devoted to the study of the basic invariance principles for a broad class of nonparametric statistics. The second part, comprising Chapters 9 through 11, deals with the theory of nonparametric statistical inference in the sequential case. Major emphasis has been laid on the development of various types of robust sequential tests and estimation procedures. Developments on sequential rank tests for location or regression encompass the central theme of Chapter 9, and the main objectives of Chapter 10 are the developments on asymptotically risk-efficient sequential point estimators and bounded-length sequential confidence intervals based on various nonparametric statistics. Some basic aspects of nonparametric life testing are studied in Chapter 11. The appendix deals with some additional results of rather technical nature whose derivations are presented there in a succinct manner. An extensive bibliography is provided at the end, and suitable exercises are appended to the various chapters.

Substantial parts of this book have been presented in lectures at the University of North Carolina, Chapel Hill; Albert-Ludwig Universität (Freiburg im Breisgau, West Germany); Calcutta University; Indian Statistical Institute, Calcutta; and University of Maryland, College Park; as well as in seminars in other places. I have benefited greatly from the comments and criticisms received from various students and colleagues and am especially grateful to Professor Hermann Witting and Dr. U. Müller-Funk (both of the Albert-Ludwig Universität) for their most critical reading of the manuscript and invaluable comments and suggestions on it. Without their unselfish help, the task of completing this work could have been much more difficult.

A substantial part of the book has cropped out of the research made by the author in collaboration with Professors B. B. Bhattacharyya (Raleigh, North Carolina), S. K. Chatterjee (Calcutta), J. C. Gardiner (East Lansing, Michigan), Malay Ghosh (Ames, Iowa), Jana Jurečková (Prague), Rupert Miller (Stanford University), H. K. Nandi (Calcutta), and Drs. H. Majumdar (New Delhi) and A. N. Sinha (Princeton). It has been a valuable collaborative experience for me, and I am grateful to all of them for their kind support. It is indeed under the inspiring guidance of Professor H. K. Nandi that I first appreciated the basic principles of statistical inference, and I am deeply indebted to Professor Nandi for all the inspiration and affection bestowed during the past 25 years. Much of the original work reported in this book evolved in meeting the requirements of (i) the Air Force Office of Scientific Research, A.F.S.C., U.S.A.F., Contract No. AFOSR-74-2736 and (ii) National Heart, Lung and Blood Institute, Contract No. NIH-NHLBI-71-2243-L from the National Institute of Health. I am grateful to both the sponsors for their support. The University of North Carolina, Chapel Hill, has been an inspirational place to undertake this task, and, in this respect, the genuine support from my colleagues there is thankfully acknowledged. Special thanks are due to Dean B. G. Greenberg and Professor J. E. Grizzle for their consistent encouragement, not only on this project but also on all other matters of professional interest. It has been a real pleasure to work with Ms. Beatrice Shube, Editor at John Wiley, in planning and completing this project, as well as others.

Finally, I express my deep sense of gratitude to my mother, Kalyani Sen, and wife, Gauri, for their constant encouragement during the writing of this book. Gauri and the children, Devadutta and Aniruddha, have been very helpful and patient during the preparation of this monograph.

PRANAB KUMAR SEN

Chapel Hill, North Carolina
April 1981

Contents

1 Introduction and General Objectives **1**

1.1 Introduction, 1

1.2 A Brief Review of the Material Covered in the Monograph, 4

PART 1 INVARIANCE PRINCIPLES

2 Some Basic Inequalities and Limit Theorems for Certain Stochastic Processes **11**

2.1 Introduction, 11

2.2 Some Basic Results in Martingale Theory, 12

2.3 Weak Convergence of Probability Measures on $C[0, 1]^p$ and $D[0, 1]^p$, 16

2.4 Weak Invariance Principles for Some Dependent Random Variables, 27

2.5 An Almost Sure Invariance Principle for Martingales, 33

2.6 Invariance Principles for Some Empirical Processes, 36

2.7 Certain Boundary-Crossing Probabilities for Wiener Processes, 41

3 Invariance Principles for *U*-Statistics and von Mises' Differentiable Statistical Functions **48**

3.1 Introduction, 48

3.2 *U*-Statistics and von Mises Functionals, 49

3.3 Weak Convergence of $\{U_n\}$ and $\{\theta(F_n)\}$, 56

3.4 Embedding of Wiener Processes for $\{U_n\}$ and $\{\theta(F_n)\}$, 64
3.5 Brownian Bridge Approximations for Finite Population Sampling, 66
3.6 Weak Convergence of Generalized U-Statistics and von Mises Functionals, 71
3.7 Strongly Consistent Estimators of the Dispersion of U-Statistics and von Mises Functionals, 80
3.8 Exercises, 83

4 Invariance Principles for Linear Rank Statistics **88**

4.1 Introduction, 88
4.2 Preliminary Notions, 88
4.3 Weak Convergence of $\{L_N\}$, 92
4.4 Skorokhod-Strassen Representation for $\{L_N\}$, 109
4.5 Asymptotic Linearity of L_N in Regression Parameter, 114
4.6 Almost Sure Convergence of $\{L_N\}$, 119
4.7 Exercises, 123

5 Invariance Principles for Signed Rank Statistics **126**

5.1 Introduction, 126
5.2 Preliminary Notions, 126
5.3 Weak Convergence of $\{S_N\}$, 129
5.4 Embedding of Wiener Processes, 132
5.5 Asymptotic Linearity of S_N in Shift Parameter, 141
5.6 Weak Convergence of Aligned Signed Rank Statistics, 145
5.7 Almost Sure Convergence of $N^{-1}S_N$, 146
5.8 Exercises, 147

6 Invariance Principles for Rank Statistics for Testing Bivariate Independence **150**

6.1 Introduction, 150
6.2 Preliminary Notions, 150
6.3 Weak Convergence of $\{M_N\}$ and $\{Q_N\}$, 155
6.4 Embedding of Wiener Processes for $\{M_N\}$ and $\{Q_N\}$, 159

6.5 A.S. Representation of M_N and Q_N in Terms of I.I.D.R.V., 163

6.6 Almost Sure Convergence of M_N and Q_N, 166

6.7 Exercises, 169

7 Invariance Principles for Linear Combinations of Functions of Order Statistics **170**

7.1 Introduction, 170

7.2 Preliminary Notions, 170

7.3 Bahadur Representation of Sample Quantiles and Invariance Principles for Type I LCFOS, 172

7.4 Weak Convergence of Type II LCFOS, 176

7.5 Embedding of Wiener Processes for Type II LCFOS, 187

7.6 A.S. Convergence of LCFOS and some Related Functionals, 191

7.7 Exercises, 198

8 Invariance Principles for *M*-Estimators and Extrema of Certain Sample Functions **203**

8.1 Introduction, 203

8.2 *M*-Estimators of Location and Regression, 204

8.3 Weak Invariance Principles for *M*-Estimators, 207

8.4 A.S. Representation of *M*-Estimators, 211

8.5 Asymptotic Equivalence of *M*-, *R*-, and *L*-Estimators, 214

8.6 Extrema of Certain Sample Functions, 217

8.7 Invariance Principles for Z_N, 221

8.8 Some General Remarks on Some Other Estimators, 224

8.9 Exercises, 227

Part 2 STATISTICAL INFERENCE

9 Asymptotic Theory of Nonparametric Sequential Tests **233**

9.1 Introduction, 233

9.2 Some Nonparametric Sequential Tests With Power One (1), 233

9.3 Nonparametric Repeated Significance Tests, 243
9.4 Sequential Rank Order and Related Tests, 252
9.5 OC Function of SROT and Related Tests, 257
9.6 ASN of SROT and Related Tests, 266
9.7 Asymptotic Efficiency of SROT and Related Tests, 272
9.8 Exercises, 276

10 Asymptotic Theory of Nonparametric Sequential Estimation **280**

10.1 Introduction, 280
10.2 Type A Sequential Confidence Regions, 281
10.3 Type B Sequential Confidence Regions, 295
10.4 Confidence Sequences for Parameters, 306
10.5 Sequential Point Estimation Based on U-Statistics, 309
10.6 Sequential Point Estimation of Location Based on Some R-, L-, and M-Estimators, 315
10.7 Some General Comments, 327
10.8 Exercises, 329

11 Nonparametric Time-Sequential Procedures **336**

11.1 Introduction, 336
11.2 Preliminary Notions, 337
11.3 Nonparametric Testing Under Progressive Censoring, 342
11.4 Asymptotic Distributions of $D_{n,r}^+$ and $D_{n,r}$, 343
11.5 Asymptotic Power of the Tests Based on $D_{n,r}^+$ and $D_{n,r}$, 348
11.6 Some Additional Problems and Related Tests, 356
11.7 Some Quasi-Nonparametric PCS Tests, 369
11.8 Exercises, 376

Appendix **385**

A.1 Introduction, 385
A.2 Proof of Theorem 2.5.1, 385
A.3 Proofs of Theorems 2.4.1, 2.4.2, 2.4.7, and Corollary 2.4.4.1, 389

A.4 Some A.S. Linearity Theorems on Nonparametric Statistics, 392

A.5 Some Selected Tables, 396

Bibliography **398**

Author Index **415**

Subject Index **419**

SEQUENTIAL NONPARAMETRICS

CHAPTER 1

Introduction and General Objectives

1.1 INTRODUCTION

During the past four decades, a steadily increasing amount of basic research has been carried out in the various aspects of *nonparametric statistical inference*. In addition, during the past 15 years, a variety of textbooks, ranging from the elementary to the advanced research level, have appeared in print. Therefore we need not stress the basic scope and advantages of nonparametric statistical procedures relative to their parametric counterparts.

By now statistical procedures based on ranks and robust statistics are quite popular in the area of classical (univariate as well as multivariate) *testing of hypotheses and estimation problems*. In some of these problems the exact distribution theory of the test statistics or the estimators is known, but usually only for small sample sizes. In the majority of the cases, for moderate or large sample sizes, the distribution theory of nonparametric statistics becomes so involved that one has to rely heavily on their asymptotic forms.

Asymptotic distribution theory of (parametric as well as) nonparametric statistics has been quite an active area of research interest during the past three decades. Some detailed accounts of the developments in this area (prior to the late 1960s) are available in Hájek and Šidák (1967), Puri and Sen (1971), and other contemporary books. In particular, the results of Hoeffding (1948a) on the asymptotic normality of the so-called U-statistics and of Chernoff and Savage (1958) on the asymptotic normality of two-sample rank order statistics deserve special mention. Hájek (1962) incorporated the idea of contiguity of probability measures in nonparametric statistics and opened a new avenue of stimulating research on the asymptotic normality of a general class of rank statistics; his earlier result (Hájek, 1961) on the permutational central limit theorem also deserves a special mention in this context. In 1968 Hájek established the most general result on the asymptotic normality of the so-called linear rank statistics. In all these works the contributions of the Czechoslovakian School under the pioneering guidance of the late Professor Jaroslav Hájek are most noteworthy. The Pyke and Shorack (1968a, b) approach based on the weak convergence of empirical processes has opened a novel direction of basic research in asymptotic nonparametrics. Shorack (1969, 1972a, b) has further advanced this method in the related area of asymptotic theory of order statistics. There are other developments that will be referred to later on. Basically these results are rich enough to provide a solid foundation of the asymptotic theory of nonparametric statistical inference in univariate as well as multivariate problems.

Invariance principles for sums of independent random variables provide useful tools for the asymptotic theory of many statistical procedures. For example, for the classical *sequential probability ratio tests* (SPRT), developed by Wald (1947), the Wiener process approximation for the likelihood ratio process greatly simplifies the expressions for the *operating characteristic* (OC) and *average sample number* (ASN) functions in many asymptotic situations. Weak convergence of partial cumulative sums of independent random variables and of the empirical distribution processes to some appropriate Gaussian functions provide most useful tools for studying the limiting distribution theory of various statistics. For an excellent review of some of these weak convergence results, we refer the reader to Billingsley (1968). In this area too an increase in the literature has taken place during the past 15 years. The Skorokhod-embedding of Wiener processes provides invariance principles stronger than the ones implied by the usual weak convergence results. Based on this embedding technique, Strassen (1964, 1967) has presented an excellent account of the almost sure behavior of sums of independent random variables and martingales (in the form of three basic invariance principles). These results are of fundamental importance in the development of the main theory covered in this monograph.

For a wide class of nonparametric problems, the underlying null hypotheses relate to the invariance of the joint distribution of the sample observations under certain finite groups of transformations that map the sample space onto itself. Motivated by the basic work of Hájek (1961), Sen and Ghosh (1971, 1972, 1973b, 1974a, b) have established the martingale property of a typical form of the usual rank statistics under such hypotheses of invariance; the reverse martingale property of U-statistics was spotted earlier by Berk (1966) (see also Kingman, 1969). As such, the weak as well as strong invariance principles developed for the martingales, reverse martingales, and related sequences of dependent random variables are all applicable (under suitable regularity conditions) for these statistics. For linear combinations of functions of order statistics, similar invariance principles have been studied by a host of workers. The results of Shorack (1969, 1972a, b) and Wellner (1977a, b) deserve special mention in this context. Strong invariance principles for such sequences have been considered by Ghosh (1972), Ghosh and Sen (1976), and Sen (1977c), among others. Sen (1978b) has provided a reverse martingale approximation to a linear combination of functions of order statistics whereby the invariance principles can be developed in a very simple and natural manner. Alternative approaches to the study of the invariance principles for linear rank statistics (without employing the martingale, reverse martingale, or submartingale properties of these statistics) are due to Braun (1976), Lai (1975a, 1978), and others. For robust estimators or the so-called M-estimators (see Huber 1973), some invariance principles have been considered by Jurečková and Sen (1981a, b) and these are incorporated in the study of the asymptotic properties of sequential estimation and testing procedures based on these estimators. In the context of bundle strength of filaments, expressible as the extremum of some sample functions, invariance principles and asymptotic behaviors of some related processes are studied by Sen (1973b, d; 1976d), Sen and Bhattacharyya (1976), Sen, Bhattacharyya, and Suh (1973), and others. In the context of progressively censored statistical inference procedures, a host of invariance principles is available in the literature. Chatterjee and Sen (1973a) considered the case of linear rank statistics (see also Sen, 1976a, b); parallel results for the likelihood ratio statistics and related functions of order statistics are due to Sen (1976f), Gardiner and Sen (1978), and Sen and Tsong (1981). For progressively censored quantile processes, invariance principles have been studied by Sen (1979b, 1981c), and others. In the context of repeated significance tests for some survival models (see Cox, 1972, 1975), similar invariance principles are studied by Sen (1981a). All these results are deeper in nature than the usual asymptotic normality results and are very useful in the context of sequential nonparametric procedures.

The principal objective of the monograph is to focus attention on the various invariance principles for all the statistics mentioned above, to unify these results in a certain fashion and to stress their fruitful applications in a general class of nonparametric inference problems with especial emphasis on the sequential (point as well as interval) estimation and testing of hypothesis problems. A more informative description is given in the next section.

1.2. A BRIEF REVIEW OF THE MATERIAL COVERED IN THE MONOGRAPH

The contents are categorized under two parts and an appendix. Part 1, Chapters 2 through 8, deals with the various invariance principles for U-statistics, generalized U-statistics, von Mises' differentiable statistical functions, linear rank statistics, signed rank statistics, rank statistics (both pure and mixed type) for testing bivariate independence, linear combinations of functions of order statistics and some extrema of sample functions. Part 2, Chapters 9, 10, 11, is devoted to the incorporation of the invariance principles, developed in Part 1, to the study of the (asymptotic) properties of some sequential testing and estimation procedures based on nonparametric statistics. The Appendix deals with certain mathematical results having direct impact on the first two parts but not treated there explicitly.

Chapter 2 is mainly expository in nature. It covers (i) the weak convergence of probability measures on separable metric spaces with especial emphasis on the $C^p[0, 1]$ and $D^p[0, 1]$ spaces (where p is a positive integer), (ii) some functional central limit theorems for martingales, reverse martingales, and related sequences of dependent random variables, developed by Loynes (1970), Brown (1971), and McLeish (1974), among others, (iii) some additional stronger forms of weak convergence results on martingales, reverse martingales, and related sequences, (iv) Strassen-type almost sure invariance principles for sums of independent random variables and martingales, and (v) weak and almost sure invariance principles for empirical distribution processes for the univariate as well as multivariate situations. Mostly, these results are presented without derivations, but, with appropriate references to facilitate a more serious investigation of these topics.

Chapter 3 is devoted to the study of U-statistics (see Hoeffding, 1948a) and Von Mises' (1947) differentiable statistical functions. Donsker-type weak invariance principles for these statistics, studied by Miller and Sen (1972), Bhattacharyya and Sen (1977), and Loynes (1978), among others, are considered. Almost sure invariance principles for these statistics, considered by Sen (1974b) and Lai (1978), are also presented. Extensions of these results to sampling from a finite population (Sen, 1972d) and to generalized U-statistics (Sen, 1974c) are incorporated. Estimation of the variance function of such a statistic and the asymptotic theory of such estimators (Sen, 1960, 1977a) and Sproule (1974) are also considered.

Chapter 4 is concerned with simple linear rank statistics. Under the hypothesis of randomness (i.e., the identity of the distributions of the basic random variables), a martingale theorem on these statistics is established and with the aid of this, weak as well as strong invariance principles are established for linear rank statistics. Almost sure convergence of these statistics to suitable centering constants is also studied. Along with an outline of the notion of *contiguity of probability measures*, the weak invariance principles are extended to local alternative hypotheses situations. Asymptotic linearity of linear rank statistics in regression parameter and related invariance principles are also considered. The results in this chapter are mainly adopted from Sen and Ghosh (1972), Sen (1975, 1978d), Jurečková (1969), and Hájek (1974).

Chapter 5 deals with the invariance principles for signed rank statistics. Under the hypotheses of sign-invariance, arising out of the assumed symmetry of the underlying distribution, a martingale theorem holds for such statistics, and this is utilized in the formulation of weak as well as strong invariance principles for these statistics. The results are also extended to local (contiguous) alternatives. Embedding of Wiener processes for signed rank statistics is considered. Asymptotic linearity of signed rank statistics in the shift parameter and related invariance principles are studied. Almost sure convergence of these statistics to suitable centering constants is considered. The results are mainly due to Sen and Ghosh (1971, 1973a) and Sen (1970a, 1974a, 1977d).

Chapter 6 is devoted to the study of invariance principles for pure rank statistics and mixed-rank statistics (or the linear combinations of induced order statistics) for testing the hypothesis of bivariate independence. Here also, a martingale theorem is developed under the null hypothesis and the same is incorporated in the formulation of the desired invariance principles. The results are extended to local alternatives as well. Almost sure convergence of these statistics is also studied under suitable regularity conditions. The results are mainly due to Sen and Ghosh (1974b) and Sen (1981e).

Chapter 7 treats the invariance principles for linear combinations of functions of order statistics along with some related almost sure convergence results. The *Bahadur (1966) representation of sample quantiles* is incorporated in the study of invariance principles for a type of linear combination of order statistics, while, for a second type, involving smooth weights, a reverse martingale approximation is utilized in the formulation of the desired invariance principles. Embedding of Wiener processes for such statistics are also considered. The results are mainly adapted from Bahadur (1966), Kiefer (1967), Sen (1977c, 1978b), and Wellner (1977a, b).

Chapter 8 is devoted to the study of the invariance principles for some miscellaneous statistics. These include the so-called M-estimators (see Huber, 1973) and some extrema of certain sample functions arising in the theory of the bundle strength of filaments. For certain stochastic processes relating to M-estimators, invariance principles considered by Jurečková and Sen (1981a, b) are incorporated here in the study of the desired results. Also, invariance principles for some other functionals of the empirical processes, studied by Sen (1973b, d; 1976d) and Sen and Bhattacharyya (1976), are incorporated here for the study of the invariance principles for the other type of statistics.

In Part 1 the major emphasis is laid on the basic invariance principles for the class of statistics described in the previous paragraphs. No attempt has been made to provide the so-called Berry–Esseen type of bounds for the errors of normal approximations or the so-called Edgeworth type of expansions for the actual distributions involved. For invariance principles, no attempt has been made to provide suitable rates of convergence. In the classical case, dealing with sums of independent random variables or vectors, a nice account of these developments is given in Bhattacharya and Rao (1976). In the nonparametric case, dealing with U-statistics and some specific form of rank statistics, some developments have taken place during the past few years, while more vigorous developments are under way. As such, it would not be proper to include these materials until they are in mature forms. For the asymptotic theory to be developed in Part 2, the rate of convergence does not appear to have any significant impact; in most of the situations, suitable simulation studies can be made to verify the adequacy of the asymptotic results in any particular setup. Some studies in this direction have already been made and they appear to be quite encouraging.

Chapter 9 in Part 2 deals with the asymptotic theory of sequential tests based on statistics studied in Chapters 3 through 8. Analogous to the Wald (1947) SPRT, a class of sequential tests based on U-statistics, von Mises' differentiable statistical functionals, Huber's M-estimators, linear rank and signed rank statistics, and linear combinations of functions of order statistics is considered and various asymptotic properties of these tests are studied with the aid of the invariance principles developed in Part 1. The Darling-Robbins (1968b) type of sequential tests (with power 1 and arbitrarily small type I error) based on these nonparametric statistics are also considered. Repeated significance tests based on these nonparametric statistics are developed; the invariance principles in Part 1 also play a vital role here. The results in this chapter are mainly adapted from Sen (1973a, b), Sen and Ghosh (1972, 1973a, 1974a, 1980), Ghosh and Sen (1976, 1977), Lai (1975a, 1978), and others.

Chapter 10 is devoted to the study of the asymptotic theory of sequential estimation based on the statistics considered in Part 1. Specifically, the problems of (i) asymptotically risk-efficient point estimation based on a general class of statistics and (ii) bounded-length (sequential) confidence intervals for suitable parametric functions based on such statistics are considered here. In both the problems, the invariance principles of Part 1 provide the key tool for the solutions. The results of this chapter are mainly adapted from Sen and Ghosh (1971, 1981), Ghosh and Sen (1972), Sen (1980e), and Jurečková and Sen (1981a, b, c).

Chapter 11 deals with nonparametric life testing under progressively censoring schemes. Based on the results of Chatterjee and Sen (1973a) and Sen (1976a, b), an invariance principle for progressively censored linear rank statistics is developed and incorporated in the study of the asymptotic properites of these time-sequential tests. Parallel results on the likelihood ratio statistics, adapted from Sen (1976f), Gardiner and Sen (1978), and Sen and Tsong (1981), are included for the motivation of the nonparametric tests. Invariance principles for progressively censored quantile processes, developed in Sen (1979b, 1981c) are also incorporated in the study of the asymptotic properties of the time-sequential tests based on them. Additional results from Hájek (1963), Sen (1981a), Majumdar and Sen (1977, 1978a, b), Sinha and Sen (1979a, b) and Cox (1972, 1975) are included in this study.

The appendix is devoted to the study of some miscellaneous results not explicitly derived in the earlier chapters but having direct relevance to the main theory. In particular, some of the proofs of the main theorems in Chapters 2, 4, and 5 are postponed to the appendix.

A bibliography is provided at the end for the convenience of readers interested in pursuing the work in this general area.

PART 1

Invariance Principles

CHAPTER 2

Some Basic Inequalities and Limit Theorems for Certain Stochastic Processes

2.1. INTRODUCTION

Various basic probability (distributional) inequalities, laws of large numbers, and (functional) central limit theorems play a fundamental role in the asymptotic theory of statistical inference. In this context invariance principles are also very useful. Though, traditionally, all these developments relate to sums (or averages) of independent random variables (r.v.), there has been a pressing need to extend their scope to sequences of dependent r.v.'s and this constitutes an area of active research interest. For the statistical inference problems considered in this monograph, typically, a realization of a *stochastic process* (i.e., a collection of r.v.'s $\{X(t), t \in T\}$ defined over some probability space $(\Omega, \mathfrak{B}, P)$ where T is an *index set* and $X(t)$ is $\mathfrak{B}$-measurable for each $t \in T$) is given (or, more generally, a certain family $\{X_\gamma(t), t \in T\}_{\gamma \in \Gamma}$ of stochastic processes is given) and one needs to construct suitable functionals of these random functions and to study various (asymptotic) properties of such *statistics* for drawing statistical inference. Indeed, in subsequent chapters, while dealing with various nonparametric statistics, we shall encounter various stochastic processes which are related to *martingales*, *reverse martingales*, and some related sequences of dependent r.v.'s and for the study of the (asymptotic) properties of these stochastic processes, we need some basic inequalities and convergence results on martingales, reverse-martingales, and submartingales. These are presented in Section 2.2.

During the past three decades, convergence of sums of independent r.v.'s to *Wiener processes* has been studied under increasing generality by a host of research workers. In the context of *weak convergence of probability measures on separable metric spaces* (and, especially, on $C[0, 1]$ and $D[0, 1]$ spaces), excellent accounts of these developments are given by Billingsley (1968) and Parthasarathy (1967), among others. The past 10 years have witnessed notable generalizations of these works in two principal directions: (i) weak convergence of probability measures on $C[0, 1]^p$ and $D[0, 1]^p$ spaces, for some $p \geqslant 1$ (see Neuhaus, 1971; Bickel and Wichura, 1971; and Straf, 1972) and (ii) weak convergence of martingales, reverse martingales and related sequences of dependent r.v.'s (see Loynes, 1970; Brown, 1971; Drogin, 1972; Scott, 1973; McLeish, 1974; Hall, 1977; and others). In Sections 2.3 and 2.4, we present a review of these works.

By the elegant use of the *Skorokhod-embedding of Wiener processes*, Strassen (1964, 1967) has nicely presented some almost sure (a.s.) invariance principles for sums of independent r.v.'s and martingales. This fundamental work has a great impact on our subsequent chapters dealing with various nonparametric statistics. We present an informative review of some of his works in Section 2.5.

Section 2.6 deals with the classical *empirical processes* (constructed from the *sample (empirical) distribution functions*, d.f.) in the general multivariate case. Invariance principles relating to such processes are considered. In the context of the results in Sections 2.5 and 2.6, the recent contributions made by the Hungarian school deserve special mention.

The concluding section deals with certain standard results on the boundary crossing probabilities for the Wiener and the tied-down Wiener processes. These results are used in Chapters 9, 10, and 11.

Most of the results in this chapter are presented without proofs; appropriate references for providing the sources of the proofs, and for deeper studies of these results are cited. A few of the basic theorems are proved in the Appendix.

2.2. SOME BASIC RESULTS IN MARTINGALE THEORY

Let $(\Omega, \mathfrak{B}, P)$ be a probability space, Γ be a linearly ordered index set and $\{\mathfrak{F}_\gamma, \gamma \in \Gamma\}$ be a system of subsigma fields of $\mathfrak{B}$. $\{\mathfrak{F}_\gamma, \gamma \in \Gamma\}$ is said to be monotone nondecreasing (or nonincreasing) if for $\gamma, \gamma' \in \Gamma$, and $\gamma < \gamma'$, $\mathfrak{F}_\gamma \subset \mathfrak{F}_{\gamma'}$ (or $\mathfrak{F}_{\gamma'} \subset \mathfrak{F}_\gamma$). Let $\{X_\gamma, \gamma \in \Gamma\}$ be a stochastic process, defined over $(\Omega, \mathfrak{B}, P)$, adapted to $\{\mathfrak{F}_\gamma, \gamma \in \Gamma\}$ in the sense that X_γ is $\mathfrak{F}_\gamma$-measurable for each $\gamma \in \Gamma$.

Definition 2.2.1. Given a probability space $(\Omega, \mathfrak{B}, P)$, a monotone increasing (or decreasing) system of subsigma fields $\{\mathfrak{F}_\gamma, \gamma \in \Gamma\}$ and a stochastic process $\{X_\gamma, \gamma \in \Gamma\}$ adapted to $\{\mathfrak{F}_\gamma, \gamma \in \Gamma\}$, if

$$E|X_\gamma| < \infty \qquad \forall \gamma \in \Gamma, \tag{2.2.1}$$

and if for every $\gamma, \gamma' \in \Gamma$ with $\gamma < \gamma'$ (or $\gamma' < \gamma$),

$$E(X_{\gamma'} \mid \mathcal{F}_\gamma) = X_\gamma \qquad \text{almost everywhere (a.e.)}, \tag{2.2.2}$$

then $\{X_\gamma, \gamma \in \Gamma\}$ is said to be a *martingale* (or *reverse martingale*). If the equality sign in (2.2.2) is replaced by $\geqslant$ (or $\leqslant$), then $\{X_\gamma, \gamma \in \Gamma\}$ is termed a *submartingale* (or *supermartingale*) when $\{\mathcal{F}_\gamma, \gamma \in \Gamma\}$ is nondecreasing and a *reverse submartingale* (or *supermartingale*) when $\{\mathcal{F}_\gamma, \gamma \in \Gamma\}$ is nonincreasing.

These sequences (of dependent r.v.'s) play a vital role in the asymptotic theory of statistical inference. There are excellent textbooks on probability theory and stochastic processes (see Loéve, 1963; Doob, 1967; Meyer, 1966; Ash, 1972; Tucker, 1967; Stout, 1974; and others) which have dealt in detail with the various properties of such sequences. For our purpose, we need a few basic inequalities and convergence theorems, presented (without proofs) below. In this context, Γ is often taken to be a discrete index set, which can equivalently be taken as the set of positive integers (or any subset of the same).

(i) If $\{X_\gamma, \mathcal{F}_\gamma; \gamma \in \Gamma\}$ is a (forward or reverse) martingale and g is an integrable, convex function (e.g., $g(x) = |x|^r$ for some $r \geqslant 1$), then $\{g(X_\gamma), \mathcal{F}_\gamma; \gamma \in \Gamma\}$ is a (forward or reverse) submartingale.

(ii) ***Kolmogorov-Hájek-Rényi-Chow Inequality*** (see Chow, 1960). Let $\{X_n, \mathcal{F}_n; n \geqslant 1\}$ be a submartingale and $\{c_n; n \geqslant 1\}$ be a nonincreasing sequence of nonnegative numbers. Then, for every $t > 0$,

$$P\Big\{\max_{1 \leqslant k \leqslant n} c_k X_k \geqslant t\Big\} \leqslant t^{-1}\Big\{c_n E X_n^+ + \sum_{i=1}^{n-1}(c_i - c_{i+1}) E X_i^+\Big\}, \tag{2.2.3}$$

where $a^+ = (a \vee 0) = \max\{a, 0\}$. In particular, (2.2.3) yields the Kolmogorov inequality

$$P\Big\{\max_{1 \leqslant k \leqslant n} X_k \geqslant t\Big\} \leqslant t^{-1} E X_n^+. \tag{2.2.4}$$

Also, if $\{X_n, \mathcal{F}_n; n \geqslant 1\}$ is a reverse submartingale and $\{c_n; n \geqslant 1\}$ is a nondecreasing sequence of nonnegative numbers, then for every $N \geqslant n \geqslant 1$ and $t > 0$,

$$P\Big\{\max_{n \leqslant k \leqslant N} c_k X_k \geqslant t\Big\} \leqslant t^{-1}\Big\{c_n E X_n^+ + \sum_{i=n+1}^{N}(c_i - c_{i-1}) E X_i^+\Big\} \tag{2.2.5}$$

and, in particular,

$$P\Big\{\sup_{k \geqslant n} X_k > t\Big\} \leqslant t^{-1} E X_n^+. \tag{2.2.6}$$

(iii) ***The Doob Upcrossing Inequality*** (see Doob, 1967, p. 316). For a sequence $\{x_1, \ldots, x_n\}$ of real numbers and for $-\infty < a < b < \infty$, let

$N_1 = \min\{i : x_i \leq a\}$, $N_2 = \min\{i : i > N_1,\ x_i \geq b\}$ and for $r \geq 1$, $N_{2r+1} = \min\{i : i > N_{2r},\ x_i \leq a\}$ and $N_{2r+2} = \min\{i : i > N_{2r+1},\ x_i \geq b\}$. Then, $U_{ab}^{(n)} = \max\{r : N_{2r} \leq n\}$ is termed the *number of upcrossing* of $[a,b]$ by $\{x_1, \ldots, x_n\}$. Let $\{X_n, \mathfrak{F}_n; n \geq 1\}$ be a submartingale sequence and let $U_{ab}^{(n)}$ be the number of upcrossings of $[a,b]$ by $\{X_1, \ldots, X_n\}$. Then

$$EU_{ab}^{(n)} \leq (b-a)^{-1}E(X_n - a)^+ \leq (b-a)^{-1}\{E|X_n| + |a|\}. \quad (2.2.7)$$

(iv) ***The Doob Moment Inequality*** (see Doob, 1967, p. 318). Let $\{X_n, \mathfrak{F}_n; n \geq 1\}$ be a nonnegative submartingale sequence. Then

$$E\left\{\max_{1 \leq k \leq n} X_k^\alpha\right\} \leq \begin{cases} [e/(e-1)][1 + E\{X_n \log^+ X_n\}], & \alpha = 1 \\ [\alpha/(\alpha-1)]^\alpha E(X_n^\alpha), & \alpha > 1, \end{cases} \quad (2.2.8)$$

where $\log^+ a = ((\log a) \vee 1)$.

(v) ***Brown's Inequality*** (see Brown, 1971, Lemma 4). Let $\{X_n, \mathfrak{F}_n; n \geq 1\}$ be a submartingale. Then, for every $t > 0$,

$$P\left\{\max_{1 \leq k \leq n} |X_k| \geq 2t\right\} \leq P\{|X_n| > t\} + E\left\{(t^{-1}|X_n| - 2)I(|X_n| \geq 2t)\right\}$$
$$\leq t^{-1}E\{|X_n| I(|X_n| \geq t)\}, \quad (2.2.9)$$

where $I(A)$ stands for the indicator function of a set A. Also, by the Hölder inequality, the right-hand side (rhs) of (2.2.9) is bounded from above by

$$t^{-1}[E(|X_n|^r)]^{1/r}[P\{|X_n| \geq t\}]^{1/s}, \qquad (r^{-1} + s^{-1} = 1) \quad (2.2.10)$$

where r and s are positive numbers and it is assumed that $E|X_n|^r < \infty$.

(vi) ***The Submartingale Convergence Theorem*** (see Doob, 1967). Let $\{X_n, \mathfrak{F}_n; n \geq 1\}$ be a submartingale and suppose that $\sup_n \mathrm{E}|X_n| < \infty$. Then there exists a r.v. X such that

$$X_n \to X \quad \text{a.s., as } n \to \infty \quad \text{and} \quad E|X| < \sup_n E|X_n|. \quad (2.2.11)$$

Also, let $\{X_n, \mathfrak{F}_n; n \geq 1\}$ be a submartingale or a reverse submartingale and assume that the X_n are uniformly integrable [i.e., $E(|X_n| I(|X_n| > c)) \to 0$ as $c \to \infty$, uniformly in n]. Then there exists a r.v. X such that $X_n \to X$ a.s., and in the first mean, as $n \to \infty$. In particular, if $\{X_n, \mathfrak{F}_n; n \geq 1\}$ is a reverse martingale, then the X_n are uniformly integrable, and hence $X_n \to X$ a.s., and in the first mean, as $n \to \infty$.

(vii) ***Moments of Randomly Stopped Martingales*** (see Chow, Robbins, and Teicher, 1965). Let $\{X_n, \mathfrak{F}_n; n \geq 1\}$ be a martingale and let $X_{n+1} - X_n = Y_{n+1}, n \geq 0$ (where $X_0 = 0$ with probability 1) be the associated martingale difference sequence. A *stopping variable* N is a positive integer valued r.v. (defined on the same probability space as of the X_n) such that the event $[N = n] \in \mathfrak{F}_n$ for every $n \geq 1$. Then we have the following:

(a) If $E|X_N| < \infty$ and $\liminf \int_{[N>n]} |X_n| dP = 0$, then

$$E(X_N \mid \mathcal{F}_n) = X_n \text{ if } N > n, \tag{2.2.12}$$

and hence,

$$EX_N = EX_1. \tag{2.2.13}$$

(b) If $EX_n^2 < \infty$, $\forall n \geqslant 1$ and N be defined as before, then

$$EX_N^2 \leqslant E\left\{ \sum_{i=1}^{N} Y_i^2 \right\}. \tag{2.2.14}$$

If any one of the following conditions

$$\liminf \int_{[N>n]} |X_n| dP = 0, \quad \liminf \int_{[N>n]} X_n^2 \, dP = 0,$$
$$E \sum_{i=1}^{N} |Y_i| < \infty, \qquad E \sum_{i=1}^{N} Y_i^2 < \infty \tag{2.2.15}$$

hold, then

$$EX_N^2 = E \sum_{i=1}^{N} Y_i^2. \tag{2.2.16}$$

If either $E\sum_{i=1}^{N} |Y_i| < \infty$ or $E\sum_{i=1}^{N} Y_i^2 < \infty$, then (2.2.12) holds.

In passing, we remark that if the Y_i, defined before (2.2.12), are all independent, then the X_n are sums of independent r.v.'s and (2.2.2) holds if $EY_n = 0$ for every $n \geqslant 1$. Hence, for centered sums of independent r.v.'s, the results presented in this section continue to hold. We conclude this section with some additional probability inequalities due to Birnbaum and Marshall (1961).

(viii) ***Birnbaum-Marshall Inequality, I.*** Let $X_1, \ldots, X_n$ be r.v.'s such that

$$E(|X_k| \,\big|\, X_1, \ldots, X_{k-1}) \geqslant \psi_k |X_{k-1}| \qquad \text{a.e.}, \tag{2.2.17}$$

where $\psi_k \geqslant 0$ for every $k \geqslant 2$. Also, let

$$a_k > 0,\, b_k = \max\left(a_k, a_{k+1}\psi_{k+1}, a_{k+2}\psi_{k+1}\psi_{k+2}, \ldots, a_n \prod_{i=k+1}^{n} \psi_i \right), k \geqslant 1, \tag{2.2.18}$$

$b_{n+1} = 0$ and let $X_0 = 0$. If $r \geqslant 1$ is such that $E|X_k|^r < \infty$, for every $k \geqslant 1$, then

$$P\left\{ \max_{1 \leqslant k \leqslant n} a_k |X_k| \geqslant 1 \right\} \leqslant \sum_{k=1}^{n} (b_k^r - \psi_{k+1}^r b_{k+1}^r) E|X_k|^r$$
$$= \sum_{k=1}^{n} b_k^r (E|X_k|^r - \psi_k^r E|X_{k-1}|^r). \tag{2.2.19}$$

It may be remarked that the Kolmogorov-Hájek-Rényi-Chow inequality in (2.2.3) and (2.2.4) is a special case of (2.2.20) where the ψ_k are all equal to 1. In the next inequality, we consider the case of separable stochastic processes, where *separable* means separable relative to the class of all closed subsets of the extended real line, although a weaker separability would suffice. Let $\{X_t, t \geqslant 0\}$ be a separable process, let f be a positive function on $[0,\infty)$ having at most countably many discontinuities and let $\tau > 0$. If S is a countable set dense in $[0,\infty)$ satisfying the definition of separability and containing the set of discontinuities of f as well as 0 and τ, then $\{\omega : \sup_{0 \leqslant t \leqslant \tau}[|X_t(\omega)|/f(t)] < 1\}$ is measureable and

$$P\left\{\sup_{0 \leqslant t \leqslant \tau} \frac{|X_t|}{f(t)} < 1\right\}$$
$$= \lim_{k\to\infty} P\left\{|X_t| \leqslant \frac{(k-1)f(t)}{k} \quad \text{for all} \quad t \in S \cap [0,\tau]\right\}. \quad (2.2.20)$$

With this, we may state the following.

(ix) ***Birnbaum-Marshall Inequality, II.*** If $\{X_t, t \geqslant 0\}$ is a separable submartingale such that $E|X_t| = \mu(t) < \infty$ for all $t \leqslant \tau$ and if f is a nondecreasing positive function on $[0,\tau]$ such that the Riemann-Stieltjes integral in the following bound exists, then

$$P\left\{\sup_{0 \leqslant t \leqslant \tau} \frac{X_t}{f(t)} \geqslant 1\right\} \leqslant \frac{\mu(0)}{f(0)} + \int_0^\tau [f(t)]^{-1} d\mu(t). \quad (2.2.21)$$

The restriction that f be monotone is not necessary; in any case, $g(t) = \inf_{t \leqslant s \leqslant \tau} f(s)$ is monotone and $g(t) \leqslant f(t)$ on $[0,\tau]$, so that f may be replaced by g on the rhs of (2.2.21).

2.3. WEAK CONVERGENCE OF PROBABILITY MEASURES ON $C[0,1]^p$ AND $D[0,1]^p$

Let S be a metric space and $\mathbb{S}$ be the class of Borel sets in S. Let $\{(\Omega_n, \mathfrak{B}_n, \mu_n); n \geqslant 1\}$ be a sequence of probability spaces and let X_n be a measurable map from $(\Omega_n, \mathfrak{B}_n)$ into $(S, \mathbb{S})$, for $n \geqslant 1$. We denote by P_n the probability measure induced in $(S, \mathbb{S})$ by the random function X_n, for $n \geqslant 1$. Also, let $(\Omega, \mathfrak{B}, \mu)$ be a probability space, X be a measurable map from $(\Omega, \mathfrak{B})$ into $(S, \mathbb{S})$ and let P be the probability measure induced in $(S, \mathbb{S})$ by X. Note that Ω is not necessarily in the same space spanned by the Ω_n.

Definition 2.3.1. The sequence $\{P_n, n \geqslant 1\}$ of probability measures on $(S, \mathfrak{S})$ is said to converge weakly to P (denoted by $P_n \Rightarrow P$), if

$$\int g\,dP_n \to \int g\,dP, \qquad \forall g \in C(S), \tag{2.3.1}$$

where $C(S)$ is the class of all bounded, continuous, real functions on S.

This definition implies that for every continuous functional h assuming values in R^k, the $k(\geqslant 1)$-dimensional Euclidean space, as $n \to \infty$,

$$h(X_n) \xrightarrow{\mathfrak{D}} h(X) \qquad \text{when} \quad P_n \Rightarrow P, \tag{2.3.2}$$

where $\xrightarrow{\mathfrak{D}}$ stands for the convergence in distribution. That is,

$$P_n \Rightarrow P \quad \Rightarrow \quad P_n h^{-1} \Rightarrow P h^{-1}, \tag{2.3.3}$$

where $P_n h^{-1}$ and Ph^{-1} are probability measures on $(R^k, \mathfrak{F}^k)$ induced by X_n and X on the Borel sets $\mathfrak{F}^k$ by the functional h. In Definition 2.3.1, one may weaken the continuity assumption on h as follows. Let $D_h (\subset S)$ be the set of discontinuities of h. Then, if h is a $\mathfrak{S}$-measurable functional such that $P\{D_h\} = 0$, that is, D_h is of null P-measure, then (2.3.3) holds. The concepts of *relative compactness* and *tightness* of probability measures are very useful in this context.

Definition 2.3.2. (Prokhorov, 1956). A family Π of probability measures on $(S, \mathfrak{S})$ is relatively compact if every sequence of elements of Π contains a weakly convergent subsequence. Explicitly, for every sequence $\{P_n\}$ in Π, there is a subsequence $\{P_{n_k}; k \geqslant 1\}$ and a probability measure P^* (on $(S, \mathfrak{S})$ but not necessarily in Π) such that $P_{n_k} \Rightarrow P^*$.

Definition 2.3.3. (LeCam 1957). A family Π of probability measures on $(S, \mathfrak{S})$ is said to be tight, if for every $\epsilon > 0$, there exists a compact set K_ϵ (depending on ϵ), such that

$$P(K_\epsilon) > 1 - \epsilon \qquad \text{for every} \quad P \in \Pi. \tag{2.3.4}$$

It follows from the basic theorems of Prokhorov (1956) that if for a metric space S, Π is tight, then it is also relatively compact. Further, if S is a separable and complete metric space, then tightness is necessary and sufficient for the relative compactness. In the next section, we restrict ourselves to certain families of separable and complete metric spaces and study the weak convergence of probability measures in more detail.

For some positive integer p, let $E^p = \{\mathbf{t} : \mathbf{0} \leqslant \mathbf{t} \leqslant \mathbf{1}\}$ be the unit cube in R^p, where $\mathbf{t} = (t_1, \ldots, t_p)$, $\mathbf{0} = (0, \ldots, 0)$, $\mathbf{1} = (1, \ldots, 1)$ and $\mathbf{a} \leqslant \mathbf{b}$ (or $\mathbf{a} < \mathbf{b}$) means that $a_j \leqslant b_j$ (or $a_j < b_j$) for every $j = 1, \ldots, p$. The space

$C[0,1]^p$ is the space of all continuous functions $Z(\mathbf{t}), \mathbf{t} \in E^p$. Thus a function $f: E^p \to R$ belongs to $C[0,1]^p$, *iff* (if and only if)

$$\lim_{\delta \to 0} \omega_f(\delta) = 0 \tag{2.3.5}$$

where the *modulus of continuity* $\omega_f(\delta)$ is defined by

$$\omega_f(\delta) = \sup\{|f(\mathbf{t}) - f(\mathbf{s})| : \mathbf{s}, \mathbf{t} \in E^p, \|\mathbf{t} - \mathbf{s}\| < \delta\}, \tag{2.3.7}$$

and $\|\mathbf{x}\|$ is the *max-norm*, that is, for $\mathbf{x} = (x_1, \ldots, x_p)$,

$$\|\mathbf{x}\| = \max\{|x_j| : 1 \leqslant j \leqslant p\} = \lim_{p \to \infty} \|\mathbf{x}\|_p^{1/p} = \lim_{p \to \infty} \left\{ \sum_{i=1}^{p} |x_i|^p \right\}^{1/p}. \tag{2.3.7}$$

Let $\mathcal{C}^p$ be the sigma-field generated by the open subsets of $C[0,1]^p$. With the space $C[0,1]^p$, we associate the *uniform topology* specified by the metric

$$\rho(x, y) = \sup_{\mathbf{t} \in E^p} |x(\mathbf{t}) - y(\mathbf{t})| \quad \text{where} \quad x, y \in C[0,1]^p. \tag{2.3.8}$$

Then, under the metric ρ, $C[0,1]^p$ is a complete and separable metric space. The space $C[0,1]^p$ is a subspace of $D[0,1]^p$, which we present below. We refer to Neuhaus (1971), Straf (1972) and Bickel and Wichura (1971) for detailed studies of the topology of $D[0,1]^p$ and the weak convergence of probability measures on it. Let

$$\mathcal{P} = \{\boldsymbol{\rho} = (\rho_1, \ldots, \rho_p) : \rho_j = 0, 1, 1 \leqslant j \leqslant p\}. \tag{2.3.9}$$

For every $\mathbf{t}_0 \in E^p$, the set of p hyperplanes $\{\mathbf{t} : t_j = t_{0j}\}$, for $j = 1, \ldots, p$, divides E^p into 2^p left closed quadrants each containing $\mathbf{t}_0$ as one of its vertices. For example, for $p = 2$, we have four quadrants where $\mathbf{t}_0 = (t_{01}, t_{02})$ is a lower corner of $[\mathbf{t}_0, \mathbf{1}]$, an upper corner of $[\mathbf{0}, \mathbf{t}_0)$ and other corners for the two other subsets of E^2. Hence, for every $\mathbf{t} \in E^p$, we have the set of 2^p quadrants

$$\{Q(\rho, \mathbf{t}) : \rho \in \mathcal{P}\}. \tag{2.3.10}$$

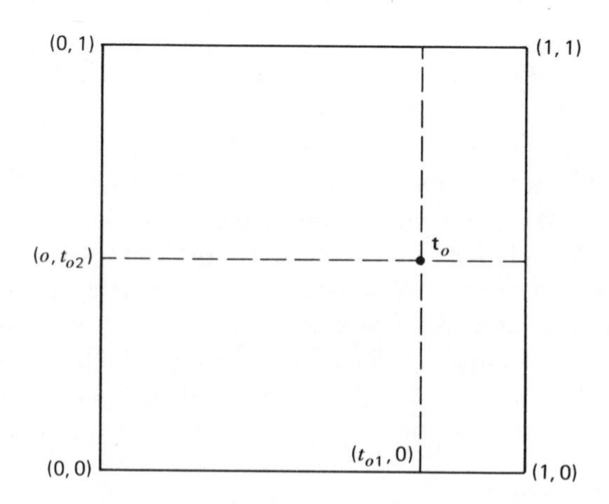

Let f be a function defined on E^p. If, for $\mathbf{t} \in E^p$ and $\rho \in \mathcal{P}$ with $Q(\rho, \mathbf{t}) \neq 0$, and every sequence $\{\mathbf{t}_n\} \in Q(\rho, \mathbf{t})$ with $\mathbf{t}_n \to \mathbf{t}$, the sequence $\{f(\mathbf{t}_n)\}$ converges, then we denote (the necessarily unique) limit by $f(\mathbf{t} + \mathbf{0}_\rho)$ and call it the ρ-limit of f at $\mathbf{t}$ (or the quadrant limit).

Definition 2.3.4. The space $D[0,1]^p$ is the space of all real-valued functions $f: E^p \to R$ for which the ρ-limit of f at $\mathbf{t}$ exists for every $\mathbf{t} \in E^p$ and $\rho \in \mathcal{P}$ with $Q(\rho, \mathbf{t}) \neq 0$ and which are continuous from above in the sense that

$$f(\mathbf{t}) = \lim_{\delta \to 0} f(\mathbf{t} + \delta \mathbf{1}) \qquad \forall \mathbf{t} \in E^p. \tag{2.3.11}$$

For $p = 1$, it means that $f(t)$ is right continuous and has a left-hand limit, that is, (i) $f(t) = \lim_{\delta \to 0} f(t + \delta)$ exists for every $0 \leqslant t < 1$ and (ii) $f(t - 0) = \lim_{\delta \to 0} f(t - \delta)$ exists for every $0 < t \leqslant 1$, though $f(t)$ and $f(t - 0)$ need not be equal everywhere. Thus f has only *discontinuity of the first kind*. Bearing this in mind, we say that $D[0,1]^p$ is the space of all real-valued functions on the unit cube E^p that have only discontinuities of the first kind. [In (2.3.11), the minor adjustments needed when $t_j = 1$ for some $j(= 1, \ldots, p)$ are understood and similar adjustments are necessary with the ρ-limits when $t_j = 0$ or 1 for some $j = 1, \ldots, p$ and $\rho \in \mathcal{P}$.]

Definition 2.3.5. A function $f \in D[0,1]^p$ has a jump at $\mathbf{t} \in E^p$, *iff*

$$H_f(\mathbf{t}) = \max\{|f(\mathbf{t} + \mathbf{0}_\rho) - f(\mathbf{t} + \mathbf{0}_{\rho'})| : \rho, \rho' \in \mathcal{P}\} > 0. \tag{2.3.12}$$

Note that $H_f(\mathbf{t})$ is the maximum possible difference of the 2^p quadrant limits of $f(\mathbf{t})$ at $\mathbf{t}$ and is termed the magnitude of jump at $\mathbf{t}$.

From the above it is clear that $C[0,1]^p$ is a subspace of $D[0,1]^p$ for which $H_f(\mathbf{t}) = 0 \forall \mathbf{t} \in E^p$. We now introduce the Skorokhod J_1-topology on $D[0,1]^p$. Let Λ denote the class of all strictly increasing, continuous mappings $\lambda: [0,1] \to [0,1]$ (i.e., from $[0,1]$ onto itself). For $\boldsymbol{\lambda} = (\lambda_1, \ldots, \lambda_p) \in \boldsymbol{\Lambda}^p = \Lambda \times \cdots \times \Lambda$ and $\mathbf{t} \in E^p$, let us denote by $\boldsymbol{\lambda}(\mathbf{t}) = (\lambda_1(t_1), \ldots, \lambda_p(t_p))$. Then, the *Skorokhod-distance* between x and y (both belonging to the $D[0,1]^p$ space) is defined by

$$d(x, y) = \inf_{\epsilon} \Big\{ \epsilon > 0; \text{ there is a } \boldsymbol{\lambda} \text{ in } \Lambda^p \text{ such that } \sup_{\mathbf{t} \in E^p} \|\boldsymbol{\lambda}(\mathbf{t}) - \mathbf{t}\| < \epsilon \text{ and } \sup_{\mathbf{t} \in E^p} |x(\mathbf{t}) - y(\boldsymbol{\lambda}(\mathbf{t}))| < \epsilon \Big\}, \tag{2.3.13}$$

where $\|\cdot\|$ is the max-norm, defined in (2.3.7). The distance function d is a metric on $D[0,1]^p$, called the *Skorokhod-metric* which generates the Skorokhod-topology on $D[0,1]^p$; we term this topology as J_1-topology. A necessary and sufficient condition for a sequence $\{f_n\} \in D[0,1]^p$ to converge to some $f \in D[0,1]^p$ in the Skorokhod J_1-topology is that there exists a sequence $\{\boldsymbol{\lambda}_n\}$ in Λ^p with

$$\lim_{n\to\infty} f_n(\boldsymbol{\lambda}_n(\mathbf{t})) = f(\mathbf{t}), \qquad \text{uniformly in } \mathbf{t} \in E^p, \tag{2.3.14}$$

$$\lim_{n\to\infty} \boldsymbol{\lambda}_n(\mathbf{t}) = \mathbf{t}, \qquad \text{uniformly in } \mathbf{t} \in E^p. \tag{2.3.15}$$

The metric space $D[0,1]^p$ with the metric d in (2.3.13) is separable. A countable, dense subset $\mathfrak{T} \subset D^p$ can be obtained as follows. Let $\mathfrak{N} = \{1, 2, \ldots\}$. Then, for $n \in \mathfrak{N}$, let $\mathfrak{R}_n$ be the partition generated by the set of points

$$T_n = \left\{\mathbf{t} = (t_1, \ldots, t_p) : t_j \in \left[0, \frac{1}{n}, \ldots, \frac{n}{n}\right], \quad j = 1, \ldots, p\right\}, \tag{2.3.16}$$

and let $\mathfrak{T}_n$ be the countable class of functions

$$\sum_{R \in \mathfrak{R}_n} I_R a_R \tag{2.3.17}$$

where a_R are all rational numbers and I_R denotes the indicator function of the set R. Then, $\mathfrak{T} = \cup_n \mathfrak{T}_n$ is countable, dense in $D[0,1]^p$.

Consider a partition of E^p formed by finitely many hyperplanes parallel to the p coordinate axes, such that each element of this partition is a left closed right open rectangle of diameter at least $\delta(>0)$. A typical rectangle is denoted by R and the partition generated by the hyperplanes is denoted by $\mathfrak{R}$. Then, for each $R \in \mathfrak{R}$, we have the modulus

$$\omega_f^0(R) = \sup\{|f(\mathbf{t}) - f(\mathbf{s})| : \mathbf{s}, \mathbf{t} \in R\}, \tag{2.3.18}$$

so that, for the partition $\mathfrak{R}$, we have a metric

$$\max\{\omega_f^0(R) : R \in \mathfrak{R}\}. \tag{2.3.19}$$

We now take the infimum of (2.3.19) with respect to all $\mathfrak{R}$ satisfying the condition that each member of $\mathfrak{R}$ has a diameter at least $\delta(>0)$ and denote this by

$$\omega_f'(\delta) = \inf_{\mathfrak{R}}\left[\max\{\omega_f^0(R) : R \in \mathfrak{R}\}\right]. \tag{2.3.20}$$

The modulus $\omega_f'(\delta)$ characterizes the space $D[0,1]^p$. In fact, a function $f: E^p \to R$ belongs to $D[0,1]^p$ iff

$$\lim_{\delta\to 0} \omega_f'(\delta) = 0. \tag{2.3.21}$$

Let $f: E^p \to R$ be a $\mathfrak{D}^p$-measurable function (where $\mathfrak{D}^p$ stands for the sigma-field generated by the open subsets of $D[0,1]^p$) and let

$$J_{\mathbf{t}} = \{f \in D[0,1]^p : H_f(\mathbf{t}) > 0\} \qquad \text{and} \qquad T_P = \{\mathbf{t} \in E^p : P(J_t) = 0\} \tag{2.3.22}$$

where P is a probability measure on $(D[0,1]^p, \mathfrak{D}^p)$. That is, T_P is the set of all $\mathbf{t} \in E^p$ for which the projection $\pi_{\mathbf{t}}$ [defined by $\pi_{\mathbf{t}} f = f(\mathbf{t})$] is continuous except on a set of P-measure 0. For a fixed P, there are countably many points $\{\mathbf{t}_n\}$ in E^p such that $P(J_{\mathbf{t}}) > 0$ implies that $\mathbf{t}$ has one component in common with some $\mathbf{t}_n$. All points $\mathbf{t}$ with $P(J_{\mathbf{t}}) > 0$ are concentrated on countably many (proper) hyperplanes. Also, T_p contains $\mathcal{P}$ and is dense in E^p. Thus for points $\mathbf{t}_1, \ldots, \mathbf{t}_k$ in T_P the projection

$$\pi_{\mathbf{t}_1, \ldots, \mathbf{t}_k}(f) = [f(\mathbf{t}_1), \ldots, f(\mathbf{t}_k)] \text{ is a.e. continuous } (P). \tag{2.3.23}$$

Further, the projection maps in (2.3.23) are measurable maps from $(D[0,1]^p, \mathfrak{D}^p)$ into $(R^k, \mathfrak{F}^k)$. For a subset A of E^p, let

$$\mathfrak{B}_A = \{\pi^{-1}_{\mathbf{t}_1, \ldots, \mathbf{t}_k}(B) : \mathbf{t}_i \in A, i = 1, \ldots, k \text{ and } B \in \mathfrak{F}^k, k = 1, \ldots, \}. \tag{2.3.24}$$

Then the $\mathfrak{B}_A$ are called the *finite-dimensional* or *cylinder sets* of $D[0,1]^p$. The collection

$$\{P\pi^{-1}_{\mathbf{t}_1, \ldots, \mathbf{t}_k} : \mathbf{t}_i \in A, \qquad i = 1, \ldots, k; \qquad k = 1, \ldots\} \tag{2.3.25}$$

is called the collection of *finite dimensional distributions* (f.d.d.) of P for points in A. The following weak convergence result has been studied independently by Neuhaus (1971), Bickel and Wichura (1971) and Straf (1972).

THEOREM 2.3.1. *Let $\{P_n\}$ and P be probability measures on $(D[0,1]^p, \mathfrak{D}^p)$, not necessarily defined on the same measure space. If $\{P_n\}$ is relatively compact and if $P_n\pi^{-1}_{\mathbf{t}_1, \ldots, \mathbf{t}_k} \Rightarrow P\pi^{-1}_{\mathbf{t}_1, \ldots, \mathbf{t}_k}$ whenever $\mathbf{t}_1, \ldots, \mathbf{t}_k$ all lie in T_p, then $P_n \Rightarrow P$.*

THEOREM 2.3.2. *A sequence $\{P_n\}$ of probability measures on $(D[0,1]^p, \mathfrak{D}^p)$ is tight (relatively compact) iff*

(a) *for every $\epsilon > 0$, there exists a $M_\epsilon (0 < M_\epsilon < \infty)$ such that*

$$P_n(\{f \in D[0,1]^p : \|f\| > M_\epsilon\}) \leqslant \epsilon \qquad \forall n \geqslant 1 \tag{2.3.26}$$

(where $\|f\| = \sup_{0 \leqslant t \leqslant 1} |f(t)|$) and

(b) *for every $\epsilon > 0$,*

$$\lim_{\delta \downarrow 0} \lim \sup_{n \to \infty} P_n(\{f \in D[0,1]^p : \omega'_f(\delta) \geqslant \epsilon\}) = 0. \tag{2.3.27}$$

Theorems 2.3.1 and 2.3.2 are also contained in Billingsley (1968) for the case of $p = 1$. Tracing his line of argument, it follows, from (2.3.6) and (2.3.8) through (2.3.20), that for any $p(>1)$,

$$\omega_f'(\delta) \leqslant \omega_f(2\delta), \qquad \text{for every} \quad 0 < \delta < \tfrac{1}{2}, \tag{2.3.28}$$

while, for continuous $f: E^p \to R$,

$$\omega_f(\delta) \leqslant 2p\omega_f'(\delta), \qquad \text{for every} \quad 0 < \delta < \tfrac{1}{2}. \tag{2.3.29}$$

Therefore, in (2.3.27), one may replace $\omega_f'(\delta)$ by $\omega_f(\delta)$, which, though more restrictive, is easier to verify in many situations. For some alternative forms of (2.3.27), we refer to Bickel and Wichura (1971). We follow the method of these authors to present some of these results.

Consider a *block* $B(\mathbf{s},\mathbf{t}], \mathbf{s} \leqslant \mathbf{t}$, defined by

$$B(\mathbf{s},\mathbf{t}] = \prod_{j=1}^{p} (s_j, t_j] \quad \text{where } \mathbf{s} = (s_1, \ldots, s_p) \quad \text{and} \quad \mathbf{t} = (t_1, \ldots, t_p) \in E^p. \tag{2.3.30}$$

The block has p faces: $B_k(\mathbf{s},\mathbf{t}), k = 1, \ldots, p$, defined by

$$B_k(\mathbf{s},\mathbf{t}] = \prod_{j=1(\neq k)}^{p} (s_j, t_j]. \tag{2.3.31}$$

Two disjoint blocks $B(\mathbf{s},\mathbf{t}]$ and $B(\mathbf{s}',\mathbf{t}']$ are called *neighbors* if they have the same kth face for some $k(=1, \ldots, p)$. For each block, we define the *increment* of $f(\epsilon D[0,1]^p)$ around the block $B(\mathbf{s},\mathbf{t}]$ by

$$f(B(\mathbf{s},\mathbf{t}]) = \sum_{j=1}^{p} \sum_{l=0,1} (-1)^{p-l_1-\cdots-l_p} f(l_j t_j + (1-l_j)s_j,\ 1 \leqslant j \leqslant p). \tag{2.3.32}$$

Let then

$$m(B(\mathbf{s},\mathbf{t}], B(\mathbf{s}',\mathbf{t}']) = \min\{|f(B(\mathbf{s},\mathbf{t}])|, |f(B(\mathbf{s}',\mathbf{t}'])|\}, \tag{2.3.33}$$

and, let μ be a finite nonnegative measure on E^p. Then, we have the following.

THEOREM 2.3.3 (Bickel and Wichura, 1971). *The hypothesis of tightness in Theorem 2.3.2 holds, if, for every pair of neighboring blocks in* $E^p, \lambda > 0$, *and* $n \geqslant 1$,

$$P_n(\{m(B(\mathbf{s},\mathbf{t}], B(\mathbf{s}',\mathbf{t}']) \geqslant \lambda\}) \leqslant \lambda^{-\gamma}\{\mu(B(\mathbf{s},\mathbf{t}] \cup B(\mathbf{s}',\mathbf{t}'])\}^{1+\beta}, \tag{2.3.34}$$

for some $\gamma > 0$ *and* $\beta > 0$ *(where* $\mathbf{s}, \mathbf{s}', \mathbf{t}$, *and* $\mathbf{t}'$ *all lie in* E^p*). A sufficient condition for* (2.3.34) *to hold is that for every pair of neighboring blocks in* E^p

and $n \geqslant 1$,

$$E\big(|f_n(B(\mathbf{s},\mathbf{t}])|^{\alpha_1}|f_n(B(\mathbf{s}',\mathbf{t}'])|^{\alpha_2}\big) \leqslant M\big\{\mu(B(\mathbf{s},\mathbf{t}])\mu(B(\mathbf{s}',\mathbf{t}'])\big\}^{(1+\beta)/2} \tag{2.3.35}$$

where the probability measure P_n *is induced by* $f_n(\in D[0,1]^p)$, α_1, α_2, β, *and* M *are positive numbers, and* $\mathbf{s}, \mathbf{s}', \mathbf{t}$, *and* $\mathbf{t}'$ *all lie in* E^p.

The case of $p = 1$ has been dealt with in detail by Billingsley (1968). For the case of the $C[0,1]^p$ space, we have the following simplified version.

THEOREM 2.3.4. *A sequence* $\{P_n\}$ *of probability measures on* $(C[0,1]^p, \mathcal{C}^p)$ *is tight if for every* $B(\mathbf{s},\mathbf{t}]$ *in* E^p, $\lambda > 0$ *and* $n \geqslant 1$,

$$P\big\{|f_n(B(\mathbf{s},\mathbf{t}])| \geqslant \lambda\big\} \leqslant \lambda^{-\gamma}\big\{\mu(B(\mathbf{s},\mathbf{t}])\big\}^{1+\beta}, \qquad \forall \mathbf{s}(\leqslant \mathbf{t}) \text{ in } E^p, \tag{2.3.36}$$

where γ *and* β *are positive numbers, independent of* $(\mathbf{s},\mathbf{t})$ *and* n. *A sufficient condition for* (2.3.36) *to hold is that for some positive* $M(<\infty)$,

$$E|f_n(B(\mathbf{s},\mathbf{t}])|^{\alpha} \leqslant M\big\{\mu(B(\mathbf{s},\mathbf{t}])\big\}^{1+\beta}, \qquad \forall \mathbf{s}, \mathbf{t}, \in E^p, \tag{2.3.37}$$

where $\alpha > 0$ *and* $\beta > 0$. *Here,* f_n *and* μ *have the same interpretations as in Theorem* 2.3.3.

For the particular case of $p = 1$ and μ being the Lebesgue measure on $[0,1]$, (2.3.37) reduces to

$$E|f_n(t) - f_n(s)|^{\alpha} \leqslant M|t-s|^{1+\beta} \qquad \text{for every} \quad 0 \leqslant s \leqslant t \leqslant 1 \quad \text{and} \quad n \geqslant 1, \tag{2.3.38}$$

(where M, α, and β are positive numbers) and is known as the *Kolmogorov condition*.

Weak convergence of probability measures on the function spaces $C[0,\infty)$, $D[0,\infty)$ and, in general, $C(R^k)$ and $D(R^k)$, where k is a positive integer and R is the real line $(-\infty,\infty)$, has also been studied by a host of workers (see Stone, 1961, 1963; Whitt, 1970; and Lindvall, 1973). Let $C = C(R)$ be the space of all continuous functions on R with values in a complete separable metric space (S,d) and let $\mathcal{C}$ be the σ-field generated by the open subsets of C when C is endowed with the topology of uniform convergence on compacta (u.c.c.). For any two functions x and y in C, let $d: C \times C \to R$ be defined as

$$d(x,y) = \sum_{j=1}^{\infty} \frac{2^{-j}d_j(x,y)}{\{1 + d_j(x,y)\}}, \tag{2.3.39}$$

where $d_j(x,y) = \sup\{d(x(t),y(t)) : |t| \leqslant j\}$, $j \geqslant 1$. Then the metric topology in (C,d) is the topology of u.c.c. Futher, Definition 2.3.1 applies to probability measures on $(C, \mathcal{C})$. Also, if for every $j \geqslant 1$, we define $r_j : C(R) \to C[-j, j]$ by letting

$$(r_j x)(t) = x(t), \qquad |t| \leqslant j \tag{2.3.40}$$

and if $\{P_n\}$ and P are probability measures on $(C, \mathcal{C})$, then $P_n \Rightarrow P$ iff $P_n r_j^{-1} \Rightarrow P r_j^{-1}$ for every $j \geqslant 1$ (ensuring that $\{P_n\}$ is tight iff $\{P_n r_j^{-1}\}$ is tight for each $j \geqslant 1$). Thus if we define $\omega_f^j(\delta)$ as in (2.3.5) where we allow s and t to lie in $[-j, j]$, $j \geqslant 1$, then, we have the following.

THEOREM 2.3.5. *If $\{P_n\}$ and P are probability measures on $(C, \mathcal{C})$, then $P_n \Rightarrow P$ iff* (i) *the f.d.d.'s of P_n converge weakly to those of P and* (ii) *for every $\epsilon > 0$ and $j \geqslant 1$,*

$$\lim_{n\to\infty} \lim_{\delta \downarrow 0} P_n \{ x \in C : \omega_x^j(\delta) > \epsilon \} = 0. \tag{2.3.41}$$

The result extends to the case of $C(R^k)$, for any $k \geqslant 1$ in a natural way. For $D = D(R^k)$, we denote by $\mathfrak{D}$ the σ-field generated by the open subsets of D and define T_p as in (2.3.22) wherein we replace E^p by R^k. Also, we define r_α as in (2.3.40) where we replace $|t| \leqslant j$ by $t \in [-\alpha, \alpha]^k$, $\alpha > 0$. Then we have the following.

THEOREM 2.3.6. *Let $\{P_n\}$ and P be probability measures on $(D, \mathfrak{D})$. Then, $P_n \Rightarrow P$ iff $P_n r_\alpha^{-1} \Rightarrow P r_\alpha^{-1}$ for every $\alpha \in T_p$.*

For the spaces $C[0, \infty)^k$ or $D[0, \infty)^k$, we need to restrict ourselves to finite intervals on $[0, \infty)^k$ and the last two theorems hold.

If two sequences $\{X_n\}$ and $\{Y_n\}$ have a common domain (e.g., E^p or R^p for some $p \geqslant 1$) and these random functions belong to a separable space S, then $\rho(X_n, Y_n)$, defined as in (2.3.8), is a r.v. and the following theorem, proved in Billingsley (1968, p. 25), provides the weak convergence of one by way of the other.

THEOREM 2.3.7. *If $X_n \xrightarrow{\mathfrak{D}} X$ and $\rho(X_n, Y_n) \xrightarrow{P} 0$, then $Y_n \xrightarrow{\mathfrak{D}} X$.*

We shall find this theorem very useful in subsequent chapters dealing with applications in nonparametric statistics. In many situations one may need some weak convergence results in metrics stronger than the usual metric in (2.3.8) or (2.3.13). For example, suppose that $\{X_n\}$ is a sequence of random functions on $D[0,1]$ and it converges in distribution to a random function X [in the Skorokhod metric in (2.3.13)] where X belongs to the $D[0, \infty)$ [or $C[0, \infty)$] space. Also, let $q = \{q(t), t \in [0, \infty)\}$ be a nonnegative function on $[0, \infty)$, and let $X_n^* = \{X_n^*(t) = X_n(t)/q(t), t \in$

$[0,\infty)\}$, and $X^* = \{X^*(t) = X(t)/q(t),\ t \in [0,\infty)\}$. We would like to know whether $\{X_n^*\}$ converges in distribution to X^*. If q is bounded away from 0 in the entire interval $[0,\infty)$, then the weak convergence of X_n to X would ensure the same for X_n^* to X^*. However, the difficulty arises when $q(t)$ approaches to 0 in certain regions inside $[0,\infty)$. We may introduce a new metric $d_q(x, y)$ by replacing $x(t)$ and $y(t)$ in (2.3.14) by $x^*(t)$ and $y^*(t)$, respectively, where $x^*(t) = x(t)/q(t)$ and $y^*(t) = y(t)/q(t)$. Then we need to show that the tightness-part of the weak convergence theorems considered earlier holds when we replace the Skorokhod metric by $d_q(x, y)$. This is termed *weak convergence in d_q metric*. Such weak convergence results have been studied by Chibisov (1964), Pyke and Shorack (1968a, b), O'Reilly (1974), and others. We shall refer to some of these in a later section.

For the space $D[0,\infty)$, we may also refer to Müller (1968, 1970) for some nonstandard proofs of invariance principles in probability theory employing some other metrics of more general nature. His proofs may need some modifications for the general case of $D[0,\infty)^p$, for $p \geqslant 1$.

In most of the applications of the theorems of this section, as we shall see later on, the probability measure P is induced by some Gaussian function on $C[0,1]^p$ or $C(R^p)$, for some $p \geqslant 1$, while $\{P_n\}$ is induced by suitable empirical processes on $D[0,1]^p$ or $D(R^p)$. In the remainder of this section, we therefore introduce some of these Gaussian functions.

Let $X(\cdot) = \{X(t), t \in [0,\infty)^p\}$ be a real-valued stochastic process defined on some probability space $(\Omega, \mathcal{C}^p)$ and denoted by W, the probability measure induced in $(C[0,\infty)^p, \mathcal{C}^p)$ by $X(\cdot)$ [where $\mathcal{C}^p$ is the σ-field generated by the open subsets of $C[0,\infty)^p$]. Then W is defined to be a Wiener measure on $(C[0,\infty)^p, \mathcal{C}^p)$, if the following hold:

(i) $W\{[X(t);\ t \in [0,\infty)^p] \in \mathrm{C}[0,\infty)^{\mathrm{p}}\} = 1$,

(ii) B and C are any two disjoint rectangles in $[0,\infty)^p$, defined as in (2.3.30), and the increment of the process $X(\cdot)$ over the rectangles, denoted by $X(B)$ and $X(C)$, respectively, are defined as in (2.3.32), then $X(B)$ and $X(C)$ are stochastically independent, and

(iii) for every $B \subset [0,\infty)^p$ and real x,

$$W\{X(B) \leqslant x\} = \left(2\pi\sigma^2\lambda(B)\right)^{1/2}\int_{-\infty}^{x} \exp\left\{-\tfrac{1}{2}u^2/\sigma^2\lambda(B)\right\}du \qquad (2.3.42)$$

where $\lambda(B)$ is the Lebesgue measure of B and $\sigma(0 < \sigma < \infty)$ is a constant. In particular, if $\sigma = 1$, we term W a *standard Wiener measure* on $C[0,\infty)^p$.

For $p = 1$, this definition is very simple. Here, for every $0 \leqslant t_1 < \cdots < t_m < \infty$ and $m \geqslant 1$, $X(t_1)$, $X(t_2) - X(t_1), \ldots, X(t_m) - X(t_{m-1})$ are all stochastically independent and normally distributed with zero means and variances $\sigma^2 t_1$, $\sigma^2(t_2 - t_1), \ldots, \sigma^2(t_m - t_{m-1})$, respectively (where $t_0 = 0$). We term $X(\cdot)$ a *Brownian motion* or *Wiener process* (standard if $\sigma = 1$) on $[0, \infty)$. Thus $X(\cdot)$ has independent increments and it satisfies

$$EX(t) = 0, \quad \sigma^{-2}E\big[X(t)X(s)\big] = (s \wedge t) \\ = \min(s,t) \qquad \forall s,t \in R^+, \tag{2.3.43}$$

where σ is a positive and finite constant.

In general, for $p \geqslant 1$, for the Wiener process described earlier, we have

$$EX(\mathbf{t}) = 0 \qquad \text{and} \quad \sigma^{-2}E\big[X(\mathbf{t})X(\mathbf{s})\big] = \mathbf{s} \wedge \mathbf{t} = \prod_{j=1}^{p} (s_j \wedge t_j), \\ \forall \mathbf{s},\mathbf{t} \in R^{+p}. \tag{2.3.44}$$

Consider now a Gaussian function $X^o(\cdot) = \{X^o(\mathbf{t}), \mathbf{t} \in E^p\}$ where

$$X^o(\mathbf{t}) = X(\mathbf{t}) - |\mathbf{t}|X(\mathbf{1}), \qquad \mathbf{t} \in E^p; \qquad |\mathbf{t}| = t_1, \ldots, t_p \tag{2.3.45}$$

and $X(\cdot) = \{X(\mathbf{t}); \mathbf{t} \in E^p\}$ is a Wiener process on E^p. This implies that X^o is a Gaussian function, defined on E^p, with $EX^o(\mathbf{t}) = 0$ for all $\mathbf{t} \in E^p$ and

$$E\big[X^o(B)X^o(C)\big] = \sigma^2\big[\lambda(B \cap C) - \lambda(B)\lambda(C)\big] \tag{2.3.46}$$

for every B and $C \in E^p$, where the increments $X^o(B)$ and $X^o(C)$ for the process $X^o(\cdot)$ are again defined by (2.3.32) and $\lambda(A)$ is the Lebesgue measure of the rectangle $A \in E^p$. The process X^o is called a *tied-down Brownian Sheet* on E^p. Note that $X^o(\mathbf{t}) = 0$ with probability 1 if $t_j = 0$ or 1 for at least one $j = 1, \ldots, p$. For $p = 1$, X^o is also called a *Brownian Bridge* where (2.3.46) simplifies to

$$E\big[X^o(s)X^o(t)\big] = \sigma^2(s \wedge t - st), \qquad \forall s,t \in [0,1]. \tag{2.3.47}$$

For $\sigma = 1$, $X^o = \{X^o(\mathbf{t}), \mathbf{t} \in E^p\}$ is termed a *standard tied-down Brownian Sheet*.

For $p = 2$, consider a Gaussian function $X^* = \{X^*(\mathbf{t}), \mathbf{t} \in E \times R^+\}$ where

$$X^*(t_1,t_2) = X(t_1,t_2) - t_1X(1,t_2), \qquad \forall (t_1,t_2) = \mathbf{t} \in E \times R^+, \tag{2.3.48}$$

and $\{X = X(\mathbf{t}), \mathbf{t} \in R^{+2}\}$ is a Brownian Sheet. Then, $EX^*(\mathbf{t}) = 0$, $\forall \mathbf{t} \in E \times R^+$ and

$$EX^*(\mathbf{s})X^*(\mathbf{t}) = (s_1 \wedge t_1 - s_1t_1)(s_2 \wedge t_2) \qquad \text{for all} \quad \mathbf{s},\mathbf{t} \in E \times R^+. \tag{2.3.49}$$

We term $X^* = \{X^*(\mathbf{t}), \mathbf{t} \in E \times R^+\}$ a *Kiefer process*. More generally, a Kiefer process can also be defined on $E^p \times R^+$ by considering a Brownian

Sheet on $R^{+(p+1)}$ and writing $\mathbf{t} = (\mathbf{t}_1, t_2)$ where $\mathbf{t}_1 \in E^p$ and $t_2 \in R^+$ and in (2.3.48), replacing t_1 by $|\mathbf{t}_1|$. Thus like the Brownian Bridge or the tied-down Brownian Sheet, the Kiefer process is also a tied-down Wiener process; it is tied-down in only one component of the time parameter and not in the other.

Finally, we define a k-parameter *Bessel process* $B = \{B(t), t \in [0, \infty)\}$ by letting

$$B(t) = \{[\mathbf{X}(t)]'[\mathbf{X}(t)]\}^{1/2}; \quad \mathbf{X}(t) = [X_1(t), \ldots, X_k(t)]', \quad t \in [0, \infty), \tag{2.3.50}$$

where the $X_j = \{X_j(t), t \in [0, \infty)\}$ are independent copies of a standard Wiener process. All these processes play a vital role in the various invariance principles to be discussed in subsequent sections and chapters.

2.4. WEAK INVARIANCE PRINCIPLES FOR SOME DEPENDENT RANDOM VARIABLES

We consider here some functional central limit theorems for martingales, reverse martingales and related sequences of dependent r.v.'s. In this context, the following result due to Dvoretzky (1972) deserves especial mention.

For a triangular array $\{(Z_{n,k}, 0 \leqslant k \leqslant k_n); n \geqslant 1\}$ of r.v.'s (not necessarily independent), we set $Z_{n,0} = 0$,

$$S_{n,k} = Z_{n,0} + \cdots + Z_{n,k} \quad \text{for} \quad k = 1, \ldots, k_n \tag{2.4.1}$$

and let $\mathcal{F}_{n,k}$ be the sigma-field generated by $S_{n,k}$ for $k \geqslant 1$ and $n \geqslant 1$ (where $\mathcal{F}_{n,0}$ is the trivial sigma-field for every $n \geqslant 1$). Assume that $EZ_{n,i}^2 < \infty$, $\forall i \geqslant 1, n \geqslant 1$. Let then

$$\mu_{n,k} = E(Z_{n,k}|\mathcal{F}_{n,k-1}) \quad \text{and} \quad \sigma_{n,k}^2 = E(Z_{n,k}^2|\mathcal{F}_{n,k-1}) - \mu_{n,k}^2 \tag{2.4.2}$$

for $1 \leqslant k \leqslant k_n, n \geqslant 1$ (these r.v.'s exist a.e.) and assume that $k_n \to \infty$ as $n \to \infty$.

THEOREM 2.4.1 (Dvoretzky, 1972). *Suppose that as* $n \to \infty$,

$$\sum_{k=1}^{k_n} \mu_{n,k} \xrightarrow{P} 0, \qquad \sum_{k=1}^{k_n} \sigma_{n,k}^2 \xrightarrow{P} 1, \tag{2.4.3}$$

and the conditional Lindeberg condition holds, that is, for every $\epsilon > 0$,

$$\sum_{k=1}^{k_n} E(Z_{n,k}^2 I(|Z_{n,k}| > \epsilon) \,|\, \mathcal{F}_{n,k-1}) \xrightarrow{P} 0 \quad \text{as} \quad n \to \infty. \tag{2.4.4}$$

Then, for every real x,

$$\lim_{n\to\infty} P\{S_{n,k_n} \leqslant x\} = \Phi(x) = \int_{-\infty}^{x} (2\pi)^{-1/2}\exp(-\tfrac{1}{2}t^2)\,dt. \qquad (2.4.5)$$

A proof of this theorem is given in the appendix. We note that (2.4.4) is implied by the usual Lindeberg condition: for every $\epsilon > 0$,

$$\sum_{k=1}^{k_n} E\left[Z_{n,k}^2 I(|Z_{n,k}| > \epsilon)\right] \to 0 \quad \text{as} \quad n \to \infty. \qquad (2.4.6)$$

In fact, (2.4.6) ensures that (2.4.4) also holds in L_1 norm.

For the same triangular array of r.v.'s, let us now define

$$W_n(t) = S_{n,k_n(t)}, \qquad 0 \leqslant t \leqslant 1; \qquad W_n = \{W_n(t); 0 \leqslant t \leqslant 1\}, \qquad (2.4.7)$$

where $\{k_n(t);\ 0 \leqslant t \leqslant 1\}$ is a sequence of integer-valued, nondecreasing, and right-continuous functions (of t) defined on $[0,1]$ with $k_n(0) = 0$. Then, for every $n(\geqslant 1)$, W_n belongs to the $D[0,1]$ space. Let $W = \{W(t);\ 0 \leqslant t \leqslant 1\}$ be a standard Wiener process on $[0,1]$. Then W belongs to the $C[0,1]$ space with probability 1.

THEOREM 2.4.2 (McLeish, 1974). *Suppose that* (2.4.4) *holds and for each* $t \in [0,1]$,

$$\sum_{k=0}^{k_n(t)} |\mu_{n,k}| \xrightarrow{P} 0 \quad \textit{and} \quad \sum_{k=0}^{k_n(t)} \sigma_{n,k}^2 \xrightarrow{P} t, \quad \textit{as} \quad n \to \infty, \qquad (2.4.8)$$

(*where* $\mu_{n,0} = \sigma_{n,0} = 0$ *with probability* 1). *Then,* W_n *converges in law* (*in the* J_1*-topology on* $D[0,1]$) *to* W.

A proof of this theorem is also given in the appendix. Note that the second condition in (2.4.8) may also be taken as $\sum_{k=1}^{k_n(t)} E(Z_{n,k}^2 | \mathfrak{F}_{n,k-1}) \to t$, in probability. In the following, we narrow our focus to martingales and reverse martingales and present some parallel results.

THEOREM 2.4.3 (Brown, 1971). *Suppose that* $\{Z_k, \mathfrak{F}_k; k \geqslant 1\}$ *is a martingale difference sequence, so that* $\{S_n = \sum_{k=1}^{n} Z_k, \mathfrak{F}_n; n \geqslant 1\}$ *is a martingale sequence and* $E(Z_k | \mathfrak{F}_{k-1}) = 0 \forall k \geqslant 1$. *Then let*

$$V_n = \sum_{k=1}^{n} E(Z_k^2 | \mathfrak{F}_{k-1}) \quad \textit{and} \quad s_n^2 = EV_n = \sum_{k=1}^{n} EZ_k^2. \qquad (2.4.9)$$

If

$$V_n / s_n^2 \to 1, \quad \textit{in probability as} \quad n \to \infty, \qquad (2.4.10)$$

and for every $\epsilon > 0$,

$$s_n^{-2}\left\{\sum_{k=1}^{n} E\big(Z_k^2 I(|Z_k| > \epsilon s_n)\big)\right\} \to 0 \qquad as \quad n \to \infty, \tag{2.4.11}$$

then, for every real x,

$$\lim_{n\to\infty} P\{S_n \leqslant x s_n\} = \Phi(x). \tag{2.4.12}$$

Since Theorem 2.4.3 is a special case of Theorem 2.4.1, its proof is not considered. We may add that for (2.4.10) to hold, it is necessary and sufficient that $E|V_n s_n^{-2} - 1| \to 0$ as $n \to \infty$. For the martingale sequence under consideration, let us now define

$$k_n(t) = \max\{k : s_k^2 \leqslant t s_n^2\} \quad \text{and} \quad W_n(t) = S_{k_n(t)}/s_n;\ t \in [0,1]. \tag{2.4.13}$$

Then by virtue of (2.4.10) and (2.4.13), $V_{k_n(t)}/s_n^2 \to t$, in probability, $\forall t \in [0,1]$, and hence, on letting $Z_{n,k} = Z_k/s_n, k \leqslant n$, we obtain from Theorem 2.4.2, the following.

THEOREM 2.4.4 (Brown, 1971). *For* $W_n = \{W_n(t);\ 0 \leqslant t \leqslant 1\}$, *defined by* (2.4.13), *the conditions* (2.4.10) *and* (2.4.11) *ensure the weak convergence of* $\{W_n\}$ *to* W.

In subsequent chapters, we will often use these theorems. Verification of (2.4.10), particularly in the context of rank statistics, usually requires elaborate manipulations. Often these can be avoided by the following projection technique. Suppose that there exists a sequence $\{Y_i, i \geqslant 1\}$ of independent r.v.'s such that $S_n = S(Y_1, \ldots, Y_n)$ for $n \geqslant 1$ (where the S_n may or may not form a martingale sequence). Let us then take

$$S_n^* = \sum_{i=1}^{n} E(S_n | Y_i) - (n-1)E(S_n) = \sum_{i=1}^{n} l_n(Y_i), \tag{2.4.14}$$

where the $l_n(Y_i)$ form a triangular array of independent r.v.'s and $E(S_n - S_n^*)^2 = V(S_n) - V(S_n^*)$. Thus if the projection in (2.4.14) satisfies the condition that

$$E(S_n - S_n^*)^2/V(S_n) \to 0 \qquad \text{as} \quad n \to \infty, \tag{2.4.15}$$

then, $(S_n - ES_n)/(V(S_n))^{1/2}$ and $(S_n^* - ES_n)/(V(S_n))^{1/2}$ both have the same asymptotic distribution, if they have any at all. On the other hand, for applying the central limit theorem to $\{S_n^*\}$, we do not require (2.4.10), and a Lindeberg-type condition on the $l_n(Y_i)$ suffices. Thus if (2.4.15) holds, simpler proofs for the asymptotic normality in Theorem 2.4.3 can usually be derived (under alternative conditions). We may further note that the proof of Theorem 2.4.4 given in Brown (1971) has a very nice feature. Namely, he uses the lemma in (2.2.9) and (2.2.10) (with $r = s = 2$) and proves the following.

Corollary 2.4.4.1. *For a martingale sequence, if W_n be defined by* (2.4.13), *then the convergence of the finite-dimensional distributions of $\{W_n\}$ to those of W ensures the tightness of $\{W_n\}$ (and hence the weak convergence of $\{W_n\}$ to W).*

A proof of this corollary is given in the appendix.

For a reverse martingale sequence $\{X_n, \mathcal{F}_n; n \geqslant 1\}$, we define $Z_{n,i} = X_{n+i-1} - X_{n+i}, i \geqslant 1$, and let

$$V_n = \sum_{i=1}^{\infty} E(Z_{n,i}^2 \mid \mathcal{F}_{n+i}) \qquad \text{and} \quad s_n^2 = EV_n = \sum_{i=1}^{\infty} EZ_{n,i}^2 = V(X_n). \tag{2.4.16}$$

The conditional Lindeberg condition in (2.4.4) then reduces to

$$s_n^{-2}\left\{\sum_{i \geqslant 1} E\left(Z_{n,i}^2 I(|Z_{n,i}| > \epsilon s_n) \mid \mathcal{F}_{n+i}\right)\right\} \to 0, \qquad \text{in probability.} \tag{2.4.17}$$

Here we consider the process $W_n = \{W_n(t); 0 \leqslant t \leqslant 1\}$ by letting

$$W_n(t) = s_n^{-1} X_{n(t)}, \qquad n(t) = \min\{k : s_k^2/s_n^2 \leqslant t\} \qquad \text{for} \quad 0 \leqslant t \leqslant 1. \tag{2.4.18}$$

THEOREM 2.4.5 (Loynes, 1970). *If $s_n^{-2}V_n \xrightarrow{p} 1$ and* (2.4.17) *holds, then W_n, defined by* (2.4.18), *converges weakly to W.*

The proof follows from Theorem 2.4.2 and hence is omitted. Loynes (1970) actually used a.s. convergence in his statement of the theorem. This, however, is not necessary.

Note that in Theorem 2.4.4., W_n is constructed from the partial sequence $\{Z_k; k \leqslant n\}$, while in Theorem 2.4.5, it is constructed from the tail sequence $\{X_k; k \geqslant n\}$. Suppose now that for the martingale $\{Z_k, \mathcal{F}_k; k \geqslant 1\}$ with $EZ_n = 0$ and $\lim_{n\to\infty} s_n^2 = \infty$, we define $W_n^* = \{W_n^*(t) = s_n Z_{n(t)} / s_{n(t)}^2;$ $0 \leqslant t \leqslant 1\}$ by letting $n(t) = \min\{k : s_n^2/s_k^2 \leqslant t\}$, $0 \leqslant t \leqslant 1$ and for the reverse martingale $\{X_k, \mathcal{F}_k;\ k \geqslant 1\}$ with $EX_n = 0$ and $EX_n^2 = s_n^2 \downarrow 0$ as $n \to \infty$, we define $W_n^* = \{W_n^*(t) = s_n s_{n(t)}^{-2} X_{n(t)}; 0 < t \leqslant 1\}$ by letting $n(t) = \max\{k :$ $s_n^2/s_k^2 \leqslant t\}, 0 < t \leqslant 1$. Then we have the following.

THEOREM 2.4.6 (Sen, 1976*c*). *Under the hypothesis of Theorem* 2.4.4 (*or* 2.4.5), *the weak convergence of $\{W_n\}$ to W ensures the weak convergence of $\{W_n^*\}$ to W.*

We refer to the proof of Theorem 2.2 of Sen (1976c) for a proof of this theorem. We conclude this section with an elegant result from Scott (1973) on a triangular array of martingales; his results are very similar to our Theorem 2.4.2. For every $n(\geqslant 1)$, let $\{S_{n,k} = \sum_{i=1}^{k} Z_{n,i}, \mathcal{F}_{n,k}; k \leqslant n\}$ be a martingale sequence and let $S_{n,0} = 0$ a.s. Let then $s_{n,k}^2 = ES_{n,k}^2$ for $0 \leqslant k \leqslant n$ and define $W_n = \{W_n(t); 0 \leqslant t \leqslant 1\}$ by letting

$$W_n(t) = \frac{S_{n,k_n(t)}}{s_{n,n}} \quad \text{where} \quad k_n(t) = \max\{k : s_{n,k}^2 \leqslant t s_{n,n}^2\}, \qquad 0 \leqslant t \leqslant 1. \tag{2.4.19}$$

THEOREM 2.4.7 (Scott, 1973). *Suppose that any one of the following equivalent sets of conditions holds (where, without any loss of generality, we take $s_{n,n}^2 = 1$):*

$$\left.\begin{aligned} &\sum_{k=1}^{k_n(t)} Z_{n,k}^2 \xrightarrow{p} t \quad \text{as} \quad n \to \infty, \quad 0 < t \leqslant 1 \quad \text{and} \\ &\sum_{k=1}^{n} Z_{n,k}^2 I(|Z_{n,k}| > \epsilon) \xrightarrow{p} 0 \quad \text{as} \quad n \to \infty, \quad \forall \epsilon > 0; \end{aligned}\right\} \quad (A) \tag{2.4.20}$$

$$\left.\begin{aligned} &\sum_{k=1}^{k_n(t)} E(Z_{n,k}^2 | \mathcal{F}_{n,k-1}) \xrightarrow{p} t \quad \text{as } n \to \infty, \quad 0 < t \leqslant 1 \quad \text{and} \\ &\sum_{k=1}^{k_n(t)} E(Z_{n,k}^2 I(|Z_{n,k}| > \epsilon) | \mathcal{F}_{n,k-1}) \xrightarrow{p} 0 \quad \text{as } n \to \infty, \quad \forall \epsilon > 0; \end{aligned}\right\} \quad (B) \tag{2.4.21}$$

$$\left.\begin{aligned} &\sum_{k=1}^{k_n(t)} Z_{n,k}^2 \xrightarrow{p} t \quad \text{as} \quad n \to \infty, \quad 0 < t \leqslant 1 \quad \text{and} \\ &\sup_{k \leqslant n} Z_{n,k}^2 \xrightarrow{p} 0 \quad \text{as} \quad n \to \infty; \end{aligned}\right\} \quad (C) \tag{2.4.22}$$

$$\left.\begin{aligned} &\sum_{k=1}^{k_n(t)} Z_{n,k}^2 \xrightarrow{p} t \quad \text{as} \quad n \to \infty, \quad 0 < t \leqslant 1 \quad \text{and} \\ &\sum_{k=1}^{n} Z_{n,k}^2 U(\epsilon^{-1}|Z_{n,k}|) \xrightarrow{p} 0 \quad \text{as} \quad n \to \infty, \quad \forall \epsilon > 0 \end{aligned}\right\} \quad (D) \tag{2.4.23}$$

$$\left.\begin{aligned} &\sum_{k=1}^{k_n(t)} E(Z_{n,k}^2 \,|\, \mathcal{F}_{n,k-1}) \xrightarrow{p} t \quad \text{as} \quad n \to \infty, \quad 0 < t \leqslant 1 \quad \text{and} \\ &\sum_{k=1}^{n} E(Z_{n,k}^2 U(\epsilon^{-1}|Z_{n,k}|) \,|\, \mathcal{F}_{n,k-1}) \xrightarrow{p} 0 \quad \text{as} \quad n \to \infty, \quad \epsilon > 0, \end{aligned}\right\} \quad (E) \tag{2.4.24}$$

where $U(\cdot)$ is any continuous, nonnegative function of bounded variation on $[0, \infty)$ for which $U(0) = 0$ and $U(x)$ converges to a positive constant as $x \to \infty$. Then, W_n, defined by (2.4.19), converges weakly to W.

We provide a proof of this theorem in the appendix. Note that if the Z_k are mutually independent (and centered at expectations) in Theorems 2.4.3 and 2.4.4., then $V_n = s_n^2$ for every $n \geqslant 1$, so that (2.4.10) holds, and hence, the Lindeberg Condition in (2.4.11) suffices for the weak convergence of $\{W_n\}$ to W; this yields the classical Donsker theorem for the weak convergence of partial sums of independent r.v.'s. Also, in Theorems 2.4.2 and 2.4.7, if, for every $n \geqslant 1$, the $Z_{n,i}$ are mutually independent (and centered at expectations), then by definition of $k_n(t)$ in (2.4.19), the first part of Condition (B) in (2.4.21) holds whenever $\max_{1 \leqslant i \leqslant n} EZ_{n,i}^2 / s_{n,n}^2 \to 0$ as $n \to \infty$ (i.e., $\{Z_{n,i}\}$ is an infinitesimal system), while the second part follows from the classical Lindeberg-type condition. Second, we have so far considered the weak convergence of suitable stochastic processes (constructed from the $Z_{n,i}$) on the unit interval $[0, 1]$. By virtue of our Theorems 2.3.5 and 2.3.6, we are now in a position to extend the domain of these processes to the semi-infinite interval $[0, \infty)$ and to derive parallel weak convergence results. This does not need any extra regularity conditions. Finally, so far we have considered the weak convergence in the Skorokhod J_1-topology. As in Theorem 2.3.7, we may also consider the weak convergence in the d_q-metric for some suitable q. Let $q = \{q(t), t \in [0, 1]\}$ be a continuous, nonnegative function on $[0, 1]$, bounded away from 0 on $[a, 1]$ and nondecreasing on $[0, a]$ for some $0 < a < 1$, such that

$$I(q) = \int_0^1 [q(t)]^{-2} dt < \infty. \tag{2.4.25}$$

Consider the metric

$$d_q = \sup\left\{ \frac{|x(t) - y(t)|}{q(t)} : t \in [0, 1] \right\}. \tag{2.4.26}$$

Define $\{W_n\}$ as in (2.4.19) and note that by definition of the $S_{n,k}$, for every $n \geqslant 1$,

$$\{S_{n,k}^2, k \geqslant 0\} \text{ is a nonnegative submartingale,} \tag{2.4.27}$$

$$ES_{n,k}^2 = s_{n,k}^2 \qquad 0 \leqslant k \leqslant n \quad \text{where } s_{n,0}^2 = 0. \tag{2.4.28}$$

Thus by the Birnbaum-Marshall inequality, II [see (2.2.21)], for every

$0 < \tau < a$, $\epsilon > 0$ and $n \geqslant 1$,

$$P\left\{\sup_{0 \leqslant t \leqslant \tau} \frac{|W_n(t)|}{q(t)} > \epsilon\right\} = P\left\{\sup_{0 \leqslant t \leqslant \tau} \frac{W_n^2(t)}{q^2(t)} > \epsilon^2\right\}$$

$$\leqslant \epsilon^{-2} \sum_{k=1}^{k_n(\tau)} \left[q\left(\frac{s_{n,k}^2}{s_{n,n}^2}\right)\right]^{-2} \frac{\left(s_{n,k}^2 - s_{n,k-1}^2\right)}{s_{n,n}^2}$$

$$\leqslant \epsilon^{-2} \int_0^{\tau} [q(t)]^{-2}\, dt,$$

$$\text{as } q \text{ is nondecreasing on } [0, \tau]. \tag{2.4.29}$$

By (2.4.25), the right-hand side of (2.4.29) can be made smaller than any preassigned $\eta(>0)$ by choosing $\tau(>0)$ adequately small. On the other hand, for $t > \tau, q(t)$ is bounded away from 0 and hence, the weak convergence in Theorem 2.4.7 ensures the same in d_q-metric as well. Hence we arrive at the following. (See Sen and Tsong, 1980.)

THEOREM 2.4.8. *For q satisfying* (2.4.25) *and the conditions stated therein, $\{W_n\}$, defined by* (2.4.19), *weakly converges to W in the d_q-metric when the hypothesis of Theorem* 2.4.7 *holds.*

A similar extension of Theorem 2.4.2 demands an inequality parallel to (2.4.29) and, in the absence of a semimartingale property, may need some extra regularity conditions.

2.5. AN ALMOST SURE INVARIANCE PRINCIPLE FOR MARTINGALES

For a sequence $\{Y_i, i \geqslant 1\}$ of independent r.v.'s (with means equal to 0 and finite variances), Skorokhod (1956) has shown that there exists a Brownian motion process $W = \{W(t), t \geqslant 0\}$ (on R^+) and a sequence $\{T_i, i \geqslant 1\}$ of nonnegative and independent r.v.'s such that for every $n \geqslant 1, W(T_1)$, $W(T_1 + T_2) - W(T_1), \ldots, W(\sum_{j=1}^n T_j) - W(\sum_{j=1}^{n-1} T_j)$ have the same joint distribution as do $Y_1, \ldots, Y_n$ and $E(T_k) = \text{Var}(Y_k), \forall k \geqslant 1$. This fundamental result allows one to examine (for various probabilistic analyses) the sequence of values of the process $\{W(\sum_{j=1}^n T_j); n \geqslant 1\}$ instead of $\{S_n = \sum_{j=1}^n Y_j; n \geqslant 1\}$ and is known as the *skorokhod embedding of Wiener processes*. Strassen (1964, 1967) has extended this elegant representation to discrete parameter martingales and his theorems have great impact on the main theorems developed in the subsequent chapters of this monograph. His main theorem is presented below.

Let $\{X_n, \mathfrak{F}_n; n \geqslant 1\}$ be a martingale, so that $\{Y_n = X_n - X_{n-1}, \mathfrak{F}_n; n \geqslant 1\}$ (with $X_0 = EX_1$) form a martingale difference sequence. We assume that

$$E(Y_n^2 \mid \mathfrak{F}_{n-1}), \qquad \text{exists a.s.,} \quad \text{for every } n \geqslant 1 \tag{2.5.1}$$

and note that $E(Y_n \mid \mathfrak{F}_{n-1}) = 0$ a.s., $\forall n \geqslant 1$. Also, let

$$V_n = \sum_{k=1}^{n} E(Y_k^2 \mid \mathfrak{F}_{k-1}), \qquad n \geqslant 1, \tag{2.5.2}$$

so that $\{V_n; n \geqslant 1\}$ is a sequence of nonnegative r.v.'s. Let, then

$$S_{V_n} = X_n, \qquad n \geqslant 1, \qquad S_0 = X_0 = 0, \qquad \text{with probability 1;} \tag{2.5.3}$$

$$S_t = S_{V_n} + \left[\frac{(t - V_n)}{(V_{n+1} - V_n)} \right] Y_{n+1} \qquad \text{for} \quad V_n \leqslant t \leqslant V_{n+1}; \quad n > 0. \tag{2.5.4}$$

Thus $S = \{S_t; t \geqslant 0\}$ is a continuous process on R^+. In (2.5.4), if $V_{n+1} = V_n$, then $Y_{n+1} = 0$ a.e., so that $S_t = S_{V_n}$. To avoid notational complications, we may assume, without any loss of generality, that $V_1 > 0$.

THEOREM 2.5.1 (Strassen, 1967). *Let $\{X_n, \mathfrak{F}_n; n \geqslant 1\}$ be a martingale and let $\{V_n; n \geqslant 1\}$ and S be defined by (2.5.2) through (2.5.4). Also, let $f = \{f(t); t > 0\}$ be a nonnegative and nondecreasing function such that*

$$f(t) \ \text{ is } \uparrow \ \text{ but } t^{-1}f(t) \ \text{ is } \downarrow \ \text{ in } \ t(>0). \tag{2.5.5}$$

Suppose now that

$$V_n \to \infty \quad \text{a.s.,} \quad \text{as} \quad n \to \infty \tag{2.5.6}$$

and

$$\sum_{n \geqslant 1} [f(V_n)]^{-1} E\{Y_n^2 I(|Y_n| > f^{1/2}(V_n)) \mid \mathfrak{F}_{n-1}\} < \infty \qquad \text{a.s.} \tag{2.5.7}$$

Then, there exists a Brownian motion process $W = \{W(t); t \geqslant 0\}$ such that

$$S_t = W(t) + o\big((\log t)[tf(t)]^{1/4}\big) \qquad \text{a.s.,} \quad \text{as } t \to \infty. \tag{2.5.8}$$

A proof of this fundamental theorem is given in the appendix. We may remark that the above result rests on the following embedding:

$$X_n = W(T_1 + \cdots + T_n) \qquad \text{a.s.,} \quad \text{for all } n \geqslant 1, \tag{2.5.9}$$

where the T_i are nonnegative r.v.'s and they satisfy certain convergence results. In fact, we need to introduce a new probability space on which W and $\{T_i;\ i \geqslant 1\}$ are defined and to consider a sequence $\{X_n^*;\ n \geqslant 1\}$ of r.v.'s on this new space such that $\{X_n; n \geqslant 1\}$ and $\{X_n^*; n \geqslant 1\}$ have the same distribution and (2.5.9) holds for the X_n^*. Actually, the X_n and the

Brownian motion process need not be defined on the same probability space.

The choice of f in (2.5.5) and (2.5.7) rests on the specific situations at hand. In the context of the law of iterated logarithm for $\{X_n\}$, it suffices to choose

$$f(t) = t(\log\log t)^2/(\log t)^4, \qquad \text{for large } t. \tag{2.5.10}$$

If we assume the existence of higher order moments of $\{Y_n\}$, then we may take $f(t) = t^{1-\epsilon}$ for some $\epsilon > 0$, so that in (2.5.8), the second term on the rhs will be $o(t^{1/2-\eta})$, for some $\eta > 0$. In subsequent chapters, we will come across some situations where this will be clarified. If the Y_n are independent and identically distributed (i.i.d.) r.v.'s with 0 mean, unit variance, and finite 4th moment, then Strassen (1967) has shown that in Theorem 2.5.1, (2.5.8) may be strengthened to

$$S_n = W(n) + O\left[(n\log\log n)^{1/4}(\log n)^{1/2}\right] \qquad \text{a.s.,} \quad \text{as } n \to \infty. \tag{2.5.11}$$

For this case of independent r.v.'s, a very interesting approach to this embedding has been developed by the Hungarian school (see Csörgö and Révész 1975a, b and Komlós, Major, and Tusnády, 1975). These authors have shown that the order of the remainder term in (2.5.11) can be improved under more stringent moment conditions. Let $\{Y_i, i \geqslant 1\}$ be a sequence of i.i.d.r.v.'s with $EY_1 = 0$, $EY_1^2 = 1$ and

$$E\{\exp(tY_1)\} < \infty, \qquad \forall |t| \leqslant t_o \qquad \text{for some} \quad t_o > 0. \tag{2.5.12}$$

Note that for i.i.d.r.v. with 0 mean and unit variance, by definition in (2.5.2), $V_n = n$, so that by (2.5.3), $S_n = X_n = Y_1 + \cdots + Y_n, \forall n \geqslant 1$. Then, under (2.5.12), there exists a positive and finite constant C_0 (depending only on the d.f. of the Y_i), such that

$$P\left\{\limsup_{n\to\infty} \frac{|S_n - W(n)|}{(\log n)} \leqslant C_o\right\} = 1, \tag{2.5.13}$$

so that we have

$$S_n = W(n) + O(\log n) \qquad \text{a.s,} \quad \text{as} \quad n \to \infty. \tag{2.5.14}$$

The novel technique of proving (2.5.13) employed by these authors depends on the independence of the Y_i and has not yet been extended to the case of martingales or other dependent sequences. We conclude this section with the remark that if F be the d.f. of Y_1 (when the Y_i are i.i.d.r.v.) and $\mathfrak{F}$ be a class of d.f.'s such that

$$\sup_{F\in\mathfrak{F}} E_F Y_1^4 < \infty \qquad \text{or} \qquad \sup_{F\in\mathfrak{F}} E_F\{\exp(tY_1)\} < \infty, \qquad \forall |t| \leqslant t_o(>0), \tag{2.5.15}$$

then (2.5.11) or (2.5.14) holds uniformly in the class of d.f.'s $\mathfrak{F}$.

2.6. INVARIANCE PRINCIPLES FOR SOME EMPIRICAL PROCESSES

Let $\{\mathbf{X}_i; i \geqslant 1\}$ be a sequence of i.i.d.r.v.'s defined on a probability space $(\Omega, \mathcal{B}, P)$ with each X_i having a continuous distribution function (d.f.) F, defined on R^p. For every $n(\geqslant 1)$, the *sample* or *empirical distribution* F_n is defined by

$$F_n(\mathbf{x}) = n^{-1} \sum_{i=1}^{n} c(\mathbf{x} - \mathbf{X}_i), \qquad \mathbf{x} \in R^p, \tag{2.6.1}$$

where $c(\mathbf{u})$ is 1 only when all the p coordinates of $\mathbf{u}$ are nonnegative, and is otherwise equal to 0. Note that

$$EF_n(\mathbf{x}) = F(\mathbf{x}), \qquad \forall x \in R^p; \tag{2.6.2}$$

$$\mathrm{Cov}\left[F_n(\mathbf{x}), F_n(\mathbf{y})\right] = n^{-1}\{F(\mathbf{x} \wedge \mathbf{y}) - F(\mathbf{x})F(\mathbf{y})\}, \qquad \forall \mathbf{x}, \mathbf{y} \in R^p. \tag{2.6.3}$$

The usual *empirical process* $V_n = \{V_n(x); x \in R^p\}$ is defined by

$$V_n(\mathbf{x}) = n^{1/2}\left[F_n(\mathbf{x}) - F(\mathbf{x})\right], \qquad \mathbf{x} \in R^p. \tag{2.6.4}$$

We shall find it convenient to consider the following equivalent (reduced) process. Let $\mathbf{X}_i = (X_{i1}, \ldots, X_{ip})'$ and let $F_{[j]}(x) = P\{X_{ij} \leqslant x\}, x \in R$ be the marginal d.f. of X_{ij} for $j = 1, \ldots, p$. Define then

$$Y_{ij} = F_{[j]}(X_{ij}), \qquad 1 \leqslant j \leqslant p; \qquad Y_i = (Y_{i1}, \ldots, Y_{ip}), \qquad i \geqslant 1. \tag{2.6.5}$$

Thus marginally, each Y_{ij} has the uniform [0, 1] d.f., while

$$G(\mathbf{t}) = P\{\mathbf{Y}_i \leqslant \mathbf{t}\} = F\left(F_{[1]}^{-1}(t_1), \ldots, F_{[p]}^{-1}(t_p)\right), \qquad \mathbf{t} \in E^p, \tag{2.6.6}$$

where $F_{[j]}^{-1}(u) = \inf\{x : F_{[j]}(x) \geqslant u\}$, $0 \leqslant u \leqslant 1$. Note that all the p univariate marginals of G are uniform [0, 1] d.f. and the Y_i belong to the E^p, with probability 1. The *reduced empirical* d.f. G_n is defined as

$$G_n(\mathbf{t}) = n^{-1} \sum_{i=1}^{n} c(\mathbf{t} - \mathbf{Y}_i), \qquad \mathbf{t} \in E^p, \tag{2.6.7}$$

and the corresponding *empirical process* $U_n = \{U_n(t); t \in E^p\}$ is defined by

$$U_n(t) = n^{1/2}\left[G_n(\mathbf{t}) - G(\mathbf{t})\right], \qquad \mathbf{t} \in E^p. \tag{2.6.8}$$

Note that, by definition,

$$U_n(\mathbf{t}) = V_n\left[F_{[1]}^{-1}(t_1), \ldots, F_{[p]}^{-1}(t_p)\right], \qquad \mathbf{t} = (t_1, \ldots, t_p) \in E^p, \tag{2.6.9}$$

so that by studying the behavior of U_n, we may conclude on parallel results on V_n.

Now $G_n(\mathbf{t})$, defined by (2.6.7), involves an average of 0 to 1 valued i.i.d.r.v.'s, on which the classical central limit theorem applies. Thus for every (fixed) $m(\geqslant 1)$ and $\mathbf{t}_1, \ldots, \mathbf{t}_m(\in E^p)$, if one considers the vector $[U_n(\mathbf{t}_1), \ldots, U_n(\mathbf{t}_m)]$, then an arbitrary linear compound,

$$\sum_{j=1}^{m} b_j U_n(\mathbf{t}_j) = n^{-1} \sum_{i=1}^{n} \left\{ \sum_{j=1}^{m} b_j c(\mathbf{t}_j - \mathbf{Y}_i) \right\} \qquad (\mathbf{b} \neq \mathbf{0}) \qquad (2.6.10)$$

also involves an average over i.i.d.r.v.'s, and hence, by the central limit theorem, one may conclude that as $n \to \infty$,

$$\left[U_n(\mathbf{t}_1), \ldots, U_n(\mathbf{t}_m) \right] \xrightarrow{\mathfrak{D}} \left[U(\mathbf{t}_1), \ldots, U(\mathbf{t}_m) \right], \qquad (2.6.11)$$

where $U = \{ U(\mathbf{t}), \mathbf{t} \in E^p \}$ is a Gaussian function with $EU(\mathbf{t}) = 0$ and $EU(\mathbf{s})U(\mathbf{t}) = G(\mathbf{s} \wedge \mathbf{t}) - G(\mathbf{s})G(\mathbf{t}), \forall \mathbf{s}, \mathbf{t} \in E^p$ and U belongs to the $C[0,1]^p$ space with probability 1. On the other hand, U_n belongs to the $D[0,1]^p$ space. If we define the rectangles B, C as in Section 2.3 [see (2.3.30)], then by direct computation, it follows that (2.3.35) holds with $\alpha_1 = \alpha_2 = 2$ and $\beta = 1$, that is,

$$EU_n^2(B)U_n^2(C) \leqslant 3\left[P(B)P(C) \right] \leqslant (3/4)\left[P(B \cup C) \right]^2, \qquad \forall n \geqslant 1, \qquad (2.6.12)$$

where $U_n(B), U_n(C)$ are defined as in (2.3.32) and $P(A) = P\{ Y \in A \}$, $\forall A \in E^p$. Thus by Theorems 2.3.1 through 2.3.3 and (2.6.11), we arrive at the following.

THEOREM 2.6.1. *For continuous F, $\{ U_n \}$ converges in law (in the Skorokhod J_1-topology on $D[0,1]^p$) to a Gaussian function U for which $EU = 0$ and*

$$EU(s)U(t) = G(s \wedge t) - G(s)G(t) \qquad \forall s, t \in E^p. \qquad (2.6.13)$$

In the particular case of $p = 1, G(t) = t$ for $t \in [0,1]$, so that U becomes a Brownian Bridge. Also, if $G(t) = t_1, \ldots, t_p$ for every $t \in E^p$, that is, $X_{i1}, \ldots, X_{ip}$ are mutually independent, then U is a tied-down Brownian Sheet.

Two possible extensions of Theorem 2.6.1 deserve mention. First, we define $U_n^* = \{ U_n^*(s, \mathbf{t}); s \in E, \mathbf{t} \in E^p \}$ by letting

$$U_n^*(s,t) = n^{-1/2}\{ n_s U_{n_s}(\mathbf{t}) \}, \qquad n_s = [ns], \qquad s \in [0,1] \quad \text{and} \quad \mathbf{t} \in E^p, \qquad (2.6.14)$$

where the $\{ U_k(\mathbf{t}) \}$ are defined by (2.6.8). Then $U_n^* \in D[0,1]^{p+1}$. Also, consider a Gaussian function $U^* = \{ U^*(s, \mathbf{t}), s \in E, \mathbf{t} \in E^p \}$ where

$EU^* = 0$ and

$$EU^*(s,\mathbf{t})U^*(s',\mathbf{t}') = (s \wedge s')\left[G(\mathbf{t} \wedge \mathbf{t}') - G(\mathbf{t})G(\mathbf{t}')\right],$$

$$(s,\mathbf{t}),(s',\mathbf{t}') \in E^{p+1}. \tag{2.6.15}$$

THEOREM 2.6.2 (Bickel and Wichura, 1971). *For every continuous F (and hence, G), U_n^* converges in law (in the J_1-topology on $D[0,1]^{p+1}$) to U^*.*

We also consider another process $U_n^0 = \{U_n^0(s,\mathbf{t}); s \in E, \mathbf{t} \in E^p\}$ by letting

$$U_n^0(s,\mathbf{t}) = n^{1/2}U_{n_s}(\mathbf{t}), \qquad n_s = \min\{k : n/k \leqslant s\}, \qquad s \in E, \mathbf{t} \in E^p, \tag{2.6.16}$$

where U_n is again defined by (2.6.8) and, by the Glivenko-Cantelli Lemma, for every fixed n, $U_n^0(0,\mathbf{t}) = 0$, with probability $1, \forall \mathbf{t} \in E^p$.

THEOREM 2.6.3 (Neuhaus and Sen, 1977). *For every continuous F (and hence, G),*

$$U_n^0 \xrightarrow{\mathfrak{D}} U^*, \qquad \text{in the } J_1\text{-topology on } D[0,1]^{p+1}, \tag{2.6.17}$$

where U^* is defined as in Theorem 2.6.2.

We now proceed to consider some a.s. invariance principles for the reduced empirical process. The following theorem (for the particular case of $p = 1$) is due to Komlos, Major, and Tusnády (1975).

THEOREM 2.6.4. *Let $Y_i, i \geqslant 1$ be i.i.d.r.v.'s with the uniform $[0,1]$ d.f. Then, one can define a Kiefer process $X^* = \{X^*(s,t), s \in E, t \in R^+\}$ [see (2.3.48)], such that for suitable positive constants A, B, and C and for all $n(\geqslant 1)$ and z,*

$$P\left\{\sup_{1 \leqslant k \leqslant n}\ \sup_{0 \leqslant t \leqslant 1} |k[G_k(t) - t] - X^*(t,k)| > (A\log n + z)\log n\right\} \leqslant Be^{-Cz}, \tag{2.6.18}$$

and hence

$$\sup_{t \in E}|n[G_n(t) - t] - X^*(t,n)| = O\left((\log n)^2\right) \qquad \text{a.s.,} \quad \text{as} \quad n \to \infty. \tag{2.6.19}$$

By virtue of Theorem 2.6.1, we have

$$\sup_{x \in E^p}|V_n(x)| = \sup_{t \in E^p}|U_n(t)| = O_p(1). \tag{2.6.20}$$

In fact, for $p = 1$, letting $U = \{U(t), t \in E\}$ be a standard Brownian Bridge, we have the following (see Doob, 1949):

$$\lim_{n\to\infty} P\left\{\sup_{x\in R} V_n(x) > \lambda\right\} = \lim_{n\to\infty} P\left\{\sup_{t\in E} U_n(t) > \lambda\right\}$$

$$= P\left\{\sup_{t\in E} U(t) > \lambda\right\} = \exp\{-2\lambda^2\}, \qquad \forall\lambda \geqslant 0; \tag{2.6.21}$$

$$\lim_{n\to\infty} P\left\{\sup_{x\in R} |V_n(x)| > \lambda\right\} = \lim_{n\to\infty} P\left\{\sup_{t\in E} |U_n(t)| > \lambda\right\}$$

$$= P\left\{\sup_{t\in E} |U(t)| > \lambda\right\}$$

$$= 2\sum_{k=1}^{\infty} (-1)^{k-1}\exp\{-2k^2\lambda^2\}, \qquad \forall\lambda \geqslant 0. \tag{2.6.22}$$

Further, Dvoretzky, Kiefer and Wolfowitz (1956) have shown that for every (finite) n, the actual probability on the left-hand side (lhs) of (2.6.21) and (2.6.22) is less than or equal to the limit specified. For $p > 1$, nothing in particular is known about the limiting distributions of $\sup_{t\in E^p} U(t)$ or $\sup_{t\in E^p}|U(t)|$. However, it follows from Kiefer (1961) that for every $\epsilon (0 < \epsilon < 2)$, there exists a positive $c(\epsilon)$, such that for every $\lambda > 0$,

$$P\left\{\sup_{t\in E^p} |U_n(t)| > \lambda\right\} \leqslant c(\epsilon)\exp\{-(2-\epsilon)\lambda^2\}, \qquad \forall n \geqslant 1, \tag{2.6.23}$$

that is, the rate of convergence is still exponential in λ^2. From (2.6.23) and the Borel-Cantelli Lemma, it follows that as $n \to \infty$,

$$\sup_{t\in E^p} |U_n(t)| = O\left((\log n)^{1/2}\right) \qquad \text{a.s.} \tag{2.6.24}$$

Invariance principles for weighted versions (with smooth weight functions) of empirical processes have also been studied in the past 10 years by a host of research workers. Weak convergence of empirical processes in the *sup-norm metric* has been studied by Chibisov (1964), Pyke and Shorack (1968a, b), Braun (1976), Wellner (1974), O'Reilly (1974) and others. Let $q = \{q(t), t \in E\}$ be a continuous, nonnegative function on [0, 1], bounded away from 0 on $[\gamma, 1-\gamma]$ for some $0 < \gamma < 1$, nondecreasing (nonincreasing) on $[0, \gamma]([1-\gamma, 1])$. Consider the metric

$$\rho_q(x, y) = \sup\{|x(t) - y(t)|/q(t) : 0 \leqslant t \leqslant 1\}. \tag{2.6.25}$$

THEOREM 2.6.5 (O'Reilly, 1974). *For q and ρ_q, as defined above, the condition*

$$\int_0^1 t^{-1}\exp(-\epsilon h_i^2(t))dt < \infty, \qquad \textit{for all} \quad \epsilon > 0, \quad i = 1, 2 \quad (2.6.26)$$

where

$$h_1(t) = t^{-1/2}q(t) \qquad \textit{and} \quad h_2(t) = t^{-1/2}q(1-t), \qquad t \in E, \quad (2.6.27)$$

is both necessary and sufficient for the weak convergence of $U_n = \{U_n(t), t \in E\}$ [in (2.6.8) for $p = 1$] to a Brownian Bridge $X^o = \{X^o(t), t \in E\}$ in the sup-norm metric $\rho_q(x, y)$ on $D[0, 1]$.

In practical applications, one often chooses

$$q(t) = \{t(1-t)\}^{(1/2)-\epsilon}, \qquad \text{for} \quad t \in E \quad \text{and some} \quad \epsilon > 0, \quad (2.6.28)$$

and this satisfies the conditions of Theorem 2.6.5. In this context, the following result due to Ghosh (1972) also deserves mention: for every $\theta > 0$, there exist $K(0 < K < \infty), \epsilon = (1 + (\theta)/2(2 + \theta)$ and n_0 (all dependent on θ), such that for every $n \geqslant n_0$,

$$P\left\{ \sup_{t \in E} \frac{|U_n(t)|}{\{t(1-t)\}^{(1/2)-\epsilon}} \geqslant K(\log n) \right\} \leqslant 2n^{-1-\theta}. \quad (2.6.29)$$

By (2.6.29) and the Borel-Cantelli Lemma, we obtain that as $n \to \infty$,

$$(\log n)^{-1} \sup_{t \in E} \frac{|U_n(t)|}{\{t(1-t)\}^{(1/2)-\epsilon}} = O(1) \qquad \text{a.s.} \quad (\text{for } \epsilon > \tfrac{1}{4}). \quad (2.6.30)$$

However, (2.6.30) is not sharp and the following result (due to Csaki, 1977) is worth mentioning: for every $\epsilon > 0$, as $n \to \infty$,

$$(\log\log n)^{-(1/2)} \sup_{t \in E} \frac{|U_n(t)|}{\{t(1-t)\}^{(1/2)-\epsilon}} = O(1) \qquad \text{a.s.} \quad (2.6.31)$$

Also Csörgö and Révész (1975c) proved the following: Let $\epsilon_n = (\log n)^4/n$. Then,

$$(\log\log n)^{-(1/2)} \sup_{\epsilon_n \leqslant t \leqslant 1-\epsilon_n} \frac{|U_n(t)|}{\{t(1-t)\}^{1/2}} \leqslant 2 \qquad \text{a.s.,} \quad \text{as} \quad n \to \infty. \quad (2.6.32)$$

For some related results, we may refer to Shorack and Wellner (1978). An excellent review of limit theorems for empirical processes has been made by Gaenssler and Stute (1979).

Wellner (1974) has extended Theorem 2.6.2 to weak convergence in

ρ_q-metric as follows. Define $U_n^*(s,t)$ as in (2.6.14) where we take $p = 1$. Let $q = \{q(s,t), (s,t) \in E^2\}$ be a nonnegative, continuous function on E^2, such that for a fixed $s(\in E), q(s,t)$ satisfies the conditions on q imposed before (2.6.25) and for every $t(\in E), q(s,t)$ is nondecreasing in s. Further, we assume that

$$\int_0^1 \int_0^1 [q(s,t)]^{-2} ds\, dt < \infty, \tag{2.6.33}$$

and define ρ_q by

$$\rho_q(x, y) = \sup\left\{ \frac{|x(s,t) - y(s,t)|}{q(s,t)} : (s,t) \in E^2 \right\}. \tag{2.6.34}$$

Then we have the following:

THEOREM 2.6.6 (Wellner, 1974). *The weak convergence in Theorem* 2.6.2 *holds in the* ρ_q*-metric when* q *satisfies the conditions stated above.*

The papers by Pyke and Shorack (1968a, b) and Braun (1976) relate to two-sample versions of Theorems 2.6.6 and will be referred to later on in Chapter 4.

2.7. CERTAIN BOUNDARY-CROSSING PROBABILITIES FOR WIENER PROCESSES

In this section, we present some results on the distribution of some functionals of the Brownian motion process and the Brownian Bridge process having useful statistical applications. As before, we denote by $X = \{X(t), t \in [0, \infty)\}$ a standard Brownian motion process on R^+ and by $X^o = \{X^o(t), t \in E\}$ a standard Brownian Bridge on [0, 1].

(i) (Doob, 1949, p. 397): For every nonnegative s, T, and λ,

$$\begin{aligned} P\left\{ \sup_{0 \leqslant t \leqslant T} (X(s+t) - X(s)) \geqslant \lambda \right\} &= 2P\{X(s+T) - X(s) \geqslant \lambda\} \\ &= 2P\{X(T) \geqslant \lambda\} \\ &= 2P\{X(1) \geqslant \lambda / T^{1/2}\} \\ &= (2/\pi)^{1/2} \int_{\lambda/T^{1/2}}^{\infty} \exp\left(-\tfrac{1}{2}x^2\right) dx. \end{aligned} \tag{2.7.1}$$

(ii) (see Parthasarathy, 1967, p. 230): For every nonnegative T and real $a_1 < a_2$,

$$P\{a_1 < X(t) < a_2, \quad \forall 0 \leqslant t \leqslant T\}$$
$$= (2\pi T)^{-(1/2)} \sum_{k=-\infty}^{\infty} \int_{a_1}^{a_2} \Big\{ \exp\Big[-\frac{1}{2T}(x + 2k(a_2 - a_1))^2\Big] - \exp\Big[-\frac{1}{2T}(x - 2a_2 + 2k(a_2 - a_1))^2\Big]\Big\}\, dx. \tag{2.7.2}$$

In particular, if $a_2 = -a_1 = a > 0$, we have

$$P\Big\{\sup_{0 \leqslant t \leqslant T} |X(t)| \leqslant a\Big\} = \sum_{k=-\infty}^{\infty} (-1)^k \Big[\Phi\big((2k+1)a/T^{1/2}\big) - \Phi\big((2k-1)a/T^{1/2}\big)\Big], \tag{2.7.3}$$

where Φ is the standard normal d.f.

(iii) (Doob, 1949, pp. 398–401): If $a \geqslant 0$ and $b > 0$,

$$P\Big\{\sup_{t \in [0,\infty)} \big[X(t) - at - b\big] \geqslant 0\Big\} = e^{-2ab}, \tag{2.7.4}$$

while if $a \geqslant 0$, $b > 0$, $\alpha \geqslant 0$, and $\beta > 0$,

$$P\Big\{\sup_{t \in [0,\infty)} \big[X(t) - at - b\big] \geqslant 0, \inf_{t \in [0,\infty)} \big[X(t) + \alpha t + \beta\big] \leqslant 0\Big\}$$
$$= \sum_{m=1}^{\infty} \Big\{ \exp\big(-2\big[m^2 ab + (m-1)^2 \alpha\beta + m(m-1)(a\beta + \alpha b)\big]\big) + \exp\big(-2\big[(m-1)^2 ab + m^2 \alpha\beta + m(m-1)(a\beta + \alpha b)\big]\big) - \exp\big(-2\big[m^2(ab + \alpha\beta) + m(m-1)a\beta + m(m+1)\alpha b\big]\big) - \exp\big(-2\big[m^2(ab + \alpha\beta) + m(m+1)a\beta + m(m-1)\alpha b\big]\big)\Big\}; \tag{2.7.5}$$

in particular, for $a = \alpha$ and $b = \beta$, (2.7.5) reduces to

$$P\Big\{\sup_{t \in [0,\infty)} \frac{|X(t)|}{a + bt} \geqslant 1\Big\} = 2 \sum_{m=1}^{\infty} (-1)^{m-1} e^{-2m^2 ab}. \tag{2.7.6}$$

If X^o is a standard Brownian Bridge, then we let

$$Y(t) = (t+1)X^o(t/(t+1)), \qquad t \in [0, \infty). \tag{2.7.7}$$

It is easy to verify that $Y = \{Y(t), t \in [0, \infty)\}$ is a standard Brownian motion on $[0, \infty)$. Hence, from the above, we obtain the following:

$$P\{X^o(t) < a(1-t) + bt, \quad \forall t \in [0,1]\} = 1 - e^{-2ab}, \quad \forall a > 0, \ b > 0; \tag{2.7.8}$$

$$P\{-a(1-t) - bt < X^o(t) < a(1-t) + bt, \quad \forall t \in [0,1]\}$$
$$= 1 - 2\sum_{m=1}^{\infty}(-1)^{m+1}\exp(-2m^2ab), \qquad \forall a > 0, \ b > 0, \tag{2.7.9}$$

and for $a = b$, we have

$$P\Big\{\sup_{t\in[0,1]} X^o(t) < a\Big\} = 1 - e^{-2a^2}, \qquad \forall a > 0; \tag{2.7.10}$$

$$P\Big\{\sup_{t\in[0,1]} |X^o(t)| < a\Big\} = 1 - 2\sum_{m=1}^{\infty}(-1)^{m+1}e^{-2a^2m^2}, \qquad \forall a > 0. \tag{2.7.11}$$

(iv) Let $g(t)$ be integrable inside $(0,1)$ and a be a real number. Then,

$$P\Big\{\int_0^1 g(t)X^o(t)\,dt \leqslant a\Big\} = \Phi(a/\sigma_0), \tag{2.7.12}$$

where

$$\sigma_0^2 = \int_0^1 [G(t)]^2\,dt - \Big(\int_0^1 G(t)\,dt\Big)^2; \qquad G(t) = \int_0^t g(s)\,ds, \qquad t \in [0,1]; \tag{2.7.13}$$

$$P\Big\{\int_0^1 g(t)X(t)\,dt \leqslant a\Big\} = \Phi(a/\sigma) \tag{2.7.14}$$

where

$$\sigma^2 = \int_0^1 [G(t)]^2\,dt + [G(1)]^2 - 2G(1)\Big(\int_0^1 G(t)\,dt\Big). \tag{2.7.15}$$

(v) (Anderson, 1960): Let $\gamma_2 < 0 < \gamma_1$ and $\delta_1 \neq \delta_2$ (not $\delta_1 = \delta_2 = 0$) be real constants. Then, for X, a Brownian motion on $[0, \infty)$,

$$P\{X(t) \geqslant \gamma_1 + \delta_1 t \quad \text{for a smaller } t \text{ than any } t \text{ for which } X(t) \leqslant \gamma_2 + \delta_2 t\}$$

$$= \sum_{r=1}^{\infty} \Big\{ \exp\big(-2\big[r^2\gamma_1\delta_1 + (r-1)^2\gamma_2\delta_2 - r(r-1)(\gamma_1\delta_2 + \gamma_2\delta_1)\big]\big) - \exp\big(-2\big[r^2(\gamma_1\delta_1 + \gamma_2\delta_2) - r(r-1)\gamma_1\delta_2 - r(r+1)\gamma_2\delta_1\big]\big)\Big\}, \quad \delta_1 \geqslant 0;$$

$$= 1 - \sum_{r=1}^{\infty} \Big\{ \exp\big(-2\big[(r-1)^2\gamma_1\delta_1 + r^2\gamma_2\delta_2 - r(r-1)(\gamma_1\delta_2 + \gamma_2\delta_1)\big]\big) - \exp\big(-2\big[r^2(\gamma_1\delta_1 + \gamma_2\delta_2) - r(r+1)\gamma_1\delta_1 - r(r-1)\gamma_2\delta_1\big]\big)\Big\}, \quad \delta_1 \leqslant 0;$$

$$= (\exp(-2\gamma_2\delta_1) - 1)/(\exp(2(\gamma_1 - \gamma_2)\delta_1) - 1), \quad \delta_1 = \delta_2 \neq 0. \tag{2.7.16}$$

Also, let $\gamma_2 < 0 < \gamma_1$ and δ_1, δ_2 be such that $\gamma_1 + \delta_1 T \geqslant \gamma_2 + \delta_2 T, T > 0$. Then,

$$P\{X(t) \geqslant \gamma_1 + \delta_1 t \quad \text{for a } t(\leqslant T) \quad \text{smaller than any } t(\leqslant T) \text{ for which } X(t) \leqslant \gamma_2 + \delta_2 t\}$$

$$= 1 - \Phi\big((\gamma_1 + \delta_1 T)/T^{1/2}\big)$$
$$+ \sum_{r=1}^{\infty} \Big\{ \exp\big(-2[r\gamma_1 - (r-1)\gamma_2][r\delta_1 - (r-1)\delta_2]\big) \times \Phi\big(T^{-(1/2)}[\delta_1 T + 2(r-1)\gamma_2 - (2r-1)\gamma_1]\big)$$
$$- \exp\big(-2\big[r^2(\gamma_1\delta_1 + \gamma_2\delta_2) - r(r-1)\gamma_1\delta_2 - r(r+1)\gamma_2\delta_1\big]\big) \times \Phi\big(T^{-(1/2)}[\delta_1 T + 2r\gamma_2 - (2r-1)\gamma_1]\big)$$
$$- \exp\big(-2[(r-1)\gamma_1 - r\gamma_2][(r-1)\delta_1 - r\delta_2]\big) \times \big[1 - \Phi\big(T^{-(1/2)}[\delta_1 T - 2r\gamma_2 + (2r-1)\gamma_1]\big)\big]$$
$$- \exp\big(-2\big[r^2(\gamma_1\delta_1 + \gamma_2\delta_2) - r(r-1)\gamma_2\delta_1 - r(r+1)\gamma_1\delta_2\big]\big) \times \big[1 - \Phi\big(T^{-(1/2)}[\delta_1 T + (2r+1)\gamma_1 - 2r\gamma_2]\big)\big]\Big\}. \tag{2.7.17}$$

(vi) (Robbins and Siegmund, 1970): Let F denote any measure on R^+ which is finite on bounded intervals. For $x \in R$, $t \in R$, and $\epsilon \in R^+$, let

$$0 < f(x,t) = \int_0^\infty \exp\left(xy - \tfrac{1}{2}y^2 t\right) dF(y) \leqslant \infty,$$
$$-\infty \leqslant A(t,\epsilon) = \inf\{x : f(x,t) \geqslant \epsilon\} < \infty \tag{2.7.18}$$

and let $\phi(x) = \Phi'(x)$ be the standard normal probability density function. Then, for X, a standard Brownian motion process on $[0,\infty)$, (a) for any b, h, ϵ such that $f(b,h) < \epsilon$,

$$P\{X(t) \geqslant A(t+h,\epsilon) - b \quad \text{for some} \quad t \geqslant 0\} = f(b,h)/\epsilon, \tag{2.7.19}$$

and (b) for any b, h, ϵ and $\tau > 0$,

$$P\{X(t) \geqslant A(t+h,\epsilon) - b \quad \text{for some} \quad t \geqslant \tau\}$$
$$= 1 - \Phi\left(\tau^{-(1/2)}\left[A(\tau+h,\epsilon) - b\right]\right) + \epsilon^{-1}\int_0^\infty \exp\left(by - \tfrac{1}{2}y^2 h\right)$$
$$\times \Phi\left(\tau^{-(1/2)}\left[A(\tau+h,\epsilon) - b\right] - y\tau^{1/2}\right) dF(y). \tag{2.7.20}$$

In particular, letting F be a measure on R that assigns measure 0 to $\{0\}$ and where $dF(y) = (2\pi)^{-(1/2)}dy$, $b = 0$ and $\epsilon = \exp(\tfrac{1}{2}a^2)$, we obtain that

$$P\left\{|X(t)| \geqslant \left[t(a^2 + \log t)\right]^{1/2} \quad \text{for some} \quad t \geqslant \tau\right\}$$
$$= 2\left\{1 - \Phi\left[(a^2 + \log\tau)^{1/2}\right] + \left[(a^2 + \log\tau)/\tau\right]^{1/2}\Phi(a)\right\} \quad (\tau > e^{-a^2}); \tag{2.7.21}$$

$$P\left\{|X(t)| \geqslant \left[(t+h)\left(a^2 + \log(t+h)\right)\right]^{1/2} \quad \text{for some} \quad t > 0\right\}$$
$$= h^{-(1/2)}e^{-(1/2)a^2} \qquad (h > e^{-a^2}). \tag{2.7.22}$$

The following result also follows from (2.7.19): for $b < a$ and $a > 0$,

$$P\left\{\sup_{0 \leqslant t \leqslant 1} X(t) \geqslant a \mid X(1) = b\right\} = P\left\{\sup_{t \geqslant 0} X(t)/t \geqslant a \mid X(1) = b\right\}$$
$$= P\left\{\sup_{t \geqslant 0}\left[b + X(t)\right]/(1+t) \geqslant a\right\}$$
$$= \exp(-2a(a-b)). \tag{2.7.23}$$

We consider next the *Bessel process* defined in (2.3.50). For this, we have the following result (see DeLong, 1980):

$$P\left\{\sup_{t \in I} B(t) > x\right\} = 1 + 2^{-q}\left(\Gamma\left(\frac{k}{2}\right)\right)^{-1}\sum_{n=1}^{\infty}(z_{\nu n})^q \exp\left(-z_{\nu n}^2/2x^2\right)\Big/ J_\nu'(z_{\nu n}), \tag{2.7.24}$$

where $q = k/2 - 1, J_\nu$ is the Bessel function with index $\nu = q + 1$ and $z_{\nu n}$ is the nth positive root of J_ν. For $k = 1$, (2.7.24) reduces to the complementary part of (2.7.3). Also, for $k = 3$, it reduces to

$$1 + 2 \sum_{n=1}^{\infty} (-1)^n \exp(-n^2\pi^2/2x^2). \tag{2.7.25}$$

Finally, if $X_j^o = \{X_j^o(t), t \in [0, 1]\}, j = 1, \ldots, q$ be independent copies of a standard Brownian Bridge and we set

$$B^0(t) = \left\{ \sum_{j=1}^{q} (X_j^o(t))^2 \right\}^{1/2}, \qquad 0 \leqslant t \leqslant 1, \tag{2.7.26}$$

then, it follows from Kiefer (1959) that for every real $x \geqslant 0$,

$$P\left\{ \sup_{0 \leqslant t \leqslant 1} B^0(t) \leqslant x \right\} = \frac{4}{\Gamma\left(\frac{q}{2}\right)(2x^2)^{q/2}} \sum_{n=1}^{\infty} (z_{\nu n})^{\nu} \{J_{q/2}(z_{\nu n})\}^{-2}$$
$$\times \exp(-z_{\nu n}^2/2x^2), \tag{2.7.27}$$

where $\nu = (q - 2)/2$ and J_h and $z_{\nu n}$ are defined as in (2.7.24). In particular, for $q = 1$, (2.7.27) reduces to (2.7.11) and for $q = 3$, it reduces to

$$1 + 4 \sum_{n=1}^{\infty} \left[\tfrac{1}{2} - 2n^2x^2\right]\exp(-2n^2x^2). \tag{2.7.28}$$

Some tabulation of the d.f. in (2.7.7) for $q \leqslant 5$ has been made in Kiefer (1959).

Finally, let us consider the case of some standardized Bessel processes when the range is restricted to a subinterval of I. Note that for arbitrary $t_2 > t_1 > 0$ with $t_2/t_1 = T(> 1)$ and $c > 0$,

$$P\left\{ \sup_{t_1 \leqslant t \leqslant 2} t^{-(1/2)}B(t) > c \right\} = P\left\{ \sup_{1 \leqslant t \leqslant T} t^{-(1/2)}B(t) > c \right\}. \tag{2.7.29}$$

Also, using (2.7.26) along with (2.7.7), we obtain that for every $0 < t_1 < t_2 < 1$ and $c \geqslant 0$, on letting $s_i = t_i/(1 - t_i)$ $i = 1, 2$ and $S = s_2/s_1(> 1)$,

$$P\left\{ \sup_{t_1 \leqslant t \leqslant t_2} \{t(1 - t)\}^{-(1/2)} B^0(t) > c \right\}$$
$$= P\left\{ \sup_{s_1 \leqslant s \leqslant s_2} s^{-(1/2)}B(s) > c \right\} = P\left\{ \sup_{1 \leqslant s \leqslant S} s^{-(1/2)}B(s) > c \right\}. \tag{2.7.30}$$

Thus for both the standardized Bessel processes, it suffices to consider the case of

$$P\left\{ \sup_{1 \leqslant t \leqslant T} t^{-(1/2)}B(t) \leqslant c \right\}, \qquad T > 1, \quad c \geqslant 0, \tag{2.7.31}$$

where $B = \{B(t), t \in [0, \infty)\}$ is a standard Bessel processes on $[0, \infty)$. For this square root boundary and truncated range, some tabulations of (2.7.31) for a range of c and q values are due to DeLong (1981). He has used the following asymptotic expansion for (2.7.31):

$$\sum_{n \geqslant 1} a_n(c,q) T^{-\{b_n(c,q)\}}, \tag{2.7.32}$$

where q is the dimension of the Bessel process, $c > 0, T > 1$ and the $b_n(c,q)$ are the roots of certain equations, while the $a_n(c,q)$ are defined in terms of the $b_n(c,q)$.

Some typical percentile points of the distributions in (2.7.24), (2.7.27) and (2.7.31), for certain specific values of k (or q), are provided at the end of the Appendix. These critical values are useful for the practical applications of the sequential procedures to be discussed in Part 2.

CHAPTER 3

Invariance Principles for U-Statistics and Von Mises' Differentiable Statistical Functions

3.1. INTRODUCTION

Consider a sequence $\{X_i; i \geqslant 1\}$ of independent and identically distributed random vectors (i.i.d.r.v.) having a $p(\geqslant 1)$-variate distribution function (d.f.) $F(x)$, $x \in R^p$, the p-dimensional Euclidean space. In problems of nonparametric statistical inference, the functional form of the d.f. F is usually unknown; it is only assumed that F belongs to some suitable family $\mathcal{F}$ of d.f.'s on R^p. In this formulation, (estimable) parameters are expressed as functionals of F, defined on $\mathcal{F}$. This chapter is concerned with such functionals of unknown distribution functions, their natural estimators, and weak as well as strong invariance principles for these estimators.

In Section 3.2, along with the preliminary notions, U-statistics and von Mises' (1947) differentiable statistical functions are introduced and their basic properties are briefly presented. Section 3.3 deals with the weak convergence of these statistics (Loynes, 1970; and Miller and Sen, 1972) and this contains earlier results of Hoeffding (1948a) and von Mises (1947) as special cases. Strong invariance principles for these statistics (Sen, 1974b) are considered in Section 3.4. Section 3.5 extends the results of Section 3.3 to the case of sampling without replacement from a finite population (Sen, 1970b, 1972d). Section 3.6 is concerned with generalized U-statistics dealing with multisample problems (Sen, 1974c). The last section is devoted to the estimation of the dispersion of U-statistics and von Mises functionals and deals with the almost sure (a.s.) convergence of these estimators (Sen, 1960,

1977a; Sproule 1974). A few problems of practical interest are presented as exercises to facilitate understanding of the theory.

3.2. U-STATISTICS AND VON MISES FUNCTIONALS

Let $\mathcal{F}$ be the space of all d.f.'s belonging to a class and for every $F \in \mathcal{F}$, consider a functional $\theta(F)$ whose domain is $\mathcal{F}$. For example, let $\mathcal{F}$ be the class of all univariate d.f.'s with finite first moment. Then $\mu = \mu(F) = \int x\, dF(x)$ is a (*linear*) functional of F. If F possesses a finite second moment, $\sigma^2 = \sigma^2(F) = \int (x-\mu)^2 dF(x) = \frac{1}{2}\int\int (x-y)^2 dF(x)\, dF(y)$ is also a (*bilinear*) functional of F. In general, moments, cumulants, and some other measures of a d.f. F are functionals of this type. In nonparametric problems, other functionals of F are also quite frequently used.

A *homogeneous statistical function of degree* $m(\geqslant 1)$ is given by

$$\begin{aligned}\theta(F) &= \int \cdots \int g(x_1, \ldots, x_m)\, dF(x_1) \cdots dF(x_m) \\ &= E\{g(X_1, \ldots, X_m)\}, \qquad \forall F \in \mathcal{F}, \qquad (3.2.1)\end{aligned}$$

where $\mathcal{F} = \{F : |\theta(F)| < \infty\}$, $g(x_1, \ldots, x_m)$ is a Borel measurable *kernel* (statistic) and the X_i are i.i.d.r.v. with the d.f. F. Without any loss of generality, we may assume that $g(x_1, \ldots, x_m)$ is symmetric in its m arguments. Thus $\theta(F)$ is termed an *estimable parameter* (or a *regular functional*) if there exists a symmetric kernel $g(X_1, \ldots, X_m)$ of degree m for which (3.2.1) holds for every $F \in \mathcal{F}$; $\mathcal{F}$ is said to be the *domain* of $\theta(F)$.

Our primary concern is to formulate some natural estimators of $\theta(F)$ and to study the properties of these estimators. For a sample $(X_1, \ldots, X_n)$ of size $n(\geqslant 1)$, consider the usual *empirical* d.f.

$$F_n(x) = n^{-1} \sum_{i=1}^{n} c(x - X_i), \qquad x \in R^p, \qquad n \geqslant 1. \qquad (3.2.2)$$

where $c(u)$ is 1 iff all the p coordinates of u are nonnegative, and is 0 otherwise. Note that F_n is an unbiased estimator of F. In fact, it is a sufficient statistic (though not necessarily the minimal sufficient statistic). Thus one way to estimate $\theta(F)$ is to replace F by F_n in (3.2.1). This leads to the following estimator, classically known as the von Mises' (1947) *differentiable statistical function*.

$$\begin{aligned}\theta(F_n) &= \int \cdots \int g(x_1, \ldots, x_m)\, dF_n(x_1) \cdots dF_n(x_m) \\ &= n^{-m} \sum_{i_1=1}^{n} \cdots \sum_{i_m=1}^{n} g(X_{i_1}, \ldots, X_{i_m}), \qquad n \geqslant 1. \qquad (3.2.3)\end{aligned}$$

That is, $\theta(F_n)$ is the same functional of the empirical d.f. F_n. We may note that though F_n unbiasedly estimates F, the same is not generally true for $\theta(F_n)$. For example, for $\theta(F) = \sigma^2(F)$, defined earlier, $m = 2$ and $E\theta(F_n) \neq \sigma^2(F)$ for every finite $n(\geqslant 1)$. (See Problem 3.2.1 in this context.) However, we shall see that for large n, the bias of $\theta(F_n)$ is not a matter of practical concern.

An unbiased estimator (called the *U-statistic*) of $\theta(F)$ may be introduced as follows. Suppose that $n \geqslant m$. Since, for all distinct indices $i_1, \ldots, i_m$, $g(X_{i_1}, \ldots, X_{i_m})$ unbiasedly estimates $\theta(F)$ [by (3.2.1)], a symmetric, unbiased estimator of $\theta(F)$ is given by

$$U_n = n^{-[m]} \sum_{1 \leqslant i_1 \neq \cdots \neq i_m \leqslant n} g(X_{i_1}, \ldots, X_{i_m})$$

$$= \binom{n}{m}^{-1} \sum_{1 \leqslant i_1 < \cdots < i_m \leqslant n} g(X_{i_1}, \ldots, X_{i_m}), \qquad n \geqslant m, \tag{3.2.4}$$

where

$$n^{-[m]} = \left(n^{[m]}\right)^{-1} = \{n \ldots (n - m + 1)\}^{-1}$$

and $\binom{n}{m} = {}^nC_m = n^{[m]}/m!$. Note that whereas $\theta(F_n)$ is defined for all $n \geqslant 1$, U_n is so for $n \geqslant m$. If $m = 1$, $\theta(F_n) = U_n = n^{-1}\sum_{i=1}^n g(X_i)$. However, for $m \geqslant 2$, these two statistics are not generally identical. Among the class of unbiased estimators, U_n possesses some optimality properties. For any unbiased estimator $T_n = T(X_1, \ldots, X_n)$ of $\theta(F)$, the corresponding U-statistic has a variance smaller than or equal to that of T_n and a similar inequality holds for the risk using any convex loss function. Further, if the vector of order statistics (in the case of univariate observations) or the matrix of orbit-collections of random vectors (in the multivariate case) is complete for the class of d.f.'s under consideration, then U_n is the unique minimum variance (as well as the minimum risk, with any convex loss function) unbiased estimator of $\theta(F)$. We refer to Halmos (1946), Hoeffding (1948a), Fraser (1957), Puri and Sen (1971) and other sources for detailed discussions of these properties of U-statistics and related von Mises functionals. Here we briefly present the principal results on U_n and $\theta(F_n)$ having relevance to their invariance principles.

Let $\mathcal{C}_n$ be the sigma field generated by the unordered collection $\{X_1, \ldots, X_n\}$ and by $X_{n+1}, X_{n+2}, \ldots$, for $n \geqslant 1$. If the X_i are real valued, $\mathcal{C}_n = \mathcal{C}(X_{n,1}, \ldots, X_{n,n}; X_{n+1}, \ldots)$ where $X_{n,1} \leqslant \cdots \leqslant X_{n,n}$ are the ordered values corresponding to $X_1, \ldots, X_n$. Then $\mathcal{C}_n$ is nonincreasing in $n(\geqslant 1)$.

THEOREM 3.2.1 (Berk-Kingman). *For every estimable $\theta(F)$, $\{U_n, \mathcal{C}_n; n \geqslant m\}$ is a reverse martingale and $U_n \to \theta(F)$ a.s., as $n \to \infty$.*

Proof. Given $\mathcal{C}_n$, the $X_{n+j}, j \geqslant 1$ are all fixed, while all possible permutations of $X_1, \ldots, X_n$ among themselves are (conditionally) equally likely. Therefore, for every $1 \leqslant j_1 < \cdots < j_m \leqslant n$,

$$E[g(X_{j_1}, \ldots, X_{j_m}) | \mathcal{C}_n] = n^{-[m]} \sum_{1 \leqslant i_1 \neq \cdots \neq i_m \leqslant n} g(X_{i_1}, \ldots, X_{i_m}) = U_n. \tag{3.2.5}$$

Consequently, for every k: $m \leqslant k \leqslant n$,

$$\begin{aligned} E[U_k | \mathcal{C}_n] &= \binom{k}{m}^{-1} \sum_{1 \leqslant j_1 < \cdots < j_m \leqslant k} E[g(X_{j_1}, \ldots, X_{j_m}) | \mathcal{C}_n] \\ &= U_n, \qquad \text{by (3.2.5).} \end{aligned} \tag{3.2.6}$$

This proves the first part of the theorem. The second part is a direct consequence of the reverse martingale convergence theorem. Q.E.D.

To derive a similar result for $\{\theta(F_n)\}$, we note that by (3.2.3) and (3.2.4), for every $n \geqslant m$,

$$\theta(F_n) = n^{-m} n^{[m]} U^*_{n,0} + \sum_{h=1}^{m-1} n^{-m} n^{[m-h]} U^*_{n,h}, \tag{3.2.7}$$

where $U^*_{n,0} = U_n$ and for h: $1 \leqslant h \leqslant m-1$,

$$\begin{aligned} U^*_{n,h} &= n^{-[m-h]} \sum\nolimits_{n,h} g(X_{i_1}, \ldots, X_{i_m}) \\ &= n^{-[m-h]} \sum\nolimits^*_{n,h} g^*_h(X_{i_1}, \ldots, X_{i_{m-h}}) \\ &= \binom{n}{m-h}^{-1} \sum_{1 \leqslant i_1 < \cdots < i_{m-h} \leqslant n} g^*_h(X_{i_1}, \ldots, X_{i_{m-h}}), \end{aligned} \tag{3.2.8}$$

the summation $\sum_{n,h}$ extends over all possible choice of $i_1, \ldots, i_m$ (over $1, \ldots, n$) for which $m-h$ of the indices are distinct, g^*_h is the symmetrized version of g involving $m-h$ distinct arguments, the summation $\sum^*_{n,h}$ extends over all $1 \leqslant i_1 \neq \cdots \neq i_{m-h} \leqslant n$ and by the last equation in (3.2.8), $U^*_{n,h}$ is also a U-statistic for every $0 \leqslant h \leqslant m-1$, where $g^*_0 = g$. Since $n^{-m} n^{[m]} = 1 + O(n^{-1})$ while $n^{-m} n^{[m-h]} = O(n^{-h})$, $h \geqslant 1$, by using (3.2.7) and Theorem 3.2.1 for every $\{U^*_{n,h}, \mathcal{C}_n; n \geqslant m\}$, $h = 0, 1, \ldots, m-1$, we obtain that *whenever* $|Eg^*_h| < \infty$ *for* $0 \leqslant h \leqslant m-1$,

$$|\theta(F_n) - U_n| = O(n^{-1}) \qquad \text{a.s.,} \quad \text{as} \quad n \to \infty, \tag{3.2.9}$$

so that by Theorem 3.2.1 and (3.2.9),

$$\theta(F_n) \to \theta(F) \qquad \text{a.s.,} \quad \text{as } n \to \infty. \tag{3.2.10}$$

Let us now assume that $Eg^2 < \infty$ and for every h: $0 \leqslant h \leqslant m$, we denote by

$$g_h(x_1, \ldots, x_h) = Eg(x_1, \ldots, x_h, X_{h+1}, \ldots, X_m), \qquad g_0 = \theta(F); \tag{3.2.11}$$

$$\zeta_h(F) = Eg_h^2(X_1, \ldots, X_h) - \theta^2(F), \qquad \zeta_0(F) = 0. \tag{3.2.12}$$

Note that the $\zeta_h(F)$ are all estimable parameters; the degree of $\zeta_h(F)$ is $\leqslant 2m - h$, $1 \leqslant h \leqslant m$. Moreover, by the Jensen inequality, $\zeta_h(F)$ is nondecreasing in h $(0 \leqslant h \leqslant m)$ and is nonnegative. Thus,

$$[\zeta_m(F) < \infty] \Rightarrow [\zeta_h(F) < \infty], \qquad \forall 0 \leqslant h \leqslant m, \tag{3.2.13}$$

and if for some d $(0 \leqslant d \leqslant m)$, $\zeta_d(F) = 0$, then $\zeta_c(F) = 0$, $\forall c \leqslant d$. In fact, Hoeffding (1948a) has obtained the following inequality:

$$0 \leqslant \zeta_c(F) \leqslant cd^{-1}\zeta_d(F), \qquad \forall 1 \leqslant c < d \leqslant m; \tag{3.2.14}$$

we leave the proof as an exercise (see Problem 3.2.8).

$\theta(F)$ is said to be *stationary of order* $d(\geqslant 0)$ at $F_0(\in \mathcal{F})$ if $\zeta_d(F_0) = 0 < \zeta_{d+1}(F_0) < \infty$. In particular, if $\zeta_1(F_0) > 0$, $\theta(F)$ is said to be stationary of order 0. As an example, let $\mu(F)$ and $\sigma^2(F)$ be defined as earlier and let $\theta(F) = \mu^2(F)$ where we assume that $0 < \sigma(F) < \infty$. Then $g(x_1, x_2) = x_1 x_2$, and hence $\zeta_1(F) = \mu^2(F)\sigma^2(F)$, $\zeta_2(F) = \sigma^4(F) + 2\mu^2(F)\sigma^2(F)$. Thus, if $\mathcal{F}_0 = \{F: \mu(F) = 0\}$, then $\theta(F)$ is stationary of order 1 on $\mathcal{F}_0$ and of order 0 on $\mathcal{F}\backslash\mathcal{F}_0$.

Note that $\mathrm{Cov}[g(X_{i_1}, \ldots, X_{i_m}), g(X_{j_1}, \ldots, X_{j_m})] = \zeta_h(F)$ whenever h of the indices $(i_1, \ldots, i_m)$ are in common with those of $(j_1, \ldots, j_m)$, $0 \leqslant h \leqslant m$. Hence, by (3.2.4), we obtain that

$$\begin{aligned} \mathrm{Var}(U_n) &= \binom{n}{m}^{-2} \sum_{h=0}^{m} \binom{n}{m}\binom{m}{h}\binom{n-m}{m-h}\zeta_h(F) \\ &= \binom{n}{m}^{-1} \sum_{h=1}^{m} \binom{m}{h}\binom{n-m}{m-h}\zeta_h(F). \end{aligned} \tag{3.2.15}$$

By virtue of (3.2.14) and (3.2.15), $m^2\zeta_1(F) \leqslant n\,\mathrm{Var}(U_n) \leqslant m\zeta_m(F)$, $\forall n \geqslant m$ and, moreover, $n\,\mathrm{Var}(U_n)$ is nonincreasing in n. The lower limit is attained as $n \to \infty$, while the upper limit is attained for $n = m$. If $\theta(F)$ is stationary of order 0, we have

$$\mathrm{Var}(U_n) = m^2 n^{-1}\zeta_1(F) + O(n^{-2}). \tag{3.2.16}$$

In general, if $\theta(F)$ is stationary of order d $(0 \leqslant d \leqslant m-1)$,

$$\mathrm{Var}(U_n) = n^{-(1+d)}(d+1)!\binom{m}{d+1}^2 \zeta_{d+1}(F) + O(n^{-2-d}). \tag{3.2.17}$$

Since in (3.2.3), the indices $i_1, \ldots, i_m$ need not be all distinct, the expression for $\mathrm{Var}(\theta(F_n))$ is more complicated. However, the following asymptotic expression can easily be obtained. We assume that

$$E\left[g_h^*(X_1,\ldots,X_{m-h})\right]^2<\infty,\qquad \forall 0\leqslant h\leqslant m-1,\tag{3.2.18}$$

where the g_h^* are defined after (3.2.8). Then by using (3.2.7), (3.2.16), and the C_r-inequality (that is, $|a+b|^r\leqslant C_r\{|a|^r+|b|^r\}$ for every $r\geqslant 0$ where $C_r=1$ or 2^{r-1} according as r is $\leqslant$ or >1), we obtain that under (3.2.18) for some C_m^*, independent of n,

$$\begin{aligned}\operatorname{Var}\left[\theta(F_n)-U_n\right]&\leqslant C_m^*\left\{(1-n^{-m}n^{[m]})^2EU_n^2\right.\\&\qquad\left.+\sum_{h=1}^{m-1}(n^{-m}n^{[m-h]})^2EU_{n,h}^{*2}\right\}\\&=O(n^{-2}),\end{aligned}\tag{3.2.19}$$

while

$$E\left[\theta(F_n)-U_n\right]=O(n^{-1}),\tag{3.2.20}$$

so that if $\theta(F)$ is stationary of order 0, under (3.2.18),

$$\operatorname{Var}(\theta(F_n))=m^2n^{-1}\zeta_1(F)+O(n^{-2}),\tag{3.2.21}$$

$$n\left[\theta(F_n)-U_n\right]^2\xrightarrow{\mathcal{L}_1}0\qquad\text{as}\quad n\to\infty.\tag{3.2.22}$$

THEOREM 3.2.2. *Let $\{c_n;\ n\geqslant m\}$ be a nondecreasing sequence of positive numbers. Then, for every $m\leqslant n\leqslant N<\infty$ and $t>0$,*

$$\begin{aligned}&P\left\{\max_{n\leqslant k\leqslant N}c_k|U_k-\theta(F)|>t\right\}\\&\qquad\leqslant t^{-2}\left\{c_n^2\operatorname{Var}(U_n)+\sum_{k=n+1}^{N}(c_k^2-c_{k-1}^2)\operatorname{Var}(U_k)\right\}.\end{aligned}\tag{3.2.23}$$

If the series on the rhs of (3.2.23) converges (as $N\to\infty$), we may replace $\max_{n\leqslant k\leqslant N}$ *by* $\sup_{k\geqslant n}$ *and N by ∞ in (3.2.23).*

Proof. By Theorem 3.2.1 and the Jensen inequality, $\{[U_n-\theta(F)]^2,\mathcal{C}_n;\ n\geqslant m\}$ is a nonnegative reverse submartingale. By reversing the index set (and thereby converting the sequence into a forward submartingale), the proof of (3.2.23) follows directly by using (2.2.5). Q.E.D.

Note that if $\theta(F)$ is stationary of order d $(0\leqslant d\leqslant m-1)$, the series on the rhs of (3.2.23) converges whenever

$$\sum_{n\geqslant m}c_n^2n^{-d-2}<\infty;\tag{3.2.24}$$

the proof is left as an exercise.

We proceed to consider now some decomposition for $\theta(F_n)$ and U_n. For this purpose, for every $1 \leqslant h \leqslant m$, we let

$$V_{n,h} = \int_{R^{ph}} \cdots \int g_h(x_1, \ldots, x_h) \prod_{j=1}^{h} d\big[F_n(x_j) - F(x_j)\big], \quad (3.2.25)$$

so that

$$V_{n,1} = n^{-1} \sum_{i=1}^{n} \big[g_1(X_i) - \theta(F) \big]. \quad (3.2.26)$$

Then writing $dF_n(x_i) = dF(x_i) + d[F_n(x_i) - F(x_i)]$, $i = 1, \ldots, m$, we obtain from (3.2.3) and (3.2.25) that

$$\theta(F_n) = \theta(F) + \sum_{h=1}^{m} \binom{m}{h} V_{n,h}, \qquad n \geqslant 1. \quad (3.2.27)$$

Similarly, we may rewrite (3.2.4) as

$$U_n = n^{-[m]} \sum_{1 \leqslant i_1 \neq \cdots \neq i_m \leqslant n} \int_{R^{pm}} \cdots \int g(x_1, \ldots, x_m) \prod_{j=1}^{m} d(c(x_j - X_{i_j})), \quad (3.2.28)$$

so that on writing $dc(x_j - X_{i_j}) = dF(x_j) + d[c(x_j - X_{i_j}) - F(x_j)]$, $1 \leqslant j \leqslant m$, we obtain that for $n \geqslant m$,

$$U_n = \theta(F) + \sum_{h=1}^{m} \binom{m}{h} U_{n,h}; \quad (3.2.29)$$

$$U_{n,h} = n^{-[h]} \sum_{1 \leqslant i_1 \neq \cdots \neq i_h \leqslant n} \int_{R^{ph}} \cdots \int g_h(x_1, \ldots, x_h) \times \prod_{j=1}^{h} d\big[c(x_j - X_{i_j}) - F(x_j)\big], \quad (3.2.30)$$

for $1 \leqslant h \leqslant m$; note that $U_{n,1} = V_{n,1}$. Further, if we write

$$g_h^0(x_1, \ldots, x_h) = g_h(x_1, \ldots, x_h) - \sum_{j=1}^{h} g_{h-1}(x_1, \ldots, x_{j-1}, x_{j+1}, \ldots, x_h) + \cdots + (-1)^h \theta(F), \qquad \forall (x_1, \ldots, x_h) \in R^{ph}, \quad 1 \leqslant h \leqslant m, \quad (3.2.31)$$

we obtain from (3.2.30) and (3.2.31) that

$$U_{n,h} = \binom{n}{h}^{-1} \sum_{1 \leqslant i_1 < \cdots < i_h \leqslant n} g_h^0(X_{i_1}, \ldots, X_{i_h}), \qquad 1 \leqslant h \leqslant m. \quad (3.2.32)$$

Thus, in the decomposition (3.2.29), all the $U_{n,h}$ are themselves U-statistics.

Moreover it follows by direct computation that $EU_{n,h} = 0, \forall 1 \leqslant h \leqslant m$ and further

$$EU_{n,h}^2 = O(n^{-h}), \qquad \text{for} \quad h = 1, 2, \ldots, m; \tag{3.2.33}$$

$$E[U_{n,h} U_{n,h'}] = 0, \qquad \forall 1 \leqslant h < h' \leqslant m. \tag{3.2.34}$$

Consequently (3.2.29) represents an orthogonal decomposition with successive terms of stochastically decreasing order of magnitudes. Finally we note that by (3.2.27) and (3.2.29),

$$\theta(F_n) - U_n = \sum_{h=2}^{m} \binom{m}{h} [V_{n,h} - U_{n,h}]. \tag{3.2.35}$$

Since the X_i are independent, by virtue of (3.2.32), for any two disjoint sets $(i_1, \ldots, i_h)$ and $(i'_1, \ldots, i'_h)$, $g_h^0(X_{i_1}, \ldots, X_{i_h})$ and $g_h^0(X_{x'_1}, \ldots, X_{i'_h})$ are independent. Also, using the definition in (3.2.31), it follows that for every positive integer s and collection $(i_{r1}, \ldots, i_{rh})$, $r = 1, \ldots, s$, if at least one of the subscripts i_j occurs exactly once in these r sets, then

$$E \prod_{r=1}^{s} g_h^0(X_{i_{r1}}, \ldots, X_{i_{rh}}) = 0. \tag{3.2.36}$$

In particular, if we take $s = 2k$ where k is a positive integer, then (3.2.36) may not hold only when none of the subscripts occurs exactly once, and thus, the total number of distinct subscripts in these $2k$ sets combined can be at most equal to kh. Thus if we assume that

$$Eg^{2k}(X_1, \ldots, X_m) < \infty, \tag{3.2.37}$$

(which ensures the same condition for the different g_h^0), it follows from (3.2.32), (3.2.36), and some routine steps that

$$E[U_{n,1}^{2k}] = n^{-k}[(2k)!/k!2^k][\zeta_1(F)]^k + O(n^{-k-1}); \tag{3.2.38}$$

$$E[U_{n,h}^{2k}] = O(n^{-kh}), \qquad h = 2, \ldots, m. \tag{3.2.39}$$

Consequently, by (3.2.27), (3.2.38), and (3.2.39), under (3.2.37),

$$\lim_{n\to\infty} n^k E[U_n - \theta(F)]^{2k} = \frac{(2k)!}{2^k k!} m^{2k} [\zeta_1(F)]^k. \tag{3.2.40}$$

In a similar manner it follows by using (3.2.36) that

$$E[V_{n,h}^{2k}] = O(n^{-kh}), \qquad 2 \leqslant h \leqslant m, \quad \text{when (3.2.37) holds}, \tag{3.2.41}$$

while $U_{n,1} = V_{n,1}$, so that (3.2.38) applies. Hence, under (3.2.37), we have

$$\lim_{n\to\infty} E[\theta(F_n) - \theta(F)]^{2k} = \frac{(2k)!}{2^k k!} m^{2k} [\zeta_1(F)]^k. \tag{3.2.42}$$

Finally, by (3.2.35), (3.2.39), and (3.2.41), under (3.2.34),

$$E[\theta(F_n) - U_n]^{2k} = O(n^{-2k}). \tag{3.2.43}$$

Note that if $n^* = [n/m]$ is the largest integer contained in n/m and if

$$\bar{g}_{n^*} = \frac{1}{n^*}\sum_{h=1}^{n^*} g(X_{(h-1)m+1}, \ldots, X_{hm}), \tag{3.2.44}$$

then

$$[U_n - \theta(F)] = E\{[\bar{g}_{n^*} - \theta(F)] \mid \mathcal{C}_n\}, \tag{3.2.45}$$

so that for every $r > 0$,

$$\begin{aligned} E|U_n - \theta(F)|^r &= E\{|E\{[\bar{g}_{n^*} - \theta(F)] \mid \mathcal{C}_n\}|^r\} \\ &\leqslant E\{E(|\bar{g}_{n^*} - \theta(F)|^r \mid \mathcal{C}_n)\} \\ &= E|\bar{g}_{n^*} - \theta(F)|^r. \end{aligned} \tag{3.2.46}$$

On the other hand, by (3.2.44), $\bar{g}_{n^*}$ involves an average of i.i.d.r.v.'s, so that whenever $E|g|^r < \infty$, for some $r > 1$, we have

$$E|\bar{g}_{n^*} - \theta(F)|^r = O([n^*]^{-r^*}) = O(n^{-r^*}); \tag{3.2.47}$$

where $r^* = r - 1$ or $r/2$ according as $r \in (1,2)$ or $r \geqslant 2$; see Problem 3.2.10 in this connection. By (3.2.46), (3.2.47), and the assumed finiteness of $E|g|^r$, we obtain that

$$E|U_n - \theta(F)|^r = O(n^{-r^*}). \tag{3.2.48}$$

For integer values of r, some results parallel to (3.2.48) have been considered by Grams and Serfling (1973). Equation 3.2.48 is due to Sen and Ghosh (1981); it remains true for the von Mises' differentiable statistical functions.

3.3. WEAK CONVERGENCE OF $\{U_n\}$ AND $\{\theta(F_n)\}$

In this section, the weak invariance principle for U_n and $\theta(F_n)$ will be considered. This principle, in turn, ensures the asymptotic normality of these statistics. First, we consider the case of U-statistics.

For every $n(\geqslant m)$, define a process $Y_n = \{Y_n(t), t \in I\}$, $I = [0,1]$, by letting $Y_n(t) = 0$, $0 \leqslant t \leqslant (m-1)/n$, $Y_n(t) = Y_n(k/n)$ for $k/n \leqslant t < (k+1)/n$, $k = m-1, \ldots, n-1$ and

$$Y_n(k/n) = k\frac{[U_k - \theta(F)]}{m[n\zeta_1(F)]^{1/2}}, \qquad k = m, \ldots, n. \tag{3.3.1}$$

Then Y_n belongs to the space $D[0,1]$, endowed with the Skorokhod J_1-topology.

THEOREM 3.3.1. *If $\theta(F)$ is stationary of order 0, then as $n\to\infty$,*

$$Y_n \xrightarrow{\mathfrak{D}} W, \qquad \text{in the } J_1\text{-topology on} \quad D[0,1], \tag{3.3.2}$$

where W is a standard Brownian motion on I.

Proof. We make use of the decomposition (3.2.29). Let us introduce a second sequence of processes $\{Y_n^{(1)}\}$ where $Y_n^{(1)} = \{Y_n^{(1)}(t), t\in I\}$, $Y_n^{(1)}(t) = 0$, $0 \leqslant t < n^{-1}$, $Y_n^{(1)}(t) = Y_n^{(1)}(k/n)$, $k/n \leqslant t < (k+1)/n$, $1\leqslant k\leqslant n-1$, and

$$\begin{aligned} Y_n^{(1)}(k/n) &= \frac{kU_{k,1}}{[n\zeta_1(F)]^{1/2}} \\ &= \frac{\sum_{i=1}^k [g_1(X_i)-\theta(F)]}{\sqrt{n}\,[\mathrm{Var}(g_1(X))]^{1/2}}, \qquad k=1,\ldots,n. \end{aligned} \tag{3.3.3}$$

By the hypothesis of stationarity of order 0, $0<\zeta_1(F)<\infty$, so that for $Y_n^{(1)}$, the regularity conditions of Theorem 2.4.4 hold (in fact, here $V_n = s_n^2$, $\forall n\geqslant 1$), and hence

$$Y_n^{(1)} \xrightarrow{\mathfrak{D}} W, \qquad \text{in the } J_1\text{-topology on} \quad D[0,1]. \tag{3.3.4}$$

We complete the proof of the theorem by using Theorem 2.3.7 and showing that

$$\rho(Y_n, Y_n^{(1)}) = \sup_{t\in I}|Y_n(t) - Y_n^{(1)}(t)| \xrightarrow{p} 0 \qquad \text{as} \quad n\to\infty. \tag{3.3.5}$$

By (3.3.1) and (3.3.3),

$$\rho(Y_n, Y_n^{(1)}) \leqslant \max\left\{\max_{1\leqslant k\leqslant m-1}\left|Y_n^{(1)}\left(\frac{k}{n}\right)\right|, \max_{m\leqslant k\leqslant n}\left|\frac{\sum_{h=2}^m \binom{m}{h} kU_{k,h}}{m[n\zeta_1(F)]^{1/2}}\right|\right\}, \tag{3.3.6}$$

Now, by (3.3.4) and the fact that $(m-1)/n\to 0$ as $n\to\infty$ (m fixed), we readily claim that as $n\to\infty$,

$$\max_{1\leqslant k\leqslant m-1}\left|Y_n^{(1)}\left(\frac{k}{n}\right)\right| \xrightarrow{p} 0, \tag{3.3.7}$$

(note that $\sup_{0\leqslant t\leqslant (m-1)/n}|W(t)| \xrightarrow{p} 0$ as $n\to\infty$). Also by (3.2.29),

$$\begin{aligned} E\left[\sum_{h=2}^m \binom{m}{h} U_{n,h}\right]^2 &= E[U_n - \theta(F)]^2 - E[U_{n,1}^2] \\ &= O(n^{-2}), \qquad \text{see (3.2.33) and (3.2.34).} \end{aligned} \tag{3.3.8}$$

Further, the $\{U_{n,h}\}$, $1 \leqslant h \leqslant m$ are all reverse martingales with respect to the common sigma-field sequence $\{\mathcal{C}_n\}$, so that $\{\sum_{h=2}^{m}\binom{m}{h}U_{n,h}, \mathcal{C}_n; n \geqslant m\}$ is also a reverse martingale, and hence, by the same method as in the proof of Theorem 3.2.2, we have for every $\epsilon > 0$,

$$P\left\{\max_{m \leqslant k \leqslant n} k\left|\sum_{h=2}^{m}\binom{m}{h}U_{n,h}\right| > \epsilon m\left[n\zeta_1(F)\right]^{1/2}\right\}$$

$$\leqslant \left[m^2\epsilon^2 n\zeta_1(F)\right]^{-1}\left\{m^2 \cdot O(m^{-2}) + \sum_{k=m+1}^{n}(2k-1)O(k^{-2})\right\}$$

$$= \epsilon^{-2}\left[O(n^{-1}\log n)\right] \rightarrow 0 \quad \text{as} \quad n \rightarrow \infty \tag{3.3.9}$$

Consequently (3.3.6) follows from (3.3.7) and (3.3.9). Q.E.D.

Let us now consider the case of $\{\theta(F_n)\}$ and introduce the sequence $\{Y_n^*\}$ of stochastic processes where $Y_n^* = \{Y_n^*(t), t \in I\}$, $Y_n^*(t) = 0$, $0 \leqslant t < n^{-1}$, $Y_n^*(t) = Y_n^*(k/n)$, $k/n \leqslant t < (k+1)/n$, $k = 1, \ldots, n-1$ and

$$Y_n^*(k/n) = \frac{k\left[\theta(F_k) - \theta(F)\right]}{m\left[n\zeta_1(F)\right]^{1/2}}, \qquad k = 1, \ldots, n. \tag{3.3.10}$$

Note that by (3.2.7), (3.3.1), and (3.3.10),

$$\rho(Y_n, Y_n^*) \leqslant \max\left\{\max_{1 \leqslant k \leqslant m-1}\left|Y_n^*\left(\frac{k}{n}\right)\right|, \max_{m \leqslant k \leqslant n}\left|Y_n^*\left(\frac{k}{n}\right) - Y_n\left(\frac{k}{n}\right)\right|\right\}, \tag{3.3.11}$$

$$\max_{m \leqslant k \leqslant n}\left|Y_n^*\left(\frac{k}{n}\right) - Y_n\left(\frac{k}{n}\right)\right| \leqslant \left[m^2 n\zeta_1(F)\right]^{-1/2}$$

$$\times\left\{\max_{m \leqslant k \leqslant n}\left[\binom{m}{2}|U_k| + \sum_{h=1}^{m-1}k^{-h+1}|U_{k,h}^*|\right]\right\}, \tag{3.3.12}$$

where by Theorem 3.2.1 and (3.2.8), $\{\binom{m}{2}|U_n| + \sum_{h=1}^{m-1}n^{-h+1}|U_{n,h}^*|, \mathcal{C}_n; n \geqslant m\}$ is a reverse submartingale on which we use (2.2.5) and claim that for every $\epsilon > 0$,

$$P\left\{\max_{m \leqslant k \leqslant n}\left|Y_n^*\left(\frac{k}{n}\right) - Y_n\left(\frac{k}{n}\right)\right| > \epsilon\right\}$$

$$\leqslant \left[\epsilon^2 m^2 n\zeta_1(F)\right]^{-1/2} E\left\{\binom{m}{2}|U_m| + \sum_{h=1}^{m-1}m^{-h+1}|U_{m,h}^*|\right\}$$

$$= O(n^{-1/2}) \rightarrow 0, \quad \text{as} \quad n \rightarrow \infty, \quad \text{under (3.2.18).} \tag{3.3.13}$$

Finally, under (3.2.18), for every $\epsilon > 0$,

$$
\begin{aligned}
P&\left\{\max_{1 \leqslant k \leqslant m-1}\left|Y_n^*\left(\frac{k}{n}\right)\right| > \epsilon\right\} \\
&= P\left\{\max_{1 \leqslant k \leqslant m-1} k|\theta(F_k) - \theta(F)| > \epsilon\left[n\zeta_1(F)\right]^{1/2}\right\} \\
&\leqslant \left[\epsilon^2 n\zeta_1(F)\right]^{-1}\left\{\sum_{k=1}^{m-1} k^2 E\left[\theta(F_k) - \theta(F)\right]^2\right\} \\
&= \left[O(n^{-1})\right]\left[O(1)\right] \to 0, \qquad \text{as} \quad n \to \infty, \quad m \text{ fixed.} \qquad (3.3.14)
\end{aligned}
$$

Thus from Theorem 3.3.1, (3.3.1), (3.3.13), and (3.3.14), we arrive at the following.

THEOREM 3.3.2. *If $\theta(F)$ is stationary of order* 0 *and* (3.2.18) *holds, then*

$$\rho(Y_n, Y_n^*) \xrightarrow{P} 0 \qquad \textit{as} \quad n \to \infty, \tag{3.3.15}$$

$$Y_n^* \xrightarrow{\mathfrak{D}} W, \qquad \textit{in the } J_1\textit{-topology on} \quad D[0,1]. \tag{3.3.16}$$

[Note that $\rho(Y_n, Y_n^*) \xrightarrow{P} 0$ as $n \to \infty$ even if $\theta(F)$ is stationary of order d ($0 \leqslant d \leqslant m-1$) and (3.2.18) holds. But, unless $d = 0$, the weak convergence to W may not hold.]

From Theorems 3.3.1 and 3.3.2 it follows immediately that under the same conditions as $n \to \infty$,

$$n^{1/2}\frac{\left[U_n - \theta(F)\right]}{m\left[\zeta_1(F)\right]^{1/2}} \xrightarrow{\mathfrak{D}} W(1) \sim \mathfrak{N}(0,1), \tag{3.3.17}$$

$$n^{1/2}\frac{\left[\theta(F_n) - \theta(F)\right]}{m\left[\zeta_1(F)\right]^{1/2}} \xrightarrow{\mathfrak{D}} W(1) \sim \mathfrak{N}(0,1). \tag{3.3.18}$$

These results were obtained by Hoeffding (1948a) and von Mises (1947) by alternative procedures. Theorems 3.3.1 and 3.3.2 are due to Miller and Sen (1972).

For each ($n \geqslant 1$), let N_n be a positive integer valued random variable defined on the same probability space as the X_i and let $\{a_n\}$ be a sequence of positive constants going to ∞ as $n \to \infty$. If

$$a_n^{-1} N_n \xrightarrow{P} \xi \qquad \text{as} \quad n \to \infty, \tag{3.3.19}$$

where ξ is a positive random variable, then as in Theorem 17.2 of

Billingsley (1968),

$$Y_{N_n}^{(1)} \xrightarrow{\mathfrak{D}} W, \qquad \text{in the } J_1\text{-topology on} \quad D[0,1]. \tag{3.3.20}$$

On the other hand, note that for every $n_0(\geqslant m)$ and $\epsilon > 0$,

$$P\left\{\frac{\max_{n\geqslant n_0}\max_{m\leqslant k\leqslant n}k|\sum_{h=2}^{m}\binom{m}{h}U_{k,n}|}{m[n\zeta_1(F)]^{1/2}} > \epsilon\right\}$$

$$\leqslant P\left\{\frac{\max_{j\geqslant 0}\max_{m\leqslant k\leqslant n_{j+1}}k|\sum_{h=2}^{m}\binom{m}{h}U_{k,h}|}{m[n_j\zeta_1(F)]^{1/2}} > \epsilon\right\}$$

$$\leqslant \sum_{j\geqslant 0} P\left\{\max_{m\leqslant k\leqslant n_{j+1}} k\left|\sum_{h=2}^{m}\binom{m}{h}U_{k,h}\right| > \epsilon m[n_j\zeta_1(F)]^{1/2}\right\}, \tag{3.3.21}$$

where

$$n_j = 2^j n_0, \qquad j = 0, 1, \dots . \tag{3.3.22}$$

Since $\sum_{j\geqslant 0}2^{-j}\cdot j < \infty$ and $n_0^{-1}\log n_0 \to 0$ as $n_0 \to \infty$, by (3.3.9), the right hand side of (3.3.21) converges to 0 as $n_0 \to \infty$. Consequently, under (3.3.19), as $n \to \infty$,

$$\frac{\max_{m\leqslant k\leqslant N_n}|\sum_{h=2}^{m}\binom{m}{h}kU_{k,h}|}{m[N_n\zeta_1(h)]^{1/2}} \xrightarrow{P} 0, \tag{3.3.23}$$

while (3.3.20) implies that $\max_{k\leqslant m-1}|Y_{N_n}^{(1)}(k/N_n)| \xrightarrow{P} 0$ as $n \to \infty$. Hence, under (3.3.19),

$$\rho\left(Y_{N_n}, Y_{N_n}^{(1)}\right) \xrightarrow{P} 0 \qquad \text{as} \quad n \to \infty. \tag{3.3.24}$$

Finally, by (3.2.9) and (3.3.19),

$$\rho(Y_{N_n}, Y_{N_n}^{*}) \xrightarrow{P} 0 \qquad \text{as} \quad n \to \infty. \tag{3.3.25}$$

Thus, from (3.3.20), (3.3.24), and (3.3.25), we obtain the following.

THEOREM 3.3.3. *For positive integer valued random sample sizes $\{N_n\}$ satisfying* (3.3.19), *the weak convergence results of Theorems* 3.3.1 *and* 3.3.2 *also hold for $\{Y_{N_n}\}$ and $\{Y_{N_n}^{*}\}$.*

By virtue of Theorem 3.2.1, $\{U_n\}$ is a reverse martingale sequence, while by (3.2.9) $\{\theta(F_n)\}$ behaves almost surely like a reverse martingale sequence. As such, motivated by Theorem 2.4.5, we may also be interested in the following invariance principle.

For every $n(\geqslant m)$, consider a stochastic process $Z_n = \{Z_n(t), t \in I\}$

where

$$Z_n(t) = \frac{\sqrt{n}\left[U_{n(t)} - \theta(F)\right]}{m\left[\zeta_1(F)\right]^{1/2}}; \qquad n(t) = \min\left\{k : \frac{n}{k} \leqslant t\right\}, \qquad t \in I. \tag{3.3.26}$$

Then Z_n belongs to the space $D[0, 1]$. Also we introduce a second stochastic process $Z_n^* = \{Z_n^*(t), t \in I\}$ by defining $n(t)$ as in (3.3.26) and letting

$$Z_n^*(t) = \sqrt{n}\,\frac{\left[\theta(F_{n(t)}) - \theta(F)\right]}{m\left[\zeta_1(F)\right]^{1/2}}, \qquad t \in I. \tag{3.3.27}$$

Note that by Theorem 3.2.1, $U_n \to \theta(F)$ a.s., as $n \to \infty$ and by (3.2.9), $\theta(F_n) \to \theta(F)$ a.s., as $n \to \infty$. Hence there is no problem in extending the definitions (3.3.26) and (3.3.27) at $t = 0$. Then we have the following.

THEOREM 3.3.4. *If $\theta(F)$ is stationary of order 0, then*

$$Z_n \xrightarrow{\mathfrak{D}} W, \qquad \text{in the } J_1\text{-topology on} \quad D[0, 1]. \tag{3.3.28}$$

If, in addition, (3.2.10) *holds, then*

$$\rho(Z_n, Z_n^*) \xrightarrow{P} 0, \qquad \text{as} \quad n \to \infty, \tag{3.3.29}$$

$$Z_n^* \xrightarrow{\mathfrak{D}} W, \qquad \text{in the } J_1\text{-topology on} \quad D[0, 1]. \tag{3.3.30}$$

Proof. For (3.3.28), we verify that $\{U_n\}$ satisfies the regularity conditions of Theorem 2.4.5. For this purpose, we rewrite (3.2.29) as

$$U_n - \theta(F) = mU_{n,1} + \tilde{U}_n; \qquad \tilde{U}_n = \sum_{h=2}^{m} \binom{m}{h} U_{n,h}, \tag{3.3.31}$$

and note that both $\{U_{n,1}\}$ and $\{\tilde{U}_n\}$ are reverse martingales with respect to the common $\{\mathcal{C}_n\}$. As such, for every $\epsilon > 0$,

$$P\left\{\sup_{k \geqslant n} \sqrt{n}\,|\tilde{U}_k| > \epsilon\right\} \leqslant n\epsilon^{-2} E\left[\tilde{U}_n^2\right] = n\epsilon^{-2}\left[O(n^{-2})\right] \to 0 \qquad \text{as} \quad n \to \infty, \quad \text{by (3.3.8).} \tag{3.3.32}$$

Thus if we define $Z_n^{(1)} = \{Z_n^{(1)}(t),\ t \in I\}$ by letting $Z_n^{(1)}(t) = \sqrt{n}\,[U_{n(t),1}] / m[\zeta_1(F)]^{1/2}$, $t \in I$, with $n(t)$ defined by (3.3.26), then

$$\rho(Z_n, Z_n^{(1)}) \xrightarrow{P} 0 \qquad \text{as} \quad n \to \infty. \tag{3.3.33}$$

On the other hand, $nU_{n,1} = \sum_{i=1}^{n} g_1^0(X_i)$; $g_1^0(x) = g_1(x) - \theta(F)$, so that

$$U_{n,1} - U_{n+1,1} = n^{-1}\{U_{n+1,1} - g_1^0(X_{n+1})\}, \qquad n \geqslant 1, \tag{3.3.34}$$

where conditionally on $\mathcal{C}_{n+1}$, $U_{n+1,1}$ is fixed and $g_1^0(X_{n+1})$ assumes any one of the $(n+1)$ values $g_1^0(X_\alpha)$, $1 \leqslant \alpha \leqslant n+1$ with equal conditional probability $(n+1)^{-1}$. Thus, for every $n \geqslant 1$,

$$\begin{aligned} E\{(U_{n,1} - U_{n+1,1})^2 \mid \mathcal{C}_{n+1}\} &= n^{-2}\left\{(n+1)^{-1}\sum_{\alpha=1}^{n+1}\left[g_1^0(X_\alpha) - U_{n+1,1}\right]^2\right\} \\ &= n^{-2}\left\{(n+1)^{-1}\sum_{\alpha=1}^{n+1}\left[g_1^0(X_\alpha)\right]^2 - U_{n+1,1}^2\right\}. \end{aligned} \tag{3.3.35}$$

By Theorem 3.2.1, $U_{n,1} \to 0$ a.s., as $n \to \infty$, while by the Kintchine strong law of large numbers, $n^{-1}\sum_{i=1}^{n}[g_1^0(X_i)]^2 \to \zeta_1(F)$ a.s., as $n \to \infty$. Hence, by (3.3.35),

$$n(n+1)E\left[(U_{n,1} - U_{n+1,1})^2 \mid \mathcal{C}_{n+1}\right] \to \zeta_1(F) \qquad \text{a.s.,} \quad \text{as} \quad n \to \infty. \tag{3.3.36}$$

To verify (2.4.17), the conditional Lindeberg condition, we observe that in this specific problem, we have for every $\epsilon > 0$,

$$\begin{aligned} &\left[n/\zeta_1(F)\right]\sum_{N \geqslant n} E\left\{\left[U_{N,1} - U_{N+1,1}\right]^2 I(|U_{N,1} - U_{N+1,1}| > \epsilon n^{-1/2}) \mid \mathcal{C}_{N+1}\right\} \\ &= \left[n/\zeta_1(F)\right]\sum_{N \geqslant n} N^{-2}\left\{(N+1)^{-1}\sum_{i=1}^{N+1}\left[g_1^0(X_i) - U_{N+1,1}\right]^2 \chi_{i,N+1}^{(n)}\right\}, \end{aligned} \tag{3.3.37}$$

where

$$\chi_{i,N}^{(n)} = \begin{cases} 1, & |g_1^0(X_i) - U_{N,1}| > \epsilon N n^{-1/2}, \\ 0, & \text{otherwise,} \quad \text{for} \quad 1 \leqslant i \leqslant N; \quad N \geqslant n. \end{cases} \tag{3.3.38}$$

Thus, to verify (2.3.36), it suffices to show that

$$n^{-1}\sum_{i=1}^{n}\left[g_1^0(X_i) - U_{n,1}\right]^2 \chi_{i,n}^{(n)} \to 0 \qquad \text{a.s.,} \quad \text{as} \quad n \to \infty. \tag{3.3.39}$$

Since (by Theorem 3.2.1), $U_{n,1} \to 0$ a.s., as $n \to \infty$, on defining

$$\chi_{i,n}^{*} = \begin{cases} 1, & |g_1^0(X_i)| > \frac{1}{2}\epsilon n^{1/2}, \\ 0, & \text{otherwise,} \quad \text{for} \quad 1 \leqslant i \leqslant n, \end{cases} \tag{3.3.40}$$

to prove (3.3.39), it suffices to show that

$$n^{-1}\sum_{i=1}^{n}\left[g_1^0(X_i)\right]^2\chi_{i,n}^* \to 0 \qquad \text{a.s.,} \quad \text{as} \quad n\to\infty. \tag{3.3.41}$$

Now, for every $\epsilon' > 0$,

$$\begin{aligned}
&P\left\{\max_{N\geq n}\max_{1\leq i\leq N}|N^{-1/2}g_1^0(X_i)| > \epsilon'\right\} \\
&\quad \leq P\left\{\left[\max_{1\leq i\leq n}|g_1^0(X_i)| > \epsilon'\sqrt{n}\right]\cup\left[\max_{N>n}N^{-1/2}|g_1^0(X_i)| > \epsilon'\right]\right\} \\
&\quad \leq P\left\{\max_{1\leq i\leq n}|g_1^0(X_i)| > \epsilon'\sqrt{n}\right\} + P\left\{\max_{N>n}N^{-1/2}|g_1^0(X_i)| > \epsilon'\right\}.
\end{aligned} \tag{3.3.42}$$

Since $n^{-1}\sum_{i=1}^{n}[g_1^0(X_i)]^2 \to \zeta_1(F)$ a.s., as $n\to\infty$ and $0<\zeta_1(F)<\infty$, it follows immediately that $n^{-1}[g_1^0(X_n)]^2\to 0$ a.s., as $n\to\infty$, so that the second term on the rhs of (3.3.42) converges to 0 as $n\to\infty$. Also note that if $P_0(x)$ stands for the d.f. of $g_1^0(X_1)$, then

$$\begin{aligned}
P\left\{\max_{1\leq i\leq n}|g_1^0(X_i)| > \epsilon'\sqrt{n}\right\} &\leq nP\left\{|g_1^0(X_1)| > \epsilon'\sqrt{n}\right\} \\
&\leq (\epsilon')^{-2}\int_{|g_1^0(x)|>\epsilon'\sqrt{n}}\left[g_1^0(x)\right]^2 dP_0 \to 0, \\
&\qquad\qquad \text{as} \quad n\to\infty,
\end{aligned} \tag{3.3.43}$$

where the last step follows again from the fact that $E[g_1^0(X_1)]^2 = \zeta_1(F) < \infty$. Consequently, (3.3.43) converges to 0 as $n\to\infty$. Thus

$$\max_{1\leq i\leq n}\chi_{i,n}^* \to 0 \qquad \text{a.s.,} \quad \text{as} \quad n\to\infty, \tag{3.3.44}$$

and this, along with the fact that as $n\to\infty$, $n^{-1}\sum_{i=1}^{n}[g_1^0(X_i)]^2\to\zeta_1(F)$ a.s., implies that (3.3.41) holds. Hence both the conditions for Theorem 2.4.5 hold for $\{Z_n\}$, and as a result, $Z_n \to W$ as $n\to\infty$. To prove (3.3.29) and (3.3.30), we note that by (3.2.9), (3.3.26) and (3.3.27),

$$\rho(Z_n, Z_n^*) = O_p(n^{-1/2}) \qquad \text{as} \quad n\to\infty, \tag{3.3.45}$$

and hence (3.3.28) and (3.3.45) imply (3.3.29) and (3.3.30). Q.E.D.

We conclude this section with the remark that by virtue of Theorem 2.3.6, the weak convergence results in Theorems 3.3.1 and 3.3.2 also hold for the $D[0,\infty)$ space. Further, by virtue of Theorem 2.4.8, defining q and d_q as in (2.4.25) and (2.4.26), it follows by some standard steps that the weak convergence results in Theorems 3.3.1, 3.3.2, and 3.3.4 also hold in the (d_q-) sup-norm as well; we leave the proofs as exercises.

3.4. EMBEDDING OF WIENER PROCESSES FOR $\{U_n\}$ AND $\{\theta(F_n)\}$

In Chapter 2 (specifically, Theorem 2.5.1), we considered the Skorokhod-Strassen embedding of Wiener processes for sums of independent random variables and martingales. In this section, we extend the result to U-statistics and von Mises functionals.

Let $S = \{S(t), t \in R^+ = [0, \infty)\}$ be a random process, where

$$S(k) = S_k = \begin{cases} 0, & k = 0, \dots, m-1, \\ k[U_k - \theta(F)]/m[\zeta_1(F)]^{1/2}, & k \geq m, \end{cases} \quad (3.4.1)$$

and $S(t) = S_k$ for $k \leq t < k+1$, $k \geq 0$; m, $\zeta_1(F)$, U_k, and $\theta(F)$ are all defined as in Section 3.2. Similarly, $S^* = \{S^*(t), t \in R^+\}$ is a random process where

$$S^*(k) = S_k^* = \begin{cases} 0, & k = 0, \\ k[\theta(F_k) - \theta(F)]/m[\zeta_1(F)]^{1/2}, & k \geq 1 \end{cases} \quad (3.4.2)$$

and $S^*(t) = S_k^*$ for $k \leq t < k+1$, $k \geq 0$. [Alternatively, $S(t)$ (and $S^*(t)$) can also be defined by linear interpolation between (S_k, S_{k+1}) (and (S_k^*, S_{k+1}^*)) for $t \in (k, k+1), k \geq 0$.] Consider now a positive and real-valued function $f: R^+ \to R^+$, such that

$$f(t) \text{ is } \uparrow \text{ but } t^{-1}f(t) \text{ is } \downarrow \text{ in } t \in R^+, \quad (3.4.3)$$

$$\sum_{n \geq 1} [f(cn)]^{-1} E\left\{[g_1^0(X_1)]^2 I\left([g_1^0(X_1)]^2 > f(cn)\right)\right\} < \infty, \quad \forall c > 0. \quad (3.4.4)$$

Note that (3.4.4) is defined by analogy with the basic condition of Theorem 2.5.1. Since $g_1^0(x) = E[g(x, X_2, \dots, X_m) - \theta(F)]$, (3.4.4) can also be stated in terms of the kernel $g(X_1, \dots, X_m)$; in applications this does not make much difference, although (3.4.4) is a little less restrictive. (See Problems 3.4.1 and 3.4.2.) Also, if we assume that $E[|g(X_1, \dots, X_m|^r] < \infty$ for some $r > 2$, then (3.4.4) holds and, further, this improves the order of approximations in the main theorem of this section. (See Problem 3.4.2.) Finally, let $W = \{W(t), t \in R^+\}$ be a standard Wiener process on R^+. Then we have the following.

THEOREM 3.4.1. *If $\theta(F)$ is stationary of order* 0 *and* (3.4.3) *and* (3.4.4) *hold, then there exists a process $\tilde{S} = \{\tilde{S}(t), t \in R^+\}$ such that*

$$|S(t) - \tilde{S}(t)| = o\left([tf(t)]^{1/4} \log t\right) \quad \text{a.s., as } t \to \infty, \quad (3.4.5)$$

$$\tilde{S}(t) = W(t) + o\left([tf(t)]^{1/4} \log t\right) \quad \text{a.s., as } t \to \infty. \quad (3.4.6)$$

If, in addition, (3.2.18) *holds, then as* $t \to \infty$,

$$|S(t) - S^*(t)| = o\left(\left[tf(t)\right]^{1/4} \log t\right) \quad a.s., \tag{3.4.7}$$

and hence, the invariance principle in (3.4.6) *also holds for* S^*.

(Note that $\tilde{S}$ and W need not be defined on the same probability space whereas S and $\tilde{S}$ are defined on the same probability space and the Skorokhod-Strassen embedding applies to $\tilde{S}$.)

Proof. We introduce another random process $S^{(1)} = \{S^{(1)}(t), t \in R^+\}$, where for $k \leqslant t < k+1$, $S^{(1)}(t) = S_k^{(1)} = S^{(1)}(k)$ and

$$S_k^{(1)} = \begin{cases} 0, & k \leqslant m-1, \\ kU_{k,1}/\left[\zeta_1(F)\right]^{1/2}, & k \geqslant m, \end{cases} \tag{3.4.8}$$

where $U_{k,1}$ is defined by (3.2.29). Further, we introduce $\tilde{U}_n$ as in (3.3.31). Then, to prove (3.4.5) and (3.4.6), it suffices to show that $\tilde{S} = S^{(1)}$, that is

$$S^{(1)}(t) = W(t) + o\left(\left[tf(t)\right]^{1/4} \log t\right) \quad \text{a.s.,} \quad \text{as} \quad t \to \infty, \tag{3.4.9}$$

$$\sup_{k \geqslant n} \left\{ |k\tilde{U}_k\left[(kf(k))^{1/4} \log k\right]^{-1} \right\} \xrightarrow{p} 0, \quad \text{as} \quad n \to \infty. \tag{3.4.10}$$

Since $kU_{k,1} = \sum_{i=1}^{k} g_1^0(X_i)$ where $0 < \zeta_1(F) = E[g_1^0(X_1)]^2 < \infty$ and (3.4.3) and (3.4.4) hold, the proof of (3.4.9) follows directly from Theorem 2.5.1. Hence we need to prove only (3.4.10). For this, we write

$$c_k = k\left[(kf(k))^{1/4} \log k\right]^{-1} = \left[(k^{1/2}/\log k)(k/f(k))^{1/4}\right], \quad k \geqslant 2, \tag{3.4.11}$$

and note that $\{c_k\}$ is a nondecreasing sequence of positive numbers with $\lim_{k\to\infty} c_k = \infty$. Further, by (3.2.29) and (3.3.31), $\{\tilde{U}_k, \mathcal{C}_k, k \geqslant m\}$ is a reverse martingale sequence, and hence,

$$\begin{aligned} E\left[\tilde{U}_k - \tilde{U}_{k+1}\right]^2 &= E\left[\tilde{U}_k^2\right] - E\left[\tilde{U}_{k+1}^2\right] = E\left[U_n - \theta(F)\right]^2 \\ &\quad - E\left[U_{n+1} - \theta(F)\right]^2 - \left\{m^2 EU_{n,1}^2 - m^2 EU_{n+1,1}^2\right\} \\ &= O(n^{-3}), \qquad \text{by (3.2.15);} \end{aligned} \tag{3.4.12}$$

$$E\tilde{U}_n^2 = E\left[U_n - \theta(F)\right]^2 - m^2 EU_{n,1}^2 = O(n^{-2}). \tag{3.4.13}$$

Thus, by (3.4.11) and (3.4.13),

$$\lim_{n\to\infty} c_n^2 EU_n^2 = 0. \tag{3.4.14}$$

From Theorem 3.2.2, (3.4.11) through (3.4.14), and some simple adjustments, it readily follows that as $n \to \infty$,

$$P\left\{ \sup_{k \geqslant n} c_k |\tilde{U}_k| > \epsilon \right\} \to 0, \qquad \forall \epsilon > 0. \tag{3.4.15}$$

Hence the proof of (3.4.5) is complete.

Note that by (3.2.9) and the definitions of S and S^*, we have

$$|S(t) - S^*(t)| = O(1) \quad \text{a.s., as} \quad t \to \infty; \tag{3.4.16}$$

on the other hand, by (3.4.3), $[tf(t)]^{1/4} \log t \to \infty$ as $t \to \infty$, and hence, (3.4.7) follows directly from (3.4.16). Finally, (3.4.5) and (3.4.7) imply (3.4.6). Q.E.D.

3.5. BROWNIAN BRIDGE APPROXIMATIONS FOR FINITE POPULATION SAMPLING

The results of Section 3.3 are extended here to sampling without replacement from a finite (but large) population. It is shown that instead of the Brownian motion processes, the allied Brownian Bridge processes provide the desired approximations.

Let $\mathbf{N} = \{N : N \geqslant 1\}$ be the set of all positive integers, and for every $N \in \mathbf{N}$, let $A_N = (a_{N1}, \ldots, a_{NN})$ be a sequence of real numbers, not necessarily all distinct. Let $X_N = (X_{N1}, \ldots, X_{NN})$ be a random vector that takes on each permutation of A_N with the common probability $(N!)^{-1}$. Associated with X_N is the sequence of samples

$$X_N^{(n)} = (X_{N1}, \ldots, X_{Nn}) \qquad \text{for} \quad 1 \leqslant n \leqslant N. \tag{3.5.1}$$

For a symmetric *kernel* $g(X_1, \ldots, X_m)$ of *degree* $m(\geqslant 1)$, we introduce the parameter

$$\theta_N = \theta(A_N) = N^{-[m]} \sum_{P_{N,m}} g(a_{Ni_1}, \ldots, a_{Ni_m}), \tag{3.5.2}$$

where $N^{-[m]} = [N \cdots (N-m+1)]^{-1}$ and $P_{N,m} = \{(i_1, \ldots, i_m) : 1 \leqslant i_1 \neq \cdots \neq i_m \leqslant N\}$. For the sample $X_N^{(n)}$, we consider the symmetric and unbiased estimator [of θ_N]

$$U_n = U_{N,n} = U(X_N^{(n)}) = n^{-[m]} \sum_{P_{n,m}} g(X_{Ni_1}, \ldots, X_{Ni_m}). \tag{3.5.3}$$

(We may refer to Nandi and Sen, 1963 and Sen, 1970b for various properties of U_n.) Also, side by side, we have the von Mises functional

$$\theta(X_N^{(n)}) = n^{-m} \sum_{i_1=1}^{n} \cdots \sum_{i_m=1}^{n} g(X_{Ni_1}, \ldots, X_{Ni_m}). \tag{3.5.4}$$

Let us now define for every h: $0 \leqslant h \leqslant m$,

$$g_h^{(N)}(X_{Ni_1}, \ldots, X_{Ni_h}) = (N-h)^{-[m-h]} \sum\nolimits_h^* g(X_{Ni_1}, \ldots, X_{Ni_m}), \quad (3.5.5)$$

where the summation $\sum_h^*$ extends over all possible (i.e., $(N-h)^{[m-h]}$) choices of distinct $(i_{h+1}, \ldots, i_m)$ from the $N-h$ units in A_N excluding the ones identified as $X_{Ni_1}, \ldots, X_{Ni_h}$; by definition, $g_0^{(N)} = \theta(A_N)$ and $g_m^{(N)} \equiv g$. Let then

$$\bar{\zeta}_{h,N} = N^{-[h]} \sum_{P_{N,h}} \left[g_h^{(N)}(a_{Ni_1}, \ldots, a_{Ni_h}) \right]^2 - \theta^2(A_N), \qquad 0 \leqslant h \leqslant m, \quad (3.5.6)$$

where $\bar{\zeta}_{0,N} = 0$. Then our basic assumptions are

$$\inf_N \bar{\zeta}_{1,N} > 0, \qquad \sup_N \bar{\zeta}_{m,N} < \infty; \quad (3.5.7)$$

$$\max_{1 \leqslant i \leqslant N} N^{-1}\left[g_1^{(N)}(a_{Ni}) - \theta_N \right]^2 / \bar{\zeta}_{1,N} \to 0 \quad \text{as} \quad N \to \infty. \quad (3.5.8)$$

Finally we rewrite $X_N = (a_{NR_{N1}}, \ldots, a_{NR_{NN}})$ where $(R_{N1}, \ldots, R_{NN})$ takes on each permutation of $(1, \ldots, N)$ with the common probability $(N!)^{-1}$. For every n: $1 \leqslant n \leqslant N$, let

$$\mathbf{R}_n^{(N)} = (R_{N1}, \ldots, R_{Nn}), \quad (3.5.9)$$

and let $\mathbf{R}_{Nn}^* = (R_{n,1}^{(N)} < \cdots < R_{n,n}^{(N)})$ be a permutation of $\mathbf{R}_n^{(N)}$ such that $R_{n,1}^{(N)} < \cdots < R_{n,n}^{(N)}$. Then let $\mathcal{C}_{n,N}$ be the sigma-field generated by $(\mathbf{R}_{Nn}^*, R_{Nn+1}, \ldots, R_{NN})$, for $1 \leqslant n \leqslant N_i$. Clearly $\mathcal{C}_n$ is $\downarrow$ in $n (1 \leqslant n \leqslant N)$.

THEOREM 3.5.1. *For every $N \in \mathbf{N}$, $\{U_{N,n}, \mathcal{C}_{n,N};\ 1 \leqslant n \leqslant N\}$ is a reverse martingale.*

The proof is similar to that of Theorem 3.2.1, and hence is left as an exercise.

Motivated by the reverse martingale property, we are led to formulate the following results. Let us consider a sequence of stochastic processes $\{Y_N\}$, where for every $N \geqslant 1$, $\{Y_N = Y_N(t), t \in I\}$, $I = [0, 1]$,

$$\begin{aligned} Y_N(t) &= Y_N([Nt]/N) \\ &= \left[m^2 N \bar{\zeta}_{1,N} \right]^{-1/2} \{ n_t [U_{N,n_t} - \theta_N] \}, \qquad t \in I, \end{aligned} \quad (3.5.10)$$

$n_t = [Nt]$ being the largest integer contained in $[Nt]$, $0 \leqslant t \leqslant 1$; for $n_t \leqslant m-1$, we let $Y_N(t) = 0$.

THEOREM 3.5.2. *Under* (3.5.7) *and* (3.5.8),

$$Y_N \xrightarrow{\mathfrak{D}} W_0, \quad \text{in the } J_1\text{-topology on} \quad D[0,1], \tag{3.5.11}$$

where $W_0 = \{W_0(t), t \in I\} = \{W(t) - tW(1), t \in I\}$ *is a standard Brownian Bridge on* I.

Proof. Let us define for every h: $0 \leqslant h \leqslant m$,

$$\begin{aligned} W_n^{(h)} &= n^{-[h]} \sum_{P_{n,h}} g_h^{(N)}(X_{Ni_1}, \ldots, X_{Ni_h}) \\ &= n^{-[h]}(N-h)^{-[m-h]} \sum_{P_{n,h}} \sum\nolimits_h^* g(X_{Ni_1}, \ldots, X_{Ni_m}), \end{aligned} \tag{3.5.12}$$

where the summation $\sum_h^*$ extends over all possible choices of $(i_{h+1}, \ldots, i_m)$ from the $(N-h)$ units of the population excluding $(i_1, \ldots, i_h)$. Naturally, $W_n^{(0)} = \theta_N$ and $W_n^{(m)} = U_n$, $\forall n \geqslant m$. Also let

$$V_n^{(h)} = \sum_{k=0}^{h} \binom{h}{k} (-1)^k W_n^{(h-k)}, \qquad 0 \leqslant h \leqslant m, \tag{3.5.13}$$

so that

$$V_n^{(1)} = W_n^{(1)} - W_n^{(0)} = \frac{1}{n} \sum_{i=1}^{n} \left[g_1^{(N)}(X_{Ni}) - \theta_N \right]. \tag{3.5.14}$$

Then, from (3.5.12) and (3.5.13), we obtain that

$$U_n - \theta_N - mV_n^{(1)} = \sum_{h=2}^{m} \binom{m}{h} V_n^{(h)} = V_n^* \text{ say.} \tag{3.5.15}$$

First, we consider the related process $Y_N^{(1)} = \{Y_N^{(1)}(t), t \in I\}$ where we let

$$Y_N^{(1)}(t) = \begin{cases} n_t \left[V_{n_t}^{(1)} \right] / \left[N \bar{\zeta}_{1,N} \right]^{1/2}, & 1/N \leqslant t \leqslant 1 \\ 0, & t < 1/N, \end{cases} \tag{3.5.16}$$

n_t being defined as in (3.5.10). Then, from (3.5.10), (3.5.15), and (3.5.16), we obtain that

$$\sup_{t \geqslant m/N} \left| Y_N(t) - Y_N^{(1)}(t) \right| = \max_{m \leqslant n \leqslant N} \left| nV_n^* / m \left[N \bar{\zeta}_{1,N} \right]^{1/2} \right|, \tag{3.5.17}$$

$$\sup_{0 \leqslant t < m/N} |Y_N(t) - Y_N^{(1)}(t)| = \sup_{t < m/N} |Y_N^{(1)}(t)|. \tag{3.5.18}$$

Next we show that as $N \to \infty$,

$$Y_N^{(1)} \xrightarrow{\mathfrak{D}} W_0, \quad \text{in the } J_1\text{-topology on} \quad D[0,1]. \tag{3.5.19}$$

Equation 3.5.19 will imply that the right hand side of (3.5.18) converges in law to a degenerate (at 0) random variable (as $N \to \infty$), so the proof of the theorem follows if one shows, in addition, that (3.5.17) converges in probability to 0 as $N \to \infty$.

To prove (3.5.19), we need to show that (i) the finite dimensional distributions (f.d.d.) of $\{Y_N^{(1)}\}$ converge to those of W_0 and (ii) $Y_N^{(1)}$ is tight. To prove (i), consider an arbitrary linear compound, say,

$$\lambda_1 Y_N^{(1)}(t_1) + \cdots + \lambda_q Y_N^{(1)}(t_q) = Z_N(\boldsymbol{\lambda}, \mathbf{t}), \tag{3.5.20}$$

where $\boldsymbol{\lambda} = (\lambda_1, \ldots, \lambda_q) \neq \mathbf{0}$, $\mathbf{t} = (t_1, \ldots, t_q)$ with $0 < t_1 < \cdots < t_q \leqslant 1$, and $q(\geqslant 1)$ is a fixed positive integer. Then let $n_j = [Nt_j]$, $1 \leqslant j \leqslant q$, so that by (3.5.16) and (3.5.20),

$$\begin{aligned} Z_N(\boldsymbol{\lambda}, \mathbf{t}) &= \sum_{j=1}^{q} \lambda_j \left(\sum_{i=1}^{n_j} \left[g_1^{(N)}(X_{Ni}) - \theta_N \right] \right) \Big/ \left[N\bar{\xi}_{1,N} \right]^{1/2} \\ &= \sum_{i=1}^{N} c_{Ni} \left\{ \left[g_1^{(N)}(X_{Ni}) - \theta_N \right] \Big/ \left[N\bar{\xi}_{1,N} \right]^{1/2} \right\} \end{aligned} \tag{3.5.21}$$

where the real constants c_{Ni} depend on $\boldsymbol{\lambda}$ and $(n_1, \ldots, n_q)$ and satisfy the following conditions:

$$\sum_{i=1}^{N} c_{Ni} = 0, \qquad \forall N \geqslant 1; \quad \left(\max_{1 \leqslant i \leqslant N} |c_{Ni}| \right) / \sqrt{N} \to 0 \quad \text{as} \quad N \to \infty; \tag{3.5.22}$$

$$\frac{1}{N} \sum_{i=1}^{N} c_{Ni}^2 \to \sum_{j=1}^{q} \sum_{l=1}^{q} \lambda_j \lambda_l \left[t_j \wedge t_l - t_j t_l \right], \tag{3.5.23}$$

where the right hand side of (3.5.23) is positive and finite for every $t_j \in I$, $1 \leqslant j \leqslant q$. Actually, for every positive integer $k(=2, 3, \ldots)$, we have

$$\frac{1}{N} \sum_{i=1}^{N} c_{Ni}^k = O(1). \tag{3.5.24}$$

Thus the vector $\mathbf{c}_N = (c_{N1}, \ldots, c_{NN})$ satisfies the classical *Wald-Wolfowitz condition* of the permutational central limit theorem. Also, by (3.5.8), the vector $\mathbf{l}_N = (l_{N1}, \ldots, l_{NN})$ with $l_{Ni} = (g_1^{(N)}(a_{Ni}) - \theta_N)/[N\bar{\xi}_{1,N}]^{1/2}$, $1 \leqslant i \leqslant N$, satisfies the *Noether condition* of the same theorem. As such, if we proceed to evaluate

$$E\left[Z_N(\boldsymbol{\lambda}, \mathbf{t}) \right]^k / \left[E Z_N^2(\boldsymbol{\lambda}, \mathbf{t}) \right]^{k/2}, \qquad k = 2, 3, \ldots \tag{3.5.25}$$

[where $(X_{N1}, \ldots, X_{NN})$ assume all possible permutations of $(a_{N1}, \ldots, a_{NN})$ with equal probability $(N!)^{-1}$], then we can show that as $N \to \infty$, (3.5.25) converges to

$$\begin{cases} 0, & \text{if } k \text{ is odd} \\ k!/2^{k/2}(k/2)!, & \text{if } k \text{ is even,} \end{cases} \tag{3.5.26}$$

which ensures that $Z_N(\boldsymbol{\lambda}, \mathbf{t})$ is asymptotically normal; the details for (3.5.25) and (3.5.26) are left as an exercise. Thus the f.d.d. of $Y_N^{(1)}$ converge to those of W_0. Now, by Theorem 3.5.1 and the fact that $V_n^{(1)}$ is also a U-statistic, we claim that $\{V_n^{(1)}, \mathcal{C}_{n,N};\ 1 \leqslant n \leqslant N\}$ is also a reverse martingale. As such, the convergence of f.d.d. and this reverse martingale property ensure the tightness of $Y_N^{(1)}$ (see Problem 3.5.4). Thus (3.5.19) holds.

By the same argument as in Theorem 3.5.1, it follows that for every h: $0 \leqslant h \leqslant m$, $\{W_n^{(h)}, \mathcal{C}_{n,N};\ 1 \leqslant n \leqslant N\}$ is a reverse martingale. Thus rewriting V_n^* in (3.5.15) as

$$\sum_{h=2}^{m} \binom{m}{h} \sum_{k=0}^{h} \binom{h}{k} (-1)^k W_n^{(h-k)}, \tag{3.5.27}$$

we claim that $\{V_n^*, \mathcal{C}_{n,N};\ 1 \leqslant n \leqslant N\}$ is also a reverse martingale.

Let us denote by

$$\nu_n = \operatorname{Var}[V_n^*] \quad \text{and} \quad \nu_n^* = \nu_n - \nu_{n+1}, \qquad m \leqslant n \leqslant N-1. \tag{3.5.28}$$

Then, by using Theorem 3.2.2 (with obvious modifications), we obtain that for every $\epsilon > 0$,

$$\begin{aligned} P&\left\{ \max_{m \leqslant n \leqslant N} n|V_n^*| > m\epsilon\left[N\bar{\zeta}_{1,N}\right]^{1/2} \right\} \\ &\leqslant \left[m^2\epsilon^2 N\bar{\zeta}_{1,N}\right]^{-1}\left\{ m^2\nu_m + \sum_{k=m+1}^{N-1} (2k+1)\nu_k \right\} \\ &= \left[m^2\epsilon^2 N\bar{\zeta}_{1,N}\right]^{-1}\left\{ m^2\nu_m + \sum_{k=m+1}^{N-1} (2k+1)\left[\operatorname{Var}[U_k] - m^2 \operatorname{Var}[V_k^{(1)}]\right] \right\} \end{aligned} \tag{3.5.29}$$

where

$$\operatorname{Var}[U_n] = \frac{m^2(N-n)}{nN}\bar{\zeta}_{1,N} + 0(n^{-2}), \qquad m \leqslant n \leqslant N; \tag{3.5.30}$$

$$\operatorname{Var}[V_n^{(1)}] = \frac{(N-n)}{n(N-1)}\bar{\zeta}_{1,N}, \qquad 1 \leqslant n \leqslant N. \tag{3.5.31}$$

Thus the right hand side of (3.5.29) is $0(N^{-1}\log N) \to 0$ as $N \to \infty$, and this proves (3.5.17). Q.E.D.

Theorem 3.5.2 is due to Sen (1972d). If we define $n_t^0 = \min\{k : (k^{-1} - N^{-1}) \leqslant t(n^{-1} - N^{-1})\}$, $Y_{Nn}^0(t) = [U_{N,n_t^0} - \theta_N]/\{\operatorname{Var}(U_{N,n})\}^{1/2}, t \in [0,1]$ and let $Y_{Nn}^0 = \{Y_{Nn}^0(t), t \in [0,1]\}$, then, using Theorem 3.5.1 and Theorems 2.4.4 through 2.4.6, we arrive at the following.

THEOREM 3.5.3. *Under* (3.5.7) *and* (3.5.8), $Y^0_{Nn} \xrightarrow{\mathfrak{D}} W$, *in the* J_1-*topology on* $D[0, 1]$, *where* W *is a standard Wiener process on* $[0, 1]$.

The proof is left as an exercise.

3.6. WEAK CONVERGENCE OF GENERALIZED U-STATISTICS AND VON MISES FUNCTIONALS

We now extend the results of Sections 3.2 and 3.3 to the multisample cases. Let $\{X_{ji};\ i \geqslant 1\}$, $j = 1, \ldots, c(\geqslant 2)$ be independent sequences of independent random vectors (irv), where X_{ji} has a d.f. $F_j(x)$, $x \in R^p$, for $j = 1, \ldots, c$. Actually the random vectors X_{ji} need not have (for different i) the same dimensions. However, for notational simplicity we shall assume that they have the common dimension $p(\geqslant 1)$. Let $G(X_{ji}, 1 \leqslant i \leqslant m_j, 1 \leqslant j \leqslant c)$ be a Borel-measurable *kernel* of *degree* $\mathbf{m} = (m_1, \ldots, m_c)$, where without any loss of generality we assume that g is symmetric in the $m_j(\geqslant 1)$ arguments (vectors) of the jth set, for $j = 1, \ldots, c$. Let $m_0 = m_1 + \cdots + m_c$, $\mathbf{F} = (F_1, \ldots, F_c)$ and consider a functional of $\mathbf{F}$

$$\theta(F) = \int_{R^{pm_0}} \cdots \int g(x_{ji}, 1 \leqslant i \leqslant m_j, 1 \leqslant j \leqslant c) \prod_{j=1}^{c} \prod_{i=1}^{m_j} dF_j(x_{ji}) \qquad (3.6.1)$$

defined on $\mathfrak{F} = \{\mathbf{F} : |\theta(\mathbf{F})| < \infty\}$; $\theta(\mathbf{F})$ is called an *estimable parameter* or a *regular functional* of $\mathbf{F}$ over $\mathfrak{F}$.

For a set of samples of sizes $\mathbf{n} = (n_1, \ldots, n_c)$ with $n_j \geqslant m_j$, $1 \leqslant j \leqslant c$, the *generalized U-statistic* for $\theta(\mathbf{F})$ is defined by

$$U(\mathbf{n}) = \prod_{j=1}^{c} \binom{n_j}{m_j}^{-1} \sum\nolimits^*_{(\mathbf{n})} g(X_{j\alpha}\alpha = i_{j1}, \ldots, i_{jm_j}, 1 \leqslant j \leqslant c), \qquad (3.6.2)$$

where the summation $\sum^*_{(\mathbf{n})}$ extends over all $1 \leqslant i_{j1} < \cdots < i_{jm_j} \leqslant n_j$, $1 \leqslant j \leqslant c$. $U(\mathbf{n})$ is a symmetric, unbiased, and optimal estimator of $\theta(\mathbf{F})$. If we define the empirical d.f.'s by

$$F_j(x, n_j) = n_j^{-1} \sum_{i=1}^{n_j} c(x - X_{ji}), \qquad x \in R^p, \quad j = 1, \ldots, c, \qquad (3.6.3)$$

then the (generalized) von Mises functional is

$$\theta(\mathbf{F}(\cdot, \mathbf{n})) = \int_{R^{pm_0}} \cdots \int g(x_{11}, \ldots, x_{cm_c}) \prod_{j=1}^{c} \prod_{i=1}^{m_j} dF_j(x_{ji}, n_j), \qquad (3.6.4)$$

where $\mathbf{F}(\cdot,\mathbf{n}) = (F_1(\cdot,n_1), \ldots, F_c(\cdot,n_c))$. Clearly

$$\theta(F(\cdot,\mathbf{n})) = \left(\prod_{j=1}^{c} n_j^{-m_j}\right)\prod_{j=1}^{c}\left\{\sum_{i_{j1}=1}^{n_j}\cdots\sum_{i_{jm_j}=1}^{n_j} g(X_{j\alpha},\right.$$

$$\left.\alpha = i_{j1}, \ldots, i_{jm_j}, 1 \leqslant j \leqslant c)\right\}. \quad (3.6.5)$$

For every d_j: $0 \leqslant d_j \leqslant m_j$, $1 \leqslant j \leqslant c$, let

$$\begin{aligned} g_{d_1 \ldots d_c}&(x_{ji}, i = 1, \ldots, d_j, 1 \leqslant j \leqslant c) \\ &= Eg(x_{j1}, \ldots, x_{jk_j}, X_{jd_j+1}, \ldots, X_{jm_j}, 1 \leqslant j \leqslant c), \end{aligned} \quad (3.6.6)$$

so that $g_{\mathbf{0}} = \theta(\mathbf{F})$ and $g_{\mathbf{m}} = g$. Then let $\mathbf{d} = (d_1, \ldots, d_c)$ and

$$\zeta_{\mathbf{d}}(\mathbf{F}) = Eg_{\mathbf{d}}^2(X_{j1}, \ldots, X_{jd_j}, 1 \leqslant j \leqslant c) - \theta^2(\mathbf{F}), \qquad \mathbf{0} \leqslant \mathbf{d} \leqslant \mathbf{m}, \quad (3.6.7)$$

so that $\zeta_{\mathbf{0}}(\mathbf{F}) = 0$. Then, by arguments similar to those in Section 3.2, it follows that for every $\mathbf{n} \geqslant \mathbf{m}$,

$$\begin{aligned} r^2(\mathbf{n}) &= \operatorname{Var}[U(\mathbf{n})] \\ &= \left[\prod_{j=1}^{c}\binom{n_j}{m_j}\right]^{-1}\left\{\sum_{d_1=0}^{m_1}\cdots\sum_{d_c=0}^{m_c}\prod_{j=1}^{c}\binom{m_j}{d_j}\binom{n_j-m_j}{m_j-d_j}\zeta_{d_1,\ldots,d_c}(F)\right\} \\ &= \sum_{j=1}^{c} n_j^{-1}\sigma_j^2\left[1 + 0(n_0^{-1})\right], \end{aligned} \quad (3.6.8)$$

where $n_0 = \min(n_1, \ldots, n_c)$ and

$$\sigma_j^2 = m_j^2\zeta_{\delta_{j1},\ldots,\delta_{jc}}(\mathbf{F}), \qquad j = 1, \ldots, c \quad (3.6.9)$$

with $\delta_{\alpha\beta} = 1$ or 0 according as $\alpha = \beta$ or not. The expression for Var[$\theta(\mathbf{F}(\cdot,\mathbf{n}))$] is evidently more complicated. However, for large $\mathbf{n}$ under a slightly more restrictive condition, namely,

$$\max_{1 \leqslant j \leqslant c}\ \max_{1 \leqslant \alpha_{j1} \leqslant \cdots \leqslant \alpha_{jm_j} \leqslant c} Eg^2(X_{ji}, i = \alpha_{j1}, \ldots, \alpha_{jm_j}, 1 \leqslant j \leqslant c) < \infty, \quad (3.6.10)$$

it can be shown that the rhs of (3.6.8) still provides an asymptotic expression for Var[$\theta(\mathbf{F}(\cdot,\mathbf{n}))$].

If one essentially assumes that

$$\lim_{n\to\infty}\frac{n_j}{n} = \lambda_j: \qquad 0 < \lambda_j < 1, \quad 1 \leqslant j \leqslant c \quad (3.6.11)$$

where $n = n_1 + \cdots + n_c$ and $n_j = n_j(n)$, $1 \leqslant j \leqslant c$, then $\{U(\mathbf{n})\}$ [or $\{\theta(\mathbf{F}(\cdot,\mathbf{n}))\}$] can be essentially regarded as a sequence in n (for given $\boldsymbol{\lambda} = (\lambda_1, \ldots, \lambda_c)$) and in that setup, the results of Section 3.3 can be extended without difficulty. On the other hand, the more general and natural case of a c-dimensional array $\{U(\mathbf{k}): \mathbf{k} \leqslant \mathbf{n}\}$ (or $\{U(\mathbf{k}): \mathbf{k} \geqslant \mathbf{n}\}$) for the von Mises functionals requires a different and more elaborate approach that will be studied in this section.

Let $E^c = [0,1]^c$ be the unit c-cube, and for every $\mathbf{t} \in E^c$ define $\mathbf{n}(\mathbf{t})$ by $\mathbf{n}(\mathbf{t}) = (n_1(t_1), \ldots, n_c(t_c))$ where for every $\mathbf{n} \geqslant \mathbf{m}$,

$$n_j(t_j) = \min\left\{k : \frac{n_j}{k} \leqslant t_j\right\}, \qquad 0 \leqslant t_j \leqslant 1, \quad 1 \leqslant j \leqslant c. \tag{3.6.12}$$

Consider then a multiparameter stochastic process $Z(\mathbf{n}) = \{Z(\mathbf{t},\mathbf{n}): \mathbf{t} \in E^c\}$ where for every $\mathbf{t} \in E^c$,

$$Z(\mathbf{t},\mathbf{n}) = \{U(\mathbf{n}(\mathbf{t})) - \theta(F)\}/r(\mathbf{n}), \tag{3.6.13}$$

and $r(\mathbf{n})$ is defined by (3.6.8). Let us also consider the related process $Z^*(\mathbf{n}) = \{Z^*(\mathbf{t},\mathbf{n}): \mathbf{t} \in E^c\}$ where

$$Z^*(\mathbf{t},\mathbf{n}) = \{\theta(F(\cdot,\mathbf{n}(\mathbf{t}))) - \theta(F)\}/r(\mathbf{n}), \qquad \mathbf{t} \in E^c. \tag{3.6.14}$$

To study the weak convergence of these processes, we introduce c independent copies of a standard Brownian motion, denoted by $W_j = \{W_j(t): t \in E\}$, $j = 1, \ldots, c$, and then let

$$\mathbf{w} = (w_1, \ldots, w_c)'; \qquad w_j = \left[\left(\frac{\sigma_j^2}{\lambda_j}\right)\left(\sum_{l=1}^{c} \frac{\sigma_l^2}{\lambda_l}\right)^{-1}\right]^{1/2}, \quad 1 \leqslant j \leqslant c; \tag{3.6.15}$$

$$\mathbf{W}(\mathbf{t}) = [W_1(t_1), \ldots, W_c(t_c)], \qquad \mathbf{t} \in E^c; \tag{3.6.16}$$

$$W^*(\mathbf{t}) = \mathbf{w}'\mathbf{W}(\mathbf{t}), \qquad t \in E^c; \qquad W^* = \{W^*(\mathbf{t}) : \mathbf{t} \in E^c\}, \tag{3.6.17}$$

where the σ_j^2 are defined by (3.6.9) and the λ_j by (3.6.11).

THEOREM 3.6.1. *If* $\zeta_{\mathbf{m}}(F) < \infty$ *and* $\max_{1 \leqslant j \leqslant c} \sigma_j^2 > 0$, *then under* (3.6.11),

$$Z(\mathbf{n}) \xrightarrow{\mathfrak{D}} W^*, \qquad \text{in the } J_1\text{-topology on} \quad D[0,1]^c. \tag{3.6.18}$$

If, in addition, (3.6.10) *holds,*

$$\rho(Z(\mathbf{n}), Z^*(\mathbf{n})) = \sup_{\mathbf{t} \in E_c} |Z(\mathbf{t},\mathbf{n}) - Z^*(\mathbf{t},\mathbf{n})| \xrightarrow{p} 0, \tag{3.6.19}$$

$$Z^*(\mathbf{n}) \xrightarrow{\mathfrak{D}} W^*, \qquad \text{in the } J_1\text{-topology on} \quad D[0,1]^c. \tag{3.6.20}$$

Before we proceed to sketch the proof of the theorem, we need a few preliminary results, which we present below. For simplicity of the proofs we specifically consider the case of $c = 2$; a similar treatment holds for general $c \geqslant 2$.

Let $\mathcal{B}_n^{(j)}$ be the sigma-field generated by $\{X_{j1}, \ldots, X_{jn}\}$ for $n \geqslant 1$ and $j = 1, 2$ and let $\mathcal{B}^*_{n_1 n_2}$ be the product sigma-field $\mathcal{B}_{n_1}^{(1)} \times \mathcal{B}_{n_2}^{(2)}$ for $\mathbf{n} \geqslant \mathbf{1}$. For every $\mathbf{i} \geqslant \mathbf{1}$, let $S_{\mathbf{i}} = S(X_{11}, \ldots, X_{1i_1}, X_{21}, \ldots, X_{2i_2})$ be a random variable such that

$$E(S_{i_1 i_2} \mid \mathcal{B}^*_{i_1' i_2}) = S_{i_1' i_2}, \qquad \forall i_1 \geqslant i_1' \geqslant 1, \quad i_2 \geqslant 1 \tag{3.6.21}$$

and on denoting by $\mathbf{S}_k^{(j)} = (S_{1j}, \ldots, S_{kj})'$, $k \geqslant 1, j \geqslant 1$, for every $j \geqslant j' \geqslant 1$, $i \geqslant 1$,

$$E(\mathbf{S}_i^{(j)} \mid \mathcal{B}^*_{ij'}) = \mathbf{S}_i^{(j')}. \tag{3.6.22}$$

Finally, assume that for every $\mathbf{i} \geqslant \mathbf{1}$,

$$E(S_{\mathbf{i}}) = 0 \qquad \text{and} \quad E(S_{\mathbf{i}}^2) = \gamma_{\mathbf{i}}^2 < \infty. \tag{3.6.23}$$

Lemma 3.6.2. Under (3.6.21), (3.6.22), and (3.6.23), for every $\mathbf{n} \geqslant \mathbf{1}$,

$$E\left\{ \max_{\mathbf{1} \leqslant \mathbf{i} \leqslant \mathbf{n}} S_{\mathbf{i}}^2 \right\} \leqslant 16 \gamma_{\mathbf{n}}^2. \tag{3.6.24}$$

Proof. By (3.6.22), $\{\mathbf{S}_{n_1}^{(j)}, \mathcal{B}^*_{n_1 j}; 1 \leqslant j \leqslant n_2\}$ has the martingale property and as $\max_{1 \leqslant i \leqslant n_1} |S_{ij}|$ is a convex function of $\mathbf{S}_i^{(j)}$, $\{\max_{1 \leqslant i \leqslant n_1} |S_{ij}|, \mathcal{B}^*_{n_1 j}; 1 \leqslant j \leqslant n_2\}$ is a submartingale. Therefore by the Doob submartingale inequality [see (2.2.8) with $\alpha = 2$], we have

$$\begin{aligned} E\left\{ \max_{\mathbf{1} \leqslant \mathbf{i} \leqslant \mathbf{n}} S_{\mathbf{i}}^2 \right\} &= E\left\{ \max_{1 \leqslant j \leqslant n_2} \left[\max_{1 \leqslant i \leqslant n_1} |S_{ij}| \right]^2 \right\} \\ &\leqslant 4E\left\{ \max_{1 \leqslant i \leqslant n_1} |S_{in_2}|^2 \right\} \\ &\leqslant 16 E S_{n_1 n_2}^2 = 16 \gamma_{\mathbf{n}}^2. \qquad \text{Q.E.D.} \end{aligned} \tag{3.6.25}$$

(For general c, we need to replace 16 by 4^c.)

Consider now a two-sample U-statistic U_{n_1, n_2} and use $r^2(n_1, n_2) = \operatorname{Var}[U_{n_1 n_2}]$; note that by (3.6.8), $r^2(n_1, n_2)$ is nonincreasing in each of n_1, n_2.

Lemma 3.6.3. For every $\mathbf{N} \geqslant \mathbf{n} \geqslant \mathbf{m}$,

$$\frac{E\left\{\max_{\mathbf{n} \leqslant \mathbf{k} \leqslant \mathbf{N}} |U_{k_1, k_2} - U_{k_1, N_2} - U_{N_1, k_2} + U_{N_1, N_2}|^2\right\}}{r^2(n_1, n_2)}$$

$$\leqslant \frac{16\left[r^2(n_1, n_2) - r_2(n_1, N_2) - r^2(N_1, n_2) + r_2(N_1, N_2) \right]}{r^2(n_1, n_2)}. \tag{3.6.26}$$

Proof. We consider the case of $\mathbf{N} > \mathbf{n}$; other cases follow trivially. For every $r \geqslant 1$, $s \geqslant 1$ and $\mathbf{N} > \mathbf{n}$, we let

$$h_{rs}^{(\mathbf{N})} = U_{N_1-r+1,N_2-s+1} - U_{N_1-r,N_2-s+1} - U_{N_1-r+1,N_2-s} + U_{N_1-r,N_2-s}, \tag{3.6.27}$$

so that for every $\mathbf{1} \leqslant \mathbf{i} \leqslant \mathbf{N} - \mathbf{n}$,

$$\begin{aligned} S_{\mathbf{i}} &= \sum_{r=1}^{i_1} \sum_{s=1}^{i_2} h_{rs}^{(\mathbf{N})} \\ &= U_{N_1-i_1+1,N_2-i_2+1} - U_{N-i_1+1,N_2} - U_{N_1,N_2-i_2+1} + U_{N_1,N_2}. \end{aligned} \tag{3.6.28}$$

Let now $\mathcal{C}_n^{(j)}$ be the sigma-field generated by the unordered collection $\{X_{j1}, \ldots, X_{jn}\}$ and by $\{X_{jn+i}, i \geqslant 1\}$, and $j = 1, 2$ and $\mathcal{C}_{\mathbf{n}}^*$ be sigma-field generated by $\mathcal{C}_{n_1}^{(1)}$ and $\mathcal{C}_{n_2}^{(2)}$, $\mathbf{n} \geqslant \mathbf{1}$. By standard arguments (as in the proof of Theorem 3.2.1) it follows that $E[U_{k_1,k_2} | \mathcal{C}_{q_1,q_2}^*] = U_{q_1,q_2}$ for every $\mathbf{q} \geqslant \mathbf{k}$. Consequently, by (3.6.28),

$$E\left[S_{\mathbf{i}} | \tilde{\mathcal{C}}_{i_1' i_2}^* \right] = S_{i_1' i_2}, \qquad \forall i_1' \leqslant i_1, \tag{3.6.29}$$

$$E\left[\mathbf{S}_{N_1-n_1}^{(j)} | \tilde{\mathcal{C}}_{N_1-n_1,j'}^* \right] = \mathbf{S}_{N_1-n_1}^{(j')}, \qquad \forall j' \leqslant j, \tag{3.6.30}$$

where $\mathbf{S}$ is defined after (3.6.21) and $\tilde{\mathcal{C}}_{\mathbf{k}}^* = \mathcal{C}_{\mathbf{N}-\mathbf{k}}^*$, $0 \leqslant k \leqslant N - n$. The latter representation makes $\mathcal{C}_{\mathbf{k}}^*$ nondecreasing in $\mathbf{k}$. Consequently, we may use Lemma 3.6.2 and obtain

$$\begin{aligned} E&\left\{ \max_{\mathbf{n} \leqslant \mathbf{k} \leqslant \mathbf{N}} |U_{\mathbf{k}} - U_{k_1,N_2} + U_{N_1,k_2} + U_{\mathbf{N}}|^2 \right\} \\ &\leqslant 16 E\left[S_{\mathbf{N}-\mathbf{n}}^2 \right] = 16 E\left[U_{\mathbf{n}} - U_{n_1,N_2} - U_{N_1,n_2} + U_{\mathbf{N}} \right]^2 \\ &= 16\left[r^2(n_1,n_2) - r^2(n_1,N_2) - r^2(N_1,n_2) + r_2(N_1,N_2) \right], \end{aligned} \tag{3.6.31}$$

where the last step follows from the easily verifiable fact that $\mathrm{Cov}[U_{k_1,k_2}, U_{k_1,k_2'}] = \mathrm{Var}[U_{k_1,k_2'}]$, $\forall k_2' \geqslant k_2$ and similar inequalities for $\{(k_1,k_2),(k_1',k_2); k_1' \geqslant k_1\}$ and $\{(k_1,k_2),(k_1',k_2'); \mathbf{k}' \geqslant \mathbf{k}\}$. Equation 3.6.26 follows from (3.6.31). Q.E.D.

We also note that for every $N_2 (\geqslant m)$, $\{U_{k,N_2}, \mathcal{C}_{k,N_2}^*; k \geqslant m_1\}$ has the reverse martingale property, so that by (2.3.11)

$$\begin{aligned} E&\left\{ \max_{n_1 \leqslant k \leqslant N_1} |U_{k,N_2} - U_{N_1,N_2}|^2 \right\} \Big/ r^2(n_1,N_2) \\ &\leqslant 4\left[r^2(n_1,N_2) - r^2(N_1,N_2) \right] \Big/ r^2(n_1,N_2). \end{aligned} \tag{3.6.32}$$

Similarly,

$$E\left\{\max_{n_2 \leqslant q \leqslant N_2} |U_{N_1,q} - U_{N_1,N_2}|^2\right\} \Big/ r^2(N_1,n_2)$$
$$\leqslant 4\left[r^2(N_1,n_2) - r^2(N_1,N_2)\right] \Big/ r^2(N_1,n_2). \qquad (3.6.33)$$

Finally,

$$E\left[U_{N_1,N_2} - \theta(\mathbf{F})\right]^2 = r^2(N_1,N_2). \qquad (3.6.34)$$

Thus from (3.6.26), (3.6.32), (3.6.33), and (3.6.34), the c_2-inequality, and the Chebyshev inequality, we obtain that for every $\epsilon > 0$, there exists a positive $\kappa_\epsilon(< \infty)$, such that for every $\mathbf{N} \geqslant \mathbf{n}$,

$$P\left\{\max_{\mathbf{n} \leqslant \mathbf{k} \leqslant \mathbf{N}} |U_{\mathbf{k}} - \theta(\mathbf{F})| > \kappa_\epsilon r(n_1,n_2)\right\} < \epsilon \qquad (3.6.35)$$

and hence, on letting $\mathbf{N} \to \infty$, we obtain that

$$P\left\{\sup_{\mathbf{k} \geqslant \mathbf{n}} |U_{\mathbf{k}} - \theta(\mathbf{F})| > \kappa_\epsilon r(n_1,n_2)\right\} < \epsilon. \qquad (3.6.36)$$

Finally, $r(n_1,n_2) \to 0$ as $\min(n_1,n_2) \to \infty$, so that

$$\limsup_{n_i \to \infty s,\, i=1,2} |U_{\mathbf{n}} - \theta(F)| = 0, \qquad \text{with probability 1.} \qquad (3.6.37)$$

The last equation justifies the definition of $Z(\mathbf{n})$ in the lower boundary of E_c, for every finite $\mathbf{n}$.

We now consider a typical decomposition for $U_{\mathbf{n}} = U(\mathbf{n})$. Let us write $U_{\mathbf{0}}^*(n) = \theta(\mathbf{F})$ and for $\mathbf{k}$: $\mathbf{0} \leqslant \mathbf{k} \leqslant \mathbf{m}$,

$$U_{\mathbf{k}}^*(\mathbf{n}) = \sum_{\mathbf{d}=\mathbf{0}}^{\mathbf{k}} (-1)^{\mathbf{d}'\mathbf{1}} \binom{k_1}{d_1}\binom{k_2}{d_2} U_{\mathbf{d}}(\mathbf{n}), \qquad (3.6.38)$$

where for every $\mathbf{0} \leqslant \mathbf{d} \leqslant \mathbf{m}$,

$$U_{\mathbf{d}}(\mathbf{n}) = \prod_{j=1}^{2} \binom{n_j}{d_j}^{-1} \sum\nolimits_{(\mathbf{n})}^{*} g_{\mathbf{d}}\left(X_{j\alpha_{ji}}, 1 \leqslant i \leqslant d_j, j = 1,2\right), \qquad (3.6.39)$$

and the summation $\sum_{(\mathbf{n})}^*$ extends over all possible $1 \leqslant \alpha_{j1} < \cdots < \alpha_{jd_j} \leqslant n_j$, $j = 1,2$. By a few routine steps, it follows that

$$U(\mathbf{n}) = \sum_{\mathbf{k}=\mathbf{0}}^{\mathbf{m}} \binom{m_1}{k_1}\binom{m_2}{k_2} U_{\mathbf{k}}^*(\mathbf{n}), \qquad \forall \mathbf{n} \geqslant \mathbf{m}, \qquad (3.6.40)$$

where each $U_{\mathbf{k}}^*(\mathbf{n})$ is also a generalized U-statistic. From (3.6.40), we obtain

$$U(\mathbf{n}) - \theta(\mathbf{F}) = U_1(n_1) + U_2(n_2) + U_3^*(\mathbf{n}), \qquad (3.6.41)$$

where

$$U_1(n_1) = \binom{n_1}{m_1}^{-1} \sum_{1 \leqslant i_1 < \cdots < i_{m_1} \leqslant n_1} g_{m_1 0}(X_{1i_1}, \ldots, X_{1im_1}) - \theta(\mathbf{F}); \tag{3.6.42}$$

$$U_2(n_2) = \binom{n_2}{m_2}^{-1} \sum_{1 \leqslant i_1 < \cdots < i_{m_2} \leqslant n_2} g_{0 m_2}(X_{2i_1}, \ldots, X_{2im_2}) - \theta(\mathbf{F}); \tag{3.6.43}$$

$$\begin{aligned} U_3^*(\mathbf{n}) &= \sum_{\mathbf{d}=\mathbf{1}}^{\mathbf{m}} \binom{m_1}{d_1}\binom{m_2}{d_2} U_{\mathbf{d}}^*(\mathbf{n}) \\ &= U(\mathbf{n}) - \theta(\mathbf{F}) - U_1(n_1) - U_2(n_2), \end{aligned} \tag{3.6.44}$$

and $U_3^*(\mathbf{n})$ is also a generalized U-statistic for which

$$\begin{aligned} \left[r^*(\mathbf{n})\right]^2 &= E\left[U_3^*(\mathbf{n})\right]^2 = r^2(\mathbf{n}) - EU_1^2(n_1) - EU_2^2(n_2) \\ &= \sum_{\mathbf{d}=\mathbf{1}}^{\mathbf{m}} \binom{m_1}{d_1}\binom{m_2}{d_2}\binom{n_1 - m_1}{m_1 - d_1}\binom{n_2 - m_2}{m_2 - d_2}\binom{n_1}{m_1}^{-1}\binom{n_2}{m_2}^{-1} \zeta_{\mathbf{d}}(\mathbf{F}) \\ &= n_1^{-1} n_2^{-1} m_1^2 m_2^2 \zeta_{11}(\mathbf{F}) + O(n^{-3}). \end{aligned} \tag{3.6.45}$$

Let us now return to the proof of Theorem 3.6.1. By (3.6.8) and (3.6.45), under (3.6.11),

$$r^*(\mathbf{n})/r(\mathbf{n}) = O(n^{-1/2}). \tag{3.6.46}$$

Thus if we use the inequality (3.6.36) but for the array $\{U_3^*(\mathbf{k}), \mathbf{n} \leqslant \mathbf{k} \leqslant \mathbf{N}\}$ (so that $r(\mathbf{n})$ has to be replaced by $r^*(\mathbf{n})$), we obtain, using (3.6.46),

$$\sup_{\mathbf{k} \geqslant \mathbf{n}} |U_3^*(\mathbf{k})| = O_p(n^{-1}), \tag{3.6.47}$$

and hence, as $n \to \infty$,

$$\frac{\sup_{k \geqslant n} |U(\mathbf{k}) - \theta(\mathbf{F}) - U_1(k_1) - U_2(k_2)|}{r(\mathbf{n})} \xrightarrow{p} 0. \tag{3.6.48}$$

Now, for each $j(=1,2)$, let us define $Z_j(n_j) = \{Z_j(t, n_j), t \in E\}$ by

$$Z_j(t, n_j) = \left[U_j(n_j(t)) - \theta(\mathbf{F})\right]/r(n_j, \infty), \qquad 0 \leqslant t_j \leqslant 1, \quad j = 1, 2, \tag{3.6.49}$$

where $n_j(t)$ is defined by (3.6.12). Then, from (3.6.48), (3.6.49), (3.6.8), and (3.6.11), it follows that as $n \to \infty$

$$\rho(Z(\mathbf{n}), w_1 Z_1(n_1) + w_2 Z_2(n_2)) \xrightarrow{p} 0. \tag{3.6.50}$$

On the other hand, for each Z_j in (3.6.49) (which are one-sample U-

statistics), Theorem 3.3.4 holds, so that

$$Z_j(n_j) \xrightarrow{\mathcal{D}} W_j, \qquad \text{in the } J_1\text{-topology on } \quad D[0,1], \tag{3.6.51}$$

for $j = 1,2$. The proof of (3.6.18) follows then from (3.6.50), (3.6.51), (3.6.16), and (3.6.17).

By the same technique as in (3.2.7) and (3.2.8), with a direct extension to the multisample case (the proof is left as an exercise), it follows that

$$\theta(F(\cdot,\mathbf{n})) = n^{-m_0} \prod_{j=1}^{c} \sum_{\mathbf{h}=\mathbf{0}}^{\mathbf{m}-\mathbf{1}} n_j^{[m_j - h_j]} U_{\mathbf{h}}^*(\mathbf{n}), \tag{3.6.52}$$

where $m_0 = m_1 + \cdots + m_c$, and for each $\mathbf{h}$: $\mathbf{0} \leqslant \mathbf{h} \leqslant \mathbf{m} - \mathbf{1}$, $U_{\mathbf{h}}^*(\mathbf{n})$ is a generalized U-statistic of degree $\mathbf{m} - \mathbf{h}$. In particular, $U_{\mathbf{0}}^*(\mathbf{n}) = U(\mathbf{n})$. So that, under (3.6.10) and (3.6.11), on using (3.6.37) for each $U_{\mathbf{h}}^*(\mathbf{n})$, we claim that

$$\theta(F(\cdot,\mathbf{n})) = U(\mathbf{n}) + O(n^{-1}) \qquad \text{a.s.,} \quad \text{as} \quad n \to \infty. \tag{3.6.53}$$

The last equation implies that $\rho(Z(\mathbf{n}), Z^*(\mathbf{n})) \xrightarrow{p} 0$ as $n \to \infty$, and this along with (3.6.18) completes the proof of the theorem. Q.E.D.

In the spirit of Theorems 3.3.1 and 3.3.2, we consider now the following. Let $Y(\mathbf{n}) = \{Y(\mathbf{t},\mathbf{n}) : \mathbf{t} \in E^c\}$ be defined by

$$Y(\mathbf{t},\mathbf{n}) = \begin{cases} \psi([\mathbf{nt}],\mathbf{n})(U([\mathbf{nt}]) - \theta(\mathbf{F})), & [\mathbf{nt}] \geqslant \mathbf{m}, \\ 0, & \text{otherwise,} \end{cases} \tag{3.6.54}$$

where $[\mathbf{nt}] = ([n_1 t_1], \ldots, [n_c t_c])$ with $[s]$, the largest integer $\leqslant s$, and for every $\mathbf{n} \geqslant \mathbf{m}$, $\mathbf{k} > \mathbf{0}$,

$$\psi(\mathbf{k},\mathbf{n}) = \frac{n^{-1/2}\left(\sum_{j=1}^{c} \sigma_j n_j^{1/2}\right)}{\left(\sum_{j=1}^{c} \sigma_j n_j^{1/2}/k_j\right)}, \tag{3.6.55}$$

$n = n_1 + \cdots + n_c$ and the σ_j are defined by (3.6.9). Thus $\psi(\mathbf{k},\mathbf{n})$ is $n^{-1/2}$ times a weighted harmonic mean of $k_1, \ldots, k_c$. Similarly, if in (3.6.53), we replace the $U(\mathbf{k})$ by $\theta(\mathbf{F}(\cdot,\mathbf{k}))$, $\mathbf{k} \leqslant \mathbf{n}$, the corresponding process is denoted by $Y^*(\mathbf{n})$. Both these processes belong to the $D[0,1]^c$ space. Finally, we define the independent Brownian movement process $W_1, \ldots, W_c$ as in the previous theorem and let

$$W^0 = \{W^0(\mathbf{t}) : \mathbf{t} \in E^c\}; \tag{3.6.56}$$

$$W^0(\mathbf{t}) = \begin{cases} \left(\sum_{j=1}^{c} \sigma_j \lambda_j^{1/2}\right)\left(\sum_{j=1}^{c} \sigma_j / \lambda_j^{1/2} t_j\right)^{-1}\left(\sum_{j=1}^{c} \sigma_j W_j(t_j) / \lambda_j^{1/2} t_j\right), & \mathbf{t} > \mathbf{0}, \\ 0, & \text{otherwise.} \end{cases} \tag{3.6.57}$$

THEOREM 3.6.4. *Under the conditions of Theorem* 3.6.1,

$$Y(\mathbf{n}) \xrightarrow{\mathfrak{D}} W^0, \quad \textit{in the } J_1\textit{-topology on} \quad D[0,1]^c, \tag{3.6.58}$$

$$\rho(Y(\mathbf{n}), Y^*(\mathbf{n})) \xrightarrow{p} 0 \quad \textit{as} \quad n \to \infty \tag{3.6.59}$$

$$Y^*(\mathbf{n}) \xrightarrow{\mathfrak{D}} W^0, \quad \textit{in the } J_1\textit{-topology on} \quad D[0,1]^c. \tag{3.6.60}$$

Proof. Here also, we use the basic decomposition (3.6.41). First we show that under the hypothesis of the theorem,

$$\max_{\mathbf{m} \leqslant \mathbf{k} \leqslant \mathbf{n}} \psi(\mathbf{k}, \mathbf{n}) |U_3^*(\mathbf{k})| \xrightarrow{p} 0 \quad \text{as} \quad \mathbf{n} \to \infty. \tag{3.6.61}$$

For this, we partition the set $A(\mathbf{n}) = \{\mathbf{k} : \mathbf{m} \leqslant \mathbf{k} \leqslant \mathbf{n}\}$ into three disjoint subsets $A_1(\mathbf{n}) = \{\mathbf{k} \in A(\mathbf{n}) : \min(k_1, k_2) \leqslant \epsilon_1 \sqrt{n}\}$, $A_2(\mathbf{n}) = \{\mathbf{k} \in A(\mathbf{n}) : \epsilon_1 \sqrt{n} < \min(k_1, k_2) \leqslant [n^{4/5}]\}$, and $A_3(\mathbf{n}) = \{\mathbf{k} \in A(\mathbf{n}) : [n^{4/5}] \leqslant \min(k_1, k_2) \leqslant \mathbf{n}\}$, where $n = n_1 + n_2$ and $\epsilon_1 (> 0)$ is arbitrarily small. It can be shown by some standard steps that

$$\left[\max_{\mathbf{k}} \psi(\mathbf{k}, n) \cdot r^*(\mathbf{k}) : \mathbf{k} \in A_l(\mathbf{n})\right] \to 0 \quad \text{as} \quad \mathbf{n} \to \infty, \quad \forall l = 1, 2, 3, \tag{3.6.62}$$

so that on using an inequality comparable to (3.6.35) for each set $A_l(\mathbf{n})$ [but involving the $U_3^*(\mathbf{k})$ and $r^*(\ )$ instead of $U(\mathbf{k})$ and $r(\ )$], (3.6.61) follows.

Now, for each of $\{U_1(k_1) : k_1 \leqslant n_1\}$ and $\{U_2(k_2) : k_2 \leqslant n_2\}$, defined by (3.6.42) through (3.6.44), one can use Theorem 3.3.1 to provide their weak convergence to appropriate (independent) Brownian motions. Hence, the rest of the proof of (3.6.58) follows by arguments similar to those in the latter part of the proof of Theorem 3.6.1.

The basic representation (3.6.52) along with the inequality (3.6.35) for each component $U_{\mathbf{h}}^*$ leads us to the following:

$$\max_{\mathbf{0} \leqslant \mathbf{h} \leqslant \mathbf{m} - \mathbf{1}} \ \max_{\mathbf{m} - \mathbf{h} \leqslant \mathbf{k} \leqslant \mathbf{n}} |U_{\mathbf{h}}^*(\mathbf{k})| = 0_p(1). \tag{3.6.63}$$

From (3.6.52) and (3.6.63), we readily obtain that

$$\max_{\mathbf{m} \leqslant \mathbf{k} \leqslant \mathbf{n}} |U(\mathbf{k}) - \theta(\mathbf{F}(\cdot, \mathbf{k}))| \psi(\mathbf{k}, \mathbf{n}) \xrightarrow{p} 0, \quad \text{as} \quad n \to \infty, \tag{3.6.64}$$

$$\max_{\mathbf{k} : \overline{\mathbf{k} \wedge \mathbf{m}} \leqslant m - 1} |\theta(\mathbf{F}(\cdot, \mathbf{k}))| \psi(\mathbf{k}, \mathbf{n}) \xrightarrow{p} 0, \quad \text{as} \quad n \to \infty \tag{3.6.65}$$

where $\overline{\mathbf{k} \wedge \mathbf{m}} = \min_{1 \leqslant j \leqslant c} [k_j \wedge m_j]$. The last two equations imply (3.6.59) and (3.6.58) through (3.6.59) imply (3.6.60). Q.E.D.

3.7. STRONGLY CONSISTENT ESTIMATORS OF THE DISPERSION OF U-STATISTICS AND VON MISES FUNCTIONALS

In (3.2.15) and (3.2.16), we have provided the expression for the variance of U_n. Similar expressions can be obtained for the covariance of two or more U_n. In actual applications, the parameters $\zeta_h(F)$, $h = 1, \ldots, m$ are often unknown and one needs to estimate them. Note that whenever $\zeta_m(F) < \infty$, these parameters are themselves estimable parameters (regular functionals of F). For example, if one considers the *kernel*

$$\phi_h(x_1, \ldots, x_{2m}) = g(x_1, \ldots, x_m) g(x_{m-h+1}, \ldots, x_{2m-h}) \\ - g(x_1, \ldots, x_m) g(x_{m+1}, \ldots, x_{2m}), \qquad (3.7.1)$$

then

$$E\phi_h(X_1, \ldots, X_{2m}) = \zeta_h(F), \qquad \forall 0 \leqslant h \leqslant m. \qquad (3.7.2)$$

Thus the U-statistic corresponding to the kernel ϕ_h is a suitable estimator of $\zeta_h(F)$ for $h = 1, \ldots, m$. In view of the fact that for the computation of the above estimator one needs $\binom{n}{2m}$ $[= O(n^{2m})]$ terms, which, for $m > 1$, becomes quite laborious, we seek to provide alternative estimators as follows. These are due to Sen (1960, 1977a) and Sproule (1974).

Looking at (3.2.16) we observe that, for large n, the parameter of interest is $\zeta_1(F)$. For every i: $1 \leqslant i \leqslant n$, we define the ith component of U_n by

$$\tilde{V}_{n,i} = \binom{n-1}{m-1}^{-1} \sum\nolimits_n^{(i)} g(X_i, X_{i_2}, \ldots, X_{i_m}), \qquad 1 \leqslant i \leqslant n, \quad n \geqslant m, \qquad (3.7.3)$$

where the summation $\sum_n^{(i)}$ extends over all possible $1 \leqslant i_2 < \cdots < i_m \leqslant n$ with $i_j \neq i$, $2 \leqslant j \leqslant m$. Note that

$$n^{-1} \sum_{i=1}^{n} \tilde{V}_{n,i} = U_n, \qquad \forall n \geqslant m. \qquad (3.7.4)$$

Let us then define

$$s_n^2 = (n-1)^{-1} \sum_{i=1}^{n} [\tilde{V}_{n,i} - U_n]^2. \qquad (3.7.5)$$

We consider s_n^2 as an estimator of $\zeta_1(F)$. Toward this, we note that $s_n^2 = [n/(n-1)]\{(1/n)\sum_{i=1}^n \tilde{V}_{n,i}^2 - U_n^2\}$, where by Theorem 3.2.1,

$$U_n^2 \to \theta^2(F) \qquad \text{a.s., as } n \to \infty.$$

Thus if we show that

$$n^{-1} \sum_{i=1}^{n} \tilde{V}_{n,i}^2 \to \zeta_1(F) + \theta^2(F) \qquad \text{a.s., as } n \to \infty \qquad (3.7.6)$$

it follows that

$$s_n^2 \to \zeta_1(F) \qquad \text{a.s.,} \quad \text{as} \quad n \to \infty. \tag{3.7.7}$$

From (3.7.3) and some simple manipulations, we obtain that

$$n^{-1}\sum_{i=1}^{n} \tilde{V}_{n,i}^2 = n^{-1}\sum_{i=1}^{n}\left\{(n-1)^{-[m-1]}\right\}^2$$

$$\times \sum\nolimits_{n,i}^{*} g(X_i, X_{i_2}, \ldots, X_{i_m})\, g(X_i, X_{i_2'}, \ldots, X_{i_m'}), \tag{3.7.8}$$

where the summation $\sum_{n,i}^{*}$ extends over all $1 \leqslant i_2 \neq \cdots \neq i_m \leqslant n$, $1 \leqslant i_2' \neq \cdots \neq i_m' \leqslant n$ with $i_j \neq i$ and $i_j' \neq i$, $2 \leqslant j \leqslant m$. Among the $2m-2$ indices $(i_2, \ldots, i_m, i_2', \ldots, i_m')$, the number of distinct units may be equal to $m - 1 + h$, $h = 0, \ldots, m-1$. Thus $\sum_{n,i}^{*}$ can be sorted into m components where the hth component extends over the $m - 1 + h$ distinct units, $h = 0, 1, \ldots, m$. This leads us to

$$n^{-1}\sum_{i=1}^{n} \tilde{V}_{n,i}^2 = \sum_{h=0}^{m-1}\left\{n^{-1}\left((n-1)^{-[m-1]}\right)^2\right\}\left\{n(n-1)^{[m-1+h]}\tilde{U}_{n,h}\right\}, \tag{3.7.9}$$

where each $\tilde{U}_{n,h}$ is a U-statistic (of degree $m - 1 + h$), $0 \leqslant h \leqslant m-1$. On simplification, we get

$$n^{-1}\sum_{i=1}^{n} \tilde{V}_{n,i}^2 = \frac{(n-1)^{[2m-2]}}{\left((n-1)^{[m-1]}\right)^2}\tilde{U}_{n,m-1}$$

$$+ \sum_{h=0}^{m-2}\frac{(n-1)^{[m-1+h]}}{\left((n-1)^{[m-1]}\right)^2}\tilde{U}_{n,h} l_{m,h}, \tag{3.7.10}$$

where the $l_{m,h}$ are the multinomial coefficients. Note that, by definition,

$$\tilde{U}_{n,m-1} = \binom{n}{2m-1}^{-1}\sum_{1 \leqslant \alpha_1 < \cdots < \alpha_{2m-1} \leqslant n} g(X_{\alpha_1}, X_{\alpha_2}, \ldots, X_{\alpha_m})$$

$$\times\, g(X_{\alpha_1}, X_{\alpha_{m+1}}, \ldots, X_{\alpha_{2m-1}}) \tag{3.7.11}$$

so that $\tilde{U}_{n,m-1}$ is a strongly consistent (by Theorem 3.2.1) estimator of $\zeta_1(F) + \theta^2(F)$. Also, by Theorem 3.2.1, each $\tilde{U}_{n,h}$ a.s. converges to its expectation $(\zeta_{m-h}(F) + \theta^2(F) < \infty)$ as $n \to \infty$, so that by (3.7.10), one gets

$$n^{-1}\sum_{i=1}^{n} \tilde{V}_{n,i}^2 = \tilde{U}_{n,m-1} + O(n^{-1}) \qquad \text{a.s.}$$

$$\to \zeta_1(F) + \theta^2(F) \qquad \text{a.s.,} \quad \text{as} \quad n \to \infty \tag{3.7.12}$$

and this completes the proof of (3.7.7). It is also possible to define the components of $\theta(F_n)$ as $\tilde{V}_{n,i}^{*} = (n-1)^{-(m-1)}\sum_n^{*(i)} g(X_i, X_{i_2}, \ldots, X_{i_m})$ where the summation extends over $i_j = 1, \ldots, i-1, i+1, \ldots, n$, $j = 2, \ldots, m$

and introduce the statistic

$$s_n^{*2} = n^{-1}\sum_{i=1}^{n}\left[\tilde{V}_{n,i}^{*} - \theta(F_n)\right]^2, \tag{3.7.13}$$

which like s_n^2 can also be expressed as a linear combination of U-statistics and also converges a.s. to $\zeta_1(F)$ as $n \to \infty$.

Let us now consider the case of generalized U-statistics. For simplicity, we again take $c = 2$. Then let

$$\tilde{V}_{10}(i,\mathbf{n}) = \binom{n_1-1}{m_1-1}^{-1}\binom{n_2}{m_2}^{-1}\sum\nolimits_{(\mathbf{n})}^{*,i} g(X_{1i}, X_{i_2}, \ldots, X_{i_{m1}}, X_{2j_1}, \ldots, X_{2j_{m_2}}), \tag{3.7.14}$$

where the summation extends over all $1 \leqslant j_1 < \cdots < j_{m_2} \leqslant n_2$ and $1 \leqslant i_2 < \cdots < i_{m_1} \leqslant n_1$ with $i_j \neq i, j = 2, \ldots, m_1$. Note that

$$n_1^{-1}\sum_{i=1}^{n_1}\tilde{V}_{10}(\mathbf{i},\mathbf{n}) = U(\mathbf{n}), \qquad \forall \mathbf{n} \geqslant \mathbf{m}. \tag{3.7.15}$$

We define then

$$s_{10}^2(\mathbf{n}) = \frac{1}{n_1 - 1}\sum_{i=1}^{n_1}\left[\tilde{V}_{10}(i,\mathbf{n}) - U(\mathbf{n})\right]^2. \tag{3.7.16}$$

In a similar way, we can define the components $\tilde{V}_{01}(i,\mathbf{n})$, $1 \leqslant i \leqslant n_2$ and the statistic $s_{01}^2(\mathbf{n})$. Then, on parallel lines, it follows that

$$s_{10}^2(\mathbf{n}) \to \zeta_{10}(\mathbf{F}), \qquad s_{01}^2(\mathbf{n}) \to \zeta_{01}(\mathbf{F}) \qquad \text{a.s., as } \mathbf{n} \to \infty. \tag{3.7.17}$$

Note that by (3.7.8) and (3.7.10) whenever $E|g(\cdot)|^{2r} < \infty$, $r > 0$ (not necessarily integer), for each $\tilde{U}_{n,h}$, $0 \leqslant h \leqslant m-1$, the kernel has a finite absolute moment of order r, so that by (3.2.48), we have, for $r \geqslant 2$,

$$E|\tilde{U}_{n,m-1} - \zeta_1(F) + \theta^2(F)|^r = 0(n^{-r/2}); \tag{3.7.18}$$

while

$$E|\tilde{U}_{n,h}|^r = 0(1), \qquad \forall 0 \leqslant h \leqslant m-2. \tag{3.7.19}$$

Also $\mathrm{Var}(V_n) = 0(n^{-1})$, so that for $r \geqslant 2$,

$$E|U_n^2 - \theta^2(F)|^r = 0(n^{-r/2}). \tag{3.7.20}$$

From (3.7.5), (3.7.10), (3.7.18), (3.7.19), and (3.7.20), we obtain, by using the c_r-inequality, that under $E|g|^{2r} < \infty$, $r \geqslant 2$,

$$E|s_n^2 - \zeta_1(F)|^r = 0(n^{-r/2}). \tag{3.7.21}$$

A similar expression can also be obtained for the $s_{10}^2(\mathbf{n})$ and $s_{01}^2(\mathbf{n})$.

Let U_{n-1}^i be the U-statistic corresponding to the kernel g based on a sample of size $n-1$ obtained from the sample of size n by deleting X_i, for $i = 1, \dots, n$. Also, let $U_{n,i} = nU_n - (n-1)U_{n-1}^i$ for $i = 1, \dots, n$ and $U_n^* = n^{-1}\sum_{i=1}^n U_{n,i}$. Then the jack-knife estimator of $m^2\zeta_1$ is

$$V_n^* = (n-1)^{-1}\sum_{i=1}^{n}(U_{n,i} - U_n^*)^2. \tag{3.7.22}$$

It is interesting to note that by definition,

$$\binom{n-1}{m}U_{n-1}^i + \binom{n-1}{m-1}V_{ni} = \binom{n}{m}U_n \qquad \text{for every} \quad i = 1, \dots, n, \tag{3.7.23}$$

and, hence it follows from (3.7.3) through (3.7.5), (3.7.22), and (3.7.23) that

$$V_n^* = m^2(n-1)^2(n-m)^{-2}s_n^2, \qquad \forall n > m. \tag{3.7.24}$$

Hence the convergence properties of s_n^2, studied in this section, ensure the same for V_n^*.

EXERCISES

3.2.1. Let $\mathbf{X}_1, \dots, \mathbf{X}_n$ be n independent p-vectors ($p \geqslant 1$) with a common d.f. $F(\mathbf{x})$, $\mathbf{x} \in R^p$, and let $\mathcal{F} = \{F : E\mathbf{X}\mathbf{X}' \text{ exists}\}$. Further, let $\boldsymbol{\mu} = E\mathbf{X}$ and $\boldsymbol{\Sigma} = E(\mathbf{X} - \boldsymbol{\mu})(\mathbf{X} - \boldsymbol{\mu})'$. Show that both $\boldsymbol{\mu}$ and $\boldsymbol{\Sigma}$ are estimable and obtain the corresponding U-statistics and von Mises functionals. Show that for the estimation of $\boldsymbol{\mu}$, both the U-statistics and the von Mises functionals are the same, but for $\boldsymbol{\Sigma}$, they are different, In fact, show that the von Mises functional for $\boldsymbol{\Sigma}$ is not unbiased for any finite n.

3.2.2. (3.2.1 continued). Is $\boldsymbol{\mu}'\boldsymbol{\mu}$ an estimable parameter? What about $\boldsymbol{\mu}'\boldsymbol{\Sigma}^{-1}\boldsymbol{\mu}$? If they are estimable, provide the corresponding U-statistics and von Mises functionals and examine whether they are the same or not.

3.2.3. Let $\theta(F)$ be estimable and of degree $m(\geqslant 2)$. Show that if $Eg(X_{i_1}, \dots, X_{i_m}) \neq \theta(F)$ when the number of distinct units in $(i_1, \dots, i_m)$ is $\leqslant m-1$, then $\theta(F_n)$ is not unbiased for $\theta(F)$ for all n.

3.2.4. Let $\mathbf{X}_i' = (X_{i1}, X_{i2})$, $1 \leqslant i \leqslant n$ be n i.i.d.r.v. with a continuous bivariate d.f. $F(x, y); (x, y) \in R^2$. Consider the parameter $\theta(F) = 4P\{\mathbf{X}_2 < \mathbf{X}_1\} - 1$. (This is known as the Kendall tau coefficient.) Show that if X_{i1}, X_{i2} are independent, $\theta(F) = 0$, while $\theta(F)$ is positive or negative according as they are positively or negatively associated. Obtain the expression for the U-statistic and von Mises functionals for $\theta(F)$ and show that they are not identical. If F is a bivariate normal d.f. with a correlation coefficient ρ, show that $\theta(F)$ is a monotonic function of ρ: $-1 < \rho < 1$.

3.2.5. (3.2.4 continued). Consider the rank correlation

$$r_n = \frac{12}{n(n^2-1)} \sum_{i=1}^{n} \left(R_i - \frac{n+1}{2}\right)\left(S_i - \frac{n+1}{2}\right),$$

where R_i = Rank of X_{i1} among $X_{11}, \ldots, X_{n1}$, $1 \leqslant i \leqslant n$ and S_i = Rank of X_{i2} among $X_{12}, \ldots, X_{m2}$, $1 \leqslant i \leqslant n$. Show that r_n is the von Mises functional corresponding to the parameter

$$\rho_g = 12 \int_{-\infty}^{\infty} \int_{-\infty}^{\infty} \left[F(x, \infty) - \tfrac{1}{2}\right]\left[F(\infty, y) - \tfrac{1}{2}\right] dF(x, y);$$

ρ_g is known as the grade correlation coefficient. Obtain the U-statistic corresponding to ρ_g and examine the closeness of this estimator and of r_n.

3.2.6. Let $X_1, \ldots, X_n$ be n independent r.v.'s from a continuous d.f. $F(x)$, $x \in R$. Consider the kernel $g(X_1, X_2) = \text{sgn}(X_1 + X_2)$ that estimates $\theta(F) = 2P\{X_1 + X_2 > 0\} - 1$. Show that (i) if F is symmetric about δ, then $\theta(F)$ is $>$, $=$, or < 0 according as δ is $>$, $=$, or < 0, (ii) the von Mises functional corresponding to $\theta(F)$ is the classical Wilcoxon signed rank statistic, and (iii) obtain the corresponding U-statistics and simplify them for the case of $\delta = 0$.

3.2.7 (3.2.6 continued). Consider the estimator

$$T_n^{(k)} = \left[\sum_{i=1}^{n} \binom{k-1}{k}\binom{n-i}{k} X_{n,i}\right] \Big/ \binom{n}{2k+1}, \qquad k = 0, 1, \ldots;$$

where $X_{n,1} < \cdots < X_{n,n}$ are the ordered random variables corresponding to $X_1, \ldots, X_n$. Show that for each k, $T_n^{(k)}$ is a U-statistic; in particular, for $k = 0$, $T_n^{(0)} = \bar{X}_n = (1/n)\sum_{i=1}^{n} X_i$ (Sen, 1964).

3.2.8. Verify the inequality (3.2.14) (Hoeffding, 1948a).

3.2.9. For Problem 3.2.2, show that if $\mathfrak{F}_0 = \{F : \boldsymbol{\mu} = \mathbf{0}\}$, $\theta(F) = \boldsymbol{\mu}'\boldsymbol{\mu}$ is stationary of order 1.

3.2.10. Show that if $Z_1, \ldots, Z_N$ are i.i.d.r.v.'s with $E|Z|^r < \infty$, for some $r > 1$, then $E|\bar{Z}_N - EZ|^r = O(N^{-r^*})$, with r^* defined by (3.2.47).

3.2.11. Show that under (3.2.24), the rhs of (3.2.23) converges.

3.3.1. Let $X_1, \ldots, X_n$ be i.i.d.r.v.'s with d.f. $F(x)$ and let $S_n = S(X_1, \ldots, X_n)$ be a Borel-measurable function. Define $\hat{S}_n = \sum_{i=1}^{n} E(S_n | X_i) - (n-1)ES_n$. Show that (i) $E\hat{S}_n = ES_n$, (ii) $E(S_n - \hat{S}_n)^2$

$= \text{Var}(S_n) - \text{Var}(\hat{S}_n)$ and if $E(S_n - \hat{S}_n)^2/\text{Var}(S_n) \to 0$ as $n \to \infty$, then $\mathcal{L}((\hat{S}_n - a_n)/b_n) \to \mathfrak{N}(0, 1) \Rightarrow \mathcal{L}((S_n - a_n)/b_n) \to \mathfrak{N}(0, 1)$.

3.3.2 (3.3.1 continued). If $X_n = U(X_1, \ldots, X_n) = U_n$ is a U-statistic, then show that $\hat{S}_n = \theta(F) + mV_{n,1} = \theta(F) + mU_{n,1}$, where the $V_{n,1}$ and $U_{n,1}$ are defined by (3.2.26) and (3.2.30).

3.3.3. Obtain the expressions for $V_{n,1}$ for the specific problems 3.2.4, 3.2.5, and 3.2.6.

3.3.4. Let $\psi(F) = \psi(\theta_1(F), \ldots, \theta_q(F))$ be a continuous and derivable function of $q(\geqslant 1)$ estimable parameters of degree $m_1, \ldots, m_q$, all being stationary of order 0. Consider the sequence of estimators $\{Z_n = \psi(U_n^{(1)}, \ldots, U_n^{(q)}), n \geqslant m_0\}$ where $m_0 = \max(m_1, \ldots, m_q)$ and $U_n^{(j)}$ is the U-statistic for $\theta_j(F)$, $1 \leqslant j \leqslant q$. Under suitable regularity conditions on ψ (to be specified) show that the invariance principles of Theorems 3.3.1, 3.3.3, and 3.3.4 also hold for $\{Z_n\}$.

3.3.5 (3.3.4 continued). For the problem 3.2.1, let $p = 2$ and consider $\psi(\theta_1(F), \theta_2(F), \theta_3(F)) = \sigma_{12}/(\sigma_{11}\sigma_{22})^{1/2}$ = correlation coefficient. Show that the invariance principles hold for the sequence of sample correlation coefficients. Obtain similar results for the regression coefficient σ_{21}/σ_{11}.

3.3.6. Show that if $X_1, \ldots, X_n$ are i.i.d.r.v.'s with $EX^2 < \infty$, then $n^{-1/2}\max_{1 \leqslant i \leqslant n}|X_i| \to 0$ a.s., as $n \to \infty$ and this implies that $\sup_{N \geqslant n} N^{-1/2}|X_N| \xrightarrow{P} 0$ as $n \to \infty$; use this in (3.3.42).

3.3.7. Show that the weak convergence in Theorems 3.3.1, 3.3.2, and 3.3.4 holds also in the d_q-norm, where d_q is defined by (2.4.25) and (2.4.26).

3.3.8. Establish the weak convergence results in Theorems 3.3.1, 3.3.2, and 3.3.4 in the sup-norm metric in (2.4.26).

3.4.1. Let $\{Z_i, i \geqslant 1\}$ be i.i.d.r.v.'s with $E|Z|^r < \infty$ for some $r > 2$, and let $f(t)$ satisfy the condition (3.4.3). Then show that (3.4.4) holds. In fact, here $f(t) = t^{1-\eta}$, $\eta = \eta(r) > 0$ satisfies both (3.4.3) and (3.4.4).

3.4.2. For the problems 3.2.4, 3.2.5, and 3.2.6, $E|g|^r < \infty$ for every finite r. Show that here $f(t)$ can be taken to be a slowly increasing function.

3.4.3. Extend the results of Problems 3.3.4 and 3.3.5 in the light of the strong invariance principles of Section 3.4.

3.5.1. Obtain the exact expression for the variance of $U_{N,n}$ in (3.5.3) in terms of the $\bar{\zeta}_{h,N}$ in (3.5.6) and show that (3.5.30) is in order.

3.5.2. Supply the proof of Theorem 3.5.1.

3.5.3. Let $Z_N = (a_{N1}, \ldots, a_{NN})$ and $B_N = (b_{N1}, \ldots, b_{NN})$ be two sequences of real numbers and let $X_N = (X_{N1}, \ldots, X_{NN})$ be a stochastic vector that takes on each permutation of the elements of B_N with equal probability $(N!)^{-1}$. Consider the random variable $L_N = \sum_{i=1}^{N} a_{Ni} X_{Ni}$ and set

$$\bar{a}_N = N^{-1} \sum_{i=1}^{N} a_{Ni} = 0 = \bar{b}_N = \frac{1}{N} \sum_{i=1}^{N} b_{Ni}$$

and

$$\sum_{i=1}^{N} a_{Ni}^2 = \sum_{i=1}^{N} b_{Ni}^2 = N.$$

Then, show that if A_N satisfy (3.3.24) and B_N satisfy the condition that $N^{-1/2} \max_{1 \leqslant i \leqslant N} |b_{Ni}| \to 0$ as $N \to \infty$, then $N^{-1/2} L_N$ is asymptotically normally distributed with 0 mean and unit variance (Hájek, 1961).

Hint: Consider $E(L_N^{2k})$ for a positive integer k. This equals $E(\sum_{i_1=1}^{N} \cdots \sum_{i_{2k}=1}^{N} a_{Ni_1} \cdots a_{Ni_1} X_{Ni_1} \cdots X_{Ni_{2k}})$. The maximum contribution arises due to the terms for which there are k pairs of equal indices: $i_1 = i_2 \neq i_3 = i_4 \neq i_5 = i_6 \neq \cdots \neq i_{2k-1} = i_{2k}$. The total contribution of these terms will be $[(2k)!/2^k k!] N^k [1 + 0(N^{-1})]$. For any other set, containing less than or more than k distinct indices, the total contribution is $0(N^{k-1})$ or lesser order. Hence $N^{-k} E(L_N^{2k}) \to (2k)!/2^k k!$ as $N \to \infty$. Similarly, $N^{-k-1/2} E(L_N^{2k+1}) \to 0$ as $N \to \infty$. The proof follows by the method of moments.

3.5.4. Apply the Brown inequality (2.3.12) to show that the convergence of the f.d.d. of $Y_N^{(1)}$ to those of W_0 and the reverse martingale property imply the tightness.

3.5.5. Provide a proof of Theorem 3.5.3.

3.6.1. Let $X_1, \ldots, X_{n_1}$ be i.i.d.r.v.'s with a continuous d.f. $F(x)$, $x \in R$, $Y_1, \ldots, Y_{n_2}$ be i.i.d.r.v.'s with a continuous d.f. $G(x)$, $x \in R$ and let $\theta_1(F, G) = P\{X < Y\}$. Show that for this parameter, the generalized U-statistics and the von Mises functionals are the same and obtain the expression for its variance. Simplify the form when $F \equiv G$.

3.6.2 (3.6.1 continued). Consider a second functional $\theta_2(F, G) = P\{|X_1 - X_2| < |Y_1 - Y_2|\}$ and show that the corresponding U-statistic and von Mises functionals are not identical, compute the variance of these estimators and show that even if $F \equiv G$, the variances are not independent of F.

3.6.3. Complete the proof of (3.6.52). Also for $m_1 = m_2 = c = 2$, obtain explicitly the expressions for the component $U_{\mathbf{h}}^*(\mathbf{n})$.

3.7.1. Let $X_1, \ldots, X_n$ be i.i.d.r.v.'s with a d.f. $F(x)$, $x \in R$ and let $\theta(F) = E(X - EX)^2$. Obtain the expression for the variance of the U-statistic for $\theta(F)$ and consider (i) the U-statistic estimator of this variance along the lines of (3.7.1) and (3.7.2) and (ii) the alternative estimator using (3.7.5).

3.7.2. Suppose $\theta(F)$ is stationary of order $d(\geqslant 0)$. Consider the components of U_n defined by

$$\tilde{V}_n(i_1, \ldots, i_{d+1}) = \binom{n-d-1}{m-d-1}^{-1} \times \sum_{n}^{(i_1, \ldots, i_{d+1})} g(X_{i_1}, \ldots, X_{i_{d+1}}, X_{i_{d+2}}, \ldots, X_{i_m}),$$

where the summation extends over all possible $i_{d+2}, < \cdots < i_m$, (excluding the particular $i_1, \ldots, i_{d+1}$). Then let

$$s_n^2 = \binom{n}{d+1}^{-1} \sum_{1 \leqslant i_1 < \cdots < i_{d+1} \leqslant n} \left[\tilde{V}_n(i_1, \ldots, i_{d+1}) - U_n\right]^2.$$

Show that $s_n^2 \to \zeta_{d+1}(F)$ a.s., as $n \to \infty$ $(0 \leqslant d \leqslant m-1)$.

3.7.3. For Problem 3.6.1, when F and G are not specified to be the same, obtain the corresponding $s_{10}^2(\mathbf{n})$ and $s_{01}^2(\mathbf{n})$. Compare these with the corresponding U-statistic estimators of $\zeta_{10}(F, G)$ and $\zeta_{01}(F, G)$.

3.7.4. For the Problem 3.6.2, derive similar results.

3.7.5. Extend the results of Problem 3.7.2 to the case of generalized U-statistics.

CHAPTER 4

Invariance Principles for Linear Rank Statistics

4.1. INTRODUCTION

Linear rank statistics play a very prominent role in the theory of nonparametric statistical inference. For multisample (univariate as well as multivariate) location, scale, or regression problems, both for estimation and tests of significance, these stastistics are indispensable. In this chapter, linear rank statistics are introduced and various invariance principles for them are considered.

Section 4.2 deals with the basic definitions and preliminary notions. Weak convergence of (partial or tail) sequences of linear rank statistics to processes of Brownian motions is studied in Section 4.3. Section 4.4 deals with the Skorokhod-Strassen embedding of Wiener processes for linear rank statistics. Asymptotic linearity (in the regression parameter) of linear rank statistics and its role in the invariance principles are studied in Section 4.5. The last section deals with almost sure convergence of linear rank statistics.

4.2. PRELIMINARY NOTIONS

Let $X_1, \ldots, X_N$ be $N(\geqslant 1)$ independent random variables (r.v.) with continuous d.f.'s $F_1(x), \ldots, F_N(x)$, respectively; all defined on the real line $(-\infty, \infty)$. For testing the hypothesis of identity of these d.f.'s (against location, scale, or regression alternatives) or for estimating parameters in a nonparametric setup, the following type of *linear rank statistics* is widely used. Let

$$L_N = \sum_{i=1}^{N} (c_i - \bar{c}_N) a_N(R_{Ni}), \qquad \bar{c}_N = \left(\sum_{i=1}^{N} c_i \right) / N, \tag{4.2.1}$$

where $\{c_1, \ldots, c_N\}$ are known constants (not all equal), $R_{Ni} = \sum_{j=1}^{N} c(X_i - X_j)$ [with $c(u) = 1$ or 0 according as u is $\geqslant$ or < 0] is the rank of X_i among $X_1, \ldots, X_N$ for $1 \leqslant i \leqslant N$ and $a_N(1), \ldots, a_N(N)$ are suitable *scores* (real numbers). We shall find it convenient to express the scores as

$$a_N(i) = E\phi(U_{Ni}) \quad \text{or} \quad \phi(i/(N+1)), \qquad 1 \leqslant i \leqslant N, \tag{4.2.2}$$

where $\phi = \{\phi(u), u \in I\}$, $I = (0, 1)$ is a suitable *score-function* and $U_{N1}, \ldots, U_{NN}$ are the ordered random variables of a sample of size N from the rectangular [0, 1] d.f. Note that $EU_{Ni} = i/(N+1), 1 \leqslant i \leqslant N$, so that for linear ϕ, the two scores in (4.2.2) are the same. But, for general ϕ, they are not necessarily the same. However, we shall see that for large N, they share the common properties under fairly general regularity conditions.

Consider now the hypothesis of invariance, that is

$$H_0 : F_1(x) \equiv \cdots \equiv F_N(x) \equiv F(x) \text{ (unknown)}. \tag{4.2.3}$$

Under H_0, the joint distribution of the vector $\mathbf{X}_N = (X_1, \ldots, X_N)$ remains invariant under any permutation of its coordinates while the continuity of F eliminates the possibility of ties among the X_i. This leads to the fact that under H_0 in (4.2.3), $\mathbf{R}_N = (R_{N1}, \ldots, R_{NN})$ takes on each permutation of $(1, \ldots, N)$ with the common probability $(N!)^{-1}$. As such,

$$E[a_N(R_{Ni}) \mid H_0] = N^{-1} \sum_{j=1}^{N} a_N(j) = \bar{a}_N, \qquad \text{say,} \quad 1 \leqslant i \leqslant N; \tag{4.2.4}$$

$$\operatorname{Cov}[a_N(R_{Ni}), a_N(R_{Nj}) \mid H_0] = \begin{cases} A_N^2, & \text{if } i = j, \\ -\dfrac{1}{N-1} A_N^2, & \text{if } i \neq j, \end{cases} \tag{4.2.5}$$

where

$$A_N^2 = \frac{1}{N} \sum_{i=1}^{N} [a_N(i) - \bar{a}_N]^2. \tag{4.2.6}$$

From (4.2.1), (4.2.4), (4.2.5), and (4.2.6), we obtain

$$E[L_N \mid H_0] = 0, \tag{4.2.7}$$

$$E(L_N^2 \mid H_0) = \operatorname{Var}(L_N \mid H_0) = \frac{N}{N-1} A_N^2 C_N^2, \tag{4.2.8}$$

where

$$C_N^2 = \sum_{i=1}^{N} (c_i - \bar{c}_N)^2, \qquad N \geqslant 1. \tag{4.2.9}$$

Actually, under suitable regularity conditions on the $\{c_i\}$ and $\{a_N(i)\}$, one can show that as $N \to \infty$,

$$\frac{L_N}{A_N C_N} \xrightarrow{\mathfrak{D}} \mathfrak{N}(0,1) \quad \text{under} \quad H_0. \tag{4.2.10}$$

We shall comment more on this in the next section. For small values of N, we can obtain the exact distribution of L_N (under H_0) with the aid of the $N!$ equally likely realizations of $\mathbf{R}_N$. As $N \to \infty$, this process becomes very cumbersome.

We are primarily interested here in the limiting behavior of the sequence $\{L_k; k \leqslant N\}$ or the tail-sequence $\{L_k; k \geqslant N\}$ when $N \to \infty$. The principal results of Sections 4.3 and 4.4 are geared in this direction. We conclude this section with a remark on two special situations where L_N is especially important. First, consider the two-sample problem, where $N = n_1 + n_2$, $n_1 \geqslant 1, n_2 \geqslant 1$; $F_1 \equiv \cdots \equiv F_{n_1} \equiv F$ and $F_{n_1+1} \equiv \cdots \equiv F_N \equiv G$, so that (4.2.3) reduces to $H_0: F \equiv G$. In this case, the constants c_i are quite simple:

$$c_i = \begin{cases} 1, & 1 \leqslant i \leqslant n_1, \\ 0, & n_1 + 1 \leqslant i \leqslant N \end{cases} \Rightarrow C_N^2 = n_1 n_2 / N. \tag{4.2.11}$$

Here L_N can also be written as $\sum_{i=1}^{n_1}[a_N(R_{Ni}) - \bar{a}_N]$. Second, consider the simple regression model

$$F_i(x) = F(x - \alpha - \beta t_i), \qquad 1 \leqslant i \leqslant N, \quad x \in (-\infty, \infty), \tag{4.2.12}$$

where $(t_1, \ldots, t_N)$ are known regression constants and α, β are unknown parameters. Here H_0: $\beta = 0$ relates to (4.2.3). For this problem, one can choose $c_i = t_i$, $1 \leqslant i \leqslant N$. The choice of the score function ϕ in (4.2.2) depends on the alternative hypothesis we have in mind as well as on other considerations of optimality (at least locally) for specific type of underlying d.f.'s; see for example Hájek and Šidák (1967).

We consider now a basic martingale property of linear rank statistics which will be used repeatedly in subsequent sections. We introduce the scores as in the first case of (4.2.2), namely,

$$a_N^0(i) = E\phi(U_{Ni}) \qquad \text{for} \quad i = 1, \ldots, N(\geqslant 1) \tag{4.2.13}$$

and, replacing $a_N(\cdot)$ by $a_N^0(\cdot)$ in (4.2.1), we define L_N^0, for $N \geqslant 1$; conventionally, we let $L_0^0 = 0$. Let $\mathfrak{B}_N$ be the sigma-field generated by $\mathbf{R}_N, N \geqslant 1$ and $\mathfrak{B}_0$ be the trivial sigma-field. Then, $\mathfrak{B}_N$ is nondecreasing in N. The following theorem, due to Sen and Ghosh (1972), is based on an idea implicit in Hájek (1961).

THEOREM 4.2.1. *Under H_0 in (4.2.3), $\{L_N^0, \mathfrak{B}_N; N \geqslant 1\}$ is a martingale whenever ϕ is integrable (with respect to the Lebesgue measure) inside I.*

Proof. By (4.2.1) and (4.2.13), for every $N \geqslant 1$, L_N^0 is invariant under any translation of the score function ($\phi^*(u) = \phi(u) + d, 0 < u < 1$ where d is any real number), and hence, without any loss of generality, we may take

$$\bar{\phi} = \int_0^1 \phi(u)\, du = 0. \tag{4.2.14}$$

Note that by (4.2.13) and (4.2.14), for every $N \geqslant 1$,

$$\begin{aligned} N^{-1}\sum_{i=1}^{N} a_N^0(i) &= N^{-1}\sum_{i=1}^{N} N\binom{N-1}{i-1}\int_0^1 \phi(u) u^{i-1}(1-u)^{N-i}\, du \\ &= \int_0^1 \phi(u)\left[\sum_{i=0}^{N-1}\binom{N-1}{i} u^i (1-u)^{N-i-1}\right] du \\ &= \int_0^1 \phi(u)\, du = 0. \end{aligned} \tag{4.2.15}$$

Also we may write

$$L_N^0 = \sum_{i=1}^{N} c_i a_N^0(R_{Ni}). \tag{4.2.16}$$

Then denoting by E_0 the expectation under H_0, we have

$$\begin{aligned} E_0\left[L_{N+1}^0 \mid \mathfrak{B}_N\right] &= \sum_{i=1}^{N} c_i E_0\left[a_{N+1}^0(R_{N+1i}) \mid \mathfrak{B}_N\right] \\ &\quad + c_{N+1} E_0\left[a_{N+1}^0(R_{N+1N+1}) \mid \mathfrak{B}_N\right]. \end{aligned} \tag{4.2.17}$$

Now, given $\mathfrak{B}_N$ (i.e., $\mathbf{R}_N$), R_{N+1N+1} can assume the values $1, \ldots, N+1$ with equal probability $1/(N+1)$, when H_0 holds. Hence

$$\begin{aligned} E_0\left[a_{N+1}^0(R_{N+1N+1}) \mid \mathfrak{B}_N\right] &= \frac{1}{N+1}\sum_{i=1}^{N+1} a_{N+1}^0(i) \\ &= 0, \qquad \text{by (4.2.15).} \end{aligned} \tag{4.2.18}$$

Further, given $\mathfrak{B}_N$, for every i: $1 \leqslant i \leqslant N$, R_{N+1i} can only assume two values: R_{Ni} and $(R_{Ni} + 1)$. The first event materializes when R_{N+1N+1} is $> R_{Ni}$ [with conditional probability $(N + 1 - R_{Ni})/(N+1)$] and the second event when R_{N+1N+1} is $\leqslant R_{Ni}$ (with conditional probability $(N+1)^{-1}R_{Ni}$). Hence, for $1 \leqslant i \leqslant N$,

$$\begin{aligned} E_0\left[a_{N+1}^0(R_{N+1i}) \mid \mathfrak{B}_N\right] &= \frac{R_{Ni}}{(N+1)} a_{N+1}^0(R_{Ni}+1) \\ &\quad + \frac{N+1-R_{Ni}}{(N+1)} a_{N+1}^0(R_{Ni}). \end{aligned} \tag{4.2.19}$$

Finally, note that by (4.2.13), for every $1 \leqslant k \leqslant N$,

$$\begin{aligned}
&\frac{k}{N+1} a_{N+1}^0(k+1) + \frac{N+1-k}{N+1} a_{N+1}^0(k) \\
&\quad = k\binom{N}{k}\int_0^1 \phi(u)u^k(1-u)^{N-k}\,du + (N+1-k)\binom{N}{k-1} \\
&\qquad \times \int_0^1 \phi(u)u^{k-1}(1-u)^{N+1-k}\,du \\
&\quad = N\binom{N-1}{k-1}\int_0^1 \phi(u)u^{k-1}(1-u)^{N-k}\,du = a_N^0(k). \qquad (4.2.20)
\end{aligned}$$

Consequently, from (4.2.17) through (4.2.20), we have for $N \geqslant 1$,

$$E_0(L_{N+1}^0 \mid \mathfrak{B}_N) = \sum_{i=1}^N c_i a_N^0(R_{Ni}) = L_N^0. \qquad \text{Q.E.D.} \qquad (4.2.21)$$

Remarks. We may observe that if we define

$$L_N^* = \sum_{i=1}^N (c_i - \bar{c}_N)\phi(F(X_i)), \qquad (4.2.22)$$

where F is defined by (4.2.3), then

$$E_0(L_N^* \mid \mathfrak{B}_N) = L_N^0, \qquad \forall N \geqslant 1. \qquad (4.2.23)$$

This representation will be very useful in future theorems. If the scores are not defined by (4.2.13), the lemma may not hold in general; an exception is for ϕ linear in u. However, if $a_N(i) = \phi(i/(N+1))$, $1 \leqslant i \leqslant N$, then in many cases, $\{L_N - L_N^0\}$ can be shown to be of lower order of magnitudes, so that $\{L_N\}$ can be approximated by a martingale up to a certain order of approximation.

4.3. WEAK CONVERGENCE OF $\{L_N\}$

Let us assume that ϕ is square integrable inside I and let

$$A^2 = \int_0^1 \phi^2(u)\,du - \bar{\phi}^2: \qquad 0 < A^2 < \infty. \qquad (4.3.1)$$

Note that by (4.2.13), (4.2.14), (4.2.15), and (4.3.1)

$$\begin{aligned}
0 \leqslant A_N^{02} &= \frac{1}{N}\sum_{i=1}^N \left[a_N^0(i) - \bar{a}_N^0\right]^2 = N^{-1}\sum_{i=1}^N \left[a_N^0(i)\right]^2 \\
&= N^{-1}\sum_{i=1}^N \left[E\phi(U_{Ni})\right]^2 \\
&\leqslant N^{-1}\sum_{i=1}^N E\phi^2(U_{Ni}) = A^2, \qquad \forall N \geqslant 1. \qquad (4.3.2)
\end{aligned}$$

On the other hand, if we define

$$\phi_N(u) = a_N^0(1 + [uN]), \qquad 0 < u < 1, \tag{4.3.3}$$

then by (4.2.13) and (4.3.3), $\phi_N(u)$ converges almost everywhere to $\phi(u)$, $u \in I$, as $N \to \infty$ and $A_N^{02} = \int_0^1 \phi_N^2(u)\,du$. Thus from (4.3.2) and the Fatou Lemma, we conclude that

$$A_N^{02} \to A^2 \qquad \text{as} \quad N \to \infty. \tag{4.3.4}$$

Let us now assume that the $\{c_i\}$ satisfy the Noether condition:

$$\max_{1 \leqslant i \leqslant N} (c_i - \bar{c}_N)^2 / C_N^2 \to 0 \qquad \text{as} \quad N \to \infty \tag{4.3.5}$$

(which also implies that $C_N^2 \to \infty$ as $N \to \infty$). Then for every $N(\geqslant 1)$, we consider a stochastic process $Y_N = \{Y_N(t), t \in I\}$, where

$$Y_N(t) = \frac{L_{N(t)}^0}{C_N A_N^0}; \qquad N(t) = \max\{k : C_k^2 \leqslant tC_N^2\}, \qquad t \in I. \tag{4.3.6}$$

(We let, conventionally, $L_0^0 = L_1^0 = 0$, so that $Y_N(0) = 0$.) Then, Y_N belongs to the $D[0, 1]$ space for every $N \geqslant 1$. Also, let $W = \{W(t), t \in I\}$ be a standard Brownian motion on I. Then the following is an improved version of an original theorem of Sen and Ghosh (1972).

THEOREM 4.3.1. *If ϕ is square integrable inside I, then under* (4.3.5) *and H_0 in* (4.2.3), *as $N \to \infty$,*

$$Y_N \xrightarrow{\mathfrak{D}} W, \qquad \textit{in the Skorokhod } J_1\textit{-topology on} \quad D[0,1]. \tag{4.3.7}$$

Proof. First, consider the convergence of the f.d.d.'s of $\{Y_N\}$ to those of W. For every $m(\geqslant 1)$ and $t_1, \ldots, t_m(\in I)$, let $\mathbf{N} = (N(t_1), \ldots, N(t_m))$ with $N(t)$ defined by (4.3.6), and let $Y_N^* = \{Y_N^*(t), t \in I\}$ be defined by (4.3.6) with $L_{N(t)}^0$ and A_N^0 being replaced by $L_{N(t)}^*$ and A, respectively, and where $L_n^*, n \geqslant 1$ are defined by (4.2.22). Then, using (4.2.23), we obtain

$$E(L_N^* - L_N^0)^2 = E(L_N^{*2}) - E(L_N^{02}) = C_N^2\left(A^2 - \frac{N}{N-1}A_N^{02}\right); \tag{4.3.8}$$

we conclude by using (4.3.4) and (4.3.5) that

$$\frac{\max_{1 \leqslant k \leqslant N} E(L_k^* - L_k^0)^2}{C_N^2 A_N^{02}} \to 0 \qquad \text{as} \quad N \to \infty, \tag{4.3.9}$$

so that by the Chebyshev and the Bonferroni inequalities, as $N \to \infty$,

$$\|(Y_N(t_1), \ldots, Y_N(t_m)) - (Y_N^*(t_1), \ldots, Y_N^*(t_m))\| \xrightarrow{p} 0. \tag{4.3.10}$$

On the other hand, $Y_N^*(t_j)$ involves a linear function of i.i.d.r.v.'s for which (4.3.5) and the assumed square integrability of ϕ ensure that the (multivariate) central limit theorem applies to $(Y_N^*(t_1), \ldots, Y_N^*(t_m))$. This completes the proof of the convergence of the f.d.d.'s of $\{Y_N^*\}$ (and hence of $\{Y_N\}$) to those of W. Comparing (2.4.13) and (4.3.6) and noting that by virtue of Theorem 4.2.1, Y_N is constructed from a martingale sequence, we appeal to the Corollary 2.4.4.1 and conclude that the convergence of the f.d.d.'s of $\{Y_N\}$ to those of W implies the tightness of $\{Y_N\}$, and hence, the proof of the theorem is complete.

Let us next consider a second weak convergence theorem concerning the tail-sequence $\{L_k^0; k \geqslant N\}$. Let us define

$$\bar{L}_k^0 = \frac{L_k^0}{C_k^2}, \qquad k \geqslant N > 1, \qquad \text{and } 0 \text{ for } k = 1. \tag{4.3.11}$$

Then let $Z_N = \{Z_N(t), t \in I\}$ be a stochastic process, where

$$Z_N(t) = \left(\frac{C_N}{A_N^0}\right) \cdot \bar{L}_{N(t)}^0; \qquad N(t) = \min\left\{k : \frac{C_N^2}{C_k^2} \leqslant t\right\}, \quad t \in I, \tag{4.3.12}$$

Thus Z_N belongs to the $D[0, 1]$ space for every N. Note that for every N, if we define $\{N_k, k \geqslant 0\}$ by letting $N_k = \max\{m : C_m^2 \leqslant 2^k C_N^2\}$, $k = 0, 1, \ldots$, then for every $\epsilon > 0$, by Theorem 4.2.1 and the Kolmogorov inequality,

$$\begin{aligned}
&P\left\{\max_{m \geqslant N_k} C_N |\bar{L}_m^0| > \epsilon\right\} \qquad (k \geqslant 0) \\
&\leqslant \sum_{l \geqslant k} P\left\{\max_{N_l \leqslant m \leqslant N_{l+1}} |L_m^0| > \frac{\epsilon C_{N_{l+1}}^2}{C_N}\right\} \\
&\leqslant \sum_{l \geqslant k} \frac{\{\epsilon^{-2} C_N^2 / C_{N_{l+1}}^4\} A_{N_{l+1}}^{02} C_{N_{l+1}}^2 N_{l+1}}{(N_{l+1} - 1)} \\
&\leqslant 2A^2 \epsilon^{-2} \sum_{l \geqslant k} \left(C_N^2 / C_{N_{l+1}}^2\right) \leqslant 2A^2 \epsilon^{-2} \sum_{l \geqslant k} 2^{-l} \\
&= A^2 \epsilon^{-2} 2^{-k+2},
\end{aligned} \tag{4.3.13}$$

which can be made to converge to 0 by letting $k \to \infty$. Thus $\lim_{t \downarrow 0} Z_N(t) = 0$, with probability 1, for every $N (\geqslant 1)$. Hence Z_N is properly defined at the lower boundary.

THEOREM 4.3.2. *Under the conditions of Theorem* 4.3.1, *as* $N \to \infty$,

$$Z_N \xrightarrow{\mathfrak{D}} W, \qquad \text{in the } J_1\text{-topology on } D[0, 1]. \tag{4.3.14}$$

Proof. By virtue of (4.3.9) and steps similar to those in the proof of Theorem 4.3.1, the convergence of the f.d.d.'s of $\{Z_N\}$ to those of W

follows readily, and hence, we need only establish the tightness of $\{Z_N\}$. Since $Z_N(0) = 0$ with probability 1, we have for every $0 < \delta < 1$,

$$\sup_{0 \leqslant t \leqslant \delta} |Z_N(t) - Z_N(0)| = \sup_{0 \leqslant t \leqslant \delta} |Z_N(t)| = \sup_{m \geqslant N(\delta)} C_N |L_m^0| / A_N^0, \quad (4.3.15)$$

where $N(\delta)$ is defined by (4.3.12). Proceeding as in (4.3.13) but using the Brown inequality (2.2.9) on the second line, we obtain that for every $\epsilon > 0$ and $\eta > 0$, there exist a $\delta : 0 < \delta < 1$ and an N_0 such that

$$P\left\{ \sup_{0 \leqslant t \leqslant \delta} |Z_N(t)| > \epsilon \right\} < \eta\delta, \qquad \forall N \geqslant N_0. \quad (4.3.16)$$

Also note that for every $N \leqslant k_1 < k_2 \leqslant N(\delta)$, $0 < \delta < 1$,

$$\max_{k_1 \leqslant k \leqslant k_2} C_N |\bar{L}_k^0 - \bar{L}_{k_1}^0| \leqslant \frac{C_N}{C_{k_1}^2} \max_{k_1 \leqslant k \leqslant k_2} |L_k^0 - L_{k_1}^0| + \frac{C_{k_2}^2 - C_{k_1}^2}{C_{k_2}^2} \frac{C_N}{C_{k_1}^2} |L_{k_1}^0|. \quad (4.3.17)$$

Then, by the Brown inequality (2.2.9) [on the first term on the rhs of (4.3.17)] and the convergence of the f.d.d. [on both $\bar{L}_{k_1}^0 C_N$ and $(\bar{L}_{k_2}^0 - \bar{L}_{k_1}^0) C_N / C_{k_1}^2$), it follows by using (4.3.17) that

$$P\left\{ \sup_{t \leqslant s \leqslant t+\delta} |Z_N(t) - Z_N(s)| > \epsilon \right\} < \eta\delta, \qquad \forall N \geqslant N_0, \quad \delta \leqslant t \leqslant 1 - \delta. \quad (4.3.18)$$

Hence the proof is complete.

For each $n (\geqslant 1)$, let N_n be a positive integer valued random variable defined on the same probability space as the X_i and let $\{a_n\}$ be a sequence of positive constants going to ∞ as $n \to \infty$. If

$$N_n / a_n \xrightarrow{P} \xi \qquad \text{as} \quad n \to \infty, \quad (4.3.19)$$

where ξ is a positive random variable and if we define $\{Y_{N_n}\}$ and $\{Z_{N_n}\}$ as in (4.3.6) and (4.3.12) with N replaced by N_n, then the following theorem holds.

THEOREM 4.3.3. *Under* (4.3.19) *and the conditions of Theorem* 4.3.1,

$$Y_N \xrightarrow{\mathfrak{D}} W \Rightarrow Y_{N_n} \xrightarrow{\mathfrak{D}} W \quad \text{and} \quad Z_N \xrightarrow{\mathfrak{D}} W \Rightarrow Z_{N_n} \xrightarrow{\mathfrak{D}} W.$$

The proof follows precisely along the line of the proof of Theorems 17.1 and 17.2 of Billingsley (1968, pp. 146–147), and hence is left as an exercise.

In passing, we note that by virtue of Theorem 2.3.6 and the basic martingale property in Theorem 4.2.1, the weak convergence results in Theorems 4.3.1 and 4.3.2 also hold for the $D[0, \infty)$ space. Further, by virtue of Theorems 2.4.8 and 4.2.1, defining q and d_q as in (2.4.25) and (2.4.26), it follows by some standard steps that the weak convergence results in Theorems 4.3.1 and 4.3.2 hold in the (d_q-) sup-norm metric as well. We leave the proofs as exercises.

For Theorems 4.3.1 through 4.3.3, we have so far restricted ourselves to the particular type of scores defined by (4.3.1). Whenever $a_N(i)$ is not equal to $E\phi(U_{Ni})$, $1 \leqslant i \leqslant N$, Theorem 4.2.1 may not hold and, as a result, the approach to proving these theorems becomes inapplicable. However, under some extra regularity conditions, all these theorems remain valid for general scores. Let us define L_N as in (4.2.1) and L_N^0 as in (4.2.13). Then note that under H_0 in (4.2.3),

$$E(L_N - L_N^0)^2 = C_N^2 \left\{ \frac{1}{N-1} \sum_{i=1}^{N} \left[a_N(i) - a_N^0(i) \right]^2 \right\}, \qquad N \geqslant 1. \tag{4.3.20}$$

As such, the convergence of f.d.d. of Theorems 4.3.1 through 4.3.3 holds whenever

$$e_n^{(1)} = n^{-1} \sum_{i=1}^{n} \left[a_n(i) - a_n^0(i) \right]^2 \to 0 \qquad \text{as} \quad n \to \infty, \tag{4.3.21}$$

and it is well known that (4.3.21) holds under very general conditions [also, in particular, for $a_n(i) = \phi(i/(n+1)), 1 \leqslant i \leqslant n$]. On the other hand, the *tightness* part requires a more restrictive condition. Since for $\{L_N\}$, Theorem 4.2.1 may not hold, one may not be able to use the Brown inequality (2.2.9) and Corollary 2.4.4.1. But

$$\begin{aligned} \max_{1 \leqslant k \leqslant N} |L_k - L_k^0| / C_N &= C_N^{-1} \max_{1 \leqslant k \leqslant N} \left| \sum_{i=1}^{k} (c_i - \bar{c}_k) \left[a_k(R_{ki}) - a_k^0(R_{ki}) \right] \right| \\ &\leqslant \max_{1 \leqslant k \leqslant N} \left\{ \max_{1 \leqslant i \leqslant k} |a_k(i) - a_i^0(i)| \right\} \left\{ C_N^{-1} \sum_{i=1}^{k} |c_i - \bar{c}_k| \right\} \\ &\leqslant \max_{1 \leqslant k \leqslant N} \left\{ \max_{1 \leqslant i \leqslant k} \left[\sqrt{k}\, |a_k(i) - a_k^0(i)| \right] \left(\frac{C_k}{C_N} \right) \right\}, \end{aligned} \tag{4.3.22}$$

and hence, if

$$e_n^{(2)} = \max_{1 \leqslant i \leqslant n} n^{1/2} |a_n(i) - a_n^0(i)| \to 0 \qquad \text{as} \quad n \to \infty, \tag{4.3.23}$$

then (4.3.22) converges to 0, so that stochastic processes Y_N or Z_N, defined on the $\{L_k\}$ and the $\{L_k^0\}$ will be asymptotically equivalent, and hence, the theorems hold. Equation 4.3.23 is also known to hold under fairly general regularity conditions, see for example, Puri and Sen (1971, Appendix), Hoeffding (1973), and Exercises 4.3.4 and 4.3.5 at the end of this chapter.

The results developed so far relate to the situation where H_0 in (4.2.3) holds. One naturally wonders what happens when H_0 may not hold. If $F_1, \ldots, F_N$ are not all identical, the rank vector $\mathbf{R}_N = (R_{N1}, \ldots, R_{NN})$ no longer assumes all the $N!$ possible permutations of $1, \ldots, N$ with equal probability $(N!)^{-1}$, so that the distribution of $\mathbf{R}_N$ (and hence, of L_N) depends on $F_1, \ldots, F_N$. Indeed, this dependence poses serious problems in evaluating the exact or large sample distribution of L_N. Also, the martingale property, considered in Theorem 4.2.1, breaks down when $F_1, \ldots, F_N$ are not equal. However, under additional regularity conditions, asymptotic distribution theory for L_N can be developed.

A variety of techniques has been employed by a host of research workers for studying the asymptotic normality of the standardized form of L_N. First, in the context of several sample problems, where the $X_i (1 \leqslant i \leqslant N)$ can be partitioned into $c(\geqslant 2)$ subsets such that within each subset the variables are i.i.d. while the distributions can vary from subset to subset, whenever the proportions of observations belonging to these subsets all converge (as $N \to \infty$) to some numbers bounded away from 0 and 1 (and adding up to 1), then, under certain additional regularity conditions on the score function, L_N can be expressed as a sum of a principal and a residual term; the principal term is a linear function of independent r.v.'s on which the central limit theorem applies, while the residual term is $o_p(N^{-1/2})$. We refer the reader to Section 3.6 of Puri and Sen (1971) where this technique (classically known as the Chernoff-Savage, 1958 representation) has been displayed in detail and extended to more complex situations. Pyke and Shorack (1968a, b) have employed the weak convergence of some two-sample empirical processes to derive similar asymptotic normality results in the two-sample as well as the one-sample case in a more fashionable way. Second, Hájek (1968) has considered the case of arbitrary (but continuous) $F_1, \ldots, F_N$ and derived a powerful variance inequality for L_N which, coupled with an elegant polynomial approximation for the score function, yields a novel and nice proof of the asymptotic normality of L_N. Here also, L_N is approximated by a function

$$\tilde{L}_N = \sum_{i=1}^{N} (c_i - \bar{c}_N) l_N(X_i), \tag{4.3.24}$$

where the $l_N(X_i)$ are independent [and the form of $l_N(\cdot)$ depends on $F_1, \ldots, F_N$, the score function, and the c_i].

The main difficulties in pursuing these approaches for the study of the invariance principles for $\{L_N\}$ are the following. First, the stochastic equivalence (in probability or in quadratic mean) of L_N and $\tilde{L}_N$ does not necessarily imply a similar equivalence of $\{L_k, 1 \leqslant k \leqslant N\}$ and $\{\tilde{L}_k, 1 \leqslant k \leqslant N\}$ or $\{L_k, k \geqslant N\}$ and $\{\tilde{L}_k, k \geqslant N\}$; indeed, the latter requires more stringent regularity conditions. Second, the functional forms of $\{l_k(\cdot), 1 \leqslant k \leqslant N\}$ or $\{l_k(\cdot), k \geqslant N\}$ may not be very convenient (for arbitrary $F_i, i \leqslant N$ or $i \geqslant 1$) and without some close relationships between the adjacent $l_k(\cdot)$, the covariance function of the process constructed from the $\tilde{L}_k$ may not be very simple and it may be difficult to apply the theorems in Section 2.4 for establishing the desired invariance principles. Third, the martingale property in Theorem 4.2.1 does not generally hold for the case where the F_i are not necessarily the same, and, in the absence of a martingale (or reverse martingale) property, verification of the tightness of the stochastic processes constructed from the L_k relies heavily on the computation of the higher order central moments, which in turn demands more restrictive regularity conditions on the score function as well as the c_i and F_i. Indeed, in general, the processes $\{Y_N\}$ and $\{Z_N\}$, even weakly convergent, need not have the structure of a drifted Brownian motion when the F_i are arbitrary. In the Pyke-Shorack approach, one not only needs to strengthen their weak convergence results to the case where the F_i may not be categorized into a finite number of possibilities (as in the case of the simple regression model with distinct $c_1, \ldots, c_N$), but also, to extend their results to include the weak convergence of these empirical processes in the sequential case. At the end of this section we briefly discuss the work of Braun (1976), where for the special case of the two-sample problem, a sequential version of the Pyke-Shorack (1968) result is treated.

In a majority of situations dealing with asymptotic solutions (as we treat in Part 2 of this monograph), we are really concerned with alternative hypotheses *close* to the null hypotheses for which the asymptotic results are well defined and have meaningful interpretations. Thus in such a case, one may consider a sequence $\{K_N\}$ of alternative hypotheses, where, under K_N, $X_1, \ldots, X_N$ (more precisely, $X_{N1}, \ldots, X_{NN}$) are independent and have d.f.'s $F_{N1}, \ldots, F_{NN}$, respectively, where each F_{Ni} converges to a common F, as $N \to \infty$. For such a sequence of alternative hypotheses, Hájek (1962) has established the asymptotic normality of L_N (without imposing any extra regularity condition on the score function or the constants c_i) by invoking the concept of *contiguity of probability measures* (demanding some mild regularity condition on the d.f. F). We intend to follow this line of approach and derive invariance principles for linear rank statistics under contiguous alternatives.

The idea of contiguity of probability measures is due to Le Cam (1960). In the area of rank tests, the concept has been highly popularized through the pioneering work of the late Professor Jaroslav Hájek (and his associates). A very nice account of it is available in Chapter VI of Hájek and Šidák (1967). We also refer the reader to the monograph of Roussas (1972) for a nice review of the topic. For the sake of completeness, we present here the basic concept of contiguity of probability measures.

Let $\{P_n\}$ and $\{Q_n\}$ be two sequences of (absolutely continuous) probability measures on measure spaces $(\Omega_n, \mathcal{Q}_n, \mu_n)$ and we denote by $p_n = dP_n/d\mu_n$ and $q_n = dQ_n/d\mu_n$. *Then, if for any sequence of events* $\{A_n\} : A_n \in \mathcal{Q}_n$,

$$[P_n(A_n) \to 0] \Rightarrow [Q_n(A_n) \to 0], \tag{4.3.25}$$

the sequence of measures $\{Q_n\}$ *is said to be contiguous to* $\{P_n\}$.

Note that if one considers the L_1-norm of $P_n - Q_n$ that is,

$$\|P_n - Q_n\| = \sup\{|P_n(A_n) - Q_n(A_n)| : A_n \in \mathcal{Q}_n\},$$

then $\|P_n - Q_n\| \to 0$ (with $n \to \infty$) implies that (4.3.25) holds, but the converse is not necessarily true. Thus contiguity is weaker than L_1-norm equivalence. The contiguity of $\{Q_n\}$ to $\{P_n\}$ [in the one-sided version in (4.3.25)] implies that for any $\mathcal{Q}_n$-measurable r.v. T_n, $T_n \to 0$ in P_n-probability ensures that $T_n \to 0$, in Q_n-probability too. In this context, we define the *likelihood ratio* L_n as q_n/p_n, 1 or ∞, according to $p_n > 0$, $p_n = q_n = 0$ and $p_n = 0 < q_n$, respectively. Let $G_n(x) = P_n\{L_n \leqslant x\}, x \in R^+$ and let $\mathfrak{N}(\mu, \sigma^2)$ stand for the normal d.f. with mean μ and variance σ^2. Then the following result due to Le Cam (1960) characterizes contiguity of $\{Q_n\}$ to $\{P_n\}$:

If $\{G_n\}$ *weakly converge to a d.f.* G *for which* $\int_0^\infty x\,dG(x) = 1$, *then* $\{Q_n\}$ *is contiguous to* $\{P_n\}$. *In particular, if* $\log L_n$ *is asymptotically* $\mathfrak{N}(-\frac{1}{2}\sigma^2, \sigma^2)$, *then* $\{Q_n\}$ *is contiguous to* $\{P_n\}$. (4.3.26)

A second result due to Le Cam (1960) is of special importance in this context. Let $\{T_n\}$ be a sequence of r.v.'s and $\mathfrak{N}_2(\mu_1, \mu_2; \sigma_1^2, \sigma_2^2, \sigma_{12})$ be the bivariate normal d.f. with means μ_1, μ_2, variances σ_1^2, σ_2^2, and covariance σ_{12}.

If $(T_n, \log L_n)$ *is asymptotically* $\mathfrak{N}_2(\mu_1, \mu_2, \sigma_1^2, \sigma_2^2, \sigma_{12})$ *under* P_n *with* $\mu_2 = -\frac{1}{2}\sigma_2^2$, *then, under* Q_n, T_n *is asymptotically* $\mathfrak{N}(\mu_1 + \sigma_{12}, \sigma_1^2)$. (4.3.27)

We refer to Chapter VI of Hájek and Šidák (1967) for the proofs of (4.3.26) and (4.3.27) along with some useful discussion on contiguity. The following result due to Behnen and Neuhaus (1975) also deserves mention in this context:

Let $\{Q_n\}$ *be contiguous to* $\{P_n\}$ *and let* $S_n = \sum_{i=1}^n d_{ni} Z_{ni}$ *where the* Z_{ni} *are independent under* P_n *with* 0 *means and unit variances and the constants* d_{ni} *satisfy the conditions that* $\max_{1 \leqslant i \leqslant n} |d_{ni}| \to 0$ *and* $\sum_{i=1}^n d_{ni}^2 = 1$. *Then, whenever the* Z_{ni}^2 *are uniformly integrable under* P_n, *there exists a sequence* $\{a_n\}$ *of real numbers, such that* $\mathcal{L}(S_n \mid P_n) \to \mathfrak{N}(0, 1)$ *and* $\mathcal{L}(S_n - a_n \mid Q_n) \to \mathfrak{N}(0, 1)$. (4.3.28)

Note that the hypothesis of $\max\{|d_{ni}| : 1 \leqslant i \leqslant n\} \to 0$ and the uniform square integrability of the Z_{ni} under P_n ensure that for each n and i, there exist measurable, real-valued and bounded functions h_{ni}, such that (i) $\int h_{ni}\, dP_{ni} = 0$, (ii) $\max_{1 \leqslant i \leqslant n} \int (h_{ni} - z)^2\, dP_{ni}(z) \to 0$ and (iii) $\max_{1 \leqslant i \leqslant n} |d_{ni}| \cdot \max_{1 \leqslant j \leqslant n} \{\sup_z h_{nj}^2(z)\} \to 0$, as $n \to \infty$. Then, a_n may be taken in the form

$$a_n = \sum_{i=1}^{n} d_{ni} \int h_{ni}\, dQ_{ni}. \tag{4.3.29}$$

Let $\{X_{N1}, \ldots, X_{NN}; N \geqslant 1\}$ be a triangular array of r.v.'s and denote by $\{P_N\}$ and $\{Q_N\}$ the sequence of probability measures pertaining to the set $\{X_{N1}, \ldots, X_{NN}\}$ under two sequences of hypotheses. Note that each of P_N and Q_N is an N-fold product measure, so that for $\{X_{N1}, \ldots, X_{Nk}\}$, the corresponding (product-) measures are P_{Nk} and Q_{Nk}, respectively, $k = 1, \ldots, N$, where $P_N = P_{NN}$ and $Q_N = Q_{NN}$. Further, the contiguity of Q_N to P_N ensures that for each $k(\leqslant N)$, Q_{Nk} is contiguous to P_{Nk}. Let now Y_N be an arbitrary stochastic process constructed from $(X_{N1}, \ldots, X_{NN})$, such that Y_N belongs to the $D[0, 1]^p$ space, for some $p \geqslant 1$. Note that Y_N is a mapping of $(X_{N1}, \ldots, X_{NN})$ into the space $D[0, 1]^p$ and we denote by P_N^* and Q_N^* the probability measures induced by P_N and Q_N respectively. Then we have the following theorem which will be used repeatedly in this and subsequent chapters.

THEOREM 4.3.4. *If $\{Q_N\}$ is contiguous to $\{P_N\}$ and Y_N is tight under the P_N-measure, then it remains tight under the Q_N-measure as well.*

Proof. Let $\|Y_N\| = \sup\{|Y_N(t)| : t \in E^p\}$ and define $\omega_f'(\delta)$ as in (2.3.20). Then let $B_{N\epsilon}(\delta) = \{(X_{N1}, \ldots, X_{NN}) : \|Y_N\| > \epsilon^{-1}$ or $\omega_{Y_N}'(\delta) > \epsilon\}, \epsilon > 0$, $\delta > 0$. Then, from Theorem 2.3.2 and the assumed tightness of $\{Y_N\}$ under $\{P_N\}$, it follows that for every $\epsilon > 0$ and $\eta > 0$, there exists an N_0 and a $\delta : 0 < \delta < 1$, such that

$$P\{B_{N\epsilon}(\delta) | P_N\} < \eta \qquad \text{for} \quad N \geqslant N_0, \tag{4.3.30}$$

where $P\{\cdot | P_N\}$ stands for the probability under the P_N-measure (and a similar notation is used for the Q_N-measure). By (4.3.30) and the contiguity of $\{Q_N\}$ to $\{P_N\}$, we obtain that

$$P\{B_{N\epsilon}(\delta) | Q_N\} \leqslant 2\eta, \qquad \text{for} \quad N \geqslant \text{some } N^*(\geqslant N_0), \tag{4.3.31}$$

and since $\eta(> 0)$ is arbitrary, (4.3.31) (along with Theorem 2.3.1) ensures the tightness of $\{Y_N\}$ under $\{Q_N\}$. Q.E.D.

Remark. For Theorem 4.3.4 to hold, only the contiguity of $\{Q_N\}$ to $\{P_N\}$ has been utilized in the proof. It is therefore not necessary to assume that for each N, $X_{N1}, \ldots, X_{NN}$ are mutually independent. Indeed, in

Chapter 11, we shall face some situations where the X_{Ni} are not independent but the aforesaid contiguity holds and therefore Theorem 4.3.4 also holds.

Let us now consider invariance principles for linear rank statistics under local (contiguous) alternatives. For this purpose, we consider a triangular array $\{(X_{N1}, \ldots, X_{NN}); N \geqslant 1\}$ of row-wise independent r.v.'s, where under the null hypothesis (H_0), the X_{Ni} are i.i.d.r.v.'s with an absolutely continuous d.f. F and let $\{K_N\}$ be a sequence of contiguous alternatives. We define as in (4.2.22)

$$L_{Nk}^* = \sum_{i=1}^{k} (c_i - \bar{c}_k)\phi(F(X_{Ni})), \qquad k = 1, \ldots, N, \tag{4.3.32}$$

and let

$$\begin{aligned} \lambda_{Nk}^* &= E(L_{Nk}^* \mid K_N), \qquad k = 1, \ldots, N \qquad \text{and} \\ \mu_N^* &= \{ \mu_N^*(t) = \lambda_{NN(t)}^* / C_N A_N^0,\ t \in I \}, \end{aligned} \tag{4.3.33}$$

where $N(t)$ is defined by (4.3.6). Finally, we assume that

$$\lim_{N\to\infty} \mu_N^* = \mu^* \qquad \text{exists and it belongs to the} \quad C[0,1] \quad \text{space.} \tag{4.3.34}$$

THEOREM 4.3.5. *If the assumptions in Theorem* 4.3.1 *hold,* $\{P_N\}$ *corresponds to* $F_1 = \cdots = F_N$ *and* $\{Q_N\}$ *is contiguous to* $\{P_N\}$, *then defining* $\{Y_N\}$ *as in* (4.3.6), *we have under* $\{Q_N\}$,

$$Y_N \xrightarrow{\mathfrak{D}} W + \mu^*, \qquad \textit{in the } J_1\textit{-topology on} \quad D[0,1]. \tag{4.3.35}$$

Proof. Defining Y_N^* as in (4.3.7), we note that the contiguity of $\{Q_N\}$ to $\{P_N\}$ (as assumed) ensures that (4.3.10) holds under $\{K_N\}$ as well. On the other hand, for Y_N^*, we may appeal to (4.3.28) and claim that under K_N, $[Y_N^*(t_1), \ldots, Y_N^*(t_m)]$ has asymptotically a multinormal distribution with mean vector $[\mu_N^*(t_1), \ldots, \mu_N^*(t_m)]$ and dispersion matrix $((t_i \wedge t_j))$. In this context note that under H_0, the $\phi^2(F(X_{Ni}))$ are i.i.d.r.v.'s, and hence are uniformly integrable when ϕ is square integrable inside I. Thus by (4.3.34) and the above, we conclude that the f.d.d.'s of Y_N^* (and hence of Y_N) converge to those of $W + \mu^*$. The tightness of the sequence $\{Y_N\}$ under $\{K_N\}$ follows directly from Theorem 4.3.4 and the fact that under P_N, Y_N is tight by Theorem 4.3.1. Q.E.D.

Let us now use Theorem 4.3.5 in a special class of problems. We assume that under K_N, $F_i = F_{Ni}$, $i = 1, \ldots, N$ relate to the model

$$F_{Ni}(x) = F(x - \alpha - \beta d_{Ni}), \quad i = 1, \ldots, N, \quad x \in (-\infty, \infty), \tag{4.3.36}$$

where α and β are unknown parameters, $\{d_{Ni} \geqslant 1\}$ are known constants and the d.f. F is assumed to be absolutely continuous (with respect to the Lebesgue measure) with a finite Fisher information

$$I(f) = \int_{-\infty}^{\infty} \left[\frac{f'(x)}{f(x)} \right]^2 dF(x) \qquad \text{where} \quad f'(x) = \frac{d}{dx} f(x) = \frac{d^2}{dx^2} F(x). \tag{4.3.37}$$

Here the null hypothesis (4.2.3) relates to $\beta = 0$ and we are interested in the case of $\beta \neq 0$. We assume that $\{K_N\}$ relates to

$$\sum_{i=1}^{N} \left(d_{Ni} - \bar{d}_N\right)^2 \to D^2: \qquad 0 < D < \infty, \quad \text{as} \quad N \to \infty \tag{4.3.38}$$

where $\bar{d}_N = N^{-1}\sum_{i=1}^{N} d_{Ni}$ and

$$\max_{1 \leqslant i \leqslant N} \left(d_{Ni} - \bar{d}_N\right)^2 \to 0 \qquad \text{as} \quad N \to \infty. \tag{4.3.39}$$

We denote by $\psi(u) = -f'(F^{-1}(u)/f(F^{-1}(u)), 0 < u < 1$ and denote by

$$\rho_1 = \rho(\psi, \phi) = \frac{\left(\int_0^1 \psi(u)\phi(u)\,du\right)}{\left[\left(\int_0^1 \psi^2(u)\,du\right)A^2\right]^{1/2}}$$

$$= \left(\int_0^1 \psi(u)\phi(u)\,du\right) \Big/ \left[I(f)A^2\right]^{1/2}, \tag{4.3.40}$$

where A^2 is defined by (4.3.1). Defining $N(t)$ as in (4.3.6), we assume that

$$\lim_{N \to \infty} \left\{ C_N^{-1} D^{-1} \sum_{i=1}^{N(t)} \left(c_i - \bar{c}_{N(t)}\right)\left(d_{Ni} - \bar{d}_N\right) \right\} = \rho_2(t) \tag{4.3.41}$$

exists for every $t \in [0, 1]$, where $\rho_2 = \{\rho_2(t); t \in [0, 1]\}$ belongs to the $C[0, 1]$ space.

THEOREM 4.3.6. *Under* (4.3.36) *through* (4.3.39), (4.3.41), *and the assumptions on the score function and the constants* $\{c_i\}$, *made in Theorem* 4.3.1, *when* $\{K_N\}$ *obtains,*

$$Y_N - \left(\beta D\left[I(f)\right]^{1/2}\right)\rho_2\rho_1 \xrightarrow{\mathfrak{D}} W \qquad \textit{as} \quad N \to \infty \tag{4.3.42}$$

Proof. For the sequence $\{K_N\}$ of alternatives, the contiguity of $\{Q_N\}$ to $\{P_N\}$ follows directly by appealing to (4.3.26) and verifying the log-normality of the likelihood ratio under the assumptions shown in (4.3.36) through (4.3.39) (see Problem 4.3.8). Having proved this, one can virtually appeal to Theorem 4.3.5 and obtain the weak convergence of $\{Y_N\}$ under

$\{K_N\}$. The rest of the proof follows by noting that under (4.3.36) through (4.3.39), we have, on using (4.3.28), (4.3.29), (4.3.32), and (4.3.33), that

$$\lim_{N\to\infty} \mu_N^*(t) = \left(\beta D[I(f)]^{1/2}\right)\rho_1\rho_2(t) \qquad \text{for every} \quad t \in [0,1]. \quad (4.3.43)$$

Consider now the particular case where

$$d_{Ni} = (c_i - \bar{c}_N)/C_N \qquad \text{for} \quad i = 1, \ldots, N, \qquad (4.3.44)$$

so that (4.3.38) holds with $D = 1$ and also (4.3.39) follows from (4.3.5). Then (4.3.41) reduces to $\rho_2(t) = t : 0 \leqslant t \leqslant 1$. Thus in this case, we have, under $\{K_N\}$,

$$\left\{Y_N(t) - \beta[I(f)]^{1/2} t\rho_1,\ t \in [0,1]\right\} \xrightarrow{\mathfrak{D}} W. \qquad (4.3.45)$$

As a special case consider the two-sample location problem where we let $c_i = 1$ if $i = 2k$ and -1 if $i = 2k-1$, $k \geqslant 1$, so that $C_N^2 \sim N$ and $N(t) \sim [Nt]$. Two-sample scale problems can be treated similarly (see Problem 4.3.9).

Let us now consider the stochastic processes $\{Z_N\}$, defined by (4.3.12) and study their weak convergence under $\{K_N\}$. Note that here, under K_N, we consider the infinite sequence $\{X_{Ni}, i \geqslant 1\}$ of r.v.'s with d.f.'s $\{F_{Ni}, i \geqslant 1\}$ and though Q_{NN} (or Q_{Nk}, $k \leqslant N$) is contiguous to P_{NN} (or P_{Nk}, $k \leqslant N$), in general, $\{Q_{Nk}, k \geqslant N\}$ and $\{P_{Nk}, k \geqslant N\}$ are not contiguous; the difficulty is caused by the fact that as $k \to \infty$, N being fixed (or as $k/N \to \infty$), Q_{Nk} need not be contiguous to P_{Nk} under $\{K_N\}$. Thus the contiguity-based arguments in the proof of Theorem 4.3.5 may not be applicable in this case. However, we may proceed as follows.

First, defining $N_\epsilon = \max\{k : C_N^2/C_k^2 \geqslant \epsilon\}$, where $0 < \epsilon < 1$, we may assume that

$$\{Q_{Nk}; N \leqslant k \leqslant N_\epsilon\} \quad \text{is contiguous to} \quad \{P_{Nk}; N \leqslant k \leqslant N_\epsilon\},$$
$$\text{for every } \epsilon > 0. \qquad (4.3.46)$$

Second, defining $N(t)$ as in (4.3.12), we let

$$\nu_N(t) = E\left\{\frac{C_N C_{N(t)}^{-2} L_{N(t)}^0}{A_N} \mid K_N\right\}, \qquad t \in (0,1] \qquad (4.3.47)$$

and assume that the following holds:

$$\lim_{N\to\infty} \nu_N(t) = \nu(t) \qquad \text{exists for every} \quad t \in (0,1], \qquad (4.3.48)$$

$$\nu(t) \qquad \text{is a continuous function of} \quad t \in (0,1] \qquad (4.3.49)$$

$$\lim_{t\downarrow 0} \nu(t) = \nu_0 \qquad \text{exists,} \qquad (4.3.50)$$

and

$$C_N C_n^{-2} L_n^0 / A_N^0 \to \nu_0 \quad \text{a.s.,} \quad \text{as} \quad n \to \infty \; (n/N \to \infty) \qquad \text{when } K_N \text{ holds.} \tag{4.3.51}$$

THEOREM 4.3.7. *Under* (4.3.46) *through* (4.3.51) *and the conditions on the score function and the constants* $\{c_i\}$ *in Theorem* 4.3.1, *as* $N \to \infty$, *under* $\{K_N\}$,

$$Z_N - \nu \xrightarrow{\mathfrak{D}} W \qquad \textit{where} \quad \nu = \{\nu(t), t \in [0,1]\}. \tag{4.3.52}$$

Proof. By (4.3.50), (4.3.51), and (4.3.12), under K_N, as $\delta \downarrow 0$,

$$\sup\{|Z_N(t) - \nu_0| : 0 \leqslant t \leqslant \delta\} \xrightarrow{P} 0 \qquad \text{as} \quad N \to \infty. \tag{4.3.53}$$

On the other hand, by virtue of (4.3.46), for $\delta \leqslant t \leqslant 1$, the proof of Theorem 4.3.5 can be adapted to show that under K_N,

$$\{Z_N(t) - \nu(t), \delta \leqslant t \leqslant 1\} \xrightarrow{\mathfrak{D}} \{W(t), \delta \leqslant t \leqslant 1\}, \tag{4.3.54}$$

and since $\delta(>0)$ is arbitrary, while $W(0) = 0$ with probability 1, the proof of the theorem follows from (4.3.53) and (4.3.54). Q.E.D.

To illustrate an application of the theorem, consider the sequence $\{K_N\}$ of alternatives, where under K_N,

$$F_{Ni}(x) = F\left(x - \alpha - \beta \frac{(c_i - \bar{c}_N)}{C_n}\right), \qquad i = 1, \dots, N, N+1, \dots, \text{ad inf}, \tag{4.3.55}$$

and assume that

$$\sup_{i \geqslant 1} |c_i| < \infty, \qquad \limsup_{N \to \infty} |\bar{c}_N| < \infty \qquad \text{and} \quad \liminf_{N \to \infty} N^{-1} C_N^2 \geqslant C_0^2 > 0, \tag{4.3.56}$$

and F is absolutely continuous with a finite Fisher information $I(f)$. Then (4.3.46) holds. Also, $\nu(t)$, defined by (4.3.47) and (4.3.48) reduces to $\beta[I(f)]^{1/2}\rho_1$, where ρ_1 is defined by (4.3.40). Thus the drift function becomes a constant for all $t \in [0,1]$. Finally (4.3.51) can be proved along the lines of the proof of Theorem 4.6.1 (to follow in Section 4.6); we leave the details as an exercise (see Problem 4.3.10). Hence the theorem holds in this model.

In the rest of this section, we confine ourselves to the classical two-sample problem [as has been characterized in (4.2.11)] and present an outline of the Pyke and Shorack (1968a) weak convergence results along with some sequential extensions by Braun (1976). In the setup of (4.2.11), that is, the two-sample case, we let the individual sample sizes be m and n, respectively, so that $N = m + n$. Let $\lambda_N = m/N$ and assume that there exists a $\lambda^* \in (0, \frac{1}{2})$, such that $0 < \lambda^* \leqslant \lambda_N \leqslant 1 - \lambda^* < 1$. The two underlying d.f.'s are denoted by F and G respectively and are assumed to be continuous a.e. Let then

$$Z_{Ni} = \begin{cases} 1, & \text{if the } i\text{th order statistic of the combined sample} \\ & \text{is from the first sample;} \\ 0, & \text{otherwise,} \quad \text{for } i = 1, \ldots, N. \end{cases} \tag{4.3.57}$$

Also, we express the scores $\{a_N(i)\}$ as

$$a_N(i) = a_N^*(i) + \cdots + a_N^*(N), \qquad \text{for } i = 1, \ldots, N. \tag{4.3.58}$$

Then, by (4.2.11) (for $n_1 = m, n_2 = n$) and (4.3.57), we have, say,

$$m^{-1}L_N = m^{-1}\sum_{i=1}^{N} a_N(i)Z_{Ni} - \bar{a}_N = T_N - \bar{a}_N, \tag{4.3.59}$$

where $\bar{a}_N = N^{-1}\sum_{i=1}^{N} a_N(i)$. Let F_m and G_n be respectively the empirical d.f. of the first and the second sample, and let

$$H_N(x) = \lambda_N F_m(x) + (1 - \lambda_N)G_n(x), \qquad x \in R. \tag{4.3.60}$$

Further, let ν_N be a signed measure which puts the mass $a_N^*(i)$ on the point i/N for $i = 1, \ldots, N$ and puts zero measure elsewhere. Then, by (4.3.59) and (4.3.60), we have

$$T_N = \int_0^1 F_m\big(H_N^{-1}(t)\big)d\nu_N(t) = \int_{1/N}^1 F_m\big(H_N^{-1}(t)\big)d\nu_N(t). \tag{4.3.61}$$

For every $\lambda \in (0, 1), x \in R$, and $t \in [0, 1]$, let

$$H_\lambda(x) = \lambda F(x) + (1 - \lambda)G(x) \qquad \text{and} \quad K_\lambda(t) = F\big(H_\lambda^{-1}(t)\big). \tag{4.3.62}$$

Further, let

$$K_N(t) = F\big(H_N^{-1}(t)\big), \qquad H = H_{\lambda_N} \qquad \text{and}$$

$$K = K_{\lambda_N} = F\big(H^{-1}(\cdot)\big); \tag{4.3.63}$$

$$\mu_N = \int_0^1 K_{\lambda_N}(t)\, d\nu_N(t), \tag{4.3.64}$$

$$Y_N(t) = N^{1/2}\big\{F_m\big(H_N^{-1}(t)\big) - F\big(H_{\lambda_N}^{-1}(t)\big)\big\}, \qquad 0 \leqslant t \leqslant 1. \tag{4.3.65}$$

Then, from the above, we have

$$N^{1/2}(T_N - \mu_N) = \int_0^1 Y_N(t)\, d\nu_N(t) = \int_{1/N}^1 Y_N(t)\, d\nu_N(t) \tag{4.3.66}$$

and, as such, for the weak convergence of $N^{1/2}(T_N - \mu_N)$, one may study the same for Y_N and ν_N. We assume that there exists another signed Lebesgue-Stieltjes measure ν on $(0,1)$ for which for every $\epsilon: 0<\epsilon<\frac{1}{2}$, $|\nu|([\epsilon, 1-\epsilon])<\infty$ (where $|\nu|=\nu^{+}+\nu^{-}$ is the total variation of ν) and it is possible to replace in (4.3.66) ν_N by ν, that is

$$\int_{1/N}^{1} Y_N(t)\,d\left[\nu_N(t)-\nu(t)\right] \xrightarrow{P} 0, \qquad \text{as} \quad N\to\infty; \tag{4.3.67}$$

later on we shall specify conditions that ensure (4.3.67). Whenever (4.3.67) holds, it suffices to consider the weak convergence of the stochastic integral, say,

$$\int_0^1 Y_n(t)\,d\nu(t) = Y_N^{(\nu)}, \tag{4.3.68}$$

and, in this context, the weak convergence of $Y_N=\{Y_N(t), 0\leqslant t\leqslant 1\}$ plays a vital role. Define the two one-sample empirical processes $U_m = \{U_m(t), 0\leqslant t\leqslant 1\}$ and $V_N=\{V_n(t), 0\leqslant t\leqslant 1\}$ by letting

$$\begin{aligned} U_m(t) &= m^{1/2}\left[F_m\left(F^{-1}(t)\right)-t\right] \quad &&\text{and} \\ V_n(t) &= n^{1/2}\left[G_n\left(G^{-1}(t)\right)-t\right], &&0\leqslant t\leqslant 1. \end{aligned} \tag{4.3.69}$$

Then we may express Y_N as

$$\begin{aligned} Y_N(t) = (1-\lambda_N)\Big\{&\lambda_N^{-1/2}B_N(t)\,U_m\left(F\left(H_N^{-1}(t)\right)\right) \\ &-(1-\lambda_N)^{-1/2}A_N(t)\,V_n\left(G_n\left(H_N^{-1}(t)\right)\right)\Big\} + \delta_N(t), \end{aligned} \tag{4.3.70}$$

for $0\leqslant t\leqslant 1$, where

$$\delta_N(t) = A_N(t)N^{1/2}\left[H_N\left(H_N^{-1}(t)\right)-t\right], \tag{4.3.71}$$

$$A_N(t) = \frac{\left[K(u_t)-K(t)\right]}{(u_t-t)}; \qquad u_t = H\left(H_N^{-1}(t)\right), \tag{4.3.72}$$

$$B_N(t): \lambda_N A_N(t) + (1-\lambda_N)B_N(t) = 1, \qquad \text{for every } t\in[0,1]. \tag{4.3.73}$$

In the above representation, for every $t\in[0,1]$, one has

$$\begin{aligned} &|A_N(t)|\leqslant 1/\lambda_N, \qquad |B_N(t)|\leqslant 1/(1-\lambda_N), \qquad \text{and} \\ &|\delta_N(t)|\leqslant \left(N^{1/2}\lambda_N\right)^{-1}. \end{aligned} \tag{4.3.74}$$

Let $U_0=\{U_0(t), 0\leqslant t\leqslant 1\}$ and $V_0=\{V_0(t), 0\leqslant t\leqslant 1\}$ be two (independent) separable Brownian Bridges and define $\overline{Y}_N=\{\overline{Y}_N(t), 0\leqslant t\leqslant 1\}$ by letting

$$\begin{aligned} \overline{Y}_N(t) = (1-\lambda_N)\Big\{&\lambda_N^{-1/2}B_N(t)\,U_0\left(F\left(H^{-1}(t)\right)\right) \\ &-(1-\lambda_N)^{-1/2}A_N(t)\,V_0\left(G\left(H^{-1}(t)\right)\right)\Big\}, 0\leqslant t\leqslant 1. \end{aligned} \tag{4.3.75}$$

Then, the following stochastic integral suggests itself as a natural analogue of $Y_N^{(\nu)}$:

$$\overline{Y}_N^{(\nu)} = \int_0^1 \overline{Y}_N(t)\,d\nu(t). \tag{4.3.76}$$

Define the weight function q and the sup-norm metric d_q as in (2.4.25) and (2.4.26). Then, by (4.3.68) and (4.3.76), we have

$$|Y_N^{(\nu)} - \overline{Y}_N^{(\nu)}| \leqslant \int_0^1 \left\{ \frac{|(Y_N(t) - \overline{Y}_N(t))|}{q(t)} \right\} q(t)\,d\nu(t). \tag{4.3.77}$$

The ingenuity of the Pyke-Shorack approach lies in the utilization of the weak convergence of the empirical processes U_m and V_n to U_0 and V_0 in the sup-norm metric and this along with the additional assumption that

$$\int_0^1 q(t)d|\nu(t)| < \infty \tag{4.3.78}$$

ensures the convergence of (4.3.77) to 0 (in probability). To obtain natural limiting processes from (4.3.76), it remains to study the limiting behavior of $A_N(t)$ and $B_N(t)$. Note that by (4.3.62), FH^{-1} and GH^{-1} are both absolutely continuous, and hence, the derivatives $a_N(t)$ and $b_N(t)$ of FH^{-1} and GH^{-1} exist a.e. on (0, 1) (with respect to the Lebesgue measure). We assume further that the functions $K_\lambda(t)$ have derivatives $a_\lambda(t)$ for all $t \in (0, 1)$, and for some λ', $a_{\lambda'}(t)$ is continuous on (0, 1) and has one-sided limits at 0 and 1. Further, we assume that as $N \to \infty$,

$$\lambda_N \to \lambda_0 \in (\lambda^*, 1 - \lambda^*). \tag{4.3.79}$$

We denote the functions a_{λ_0} and b_{λ_0} by a_0 and b_0, respectively. Then we have the following theorem due to Pyke and Shorack (1968a).

THEOREM 4.3.8. *Under* (4.3.67), (4.3.78), *and the assumptions made above,* $Y_N^{(\nu)}$ *or* $N^{1/2}(T_N - \mu_N)$ *is asymptotically normal with mean zero and variance*

$$\begin{aligned}\sigma_0^2 = 2(1-\lambda_0)^2 \Big\{ & \lambda_0^{-1} \int_0^1 \int_0^v b_0(u) b_0(v) F\big(H_{\lambda_0}^{-1}(u)\big) \\ & \big[1 - F\big(H_{\lambda_0}^{-1}(v)\big)\big]\,d\nu(u)\,d\nu(v) \\ & + (1-\lambda_0)^{-1} \int_0^1 \int_0^v a_0(u) a_0(v) G\big(H_{\lambda_0}^{-1}(u)\big) \\ & \big[1 - G\big(H_{\lambda_0}^{-1}(v)\big)\big]\,d\nu(u)\,d\nu(v) \Big\}.\end{aligned} \tag{4.3.80}$$

We refer to Pyke and Shorack (1968a, b) and Pyke (1970) for a proof of this theorem along with some related results of interest. If $F = G$ or if F and G

are both absolutely continuous, then the derivative assumption on $K_\lambda(t)$ holds. Further, if we set $a_N(i) = \phi_N(t)$, for $(i-1)/N < t \leqslant i/N, i = 1, \ldots, N$ and $\phi_N(0) = \phi_N(0+)$ and if ϕ denotes a nonconstant function of bounded variation on $(\epsilon, 1-\epsilon)$ for all $\epsilon: 0 < \epsilon < \frac{1}{2}$, (which includes the Lebesgue-Stieltjes measure ν), then the following Theorem due to Pyke and Shorack (1968a) defends the assumptions in (4.3.67) and (4.3.78).

THEOREM 4.3.9. *Suppose that* (i) $N^{1/2}\int_0^1 |\phi_N(t) - \phi_N(t-)| dH_N(H_N^{-1}(t)) = o(1)$, (ii) $\phi_N(0) = o(N^{1/2}) = \phi_N(1)$, (iii) $|\phi(t)| \leqslant K\{t(1-t)\}^{-1/2+\delta}(0 < t < 1)$ *for some constants* K *and* $\delta > 0$, *and* (iv) $N^{1/2}\int_0^1 |\phi_N(t) - \phi(t)| dt = o(1)$, *then* (4.3.67) *and* (4.3.78) *hold. It is also possible to replace* (iv) *by* (iv)$'$ *that* $\phi = \phi_d + \phi_c$ *where* ϕ_d *is a saltus function taking only a finite number of jumps and where* ϕ_c *has a continuous derivative* ϕ_c' *on intervals* $(0, a_1), (a_1, a_2), \ldots, (a_s, 1)$ *which satisfies* $|\phi_c'(t)| \leqslant K\{t(1-t)\}^{-3/2+\delta}$ *for* $t \neq a_i, i = 1, \ldots, s$.

For the proof, we again refer to Pyke and Shorack (1968a).

Braun (1976) has extended the Pyke-Shorack results in the sequential case. Define $W_N = \{W_N(t), 0 \leqslant t \leqslant 1\}$ by letting

$$W_N(t) = \frac{[Nt](T_{[Nt]} - \mu_{[Nt]})}{(N^{1/2}\sigma_0)}, \qquad t \in [0, 1], \tag{4.3.81}$$

where T_n, μ_n and σ_0 are defined by (4.3.61), (4.3.64), and (4.3.80), respectively and $[s]$ stands for the largest integer contained in s.

THEOREM 4.3.10. *Assume that* (i) *the functions* $F(H_\lambda^{-1}(t))$ *have derivatives* $a_\lambda(t)$ *with respect to* t *for all* $t \in (0,1)$ *and for some* $\lambda' \in (0,1)$, $a_{\lambda'}$ *is continuous on* $(0,1)$ *and has one-sided limits at* 0 *and* 1, (ii) *there exists a Lebesgue-Stieltjes measure* ν *on* $(0,1)$ *for which* $|\nu|([\epsilon, 1-\epsilon]) < \infty$ *for all* $\epsilon > 0$, *such that* $\int_0^1 Y_N(t) d(\nu_N(t) - \nu(t)) \to 0$ *a.s. and* $\int_0^1 d|\nu(t)| \to \infty$, *and* (iii) $\lambda_N \to \lambda_0$ *a.s., as* $N \to \infty$. *Then,* W_N *converges weakly to a standard Wiener process.*

We refer to Braun (1976) for the proof of this theorem. His approach is based on an extension of the Pyke-Shorack (1968) weak convergence to the sequential case. For further work in this direction, we may also refer to Wellner (1974). It may be remarked that the a.s. convergence of $\int_0^1 Y_N(t) \cdot d(\nu_N(t) - \nu(t))$, needed in Theorem 4.3.10, holds under the hypothesis of Theorem 4.3.9. The last three theorems can also be extended to the case of $c(\geqslant 2)$ samples by assuming that the individual sample size ratios (to the total sample size) are all bounded away from 0 and 1 and imposing the other regularity conditions as in Theorems 4.3.8 through 4.3.10. It remains an open problem to extend this approach to the case of completely arbitrary sequence of independent r.v.'s. It is quite clear from the above discussion that in one hand, the Pyke-Shorack approach is not limited to the case of local alternatives, though, on the other hand, it rests on comparatively more stringent regularity conditions and relates to more restrictive designs.

4.4. SKOROKHOD-STRASSEN REPRESENTATION FOR $\{L_N\}$

Under the hypothesis of invariance in (4.2.3), the Shorokhod-Strassen embedding of Wiener processes for linear rank statistics is considered in this Section. For this purpose, Theorems 2.5.1 and 4.2.1 are incorporated to yield the stronger invariance principle; we need more stringent regularity conditions too. Let us define

$$q_1(n) = \left\{ \frac{\max_{1 \leqslant i \leqslant n}(c_i - \bar{c}_n)^2}{C_n^2} \right\}, \qquad n \geqslant 2; q_1(1) = 0; \tag{4.4.1}$$

$$q_2(n) = n^{-1} \sum_{i=1}^{n} E\left[\phi(U_{ni}) - \phi\left(\frac{i}{(n+1)}\right)\right]^2, \qquad n \geqslant 1; \tag{4.4.2}$$

$$q(n) = q_1(n)q_2(n) \quad \text{and} \quad Q(n) = \sum_{k=1}^{n} q(k), \qquad n \geqslant 1, \tag{4.4.3}$$

where the U_{ni} are defined as in (4.2.2). We assume that ϕ and the c_i satisfy the condition that

$$\lim_{n \to \infty} Q(n) = Q : 0 \leqslant Q < \infty. \tag{4.4.4}$$

Note that by the Noether-condition, $q_1(n) \to 0$ as $n \to \infty$, while, by definition, $q_1(n) \geqslant n^{-1}, \forall n \geqslant 1$. In most of the common problems, $q_1(n) = O(n^{-1})$. Also, if ϕ has a bounded first derivative inside I, then $q_2(n) = O(n^{-1})$ (see Problem 4.4.1), if $\phi \in L_r$, for some $r > 2$, then (see Problem 4.4.2), $q_2(n) = O(n^{1-r/2})$, and if ϕ satisfies the Hoeffding (1973) condition that

$$\int_0^1 |\phi(u)| \{\log(1 + |\phi(u)|)\}^{1+\delta} du < \infty, \qquad \text{for some} \quad \delta > 0, \tag{4.4.5}$$

then (see Problem 4.4.3), $q_2(n) = O((\log n)^{-1-\eta}), \eta > 0$. Thus, depending on the nature of $q_1(n)$, appropriate scores can be chosen so that (4.4.4) holds.

As in Section 4.3, we define L_N^0, $\mathcal{B}_N$, and so on, and let then

$$V_N = \sum_{k=2}^{N} \text{Var}(L_k^0 \mid \mathcal{B}_{k-1}), \qquad N \geqslant 2, \qquad V_1 = V_0 = 0; \tag{4.4.6}$$

$$S_{V_N}^0 = L_N^0, N \geqslant 2, S_{V_1}^0 = S_{V_0}^0 = L_1^0 = L_0^0 = 0 \tag{4.4.7}$$

and by linear interpolation between (V_N, V_{N+1}), we complete the definition of S_t^0, $V_n \leqslant t \leqslant V_{N+1}$ for $N \geqslant 0$. The following theorem is an improved version of Theorem 1.2 of Sen and Ghosh (1972) (see Sen, 1978d).

THEOREM 4.4.1. *Suppose that* (i) (4.2.3) *and* (4.4.4) *hold,* (ii) $\phi(u)$ *is the difference of two nondecreasing and absolutely continuous functions* (*inside* I) *and* (iii) $f = \{f(t), t > 0\}$ *is a nondecreasing function on* R^+ *such that* $f(t)$ *is* $\uparrow$ *but* $t^{-1}f(t)$ *is* $\downarrow$ *in* t *and*

$$\lim_{N\to\infty}\left\{\frac{C_N^2 q_1(N)}{f(A_N^2 C_N^2)}\right\}\left[\max_{1\leqslant i\leqslant N}\left\{\left[E\phi(U_{Ni})\right]^2\right\}\right] = 0. \tag{4.4.8}$$

Then, there exists a standard Wiener process $W = \{W(t), t \in R^+\}$, *such that*

$$S_t^0 = W(t) + o\left(\left[tf(t)\right]^{1/4}\log t\right) \quad a.s., \quad as\ t\to\infty. \tag{4.4.9}$$

Proof. By virtue of Theorem 4.2.1, we need to verify (2.5.6) and (2.5.7). For this reason, we first consider the following.

Lemma 4.4.2. Under the hypothesis (4.2.3) and (4.4.4), $V_n/C_n^2A_n^2 \to 1$ a.s. as $n\to\infty$.

Proof. Define L_N^* as in (4.2.22). Then, note that

$$\begin{aligned}\mathrm{Var}(L_k^0 \mid \mathfrak{B}_{k-1}) &= E\left[\left(L_k^0 - L_{k-1}^0\right)^2 \mid \mathfrak{B}_{k-1}\right] \\ &= E\left\{\left[W_k^{(1)} + W_k^{(2)} + W_k^{(3)}\right]^2 \mid \mathfrak{B}_{k-1}\right\}, \quad k \geqslant 2 \end{aligned}\tag{4.4.10}$$

where

$$W_k^{(1)} = (c_k - \bar{c}_{k-1})\phi(F(X_k)), \qquad k \geqslant 2, \tag{4.4.11}$$

$$W_k^{(2)} = (c_k - \bar{c}_{k-1})\left[a_k^0(R_{kk}) - \phi(F(X_k))\right], \qquad k \geqslant 2 \tag{4.4.12}$$

$$W_k^{(3)} = \sum_{i=1}^{k-1}(c_i - \bar{c}_{k-1})\left[a_k^0(R_{ki}) - a_{k-1}^0(R_{k-1i})\right], \qquad k \geqslant 2. \tag{4.4.13}$$

Thus it suffices to show that as $n\to\infty$,

$$C_n^{-2}A_n^{-2}\sum_{k=2}^{n} E\left(\left[W_k^{(j)}\right]^2 \mid \mathfrak{B}_{k-1}\right) \to \delta_{1j} \qquad \text{a.s.}, \quad j = 1,2,3, \tag{4.4.14}$$

where δ_{rs} is the usual Kronecker delta. Now, $\phi(F(X_k))$ is independent of $\mathfrak{B}_{k-1}$, so that

$$E\left(\left[W_k^{(1)}\right]^2 \mid \mathfrak{B}_{k-1}\right) = E\left[W_k^{(1)}\right]^2 = (c_k - \bar{c}_{k-1})^2 A^2, \qquad k \geqslant 1, \tag{4.4.15}$$

where A^2 is defined by (4.3.1). Thus, on noting that for every $n \geqslant 2$, $C_n^2 = \sum_{k=2}^n[(k-1)/k](c_k - \bar{c}_{k-1})^2$, we obtain from (4.4.15)

$$\sum_{k=2}^n E\left(\left[W_k^{(1)}\right]^2 \mid \mathfrak{B}_{k-1}\right)/C_n^2A_n^2$$
$$= \frac{\sum_{k=2}^n(c_k - \bar{c}_{k-1})^2A^2}{\left(\sum_{k=2}^n((k-1)/k)(c_k - \bar{c}_{k-1})^2\right)A_n^2} \to 1, \tag{4.4.16}$$

by the convergence of A_N^2 to A^2 [viz. (4.3.4)] and the fact that $C_N^2 \to \infty$ as $N \to \infty$. Second, given $\mathfrak{B}_{k-1}$, R_{kk} assumes the values $1, \dots, k$ with equal conditional probability k^{-1}. Hence

$$E\left[\left\{W_k^{(2)}\right\}^2 \mid \mathfrak{B}_{k-1}\right] = (c_k - \bar{c}_{k-1})^2 E\left\{\frac{1}{k}\sum_{i=1}^k\left[a_k^0(i) - \phi(U_{ki})\right]^2\right\}$$
$$= (c_k - \bar{c}_{k-1})^2 q_2(k) \leqslant q_1(k)C_k^2q_2(k)$$
$$= C_k^2 q(k), \qquad k \geqslant 2. \tag{4.4.17}$$

Therefore,

$$\sum_{k=2}^n E\left\{\left[W_k^{(2)}\right]^2 \mid \mathfrak{B}_{k-1}\right\} \leqslant \sum_{k=2}^n C_k^2 q(k), \tag{4.4.18}$$

and hence, by (4.4.4), $\sum_{k=2}^n C_k^2 q(k)/C_n^2 \to 0$ as $n \to \infty$, so that (4.4.14) holds for $j = 2$. Finally, given $\mathfrak{B}_{k-1}$, R_{ki} can only assume the two values R_{k-1i}, and $(R_{k-i} + 1)$ with respective conditional probabilities $k^{-1}(k - R_{k-1i})$ and $k^{-1}R_{k-1i}$, and similarly, given $R_{k-1i} < R_{k-1j}$, (R_{ki}, R_{kj}) can be (R_{k-1i}, R_{k-1j}), $(R_{k-1i}, R_{k-1j} + 1)$ or $(R_{k-1i} + 1, R_{k-1j} + 1)$ with respective conditional probabilities $k^{-1}(k - R_{k-1j}), k^{-1}(R_{k-1j} - R_{k-1i})$ and $k^{-1}R_{k-1i}$. Hence, if we denote by $g_k(i,j) = g_k(j,i)$, where

$$g_k(i,j) = \frac{i(k-j)}{k^2}\left[a_k^0(i+1) - a_k^0(i)\right]\left[a_k^0(j+1) - a_k^0(j)\right],$$
$$1 \leqslant i \leqslant j \leqslant k-1, \tag{4.4.19}$$

we obtain by a few standard steps that

$$E\left\{\left[W_n^{(3)}\right]^2 \mid \mathfrak{B}_{k-1}\right\} = \sum_{i=1}^{k-1}(c_i - \bar{c}_{k-1})^2 g_k(R_{k-1i}, R_{k-1j})$$
$$+ \sum_{i \neq j = 1}^{k-1}(c_i - \bar{c}_{k-1})(c_j - \bar{c}_{k-1})g_k(R_{k-1i}, R_{k-1j})$$
$$\leqslant q_1(k-1)C_{k-1}^2\left[\sum_{i=1}^{k-1}\sqrt{\frac{i(k-i)}{k^2}}\,|a_k^0(i+1) - a_k^0(i)|\right]^2,$$
$$k \geqslant 2. \tag{4.4.20}$$

Also,

$$\begin{aligned}
|a_k^0(i+1) - a_k^0(i)| &= |E[\phi(U_{ki+1}) - \phi(U_{ki})]| \\
&= \frac{k^2}{i(k-i)} \left| \frac{(k-1)!}{(i-1)!(k-1)!} \right. \\
&\quad \times \left. \int_0^1 \left[\phi(u) - \phi\left(\frac{i}{k}\right)\right]\left[u - \frac{i}{k}\right] u^{i-1}(i-u)^{k-i} du \right| \\
&\leqslant \frac{k^2}{i(k-i)} \left(E\left|U_{k-1i} - \frac{i}{k}\right|^2 E\left[\phi(U_{k-1i}) - \phi\left(\frac{i}{k}\right)\right]^2 \right)^{1/2} \\
&\leqslant \frac{k}{\sqrt{i(k-i)}} \cdot \frac{1}{\sqrt{k}} \left(E\left[\phi(U_{k-1i}) - \phi\left(\frac{i}{k}\right)\right]^2 \right)^{1/2}, \\
&\quad 1 \leqslant i \leqslant k-1, \qquad (4.4.21)
\end{aligned}$$

so that the rhs of (4.4.20) is bounded from above by

$$\begin{aligned}
q_1(k-1)C_{k-1}^2 &\left[\frac{1}{\sqrt{k}} \sum_{i=1}^{k-1} \left(E\left[\phi(U_{k-1i}) - \phi\left(\frac{i}{k}\right)\right]^2 \right)^{1/2} \right]^2 \\
&< C_{k-1}^2 q_1(k-1) \frac{1}{k-1} \sum_{i=1}^{k-1} E\left[\phi(U_{k-1i}) - \phi\left(\frac{i}{k}\right)\right]^2 \\
&= C_{k-1}^2 q_1(k-1) q_2(k-1) = C_{k-1}^2 q(k-1), \qquad k > 2. \qquad (4.4.22)
\end{aligned}$$

Hence, as in (4.4.18),

$$\frac{\sum_{k=2}^n E\left\{ \left[W_k^{(3)}\right]^2 | \mathfrak{B}_{k-1} \right\}}{C_n^2} \to 0 \quad \text{as } n \to \infty. \qquad \text{Q.E.D.} \qquad (4.4.23)$$

Returning now to the proof of Theorem 4.4.1, we note that Lemma 4.4.2 implies that $V_n \to \infty$ a.s., as $n \to \infty$. Further, by definition, for $k \geqslant 2$ and nondecreasing ϕ,

$$\begin{aligned}
|Z_k| &= |L_k - L_{k-1}^0| \\
&= \left| (c_k - \bar{c}_{k-1}) a_k^0(R_{kk}) + \sum_{i=1}^{k-1} (c_i - \bar{c}_{k-1}) \left[a_k^0(R_{ki}) - a_{k-1}^0(R_{k-1i}) \right] \right| \\
&\leqslant C_k q_1^{1/2}(k) |a_k^0(R_{kk})| + \sum_{i=1}^{k-1} |c_i - \bar{c}_{k-1}| \, |a_k^0(R_{k-1i}+1) - a_k^0(R_{k-1i})| \\
&\leqslant C_k q_1^{1/2}(k) \left\{ \max_{1 \leqslant i \leqslant k} |a_k^0(i)| + \sum_{i=1}^{k-1} |a_k^0(i+1) - a_k^0(i)| \right\} \\
&\leqslant 3 C_k q_1^{1/2}(k) \left\{ \max_{1 \leqslant i \leqslant k} |a_k^0(i)| \right\}, \qquad (4.4.24)
\end{aligned}$$

and a similar inequality holds when ϕ is the difference of two nondecreasing functions. Thus, by Lemma 4.2.2, (4.4.8), and (4.4.24),

$$P\{Z_N^2 \geqslant f(V_N) \mid \mathfrak{B}_{N-1}\} = 0 \quad \text{a.s., as } N \to \infty, \tag{4.4.25}$$

and this implies that (2.5.10) holds. Q.E.D.

So far, we have considered the case of $\{L_N^0\}$. Let us now consider the case of $a_n(i) = \phi(i/(n+1))$, $1 \leqslant i \leqslant n; n \geqslant 1$. Note that for every $N \geqslant 2$,

$$(L_N^0 - L_N)^2 \leqslant C_N^2 \sum_{i=1}^{N} \left[E\phi(U_{Ni}) - \phi\left(\frac{i}{N+1}\right) \right]^2, \tag{4.4.26}$$

so that whenever, for $N \to \infty$,

$$\sum_{i=1}^{N} \left[E\phi(U_{Ni}) - \phi\left(\frac{i}{(N+1)}\right) \right]^2 = o\left\{ (\log C_N^2 A_N^2)^2 \left[\frac{f(C_N^2 A_N^2)}{C_N^2 A_N^2} \right]^{1/2} \right\}, \tag{4.4.27}$$

we may replace $\{L_N^0\}$ by $\{L_N\}$ in the definition of S^0 in (4.4.7) as well as in (4.4.9). After (4.3.23), we have considered a similar condition for the weak invariance principle. It appears that whenever $(t^{-1/2}[tf(t)]^{1/4}\log t)$ is nondecreasing in $t(\geqslant t_0)$, the same condition implies (4.4.27). However, if $[t^{-1}f(t)]^{1/4}\log t \to 0$ as $t \to \infty$, then we need a more precise rate of convergence of the lhs of (4.4.27). In the context of the law of iterated logarithm, we take $f(t) = t(\log\log t)^2/(\log t)^4$ ($\uparrow$ in $t \geqslant 8$), so that (4.3.23) ensure (4.4.27). In this context, note that if either (a) ϕ is square integrable and $q_1(t)/(\log t)^4$ is bounded as $t \to \infty$, or (b) $\phi \in L_r$ for some $r > 2$, then (4.4.8) holds, so that the law of iterated logarithm holds for $\{L_N\}$ (and hence $\{L_N\}$ under (4.3.23); see Problem 4.4.4 in this context). We may also remark that whenever $|V_n/C_n^2A_n^2 - 1| < (\log\log C_n^2A_n^2)/(\log C_n^2A_n^2)$ a.s., as $n \to \infty$, then by virtue of Lemma 2.5.5 and (4.4.3), we have

$$L_n^0/A_n^0 = W(C_n^2) + o\left([C_n^2 f(C_n^2)]^{1/4} \log C_n^2 \right) \quad \text{a.s., as } n \to \infty; \tag{4.4.28}$$

Problem 4.4.5 is to show that the above holds under fairly general conditions.

So far we have considered the case of the null hypothesis in (4.2.2). When (4.2.2) may not hold, one may still venture to express the linear rank statistic as a sum of a principle term (involving independent summands) and a remainder term, namely, $L_n = \sum_{i=1}^{n} l_{ni}(X_i) + \xi_n$ where the $l_{ni}(X_i)$, $1 \leqslant i \leqslant n; n \geqslant 1$ form a triangular array of row wise independent r.v.'s and ξ_n is the remainder term. In that context, one may even proceed to show that there exist two positive numbers α and β, where (typically) α is less than $\frac{1}{2}$

and $\beta > 1$, such that for n sufficiently large,

$$P\{|\xi_n| > n^{\alpha}\} \leqslant n^{-\beta}, \tag{4.4.29}$$

which provides an a.s. rate of convergence for the remainder term, and then to study suitable limiting behavior for the principal term. In general, this method may not meet the full success for several reasons. First, for arbitrary $F_i, i \geqslant 1$, the $l_{ni}(X_i)$ need not have a limiting form and for this reason, the computation of the covariance of L_n and $L_{n'}$ when n and n' are not sufficiently close to each other may not be very handy or may not even converge to some simple limit. Indeed, in general, the standard Wiener process approximation may not be available. Moreover, for an arbitrary triangular array of r.v.'s, study of the limiting behavior of the row-wise sums, in general, requires more restrictive regularity conditions. Finally, the proof of (4.4.29), in general, may be quite involved and may not follow any general pattern (to sketch). In the particular case of the two-sample problem (as has been discussed in Section 4.3), when the c_i are either 0 or 1 and when the ratio of the two sample sizes is held fixed while their sum is allowed to increase, then L_n becomes a functional of the two empirical distributions and in this case, (4.4.29) can be verified in a more convenient way. Lai (1975a) has considered a general Chernoff-Savage (1958) type decomposition for a class of rank statistics. His results are particularly applicable in this context of the two sample problem where the l_{ni} also simplifies and the Wiener process approximation holds under certain additional regularity conditions. For the one sample problem, this method has been considered earlier by Sen and Ghosh (1973) and the same is considered in detail in Section 5.4. Since the treatment is essentially the same for both the one sample and several sample problems, we shall not consider the Lai's result here. Extending such a theorem for the general case of arbitrary F_i remains an open problem.

4.5. ASYMPTOTIC LINEARITY OF L_N IN REGRESSION PARAMETER

We study here some monotonicity and linearity (both in regression parameter) properties of L_N. First, we consider the following.

Let $a_N(1) \leqslant \cdots \leqslant a_N(N)$ be N real scores (not all equal) and $c_1 \cdots c_N$ be constants (not all equal). Let $X_1, \ldots, X_N$ be independent r.v.'s with continuous d.f.'s $F_1, \ldots, F_N$, respectively (not necessarily identical). Further, let $Z_i(b) = X_i - bc_i$, $1 \leqslant i \leqslant N$, $-\infty < b < \infty$ and let $R_{Ni}(b) = \sum_{j=1}^{N} c(Z_i(b) - Z_j(b))$ be the rank of $Z_i(b)$ among $Z_1(b), \ldots, Z_N(b)$,

$1 \leqslant i \leqslant N$. Finally, let

$$L_N(b) = \sum_{i=1}^{N} (c_i - \bar{c}_N) a_N(R_{Ni}(b)), \qquad -\infty < b < \infty. \tag{4.5.1}$$

THEOREM 4.5.1. *Under the above setup, $L_N(b)$ is $\searrow$ in b: $-\infty < b < \infty$.*

Proof. Let us rewrite (4.5.1) as

$$L_N(b) = \sum_{i=1}^{N} c_i^* \left[a_N(R_{Ni}(b)) - \bar{a}_N \right]; c_i^* = c_i - c_1 (\geqslant 0), \qquad 1 \leqslant i \leqslant N; \tag{4.5.2}$$

$$Z_i^*(b) = X_i - bc_i^*, \qquad 1 \leqslant i \leqslant N, \qquad b \in R. \tag{4.5.3}$$

Now (4.5.3) represents N straight lines (in b). The (i, i')th lines are either parallel (if $c_i = c_{i'}$) or they intersect at a point $(b_{ii'})$ (if $c_i \neq c_{i'}$), $1 \leqslant i < i' \leqslant N$. Let $N^* = \#\{(i, i') : c_i \neq c_{i'};\ 1 \leqslant i < i' \leqslant N\}$, so that

$$N - 1 \leqslant N^* \leqslant \binom{N}{2}.$$

We denote the ordered values of the N^* points of intersection of the N lines in (4.5.3) by $b_1 < \cdots < b_{N^*}$; since the F_i are all continuous, ties among the X_i and hence among the b_i can be neglected, in probability. Also, let $b_0 = -\infty$ and $b_{N^*+1} = +\infty$. Then, for $b_\rho < b < b_{\rho+1}$, no two lines in (4.5.3) intersect, the ranks of $Z_1^*(b), \ldots, Z_N^*(b)$ are the same as those of $Z_1^*(b_\rho + 0), \ldots, Z_N^*(b_\rho + 0)$, so that $L_N(b) = L_N(b_\rho + 0)$, $\forall b_\rho < b < b_{\rho+1}$, $\rho = 0, \ldots, N^*$. It remains to show that

$$L_N(b_\rho - 0) \geqslant L_N(b_\rho) \geqslant L_N(b_\rho + 0), \qquad \rho = 1, \ldots, N^*. \tag{4.5.4}$$

At $b = b_\rho$, let the two intersecting lines have the indices (k, q) and let $R_k(b_\rho - 0) = s_1$ and $R_q(b_\rho - 0) = s_2$. As $k < q$, $c_k < c_q$ (otherwise, the two lines do not intersect), we must have $s_1 = s_2 - 1$, $R_k(b_\rho + 0) = s_2 = 1 + R_q(b_\rho + 0)$ and $R_i(b_\rho + 0) = R_i(b_\rho - 0)$, $i = 1, \ldots, N (\neq k \neq q)$. Hence $L_N(b_\rho + 0) = L_N(b_\rho - 0) - (c_q^* - c_k^*)(a_N(s_2) - a_N(s_2 - 1)) = L_N(b_\rho - 0) - (c_q - c_k)[a_N(s_2) - a_N(s_{2-1})] \leqslant L_N(b_\rho - 0), 1 \leqslant \rho \leqslant N^*$. On the other hand, $Z_k^*(b_\rho) = Z_q^*(b_\rho)$, so according to the usual convention for mid-ranks for tied observations, $a_N(R_{Nk}(b_\rho)) = a_N(R_{Nq}(b_\rho)) = \frac{1}{2}[a_N(s_2) + a_N(s_2 - 1)]$, and hence $L_N(b_\rho) = L_N(b_\rho - 0) - (c_q^* - c_k^*)\frac{1}{2}[a_N(s_2) - a_N(s_2 - 1)] = \frac{1}{2}[L_N(b_\rho - 0) + L_N(b_\rho + 0)]$. Thus $L_N(b_\rho - 0) \geqslant L_N(b_\rho) \geqslant L_N(b_\rho + 0)$, for every $1 \leqslant \rho \leqslant N^*$. Q.E.D.

Remark. Note that $L_N(b) > 0$ for $b < b_1$ and $L_N(b) < 0$ for $b > b_{N^*}$. Also, the above monotonicity result holds even if one takes $Z_i(b) = X_i - bd_i$, $1 \leqslant i \leqslant N$ where $d_1 \leqslant \cdots \leqslant d_N$; the proof is left as an exercise.

We now proceed to consider the asymptotic linearity of $L_N(b)$ in b. For this purpose, let us define $c^*_{Ni} = (c_i - \bar{c}_N)/C_N$, $1 \leq i \leq N$ (so that $\sum_{i=1}^N (c^*_{Ni})^2 = 1$) and let

$$R^0_{Ni}(b) = \text{Rank of } X_i - bc^*_{Ni} \quad \text{among} \quad X_\alpha - bc^*_{N\alpha}, \quad 1 \leq \alpha \leq N; 1 \leq i \leq N; \tag{4.5.5}$$

$$\tilde{L}_N(b) = \sum_{i=1}^N c^*_{Ni} a_N\big(R^0_{Ni}(b)\big), \qquad -\infty < b < \infty. \tag{4.5.6}$$

Finally, define A^2, $I(f)$ and $\rho_1 = \rho(\phi, \psi)$ as in (4.3.1), (4.3.37), and (4.3.40), respectively.

THEOREM 4.5.2. *If the X_i are i.i.d.r.v. with a continuous d.f. F for which $I(f) < \infty$, ϕ is nondecreasing and square integrable inside I and* (4.3.5) *holds, then for every positive $C(< \infty)$, as $N \to \infty$,*

$$\sup_{b:|b| \leq C} |\tilde{L}_N(b) - \tilde{L}_N(0) + bAI^{1/2}(f)\rho_1| \xrightarrow{P} 0. \tag{4.5.7}$$

Proof. For every $\epsilon > 0$ and $C(< \infty)$, there exists a positive integer $K_\epsilon(< \infty)$ such that

$$AI^{1/2}(f)\rho_1 C/K_\epsilon < \tfrac{1}{2}\epsilon. \tag{4.5.8}$$

Define then a set of points

$$b_j = -C + jK_\epsilon^{-1}C, \qquad \text{for } j = 0, 1, \ldots, 2K_\epsilon. \tag{4.5.9}$$

Then, by Theorem 4.5.1, for $b_j \leq b \leq b_{j+1}$, $\tilde{L}_N(b_j) \geq \tilde{L}_N(b) \geq \tilde{L}_N(b_{j+1})$, and hence, by (4.5.8) and (4.5.9),

$$\sup_{b:|b| \leq C} |\tilde{L}_N(b) - \tilde{L}_N(0) + bAI^{1/2}(f)\rho_1| \leq \max_{0 \leq j \leq 2K_\epsilon} |\tilde{L}_N(b_j) - \tilde{L}_N(0) + b_j AI^{1/2}(f)\rho_1| + \tfrac{1}{2}\epsilon. \tag{4.5.10}$$

Since $\epsilon(> 0)$ and $K_\epsilon(< \infty)$ are given, to prove (4.5.7), it suffices to show that for every $\eta > 0$, there exists an N_η, such that for $N \geq N_\eta$ (and uniformly in $b \in [-C, C]$),

$$P\left\{|\tilde{L}_N(b) - \tilde{L}_N(0) + bAI^{1/2}(f)\rho_1| > \tfrac{1}{2}\epsilon\right\} \leq \frac{\eta}{2K_\epsilon}. \tag{4.5.11}$$

Let $K_N^{(b)}$ denote the hypothesis that X_i has the d.f. $F(x + bc^*_{Ni}), i = 1, \ldots, N$, then $\tilde{L}_N(b)$ (under H_0) has the same distribution as of $\tilde{L}_N(0)$ under $K_N^{(b)}$. Let us now consider a sequence $\{\phi^{(k)}(u), 0 \leq u \leq 1\}$ of score functions such that for each k, $\phi^{(k)}(u)$ is bounded and $\nearrow$ on the interval $[0, 1]$ and $\int_0^1 \{\phi^{(k)}(u) - \phi(u)\}^2 du < \epsilon_k$, where $\epsilon_k \to 0$ as $k \to \infty$. We denote by $\tilde{L}_N^*(b) = \sum_{i=1}^N c_{Ni}^* \phi(F(X_i - bc_{Ni}^*))$ and in (4.5.6) and above, we replace ϕ by $\phi^{(k)}$ and denote the resulting quantities by $\tilde{L}_{Nk}(b)$ and $\tilde{L}_{Nk}^*(b)$, respectively. Note that the hypothesis of the theorem ensures contiguity of the probability measure under $K_n^{(b)}$ to that under H_0, and under H_0, $\tilde{L}_{Nk}(0)$ and $\tilde{L}_{Nk}^*(0)$ are asymptotically equivalent in mean square (and hence in probability). Thus, for every k,

$$\tilde{L}_{Nk}(0) - \tilde{L}_{Nk}^*(0) \xrightarrow{p} 0 \quad \text{and} \quad \tilde{L}_{Nk}(b) - \tilde{L}_{Nk}^*(b) \xrightarrow{p} 0, \qquad \text{as } N \to \infty. \tag{4.5.12}$$

On the other hand, under H_0, $E[\tilde{L}_N^*(0) - \tilde{L}_{Nk}^*(0)]^2 = \int_0^1 \{\phi^{(k)}(u) - \phi(u)\}^2 du < \epsilon_k$ and a similar variance inequality holds for $\tilde{L}_N(0) - \tilde{L}_{Nk}(0)$ (under H_0). Hence it suffices to show that (by choosing k adequately large) for every $\epsilon > 0$,

$$P\left\{|\tilde{L}_{Nk}^*(b) - \tilde{L}_{Nk}^*(0) + bAI^{1/2}(f)\rho_1| > \epsilon\right\} \to 0 \qquad \text{as} \quad N \to \infty. \tag{4.5.13}$$

For this note that

$$\tilde{L}_{Nk}^*(b) - \tilde{L}_{Nk}^*(0) = \sum_{i=1}^N c_{Ni}^* \left[\phi^{(k)}(F(X_i - bc_{Ni}^*)) - \phi^{(k)}(F(X_i))\right] \tag{4.5.14}$$

involves independent summands, $\max_{1 \leq i \leq N} |c_{Ni}^*| \to 0$, $\phi^{(k)}$ is bounded and $I(f) < \infty$ ensures that $E[\tilde{L}_{Nk}(b) - \tilde{L}_{Nk}(0) + bAI^{1/2}(f)\rho_1] \to 0$ as k increases. Thus (4.5.13) follows by using the Chebyshev inequality along with the fact that $E[\phi^{(k)}(F(X_i - bc_{Ni}^*)) - \phi^{(k)}(F(X_i))]^2 \to 0$ as $N \to \infty$. Q.E.D.

Remark. In (4.5.5), instead of working with $\{c_{Ni}^*\}$, one could have worked with $\{d_{Ni}\}$ where the d_{Ni} are concordant to the c_{Ni}^* and $\sup_N \sum d_{Ni}^2 < \infty$. Then a similar result holds with $bAI^{1/2}(f)\rho_1$ being replaced by $bAI^{1/2}(f)\rho_1(\sum_{i=1}^N c_{Ni}^* d_{Ni})$; we pose this as an exercise (see Problem 4.5.2). The above results are due to Jurečková (1969).

Suppose now parallel to (4.3.6), we define for each b the stochastic process $Y_{N,b} = \{Y_{N,b}(t), t \in I\}$ where

$$Y_{N,b}(t) = \frac{C_{N(t)} \tilde{L}_{N(t)}(b)}{C_N A_N^0} = \left(C_N A_N^0\right)^{-1} \sum_{i=1}^{N(t)} (c_i - \bar{c}_{N(t)}) a_{N(t)}(R_{N(t)i}^*(b)), \quad t \in I \tag{4.5.15}$$

$N(t)$ is defined by (4.3.6) and $R_{N(t)i}^*(b) = \text{Rank of } X_i - b(c_i - \bar{c}_{N(t)})/C_N$ among $X_\alpha - b(c_\alpha - \bar{c}_{N(t)})/C_N$, $1 \leq \alpha \leq N(t)$, for $1 \leq i \leq N(t)$. Then, the following theorem strengthens the results of Theorems 4.5.2 in the light of the invariance principle of Theorem 4.3.2.

THEOREM 4.5.3. *Under the conditions of Theorem* 4.5.2, *as* $N \to \infty$,

$$\sup\left\{|Y_{N,b}(t) - Y_{N,0}(t) + btAI^{1/2}(f)\rho_1 : t \in I, |b| \leqslant C\right\} \xrightarrow{p} 0. \quad (4.5.16)$$

Proof. As in (4.5.10), it follows that the lhs of (4.5.16) is bounded from above by

$$\max_{0 \leqslant j \leqslant 2k_\epsilon}\left\{\sup_{t \in I}|Y_{N,b_j}(t) - Y_{N,0}(t) + b_j tAI^{1/2}(f)\rho_1|\right\} + \tfrac{1}{2}\epsilon. \quad (4.5.17)$$

By the same arguments as in (4.5.12) through (4.5.14), it follows that for every (fixed) $m(\geqslant 1)$ and $t_1, \ldots, t_m(\in I)$, the joint distribution of ($[Y_{N},b(t_j) - Y_{N,0}(t_j) + bt_jAI^{1/2}(f)\rho_1]$, $1 \leqslant j \leqslant m$) is asymptotically a degenerate multinormal distribution (with null mean vector and null dispersion matrix). If we define $\omega_\delta(x)$ as in (2.3.7) and note that $\omega_\delta(x \pm y) \leqslant \omega_\delta(x) + \omega_\delta(y)$, then on using Theorems 4.3.2 and 4.3.6 (ensuring the tightness of $Y_{N,0}$ and $Y_{N,b}$), we obtain that for every $\epsilon > 0$, $\eta'' > 0$, there exists a $\delta(>0)$, sufficiently small, so that

$$\begin{aligned}
&P\left\{\omega_\delta\left(\left[Y_{N,b}(t) - Y_{N,0}(t) + btAI^{1/2}(f)\rho_1\right], t \in I\right) > \tfrac{1}{2}\epsilon\right\} \\
&\quad \leqslant P\left\{\omega_\delta\left(\left[Y_{N,b}(t) + btAI^{1/2}(f)\rho_1\right], t \in I\right) > \tfrac{1}{4}\epsilon\right\} \\
&\qquad + P\left\{\omega_\delta(Y_{N,0}) > \tfrac{1}{4}\epsilon\right\} \\
&\quad \leqslant \eta'' \quad \text{for} \quad N \geqslant N_0;\ N_0 \ \text{depends on} \ \epsilon, \eta'', \ \text{and} \ \delta. \quad (4.5.18)
\end{aligned}$$

Thus, from the above we conclude that for every $\epsilon > 0$ and $\eta' > 0$, there exist a $\delta(0 < \delta < 1)$ and an N_0, such that for $N \geqslant N_0$,

$$P\left\{\sup_{t \in I}|Y_{N,b}(t) - Y_{N,0}(t) + btAI^{1/2}(f)\rho_1| > \tfrac{1}{2}\epsilon\right\} < \eta', \quad (4.5.19)$$

uniformly in $b \in [-c, c]$. Hence, for every η (>0), choosing $\eta' = \eta/2K_\epsilon$, the desired result follows from (4.5.17) and (4.5.19). Q.E.D.

These asymptotic linearity results are of fundamental importance in the study of the asymptotic properties of estimators of regression slopes based on linear rank statistics. While the details of these estimators will be provided in Chapter 8 (Section 8.5) and in Chapters 9 and 10, we like to introduce the same here and briefly mention the asymptotic properties. Consider the regression model in (4.2.12) and define $L_N(b)$ as in (4.5.1). Under the null hypothesis $H_0: \beta = 0$, $L_N(0)$ has location 0 and, under β, $L_N(\beta)$ has the same distribution as $L_N(0)$ under H_0. Further, by Theorem 4.5.1, $L_N(b)$ is $\searrow$ in b. Hence one way to estimate β is to align the observations X_i (by $X_i - bc_i$) in such a way that the resulting $L_N(b)$ is closest to the location 0. With this in mind, we define

$$\hat{\beta}_N = \tfrac{1}{2}(\hat{\beta}_{N1} + \hat{\beta}_{N2}); \tag{4.5.20}$$

$$\hat{\beta}_{N1} = \sup\{b : L_N(b) > 0\} \quad \text{and} \quad \hat{\beta}_{N2} = \inf\{b : L_N(b) < 0\}. \tag{4.5.21}$$

Then, $\hat{\beta}_N$ may be considered as an estimator of β in the model in (4.2.12). This estimator is due to Adichie (1967). It is a translation-invariant, robust and consistent estimator, and whenever $L_N(0)$ has a distribution symmetric about 0, $\hat{\beta}_N$ has also a distribution symmetric about β. This is the case when the underlying d.f. F in (4.2.12) is symmetric about 0. Note that for every real t, by (4.5.20) and (4.4.21),

$$[L_N(t) > 0] \subseteq [\hat{\beta}_N \geqslant t] \subseteq [L_N(t) \geqslant 0]. \tag{4.5.22}$$

Also, defining $\tilde{L}_N(b)$ as in (4.5.6) and noting that $\tilde{L}_N(0)$ is bounded in probability, we conclude from (4.5.7) and (4.5.22) that by virtue of (4.5.20) and (4.5.21),

$$C_N(\hat{\beta}_N - \beta)AI^{1/2}(f)\rho_1 - \tilde{L}_N(0) \to 0, \qquad \text{in probability}, \tag{4.5.23}$$

where (without any loss of generality) we have taken $\beta = 0$, for $\tilde{L}_N(0)$. The asymptotic normality of $C_N(\hat{\beta}_N - \beta)$ follows directly from (4.2.10) and (4.5.23). Actually, Theorem 4.5.3 extends (4.5.23) to the sequential case. Defining $N(t)$ as in (4.3.6), we obtain from (4.5.16) (by letting $\beta = 0$) and (4.5.20) and (4.5.21) that for every $\epsilon > 0$, as $N \to \infty$,

$$\sup_{\epsilon \leqslant t \leqslant 1} |C_N(\hat{\beta}_{N(t)} - \beta)AI^{1/2}(f)\rho_1 - Y_{N,0}(t)| \xrightarrow{p} 0. \tag{4.5.24}$$

(4.5.24) and (4.3.7) provide an invariance principle for the estimators.

Though in most of the cases, Theorems 5.5.2 and 4.5.3 suffice the purpose, there are certain situations where the range of b in (4.5.7) may depend on N and we may need to provide an a.s. result (instead of the "in probability" one). Such a stronger result may naturally require more stringent regularity conditions and will be considered in the appendix.

4.6. ALMOST SURE CONVERGENCE OF $\{L_N\}$

It follows from Theorem 4.4.1 that if the F_i are all identical and if $g(N)$ is a nonnegative, nondecreasing function of N satisfying the condition that

$$\overline{\lim_{N\to\infty}} \left(C_N\sqrt{2\log\log C_N^2}\right)/g(N) = 0, \tag{4.6.1}$$

then under the conditions of Theorem 4.4.1

$$[g(N)]^{-1}L_N \to 0 \quad \text{a.s.}, \quad \text{as} \quad N \to \infty. \tag{4.6.2}$$

In this section, we are primarily interested in the a.s. convergence of $\{L_N\}$ when the F_i are not necessarily identical. For this purpose, we define first

$$\check{L}_N = N^{-1/2}A_N^{-1}C_N^{-1}L_N, \tag{4.6.3}$$

$$\check{\lambda}_N = N^{-1/2}A_N^{-1}\sum_{i=1}^{N} c_{Ni}^* \int_{-\infty}^{\infty} \phi\big(\bar{F}_{(N)}(x)\big)\,dF_i(x), \tag{4.6.4}$$

where $c_{Ni}^* = (c_i - \bar{c}_N)/C_N$, $1 \leqslant i \leqslant N$ $(\Rightarrow \sum_{i=1}^{N}(c_{Ni}^*)^2 = 1)$, $\bar{F}_{(N)} = N^{-1} \cdot \sum_{i=1}^{N} F_i$ and A_N^2 is defined by (4.2.6). Our basic assumptions are the following:

$$\max_{1 \leqslant i \leqslant N} |c_{Ni}^*| = O(N^{-1/2}); \tag{4.6.5}$$

$$\phi(u) \text{ is of bounded variation in closed subintervals of } (0,1) \text{ with } \int_0^1 |\phi(u)|\,du < \infty; \tag{4.6.6}$$

$$\int_0^1 |a_N(1+[Nu]) - \phi(u)|\,du \to 0 \quad \text{as} \quad N \to \infty. \tag{4.6.7}$$

Note that we allow $F_1, \ldots, F_n, \ldots$ to be arbitrary. Then, we have the following* theorem.

THEOREM 4.6.1. *Under* (4.6.5), (4.6.6), *and* (4.6.7)

$$\check{L}_N - \check{\lambda}_N \to 0 \quad a.s., \quad as \quad N \to \infty.$$

Proof. Here also we work, for simplicity, with the scores $a_N(i) = E\phi(U_{Ni})$, $1 \leqslant i \leqslant N$, so that

$$\int_0^1 a_N(1+[Nu])\,du = \int_0^1 \phi(u)\,du, \qquad \forall N \geqslant 1.$$

Thus by (4.6.6) and (4.6.7), we claim that for every $\delta > 0$, there exists a $K(=K_\delta < \infty)$ and an integer N_δ such that on defining

$$a_N^*(i) = \begin{cases} a_N(i), & \text{if } |a_N(i)| \leqslant K \\ 0, & \text{otherwise;} \end{cases} \tag{4.6.8}$$

$$\phi^*(u) = \begin{cases} \phi(u), & \text{if } |\phi(u)| \leqslant K, \\ 0, & \text{otherwise,} \end{cases} \tag{4.6.9}$$

*Proved in an unpublished work of M. Ghosh and the author. A somewhat different theorem, without requiring (4.6.6) but assuming $\int_0^1 |\phi(u)|^r\,du < \infty$ for some $r > 2$, is proved in Sen and Ghosh (1972). Theorem 4.6.1 is a direct extension of a two-sample result of Hájek (1974).

we have

$$\int_0^1 |\phi^*(u) - \phi(u)|\, du < \delta. \tag{4.6.10}$$

$$N^{-1} \sum_{i=1}^{N} |a_N^*(i) - a_N(i)| < \delta, \qquad \forall N \geqslant N_\delta. \tag{4.6.11}$$

Let us also introduce the following functions:

$$F_N^*(x) = N^{-1/2} A_N^{-1} \sum_{i=1}^{N} c_{Ni}^* F_i(x), \qquad x \in R; \tag{4.6.12}$$

$$S_N^*(x) = N^{-1/2} A_N^{-1} \sum_{i=1}^{N} c_{Ni}^* c(x - X_i), \qquad x \in R; \tag{4.6.13}$$

$$\bar{S}_N(x) = N^{-1} \sum_{i=1}^{N} c(x - X_i), \qquad x \in R. \tag{4.6.14}$$

Then, it follows that under (4.6.5), $|dF_N^*(x)| \leqslant [d\bar{F}_{(N)}(x)]0(1)$, $|dS_N^*(x)| \leqslant [d\bar{S}_N(x)]0(1)$, $\forall N \geqslant 1$ and $x \in R$. Thus, rewriting $\check{L}_N$ and $\check{\lambda}_N$ as $\int_{-\infty}^{\infty} a_N(1 + N\bar{S}_N(x))\, dS_N^*(x)$ and $\int_{-\infty}^{\infty} \phi(\bar{F}_{(N)}(x))\, dF_N^*(x)$, respectively, and using (4.6.11) and (4.6.12), it follows that one may replace a_N and ϕ by a_N^* and ϕ^*, respectively, with a maximum error bounded by δ' (for $N \geqslant N_\delta$), where $\delta'(>0)$ goes to 0 as $\delta \to 0$. Thus it suffices to prove the theorem in the case where ϕ has bounded variation over all [0, 1]. Note then

$$\begin{aligned} \check{L}_N - \check{\lambda}_N &= \int_{-\infty}^{\infty} a_N\big(1 + N\bar{S}_N(x)\big)\, dS_N^*(x) - \int_{-\infty}^{\infty} \phi\big(\bar{F}_{(N)}(x)\big)\, dF_N^*(x) \\ &= \int_0^1 a_N(1 + [Nu])\, d\big[G_N^*(u) - H_N^*(u)\big] \\ &\quad + \int_0^1 \big[a_N(1 + [Nu]) - \phi(u)\big]\, dH_N^*(u), \end{aligned} \tag{4.6.15}$$

where

$$G_N^*(u) = S_N^*\big(\bar{S}_N^{-1}(u)\big), \qquad H_N^*(u) = F_N^*\big(\bar{F}_{(N)}^{-1}(u)\big), \qquad 0 < u < 1, \tag{4.6.16}$$

Note that, by definition,

$$\begin{aligned} &\left| \int_0^1 \big[a_N(1 + [Nu]) - \phi(u)\big]\, dH_N^*(u) \right| \\ &\quad \leqslant N^{-1/2} A_N^{-1} \Big(\max_{1 \leqslant i \leqslant N} |c_{Ni}^*| \Big) \\ &\qquad \times \sum_{i=1}^{N} \int_0^1 |a_N(1 + [Nu]) - \phi(u)|\, dF_i\big(\bar{F}_{(N)}^{-1}(u)\big) \\ &\quad = A_N^{-1} \Big\{ N^{1/2} \max_{1 \leqslant i \leqslant N} |c_{Ni}^*| \Big\} \int_0^1 |a_N(1 + [Nu]) - \phi(u)|\, du \\ &\quad \to 0, \qquad \text{as} \quad N \to \infty, \quad \text{by (4.6.5) and (4.6.7).} \end{aligned} \tag{4.6.17}$$

For the first term on the rhs of (4.6.15), on denoting by V the variation of a_N (as well as ϕ) over $[0,1]$, $(V<\infty$, by assumption), for every $\eta(0<\eta\leqslant\frac{1}{2})$, $|\int_0^\eta a_N(1+[Nu])\,dG_N^*(u)| = |\int_0^\eta G_N^*(u)\,da_N(1+[Nu])| \leqslant V\{N^{-1/2}A_N^{-1}\max_{1\leqslant i\leqslant N}|c_{Ni}^*|\}|N\bar{S}_N(\bar{S}_N^{-1}(\eta))| \leqslant 2VKA_N^{-1}\eta$, for $N\geqslant N_\delta$. Similar bounds hold for $\int_{1-\eta}^1 a_N(1+[Nu])\,dG_N^*(u)$, $\int_0^\eta a_N(1+[Nu])\,dH_N^*(u)$, and $\int_{1-\eta}^1 a_N(1+[Nu])\,dH_N^*(u)$. Thus, choosing $\eta(>0)$ arbitrarily small, the tails $[(0,\eta)\cup(1-\eta,1)]$ can be neglected. Further, for every $\eta(0<\eta<\frac{1}{2})$,

$$\Big|\int_\eta^{1-\eta} a_N(1+[Nu])\,d[G_N^*(u)-H_N^*(u)]\Big| \leqslant V\Big\{\sup_{\eta\leqslant u\leqslant 1-\eta}|G_N^*(u)-H_N^*(u)|\Big\}, \tag{4.6.18}$$

and hence it suffices to show that as $N\to\infty$,

$$\sup_x|S_N^*(x)-F_N^*(x)|\to 0 \quad \text{a.s.} \tag{4.6.19}$$

For this purpose, note that

$$\begin{aligned} S_N^*(x)-F_N^*(x) &= A_N^{-1}N^{-1/2}\sum_{i=1}^N c_{Ni}^*[c(x-X_i)-F_i(x)] \\ &= Q_{N1}(x)+Q_{N2}(x) \end{aligned} \tag{4.6.20}$$

where

$$Q_{Nj}(x) = N^{-1/2}A_N^{-1}\sum\nolimits_N^{(j)} c_{Ni}^*[c(x-x_i)-F_i(x)], \qquad j=1,2 \tag{4.6.21}$$

and the summations $\sum_N^{(1)}$ and $\sum_N^{(2)}$ range over the set of positive and negative c_{Ni}^*, respectively. Let $b_{N,r}$, $0\leqslant r\leqslant N$ be defined by $\bar{F}_{(N)}(b_{N,r}) = r/N$, $0\leqslant r\leqslant N$. Then, for every $x\in[b_{N,r},b_{N,r+1}]$, by (4.6.21), $\sum_N^{(1)}c_{Ni}^*[c(b_{N,r}-X_i)-F_i(b_{N,r+1})] \leqslant A_N N^{1/2}Q_{N1}(x) \leqslant \sum_N^{(1)}c_{Ni}^*[c(b_{N,r+1}-x_i)-F_i(b_{N,r})]$, where $\sum_N^{(1)}c_{Ni}^*[F_i(b_{N,r+1})-F_i(b_{N,r})] \leqslant \{\max_{1\leqslant i\leqslant N}c_{Ni}^*|\}N|\bar{F}_{(N)}(b_{N,r+1})-\bar{F}_{(N)}(b_{N,r})| = \max_{1\leqslant i\leqslant N}|c_{Ni}^*|$. Thus, by a few standard steps,

$$\sup_x|Q_{N1}(x)| \leqslant \max_{0\leqslant r\leqslant N}|Q_{N1}(b_{N,r})| + O(N^{-1}). \tag{4.6.22}$$

Finally, for each $r(0\leqslant r\leqslant N)$, $Q_{N1}(b_{N,r})$ involves an average of N independent and bounded valued (as $N^{1/2}|c_{Ni}^*| = O(1)$, $1\leqslant i\leqslant N$) random variables. Hence, for every $\epsilon>0$,

$$\begin{aligned} &P\Big\{\max_{0\leqslant r\leqslant N}|Q_{N1}(b_{N,r})|>\epsilon\Big\} \\ &\quad\leqslant \sum_{r=0}^N P\{|Q_{N1}(b_{N,r})|>\epsilon\} \\ &\quad\leqslant \sum_{r=0}^N\Big[\inf_{t>0} e^{-Nt\epsilon}\prod_{i=1}^N E\big\{\exp\big(tN^{1/2}c_{Ni}^*[c(b_{N,r}-X_i)-F_i(b_{N,r})]/A_N\big)\big\}\Big] \\ &\quad\leqslant (N+1)\exp\{-N\epsilon^2/4K^2\};\ K=\max_{1\leqslant i\leqslant n}N^{1/2}|c_{Ni}^*|. \end{aligned} \tag{4.6.23}$$

Since the rhs of (4.6.23) converges exponentially to 0, using the Borel-Cantelli lemma, we conclude from (4.6.23) and (4.6.24) that $\sup_x |Q_{N1}(x)| \to 0$ a.s., as $N \to \infty$. A similar treatment holds for $\sup_x |Q_{N2}(x)|$. Q.E.D.

EXERCISES

4.3.1.

(a) Let $\{Z_{Ni}, 1 \leqslant i \leqslant N; N \geqslant 1\}$ be a triangular array of independent r.v.'s with $EZ_{Ni} = 0$, $EZ_{Ni}^2 = 1$ and let $\{d_{Ni}, 1 \leqslant i \leqslant N; N \geqslant 1\}$ be another triangular array of nonstochastic positive constants with $\sum_{i=1}^N d_{Ni}^2 = 1$, $\forall N \geqslant 1$. If then (i) $\max_{1 \leqslant i \leqslant N} d_{Ni}^2 \to 0$ as $N \to \infty$ and (ii) $\max_{1 \leqslant i \leqslant N} \cdot E\{Z_{Ni}^2 I(|Z_{Ni}| > \lambda)\} \to 0$ as $\lambda \to \infty$, then show that $\sum_{i=1}^N d_{Ni} Z_{Ni} \sim \mathfrak{N}(0, 1)$ (Hájek and Sidák, 1967).

(b) Apply (a) to establish the convergence of the f.d.d.'s of $\{Y_N^*\}$ in Theorem 4.3.1.

4.3.2. Use (2.2.9) and Theorem 4.2.1 to prove the tightness part of Theorem 4.3.1.

4.3.3. Provide the proof of Theorem 4.3.3.

4.3.4. Show that if ϕ has a bounded second derivative inside I, then $\max_{1 \leqslant i \leqslant N} |\phi(i/(N+1)) - E\phi(U_{Ni})| = 0(n^{-1})$ and hence (4.3.23) holds for $a_N(i) = \phi(i/(N+1))$, $1 \leqslant i \leqslant N$ (Hájek, 1968).

4.3.5. Suppose that ϕ has a continuous first order derivative $\phi^{(1)}$ and $|\phi^{(r)}(u)| \leqslant K[u(1-u)]^{-(1/2)-r+\delta}$, $r = 0, 1$, for some $\delta > 0$, $K < \infty$, $\forall u \in I$, prove that then $\max_{1 \leqslant i \leqslant N} N^{1/2} |\phi(i/(N+1)) - E\phi(U_{Ni})| = 0(N^{-\eta})$, $\eta > 0$ (Puri and Sen, 1971).

4.3.6. Show that (4.3.23) holds for the following scores:

(a) $a_N(i) = i/(N+1)$, $1 \leqslant i \leqslant N$, (Wilcoxon scores);

(b) $a_N(i) = \phi^{-1}(i/(N+1))$, $1 \leqslant i \leqslant N$ where ϕ is the standard normal d.f. (Van der Wearden scores), and

(c) $a_N(i) = [i/(N+1) - \frac{1}{2}]^2$, $1 \leqslant i \leqslant N$ (Mood scores).

4.3.7. Show that if $a_N(i) = \mathrm{sgn}(i - (N+1)/2)$, $1 \leqslant i \leqslant N$ ($\phi(u) = \mathrm{sgn}(u - t)$), (4.3.34) does not hold but $\max_{1 \leqslant k \leqslant N} |L_k^0 - L_k| / C_N \to 0$.

4.3.8. Show that under (4.3.36) through (4.3.39), $\{Q_N\}$ is contiguous to $\{P_N\}$ (Hájek and Sidák, 1967).

4.3.9. Consider the model $F_{2k+1}(x) = F(x)$, $F_{2k+2}(x) = G(x)$, $k \geqslant 0$, where $G(x) = F(\theta x)$, $\theta > 0$ and let $H_0: \theta = 1$, $K_N: \theta = 1 + N^{-1/2}\gamma$. Establish the weak convergence of $\{Y_N\}$ under $\{K_N\}$ and obtain the expression for the drift function.

4.3.10. Show that under (4.3.55) and (4.3.56), (4.3.51) holds with $\nu_0 = \rho_1 \beta I^{1/2}(f)$.

Hint. Use the method of Theorem 4.6.1.

4.3.11. Establish the weak convergence results in Theorems 4.3.1 and 4.3.2 under the (d_q-) sup-norm metric in (2.4.26).

4.4.1. Show that if ϕ has a bounded first derivative inside I, then $q_2(n) = O(n^{-1})$.

4.4.2. If $\phi \in L_r$ for some $r > 2$, show that $q_2(n) = O(n^{1-r/2})$.

4.4.3. If ϕ satisfies (4.4.5), show that $q_2(n) = O((\log n)^{-1-\eta})$ for some $\eta > 0$ (Hoeffding, 1973).

4.4.4. Show that the law of iterated logarithm holds for $\{L_N^0\}$ when $\phi \in L_r$ for some $r > 2$ and $q_1(r) = O(n^{-1})$ (Sen and Ghosh, 1972).

4.4.5. Show that if $\phi \in L_r$, $r > 2$ and (i) $\liminf N^{-1}C_N^2 \geqslant C_0 > 0$ and (ii) $\limsup N^{-h}C_N^2 \leqslant C^0 < \infty$ for some $h(\geqslant 1)$, then (4.4.28) holds with $f(t) = t^{1-\eta}$ for some $\eta > 0$.

4.5.1. Establish the monotonicity of $L_N(b)$ when the alignment is made by regression constants $\{d_{Ni}\}$ where $(c_i - c_j)(d_{Ni} - d_{Nj}) \geqslant 0\ \forall i \neq j = 1, \ldots, N$ (with at least one strict inequality) (Jurečková, 1969).

4.5.2 (4.5.1 continued). Show that in (4.5.7), the drift has to be changed then to $bAI^{1/2}(f)\rho_1(\sum_{i=1}^N c_{Ni}^* d_{Ni})$ (Jurečková, 1969).

4.6.1 (A sequential version of the LRS). With the same notations as in (4.2.1) and (4.2.2), consider the rank statistics

$$T_n^* = \sum_{i=1}^{n} (c_i - \bar{c}_i)\left[a_i(R_{ii}) - \bar{a}_i\right], \qquad n \geqslant 1.$$

(a) Show that the sequential rankings $R_{kk}, k \geqslant 1$ are stochastically independent under the null hypothesis in (4.2.3).

(b) Verify that $E(T_N^* \mid \mathfrak{B}_{n-1}) = T_{n-1}^*$ a.e., for every $n \geqslant 1$, when (4.2.3) holds.

(c) Obtain results parallel to those in Theorems 4.3.1, 4.3.5, 4.4.1, and 4.6.1.

4.6.2. Suppose that in (4.2.1), you replace the c_i (and $\bar{c}_n$) by q-vectors $\mathbf{c}_i$ (and $\bar{\mathbf{c}}_n$) and denote the resulting q-vector by $\mathbf{L}_N$. Let then $\mathbf{C}_N = \sum_{i=1}^N (\mathbf{c}_i - \bar{\mathbf{c}}_N)(\mathbf{c}_i - \bar{\mathbf{c}}_N)'$.

(a) Show that under H_0 in (4.2.3), Theorem 4.2.1 holds for the vector $\mathbf{L}_N$ as well.

(b) Under the assumptions of Theorem 4.3.1 and condition that $\text{Max}\{(\mathbf{c}_i - \bar{\mathbf{c}}_N)'\mathbf{C}_N^{-1}(\mathbf{c}_i - \bar{\mathbf{c}}_N) : 1 \leqslant i \leqslant N\} \to 0$ as $N \to \infty$, show that $A_N^{-1}\mathbf{C}_N^{-1/2}\mathbf{L}_N$ is asymptotically normal with mean $\mathbf{0}$ and dispersion matrix $\mathbf{I}_q$.

(c) Under the assumptions in (b), show that whenever $N^{-1}\mathbf{C}_N \to \mathbf{C}^*$ (positive definite) as $N \to \infty$, $\{A_N^{-1}\mathbf{C}_N^{-1/2}\mathbf{L}_{[Nt]}, t \in [0, 1]\} \xrightarrow{\mathcal{D}} \{\mathbf{W}(t), t \in [0, 1]\}$ where $\mathbf{W}(t) = (W_1(t), \ldots, W_q(t))'$ has independent copies of a standard Wiener process.

(d) Under the same assumptions as in (c), show that

$$\left\{A_N^{-1} \cdot \left(\mathbf{L}'_{[Nt]}\mathbf{C}_N^{-1}\mathbf{L}_{[Nt]}\right)^{1/2}, \quad t \in [0, 1]\right\}$$

converges weakly to a Bessel process $B = \{B(t), t \in [0, 1]\}$, where $B(t)$ is defined by (2.3.50) for $k = q$.

4.6.3. Under the conditions in Theorem 4.3.5 or 4.3.6 [with possibly a vector in (4.3.36)], extend the results in (b), (c), and (d) in Problem 4.6.2 to that of a drifted Bessel process.

4.6.4. Extend the results in Problem 4.6.1 to the case where the $\mathbf{c}_i$ are q-vectors for some $q \geqslant 1$.

CHAPTER 5

Invariance Principles for Signed Rank Statistics

5.1. INTRODUCTION

Signed rank statistics are commonly employed for tests of significance and estimation of location parameter of a distribution. This chapter is devoted to the study of invariance principles for such statistics.

Signed rank statistics are introduced in Section 5.2. Weak convergence to Wiener processes is studied in Section 5.3. The Skorokhod-Strassen embedding of Wiener process is considered in Section 5.4. Asymptotic linearity (in the shift parameter) of signed rank statistics is considered in Section 5.5. Section 5.6 deals with tied-down Wiener process approximation for aligned signed rank statistics. Almost sure convergence of such statistics is studied in the last section.

5.2. PRELIMINARY NOTIONS

Let $\{X_i, i \geqslant 1\}$ be a sequence of independent and identically distributed random variables (i.i.d.r.v.) with a continuous d.f. $F(x)$, $x \in R$, the real line $(-\infty, \infty)$. For testing the hypothesis of symmetry of F (around a specific median, say, 0) or for estimating the unknown median of F, a general class of nonparametric procedures rests on the following type of *signed rank statistics*:

$$S_N = S(X_1, \dots, X_N) = \sum_{i=1}^{N} \operatorname{sgn} X_i a_N(R_{Ni}^+), \qquad N \geqslant 1, \qquad (5.2.1)$$

where $R_{Ni}^+ = \sum_{j=1}^{N} c(|X_i| - |X_j|)$ is the rank of $|X_i|$ among $|X_1|, \dots, |X_N|$, $1 \leqslant i \leqslant N$ and the scores $a_N(1), \dots, a_N(N)$ are defined as in (4.2.2).

Consider now the hypothesis of sign-invariance, namely,

$$H_0 : F(x) + F(-x) = 1, \qquad \forall x \geqslant 0, \qquad F \text{ continuous.} \tag{5.2.2}$$

Under H_0, $(\operatorname{sgn} X_1, \dots, \operatorname{sgn} X_N)$ and $(R_{N1}^+, \dots, R_{NN}^+)$ are stochastically independent; the first vector assumes all possible 2^N sign-inversions with the common probability 2^{-N}, and the second vector assumes all possible permutations of $(1, \dots, N)$ with the common probability $(N!)^{-1}$. Thus the distribution of S_N (under H_0) is generated by the $2^N N!$ equally likely realizations of signs and ranks and is independent of F. Note that

$$E\left[a_N(R_{Ni}^+) \mid H_0\right] = \frac{1}{N}\sum_{\alpha=1}^{N} a_N(\alpha) = \bar{a}_N, \qquad \text{say;} \tag{5.2.3}$$

$$E\left[a_N(R_{Ni}^+)a_N(R_{Ni'}^+) \mid H_0\right] = \begin{cases} A_N^2, & \text{if } i = i' \\ \dfrac{1}{N-1}\left\{N\bar{a}_N^2 - A_N^2\right\}, & \text{if } i \neq i', \end{cases} \tag{5.2.4}$$

where

$$A_N^2 = N^{-1}\sum_{\alpha=1}^{N} a_N^2(\alpha). \tag{5.2.5}$$

Further, $\operatorname{sgn} X_i$, $1 \leqslant i \leqslant N$ are i.i.d.r.v. with $P\{\operatorname{sgn} X_i = \pm 1\} = \frac{1}{2}$, and hence we have

$$E(S_N \mid H_0) = 0 \qquad \text{and} \qquad V(S_N \mid H_0) = NA_N^2. \tag{5.2.6}$$

Actually, conditionally on $(R_{N1}^+, \dots, R_{NN}^+)$, S_N is a linear function of i.i.d.r.v., and hence, whenever

$$\max_{1 \leqslant i \leqslant N}\left[a_N^2(i)/NA_N^2\right] \to 0 \qquad \text{as} \quad N \to \infty, \tag{5.2.7}$$

$$S_N/\left(NA_N^2\right)^{1/2} \xrightarrow{\mathfrak{D}} \mathfrak{N}(0,1) \qquad \text{under} \quad H_0. \tag{5.2.8}$$

In subsequent sections, we shall extend this result in different directions. In passing, we may remark that the classical sign-statistic is a special case of (5.2.1) where $a_N(i) = 1$, $\forall N \geqslant i \geqslant 1$, that is, $S_N = \sum_{i=1}^N \operatorname{sgn} X_i, N \geqslant 1$. Since this involves a sum of i.i.d.r.v.'s, the results on U-statistics in Chapter 3 hold. Hence, in the sequel, we exclude this special case and assume that the $a_N(i)$ are not a constant for every i and N.

As in (4.2.13), we introduce the scores $a_N^0(i) = E\phi(U_{Ni})$, $1 \leqslant i \leqslant N$ and define the corresponding S_N by S_N^0, $N \geqslant 1$. Let $\mathfrak{B}_N$ be the sigma field generated by $(\operatorname{sgn} X_1, \dots, \operatorname{sgn} X_N)$ and $(R_{N1}^+, \dots, R_{NN}^+)$, $N \geqslant 1$. Thus $\mathfrak{B}_N$ is nondecreasing in N.

THEOREM 5.2.1. *Under H_0 in (5.2.2), $\{S_N^0, \mathfrak{B}_N; N \geqslant 1\}$ is a martingale whenever ϕ is in L_1 space.*

Proof. Note that by (5.2.1), for every $N \geqslant 1$,

$$E_0(S_{N+1}^0 \mid \mathcal{B}_N) = \sum_{i=1}^{N} \operatorname{sgn} X_i E_0\big(a_{N+1}^0(R_{N+1i}) \mid \mathcal{B}_N\big)$$
$$+ E_0\big[\operatorname{sgn} X_{N+1} a_{N+1}^0(R_{N+1N+1}) \mid \mathcal{B}_N\big], \qquad (5.2.9)$$

where E_0 stands for the expectation under H_0 in (5.2.2). Now, conditionally on $\mathcal{B}_N$, R_{N+1i}^+ $(1 \leqslant i \leqslant N)$ can only have two possible values, R_{Ni}^+ and $(R_{Ni}^+ + 1)$ with respective conditional probabilities $(N+1)^{-1}(N+1-R_{Ni}^+)$ and $(N+1)^{-1}R_{Ni}^+$, so that the first term on the rhs of (5.3.1) is equal to

$$\sum_{i=1}^{N} \operatorname{sgn} X_i\big[(N+1)^{-1}R_{Ni}^+ a_{N+1}^0(R_{Ni}^+ + 1)$$
$$+ (N+1)^{-1}(N+1-R_{Ni}^+)a_{N+1}^0(R_{Ni}^+)\big]$$
$$= \sum_{i=1}^{N} \operatorname{sgn} X_i a_N^0(R_{Ni}^+) = S_N \qquad \big[\text{by } (4.2.20)\big]. \qquad (5.2.10)$$

On the other hand, given $\mathcal{B}_N$, $\operatorname{sgn} X_{N+1}$ and R_{N+1N+1}^+ are (conditionally) independent; $\operatorname{sgn} X_{N+1}$ assumes the values ± 1 with equal conditional probability $\frac{1}{2}$ and R_{N+1N+1}^+ can assume the values $1, \ldots, N+1$ with equal probability $(N+1)^{-1}$. Hence the second term on the rhs of (5.2.9) is equal to

$$\left\{\frac{1}{2}\sum_{j=1}^{2}(-1)^j\right\}\left\{\frac{1}{N+1}\sum_{\alpha=1}^{N+1} a_N(\alpha)\right\} = 0. \qquad (5.2.11)$$

Hence, from (5.2.10) and (5.2.11), $E_0(S_{N+1} \mid \mathcal{B}_N) = S_N$, $\forall N \geqslant 1$, while $E_0(S_1^0) = a_N^0(1)E_0(\operatorname{sgn} X_1) = 0$. Q.E.D.

If we conceive of a score function $\phi^*(u)$, $0 < u < 1$, such that ϕ^* is skew-symmetric, that is,

$$\phi^*(u) + \phi^*(1-u) = 0, \qquad 0 < u < 1 \qquad (5.2.12)$$

and relate ϕ to ϕ^* by

$$\phi(u) = \phi^*\left(\frac{1+u}{2}\right), \qquad 0 \leqslant u < 1, \qquad (5.2.13)$$

then, we note that on writing

$$S_N^* = \sum_{i=1}^{N} \phi^*(F(X_i)) \qquad (5.2.14)$$

we have the following:

$$S_N^* = \sum_{i=1}^{N} \operatorname{sgn} X_i \phi(F(|X_i|) - F(-|X_i|)), \tag{5.2.15}$$

$$E_0(S_N^* \mid \mathcal{B}_N) = S_N, \qquad \forall N \geqslant 1. \tag{5.2.16}$$

Both (5.2.15) and (5.2.16) are useful for later manipulations.

5.3. WEAK CONVERGENCE OF $\{S_N\}$

Let us assume that ϕ is square integrable inside $I = [0, 1]$ and set

$$A^2 = \int_0^1 \phi^2(u)\,du = \int_0^1 \left[\phi^*(u)\right]^2 du : 0 < A < \infty. \tag{5.3.1}$$

Note that as in (4.3.2) through (4.3.4), A_N^2, defined by (5.2.5) converges to A^2 as $N \to \infty$. For every $N(\geqslant 1)$, we define a process $Y_N = \{Y_N(t), t \in I\}$ by letting

$$\begin{aligned} &Y_N(t) = S_{N(t)}^0 / N^{1/2} A_N^0; \\ &N(t) = \max\{k : k/N \leqslant t\}, \qquad t \in I \quad \text{and} \quad S_0^0 = 0. \end{aligned} \tag{5.3.2}$$

Then, we have the following theorem due to Sen and Ghosh (1973b) and Sen (1974a).

THEOREM 5.3.1. *If ϕ is square integrable inside I, then under H_0 in* (5.2.2),

$$Y_N \xrightarrow{\mathfrak{D}} W \qquad \text{in the } J_1\text{-topology on} \quad D[0,1] \qquad (\text{as } N \to \infty), \tag{5.3.3}$$

where W is a standard Brownian motion on I.

Proof. By virtue of (5.2.6), (5.2.15), and (5.2.16),

$$E\left[\left(S_N^* - S_N^0\right)^2 \mid H_0\right] = E\left[S_N^{*2} \mid H_0\right] - E\left[S_N^{02} \mid H_0\right] = N(A^2 - A_N^2), \tag{5.3.4}$$

and hence for every $t \in I$, $N^{-1}E[(S_{[Nt]}^* - S_{[Nt]}^0)^2] \to 0$ as $N \to \infty$. On the other hand, for every (fixed) $m(\geqslant 1)$ and $t_1, \ldots, t_m(\in I)$, $S_{[Nt_j]}^*$ involves a sum of i.i.d.r.v.'s, so that by the classical central limit theorem, the joint distribution of $N^{-1/2}(S_{[Nt_1]}^*, \ldots, S_{[Nt_m]}^*)$ is asymptotically multinormal with null mean vector and covariance matrix $A^2((t_r \wedge t_s))_{r,s=1,\ldots,m}$. Thus the convergence of the f.d.d.'s of $\{Y_N\}$ to those of W follows. By virtue of the martingale property in Theorem 5.2.1, we now use Corollary 2.4.4.1 and conclude that the convergence of the f.d.d.'s implies the tightness of $\{Y_N\}$.

Q.E.D.

Suppose now we consider arbitrary scores $a_N(1), \dots, a_N(N)$ satisfying the condition (4.3.23). Then, note that

$$\begin{aligned}
&\max_{1 \leqslant k \leqslant N} N^{-1/2}|S_k - S_k^0| \\
&= N^{-1/2} \max_{1 \leqslant k \leqslant N} \left| \sum_{i=1}^{k} \operatorname{sgn} X_i \left[a_k(R_{ki}^+) - a_k^0(R_{ki}^+) \right] \right| \\
&\leqslant N^{-1/2} \max_{1 \leqslant k \leqslant N} \left\{ \sum_{i=1}^{k} |a_k(R_{ki}^+) - a_k^0(R_{ki}^+)| \right\} \\
&\leqslant \max_{1 \leqslant k \leqslant N} \left\{ (k/N)^{1/2} \left[\max_{1 \leqslant i \leqslant k} k^{1/2} |a_k(i) - a_k^0(i)| \right] \right\} \\
&\to 0 \quad \text{as} \quad N \to \infty.
\end{aligned} \tag{5.3.5}$$

Hence stochastic processes defined on $\{S_k\}$, instead of $\{S_k^0\}$, have the same asymptotic behavior when (4.3.23) holds.

Let us now consider the second weak convergence result where we define $Z_N = \{Z_N(t), t \in I\}$ by letting

$$Z_N(t) = \sqrt{N}\, S_{N(t)}^0 / (A_N N(t));\; N(t) = \min\{k : N/K \leqslant t\} \tag{5.3.6}$$

Then the following theorem holds; the proof follows along the lines of (4.3.15) through (4.3.18) and hence is left as an exercise (see Problem 5.3.1).

THEOREM 5.3.2. *Under the hypothesis of Theorem* 5.3.1, *as* $N \to \infty$

$$Z_N \to W, \qquad \text{in the } J_1\text{-topology on} \quad D[0,1]. \tag{5.3.7}$$

Further, if one considers stochastic sample sizes $\{N_n\}$ satisfying (4.3.19), then the conclusions of Theorem 4.3.3 also hold for $\{Y_{N_n}\}$ and $\{Z_{N_n}\}$, defined by (5.3.2) and (5.3.6) with N being replaced by N_n; the proof is again left as an exercise (see Problem 5.3.2). Finally, for both Theorems 5.3.1 and 5.3.2, the weak convergence in the Skorokhod-metric can be strengthened to the sup-norm metric d_q in (2.4.26); the proof is also left as an exercise (see Problem 5.3.3).

In the same fashion as in Chapter 4, we consider here invariance principles for contiguous alternatives. We assume that F possesses (almost everywhere) an absolutely continuous density function f with a finite Fisher information $I(f)$, defined by (4.3.37), and the score function ϕ satisfies (5.2.12) and (5.2.13). Further, we consider the sequence of alternative hypotheses $\{K_N\}$ where

$$K_N : F(x) \qquad \text{is symmetric about} \quad N^{-1/2}\theta. \tag{5.3.8}$$

Also, we define $\psi(u)$, $0 < u < 1$ as in before (4.3.40) and φ^* as in (5.2.12). Then, in (4.3.40), we replace ϕ by ϕ^* and denote the corresponding quantity by ρ_1^*.

THEOREM 5.3.3. *For F possessing an absolutely continuous density function with a finite Fisher information and for square integrable ϕ^*, under $\{K_N\}$,*

$$Y_N \xrightarrow{\mathfrak{D}} \left\{ W(t) + t\theta\rho_1^*\left[I(f)\right]^{1/2}, t \in I \right\}. \tag{5.3.9}$$

Proof. Let us denote the joint d.f. of $(X_1, \ldots, X_N)$ under K_N by $P_N^{(\theta)}$, so that $P_N^{(0)}$ relates to the null case. Now, as in the proof of Theorem 4.3.6, under the hypothesis of this theorem, $\{P_N^{(\theta)}\}$ is contiguous to $\{P_N^{(0)}\}$ (see Problem 5.3.4). Further, the tightness of $\{Y_N\}$ under $\{P_N^{(0)}\}$ has been established in Theorem 5.3.1. Hence, by an appeal to Theorem 4.3.4, we conclude that $\{Y_N\}$ is tight under $\{K_N\}$ as well. Hence it suffices to establish only the convergence of the f.d.d.'s in (5.3.9). By (5.3.4), the Chebyshev inequality and the contiguity of $\{P_N^{(\theta)}\}$ to $\{P_N^{(0)}\}$, we obtain that for every fixed $m(\geqslant 1)$ and $t_1, \ldots, t_m (\in I)$, as $N \to \infty$,

$$\max_{1 \leqslant j \leqslant m} \left\{ N^{-1/2} \left| S_{[Nt_j]}^0 - S_{[Nt_j]}^* \right| \right\} \xrightarrow{p} 0, \qquad \text{under} \quad \{K_N\}. \tag{5.3.10}$$

Also, under K_N, S_N^* involves a sum of a triangular array of row-wise i.i.d.r.v.'s, so that by the (multivariate) central limit theorem, $N^{-1/2}(S_{[Nt_1]}^*, \ldots, S_{[Nt_m]}^*)$ is asymptotically normally distributed with mean vector

$$\theta A\left[I(f)\right]^{1/2}\rho_1^*(t_1, \ldots, t_m) \tag{5.3.11}$$

and dispersion matrix $A^2((t_r \wedge t_s))_{r,s=1,\ldots,m}$. Thus, the convergence of f.d.d.'s of $\{Y_N\}$ to those of $\{W(t) + t\theta[I(f)]^{1/2}\rho_1^*, t \in I\}$ follows from (5.3.10) and (5.3.11), and the proof of the theorem is complete.

For the time being let us assume that when $\{K_N\}$ holds,

$$N^{1/2}S_n/n \to A\left[I(f)\right]^{1/2}\rho_1^*\theta \qquad \text{a.s.,} \quad \text{as} \quad n/N \to \infty, \quad N \to \infty; \tag{5.3.12}$$

later on in Section 5.7, we shall see that (5.3.12) holds under fairly general regularity conditions. Finally, let $\mathbf{1} = \{1; t \in I\}$.

THEOREM 5.3.4. *Under* (5.3.12) *and the assumptions of Theorem* 5.3.3,

$$Z_N - AI^{1/2}(f)\rho_1^*\theta\mathbf{1} \xrightarrow{\mathfrak{D}} W. \tag{5.3.13}$$

The proof is similar to that of Theorem 4.3.7 and hence is left as an exercise (see Problem 5.3.5).

5.4. EMBEDDING OF WIENER PROCESSES

Under H_0 in (5.2.2), the Skorokhod-Strassen embedding of Wiener processes for signed rank statistics holds under very general conditions and will be considered first. Subsequently, the general case of arbitrary F (not necessarily symmetric) will be considered and similar embedding of Wiener processes will be established under comparatively more stringent regularity conditions.

Define $\{S_N^0, \mathfrak{B}_N; N \geqslant 1\}$ as in Section 5.2 and let

$$V_N = \sum_{k=1}^{N} \mathrm{Var}\left[S_k^0 \mid \mathfrak{B}_{k-1}\right], \qquad N \geqslant 1, \qquad V_0 = 0; \tag{5.4.1}$$

$$\tilde{S}_{V_N}^0 = S_N^0 \quad \text{for } N \geqslant 1, \qquad \tilde{S}_{V_0}^0 = S_0^0 = 0, \tag{5.4.2}$$

and finally, we complete the definition of $\tilde{S}^0 = \{\tilde{S}_t^0, t \geqslant 0\}$ by linear interpolation between (V_N, S_N^0) and (V_{N+1}, S_{N+1}^0), $V_N \leqslant t \leqslant V_{N+1}$, $N \geqslant 0$. The following theorem is due to Sen and Ghosh (1973b).

THEOREM 5.4.1. *Suppose that $\phi = \phi_1 - \phi_2$ where the ϕ_j are nondecreasing and $\in L_r$ for some $r > 2$. Then, under H_0 in* (5.2.2), *there exists a standard Brownian motion $W = \{W(t), t \geqslant 0\}$ such that for $s = (2+r)/4r(<\frac{1}{2})$,*

$$S_t^0 = W(t) + o\left(\left[t^s(\log t)\right]\right) \quad \text{a.s., as } t \to \infty. \tag{5.4.3}$$

Proof. Let use denote by $Z_k = S_k^0 - S_{k-1}^0$, $k \geqslant 1$, so that by Theorem 5.2.1, under H_0, $E_0(Z_k \mid \mathfrak{B}_{k-1}) = 0$, $\forall k \geqslant 1$, $E_0(Z_1^2) = [a_1^0(1)]^2$ and for $k \geqslant 2$,

$$\begin{aligned} E_0\left(Z_k^2 \mid \mathfrak{B}_{k-1}\right) &= E_0\left(\left[a_k^0(R_{kk}^+)\right]^2 \mid \mathfrak{B}_{k-1}\right) \\ &+ E_0\left[\left[\sum_{j=1}^{k-1} \mathrm{sgn}\, X_j \left\{a_k^0(R_{kj}^+) - a_{k-1}^0(R_{k-1j}^+)\right\}\right]^2 \mid \mathfrak{B}_{k-1}\right] \\ &+ 2E_0\left(\mathrm{sgn}\, X_k a_k^0(R_{kk}^+)\left[\sum_{j=1}^{k-1} \mathrm{sgn}\, X_j \left\{a_k^0(R_{kj}^+) - a_{k-1}^0(R_{k-1j}^+)\right\}\right] \mid \mathfrak{B}_{k-1}\right). \end{aligned} \tag{5.4.4}$$

Since given $\mathcal{B}_{k-1}$, $\operatorname{sgn} X_k$ is independent of R_{kj}^+, $1 \leqslant j \leqslant k$, the third term on the rhs of (5.4.4) is equal to 0. Also, the first term on the rhs of (5.4.4) is equal to

$$k^{-1}\sum_{i=1}^{k}\left[a_k^0(i)\right]^2 = A_k^{02}(\rightarrow A^2 \quad \text{as} \quad k\rightarrow\infty). \tag{5.4.5}$$

Thus

$$V_N \geqslant \sum_{k=1}^{N} E_0\Big(\big[a_k^0(R_{kk}^+)\big]^2 \mid \mathcal{B}_{k-1}\Big) = \sum_{k=1}^{N} A_k^{02} \rightarrow \infty \quad \text{as} \quad N\rightarrow\infty. \tag{5.4.6}$$

On the other hand, by steps similar to those in the proof of Lemma 4.4.2, it follows that as $N\rightarrow\infty$,

$$\left(NA_N^{02}\right)^{-1}\sum_{k=1}^{N} E_0\left[\left[\sum_{j=1}^{k-1}\operatorname{sgn} X_j\left\{a_k^0(R_{kj}^+)\right.\right.\right.$$

$$\left.\left.\left. - a_{k-1}^0(R_{k-1j}^+)\right\}\right]^2 \mid \mathcal{B}_{k-1}\right] \rightarrow 0 \qquad \text{a.s.,} \tag{5.4.7}$$

the proof of (5.4.7) is left as an exercise (see Problem 5.4.1). Thus

$$V_N/NA^2 \rightarrow 1 \qquad \text{a.s.,} \quad \text{as} \quad N\rightarrow\infty. \tag{5.4.8}$$

Note that if $X_1^*, \ldots, X_N^*$ are i.i.d.r.v. with a d.f. G such that $\int|x|^r\,dG(x) < \infty$ for some $r \geqslant 1$, then $(1/N)\max_{1\leqslant i\leqslant N}|X_i^*|^r \rightarrow 0$ in L_1-norm (see Problem 5.4.2). As such, if ϕ is nondecreasing and $\in L_r$, then

$$\begin{aligned}\max_{1\leqslant i\leqslant N} N^{-1}|a_N^0(i)|^r &\leqslant \max_{1\leqslant i\leqslant N} E\left\{N^{-1}|\phi(U_{Ni})|^r\right\}\\ &\leqslant E\left\{\max_{1\leqslant i\leqslant N} N^{-1}|\phi(U_i)|^r\right\} \rightarrow 0.\end{aligned} \tag{5.4.9}$$

The same result holds if $\phi = \phi_1 - \phi_2$ with $\phi_j \uparrow$ and $\in L_r$. Further, by definition, for nondecreasing ϕ,

$$\begin{aligned}|Z_k| &\leqslant |a_k^0(R_{kk}^+)| + \left|\sum_{i=1}^{k-1}\operatorname{sgn} X_i\left\{a_k^0(R_{ki}^+) - a_{k-1}^0(R_{k-1i}^+)\right\}\right|\\ &\leqslant \max_{1\leqslant i\leqslant k}|a_k^0(i)| + \sum_{i=1}^{k-1}|a_k^0(R_{k-1i}^+ + 1) - a_k^0(R_{k-1i}^+)|\\ &\leqslant 3\max_{1\leqslant i\leqslant k}|a_k^0(i)| = o(k^{1/r}), \qquad \text{by} \quad (5.4.9),\end{aligned} \tag{5.4.10}$$

and the same conclusion holds when $\phi = \phi_1 - \phi_2$ with $\phi_j \uparrow$ and $\in L_r$. Thus, from (5.4.8) and (5.4.10), we conclude that

$$P\{Z_N^2 > V_N^{2/r} \mid \mathfrak{B}_{N-1}\} = 0 \quad \text{a.s.,} \quad \text{for} \quad N \geqslant N_0. \tag{5.4.11}$$

Thus, by (5.4.8) and (5.4.11), we conclude that both (2.5.9) and (2.5.10) hold in our case with $f(t) = t^{2/r}$ and hence, the proof of the theorem follows. Q.E.D.

The law of iterated logarithm for $\{S_N^0\}$ follows readily from (5.4.3) and (5.4.8). However, for this special purpose, we may replace the condition that $\phi \in L_r$, $r > 2$ by $\phi \in L_2$; we refer to Problems 5.4.3 and 5.4.4 in this context.

So far, we considered the case of $a_N^0(i) = E\phi(U_{Ni})$, $1 \leqslant i \leqslant N$. The case of $a_N(i) = \phi(i/(N+1))$, $1 \leqslant i \leqslant N$ follows by using (5.3.5), so that $(S_N - S_N^0) = o(N^{1/2})$ a.s., and hence (5.4.3) holds with the second term on the rhs being replaced by $o(t^{1/2})$.

We may remark that even if H_0 in (5.2.2) is not true, when the X_i are i.i.d.r.v. with a d.f. F, $N^{-1}S_N$ (or $N^{-1}S_N^0$) converges a.s. to a limit, say, $\mu(F, \phi)$, under fairly general conditions (see Section 5.7). We like to show now that under more stringent regularity conditions on ϕ, we have a similar embedding theorem for $\{S_N - N\mu(F,\phi), N \geqslant 1\}$ where

$$\begin{aligned} \mu(F,\phi) &= \int_0^\infty \phi(H(x))dF(x) - \int_{-\infty}^0 \phi(H(-x))dF(x); \\ H(x) &= F(x) - F(-x), \qquad x \geqslant 0. \end{aligned} \tag{5.4.12}$$

Toward this goal, we consider first an a.s. representation of S_N in terms of a sum of independent r.v.'s. For convenience of manipulations, we work here with the equivalent form of S_N, namely $T_N = \frac{1}{2}[S_N + \sum_{i=1}^N a_N(i)]$, so that

$$T_N = \sum_{i=1}^N c(X_i) a_N(R_{Ni}^+) = N\int_0^\infty a_N(NH_N(x))\,dF_N(x), \tag{5.4.13}$$

where F_N is the empirical d.f. and $H_N(x) = F_N(x) - F_N(-x), x \geqslant 0$. Concerning the score function ϕ^*, we assume that ϕ^* is twice differentiable inside I and there exist positive constants $K(< \infty)$ and $\delta(0 < \delta \leqslant \frac{1}{2})$, such that

$$|(d^r/du^r)\phi^*(u)| \leqslant K[u(1-u)]^{-(1/2)-r+\delta}, \qquad u \in I, \qquad r = 0,1,2. \tag{5.4.14}$$

Finally, we assume that F is absolutely continuous with a density function $f(x)$ where

$$\sup_x f(x) = f_0 < \infty. \tag{5.4.15}$$

THEOREM 5.4.2. *Under* (5.4.14) *and* (5.4.15),

$$T_N = \sum_{i=1}^{N} B(X_i) + R_N, \tag{5.4.16}$$

where

$$B(X_i) = c(X_i)\phi(H(|X_i|)) + \int_0^\infty \left[c(x - |X_i|) - H(x)\right]\phi'(H(x))\,dF(x), \qquad i \geqslant 1 \tag{5.4.17}$$

and for some $\eta > 0$,

$$N^{-1/2}R_N = 0(N^{-\eta}) \qquad \text{a.s.,} \quad \textit{as} \quad N \to \infty. \tag{5.4.18}$$

Proof. We know (see Exercise 4.3.5) that under (5.5.14),

$$\left|\sum_{i=1}^{N} c(X_i)\left[\phi(R_{Ni}^{+}/(N+1)) - E\phi(U_{NR_{Ni}^{+}})\right]\right| \leqslant \sum_{i=1}^{N} |\phi(i/(N+1)) - E\phi(U_{Ni})| = O(N^{(1/2)-\eta}), \qquad \eta > 0. \tag{5.4.19}$$

Hence it suffices to work with $a_N(i) = \phi(i/(N+1))$, $1 \leqslant i \leqslant N$. Let us define a_N by $H(a_N) = 1 - N^{-1+\delta}$, $\delta(>0)$ being defined by (5.4.14). Then, using the integral representation in (5.4.13), we obtain by expansion that

$$R_N = \sum_{j=1}^{5} R_N^{(j)}; \tag{5.4.20}$$

$$N^{-1}R_N^{(1)} = \int_{a_N}^{\infty} \left[\phi\left(N(N+1)^{-1}H_N\right) - \phi(H)\right] dF_N, \tag{5.4.21}$$

$$N^{-1}R_N^{(2)} = \int_0^{a_N} \left[\phi\left(N(N+1)^{-1}H_N\right) - \phi(H) - \left(\frac{N}{N+1}H_N - H\right)\phi'(H)\right] dF_N, \tag{5.4.22}$$

$$N^{-1}R_N^{(3)} = -(N+1)^{-1}\int_0^{a_N} H_N\phi'(H)\,dF_N, \tag{5.4.23}$$

$$N^{-1}R_N^{(4)} = \int_0^{a_N} [H_N - H]\phi'(H)\,d(F_N - F), \tag{5.4.24}$$

$$N^{-1}R_N^{(5)} = -\int_{a_N}^{\infty} [H_N - H]\phi'(H)\,dF, \tag{5.4.25}$$

and H, H_N, F_N, and F stand for $H(x)$, $H_N(x)$, $F_N(x)$, and $F(x)$, respectively. Note that

$$|N^{-1}R_N^{(1)}| \leqslant \left|\int_{a_N}^{\infty}\phi\big(N(N+1)^{-1}H_N\big)\,dF_N\right| + \left|\int_{a_N}^{\infty}\phi(H)\,dF_N\right|$$

$$= \left|\int_{a_N}^{\infty}\phi\big(N(N+1)^{-1}H_N\big)\,dF_N\right| + \left|\int_{a_N}^{\infty}[1-F_N]\phi'(H)\,dH + \big[1-F_N(a_N)\big]\phi(H(a_N))\right|$$

$$\leqslant \int_{a_N}^{\infty}\left|\phi\Big(\frac{N}{N+1}H_N\Big)\right|dH_N + \big[1-H_N(a_N)\big]|\phi(H(a_N))| + \int_{a_N}^{\infty}|1-H_N|\,|\phi'(H)|\,dH. \tag{5.4.26}$$

We assume that N is so large that $H(a_N) > \frac{1}{2}$. Using them (2.6.28) and (2.6.31), it follows that

$$|H_N(a_N) - H(a_N)| = O(N^{-1+\delta/2}\log N) \quad \text{a.s., as } N\to\infty. \tag{5.4.27}$$

As such, on denoting by $N^* = N[1-H_N(a_N)]$, we obtain from (5.4.14) and (5.4.27) that the first term on the rhs of (5.4.26) is a.s. bounded by

$$K\sum_{i\leqslant N^*}\big[i(N+1-i)/N^2\big]^{-(1/2)+\delta} = O\big((N^*/N)^{(1/2)+\delta}\log N\big)$$
$$= O(N^{-1/2-\delta(1-2\delta)/2}\log N) \quad \text{a.s.}$$
$$= O(N^{-(1/2)-\eta}) \quad \text{a.s., where } 0<\eta<\tfrac{1}{2}\delta(1-\delta). \tag{5.4.28}$$

A similar treatment holds for the second term on the rhs of (5.4.26). For the third term we use (2.6.31) so that $1-H_N(x) \leqslant [1-H(x)] + O(N^{-(1/2)}\log N)[H(x)\{1-H(x)\}]^{(1/2)-\epsilon})$ a.s., for every $x(\geqslant a_N)$ and hence, by (5.4.14) and (5.4.27), with probability 1, as $N\to\infty$,

$$\int_{a_N}^{\infty}[1-H_N]|\phi'(H)|dH \leqslant \int_{a_N}^{\infty}[1-H]|\phi'(H)|dH + O(N^{-1/2}\log N) \times \int_{a_N}^{\infty}\big[H(1-H)\big]^{(1/2)-\epsilon}|\phi'(H)|dH$$
$$\leqslant K\int_{a_N}^{\infty}H^{-(3/2)+\delta}(1-H)^{-(1/2)+\delta}\,dH + O(N^{-1/2}\log N)K\int_{a_N}^{\infty}\big[H(1-H)\big]^{-(1+\delta)-\epsilon}\,dH$$
$$= O\big(\big[1-H(a_N)\big]^{(1/2)+\delta}\big) + \big[O(N^{-1/2}\log N)\big]\big[O\big[1-H(a_N)\big]^{\delta-\epsilon}\big]$$
$$= O(N^{-(1/2)-\eta}) \quad \text{a.s., where we let } 0<\epsilon<\delta/2. \tag{5.4.29}$$

Hence, from (5.4.26) through (5.4.29), we obtain

$$N^{-1/2}|R_N^{(1)}| = O(N^{-\eta}) \qquad \text{a.s.,} \quad \text{as} \quad N \to \infty, \tag{5.4.30}$$

Next, using the fact that $dF_N \leqslant dH_N$ and the mean value theorem, we obtain

$$N^{-1}|R_N^{(2)}| \leqslant \frac{1}{2}\int_0^{a_N}\left[\frac{N}{N+1}H_N - H\right]^2 |\phi''(\tilde{H}_N)|\,dH_N \tag{5.4.31}$$

where $\tilde{H}_N = \theta N H_N/(N+1) + (1-\theta)H$, $0<\theta<1$. Since $H(a_N) = 1 - N^{-1+\delta}, \delta > 0$, and, in (2.6.31), we choose $0 < \epsilon < \delta/2$, we have for all $x \leqslant a_N$,

$$\begin{aligned}
1 - \tilde{H}_N &= (1-H) + (H - \tilde{H}_N)\\
&= (1-H) + O(N^{-1/2}\log N)\left[H(1-H)\right]^{(1/2)-\epsilon}\\
&= (1-H)\left\{1 + \left[O(N^{-1/2}\log N)\right]H^{(1/2)-\epsilon}(1-H)^{-(1/2)-\epsilon}\right\}\\
&= (1-H)\left\{1 + \left[O(N^{-1/2}\log N)\right]\left[O(N^{((1/2)+\epsilon)(1-\delta)})\right]\right\}\\
&= (1-H)\left\{1 + O(N^{-\delta/2+\epsilon-\delta\epsilon}\log N)\right\}\\
&= (1-H)\left[1 + o(1)\right] \qquad \text{a.s.;}
\end{aligned} \tag{5.4.32}$$

the same inequality holds for $1 - H_N$ as well. Hence, by (5.4.14) and (5.4.32), the rhs of (5.4.31) is a.s. bounded by

$$\begin{aligned}
&\frac{1}{2}\int_0^{a_N}\left[0\left(N^{-1}(\log N)^2\right)\right]\left[H(1-H)\right]^{1-2\epsilon}K\left[\tilde{H}_N(1-\tilde{H}_N)\right]^{-(5/2)+\delta}dH_N\\
&\qquad \leqslant O\left(N^{-1}(\log N)^2\right)\int_0^{a_N}\left[H_N(1-H_N)\right]^{-(3/2)+\delta}dH_N\\
&\qquad\qquad\qquad\qquad \text{a.s.} \quad [\text{by } (5.4.32)]\\
&\quad = \left[O\left(N^{-1}(\log N)^2\right)\right]\left[O\left([1-H_N]^{(1/2)+\delta}\right)\right]\\
&\quad = \left[O\left(N^{-1}(\log N)^2\right)\right]\left[O\left((N^*/N)^{-(1/2)+\delta}\right)\right]\\
&\quad = \left[O\left(N^{-1}(\log N)^2\right)\right]\left[O(N^{(1/2-\delta+2\epsilon)(1-\delta)})\right] \qquad \text{a.s.} \quad [\text{by } (5.4.27)]\\
&\quad = O(N^{-(1/2)-\eta}) \qquad \text{a.s., where} \quad \eta > 0.
\end{aligned} \tag{5.4.33}$$

Thus $N^{-1/2}|R_N^{(2)}| = O(N^{-\eta})$ a.s., as $N \to \infty$. Again, by (5.4.14) and (5.4.23), we obtain on using the fact that $dF_N \leqslant dH_N$ and (5.4.27)

$$\begin{aligned}
N^{-1}|R_N^{(3)}| &\leqslant (N+1)^{-1}\int_0^{a_N}\left[1-H\right]^{-(3/2)+\delta}dH_N\\
&\leqslant (N+1)^{-1}\int_0^{a_N}\left[1-H_N\right]^{-(3/2)+\delta}(1+o(1))\,dH_N \qquad [\text{by } (5.4.32)]\\
&= (N+1)^{-1}\left[0\left(\left[1-H_N(a_N)\right]^{-(1/2)+\delta}\right)\right]\\
&= O(N^{-(1/2)-\eta}) \qquad \text{a.s.,} \quad [\text{by } (5.4.27)].
\end{aligned} \tag{5.4.34}$$

Similarly, on using (5.4.14) and (2.6.31),

$$
\begin{aligned}
N^{-1}|R_N^{(5)}| &\leqslant O(N^{-1/2}\log N) \\
&\quad \times \int_{a_N}^{\infty}[H(1-H)]^{(1/2)-\epsilon}K[H(1-H)]^{-(3/2)+\delta}\,dH \\
&= O(N^{-1/2}\log N)\int_{a_N}^{\infty}[H(1-H)]^{-1+\delta-\epsilon}\,dH \\
&= O(N^{-1/2}\log N)O([1-H(a_N)]^{\delta-\epsilon}) \\
&= [O(N^{-1/2}\log N)][O(N^{-(1-\delta)(\delta-\epsilon)})] \\
&= O(N^{-(1/2)-\eta}) \quad \text{a.s.}
\end{aligned}
\tag{5.4.35}
$$

Let us finally consider $R_N^{(4)}$. Let us introduce a set $\{t_{Nj}, j=1,\ldots,N_0\}$ points on $[0, a_N]$ where

$$
t_{Nj} \text{ is defined by } H(t_{Nj}) = jN^{-1/2}, j=0,1,\ldots,N_0; \tag{5.4.36}
$$

so that $N_0N^{-1/2} \geqslant H(a_N) - H(0) > (N_0-1)N^{-1/2}$, that is, $N_0 = O(N^{1/2})$. Let I_{Nj} $\{x : t_{Nj-1} \leqslant x \leqslant t_{Nj}\}$, $1 \leqslant j \leqslant N_0$. Then

$$
N^{-1}R_N^{(4)} = \sum_{j=1}^{N_0}\int_{I_{Nj}}(H_N-H)\phi'(H)\,d(F_N-F). \tag{5.4.37}
$$

Let $B_n = \{(u,v) : 0 \leqslant u \leqslant v \leqslant 1 \text{ with } (v-u) \leqslant n^{-1/2}\log n\}$. Then, we have the following result due to Bahadur (1966) and Kiefer (1967). The proof of this result follows from the proof of Theorem 7.3.2 and hence is omitted. We define the empirical d.f.'s G_n and the empirical process U_n as in (2.6.7) and (2.6.8).

$$
\sup_{(u,v)\in B_n}|U_n(u)-U_n(v)| = O(n^{-1/4}\log n) \quad \text{a.s., as } n\to\infty. \tag{5.4.38}
$$

By virtue of (5.4.15), (5.4.38), and the definitions of H_N and H, for every $x \in I_{Nj}$,

$$
H_N(x) - H(x) = H_N(t_{Nj}) - H(t_{Nj}) + O(N^{-3/4}\log N) \quad \text{a.s.,} \tag{5.4.39}
$$

$$
F_N(x) - F(x) = F_N(t_{Nj}) - F(t_{Nj}) + O(N^{-3/4}\log N) \quad \text{a.s.} \tag{5.4.40}
$$

and by (5.4.14) and (5.4.32),

$$
\phi'(H(x)) = \phi'(H(t_{Nj})) + [O(N^{-1/2})][O([1-H(t_{Nj})]^{-(5/2)+\delta})]. \tag{5.4.41}
$$

Consequently,

$$\left|\int_{I_{Nj}}(H_N - H)\phi'(H)d(F_N - F)\right|$$
$$\leqslant |H_N(t_{Nj}) - H(t_{Nj})||\phi'(H(t_{Nj}))|$$
$$\times |F_N(t_{Nj}) - F_N(t_{Nj-1}) - F(t_{Nj}) + F(t_{Nj-1})|$$
$$+ O(N^{-3/4}\log N)|\phi'(H(t_{Nj}))|\int_{I_{Nj}}|d[F_N - F]|$$
$$+ O(N^{-1/2})\int_{I_{Nj}}|H_N(t_{Nj}) - H(t_{Nj})|[1 - H(t_{Nj})]^{-(5/2)+\delta}|d[F_N - F]|$$
$$+ O(N^{-5/4}\log N)\int_{I_{Nj}}[1 - H(t_{Nj})]^{-(5/2)+\delta}|d[F_N - F]|, \qquad (5.4.42)$$

where, by (2.6.29), (5.4.14), and (5.4.40), the first term on the rhs of (5.4.42) is $O(N^{-5/4}(\log N)^2)([1 - H(t_{Nj})]^{-1+\delta-\epsilon})$ a.s. and noting that

$$\int_{I_{Nj}}|d(F_N - F)| \leqslant F_N(t_{Nj}) - F_N(t_{Nj-1}) + F(t_{Nj}) - F(t_{Nj-1}) = O(N^{-1/2})$$

$+ O(N^{-3/4}\log N)$ a.s. as $N \to \infty$ [by (5.4.40)], the second term is also

$$O\left(N^{-5/4}(log\, N)^2\right)\left(\left[1 - H(t_{Nj})\right]^{-1+\delta-\epsilon}\right) \qquad \text{a.s., as } N \to \infty.$$

Similarly, the third and the fourth terms are a.s. (as $N \to \infty$) $[O(N^{-3/2} \cdot \log N)]$, $[1 - H(t_{Nj})]^{-2+\delta}$, and $[O(N^{-7/4}\log N)][1 - H(t_{Nj})]^{-5/2+\delta}$, respectively. Finally, for $r > 1$,

$$\sum_{j=1}^{N_0}[1 - H(t_{Nj})]^{-r} = \sum_{j=1}^{N_0}(1 - jN^{-1/2})^{-r}$$
$$= N^{r/2}\sum_{j=1}^{N_0}(N^{1/2} - j)^{-r} = O(N^{r/2})$$

and for $0 < r < 1$,

$$\sum_{j=1}^{N_0}[1 - H(t_{Nj})]^{-r} = \sqrt{N}\left\{(1/\sqrt{N})\sum_{j=1}^{N_0}[1 - H(t_{Nj})]^{-r}\right\}$$
$$\leqslant \sqrt{N}\int_0^{t_{NN_0}} O[1 - H(x)]^{-r}dH(x)$$
$$= \sqrt{N}\left(O\left(\left[1 - H(t_{NN_0})\right]^{-r+1}\right)\right)$$
$$= O(N^{(r-(1/2))(1-r)(\delta-\epsilon)}).$$

Hence, using the bounds for the four terms in (5.4.42) and summing over $j(=1,\ldots,N_0$ where $N_0 = O(\sqrt{N}))$, we obtain

$$\begin{aligned}|N^{-1}R_N^{(4)}| &\leqslant \left[O\left(N^{-5/4}(\log N)^2\right)\right]\left[O(N^{1/2-(\delta-\epsilon)(1-\delta)}\right]\\ &\quad + \left[O(N^{-3/2}\log N)\right]\left[O(N^{1-\delta/2})\right]\\ &\quad + \left[O(N^{-7/4}\log N)\right]\left[O(N^{5/4-\delta/2})\right] \quad \text{a.s.}\\ &= O\left(N^{-3/4-(\delta-\epsilon)(1-\delta)}(\log N)^2\right) + O\left(N^{-1/2-\delta/2}\log N\right) \quad \text{a.s.}\\ &= O(N^{-1/2-\eta}) \quad \text{a.s.} \quad \text{for some} \quad \eta > 0. \end{aligned} \tag{5.4.43}$$

Thus $N^{-1/2}|R_N| = O(N^{-\eta})$ a.s., as $N \to \infty$. Q.E.D.

In the preceding proof, (2.6.31) plays a vital role. Though an a.s. convergence result is contained in (2.6.31), in many cases a stronger result is required. The following result, due to Ghosh (1972), is useful in this direction. [See (2.6.30).]

For every $k > 1$, there exist a positive constant C and an n_0, such that

$$P\left\{\sup_{x\in R} n^{1/2}|F_n(x) - F(x)|/\{F(x)[1-F(x)]\}^{1/2-\epsilon} > C\log n\right\} \leqslant n^{-k}, \qquad \forall n \geqslant n_0, \tag{5.4.44}$$

where

$$\left(\tfrac{1}{2} >\right)\epsilon = k/(2k+2)\left(> \tfrac{1}{4}\right). \tag{5.4.45}$$

We refer to Ghosh (1972) for a simple proof of (5.4.44).

Now, in the proof of Theorem 5.4.2, we replace the use of (2.6.31) by (5.4.44) and thereby arrive at the following.

THEOREM 5.4.3 (Müller-Funk 1979). *For every $k > 1$, there exist positive numbers C, η, $\epsilon = k/(2k+2)$ and N_0, such that if* (5.4.14) *holds with $\delta = \epsilon$, then under* (5.4.15), *for the decomposition in* (5.4.16),

$$P\{N^{-1/2}|R_N| > CN^{-\eta}\} \leqslant N^{-k}, \qquad \forall N \geqslant N_0, \tag{5.4.46}$$

uniformly in the underlying d.f. F.

Let us now denote by

$$T_N^* = \sum_{i=1}^{N} B(X_i) \qquad \text{and} \qquad \mu_N^* = N\mu^* = N\int_0^\infty \phi(H(x))\,dF(x) \tag{5.4.47}$$

where the $B(X_i)$ are defined by (5.4.17), and let

$$\sigma^2 = \text{Var}\left[B(X_1)\right] \qquad \text{and assume that} \quad \sigma > 0. \tag{5.4.48}$$

Note that under (5.4.14), by a direct use of the Fubini theorem, we have

$$E|B(X_1)|^{2+\delta} < \infty (\Rightarrow \sigma < \infty). \tag{5.4.49}$$

On the other hand, T_N^* involves a sum of i.i.d.r.v.'s on which Theorem 2.5.1 applies. Hence we arrive at the following.

THEOREM 5.4.4. *Under the assumptions of Theorem* 5.4.2 (*or* 5.4.3) *and* (5.4.48),

$$\sigma^{-1}(T_N^* - \mu_N^*) = W(N) + o(N^s \log N) \qquad a.s., \quad as \quad N \to \infty, \tag{5.4.40}$$

where $s(<\frac{1}{2})$ *is a positive number.*

By virtue of (5.4.50), the law of iterated logarithm holds for the T_N^* and hence for the T_N (or S_N).

5.5. ASYMPTOTIC LINEARITY OF S_N IN SHIFT PARAMETER

We study here the monotonicity and asymptotic linearity (in the shift parameter) of aligned forms of S_N, where in (5.2.1), we replace X_i by $X_i - b$ and denote the corresponding ranks (of the absolute values) by $R_{Ni}^+(b), i = 1, \ldots, N$ and the resulting statistic by $S_N(b)$. Thus, for every real b,

$$S_N(b) = S_N(X_1 - b, \ldots, X_N - b) = \sum_{i=1}^{N} \text{sgn}(X_i - b) a_N(R_{Ni}^+(b)). \tag{5.5.1}$$

THEOREM 5.5.1. *If* $a_N(i)$ *is* $\nearrow$ *in* $i: 1 \leqslant i \leqslant N$, $S_N(b)$ *is* $\searrow$ *in* $b: -\infty < b < \infty$.

Proof. Consider any particular $b_0(-\infty < b_0 < \infty)$ and set $\sum_j = \{X_i - b_0 : \text{sgn}(X_i - b_0) = (-1)^j, 1 \leqslant i \leqslant N\}, j = 1, 2$. If, then, we let $b = b_0 + d$, $d > 0$, we note that (i) $\text{sgn}(X_i - b_0 - d)$ is $\searrow$ in d, (ii) for every $i \in \sum_1$, $|X_i - b_0 - d|$ is $\uparrow$ in d (but their relative ranks are invariant while for $d > X_i - b_0$, the element has a negative sign. As such, $\text{sgn}(X_i - b_0 - d)$ $a_N(R_{Ni}^+(b_0 + d))$ is $\searrow$ in d for every $i(= 1, \ldots, N)$, and the proof follows from (5.5.1). Q.E.D.

Remark. Constance van Eeden (1972) has established a more general result where she allows the shift bc_i, $1 \leqslant i \leqslant N$, with suitable regularity conditions on the c_i; see also Jurečková (1971a) in this context. We pose these as exercises (see Problems 5.5.1 and 5.5.2).

We now proceed to consider the asymptotic linearity of $S_N(b)$ and for this purpose, we define A^2, $I(f)$ and ρ_1^* as in Section 5.3.

THEOREM 5.5.2. *If this X_i are i.i.d.r.v. with an absolutely continuous d.f. F, symmetric about* 0 *and having a finite Fisher information $I(f)$, and if ϕ is nondecreasing and square integrable inside I, then for every positive $C(<\infty)$, as $N\to\infty$,*

$$\sup_{b\,:\,|b|\leqslant C} |N^{-1/2}\{S_N(N^{-1/2}b) - S_N(0)\} + bAI^{1/2}(f)\rho_1^*| \xrightarrow{p} 0. \quad (5.5.2)$$

The proof again runs on parallel lines to that of Theorem 4.5.2 and hence is left as an exercise (Problem 5.5.3). In this context, we also refer to van Eeden (1972) and Jurečková (1971a) for some related results.

Suppose now in (5.3.10), we replace $S_{N(t)}^0$ by $S_{N(t)}(N^{-1/2}b)$, $t\in I$ and denote the resulting process by $\{Y_{N,b}(t), t\in I\} = Y_{N,b}$. Then as an extension of Theorem 5.5.2, we have the following.

THEOREM 5.5.3. *Under the conditions of Theorem* 5.3.2, *as $N\to\infty$,*

$$\sup\{|Y_{N,b}(t) - Y_{N,0}(t) + btI^{1/2}(f)\rho_1^*| : t\in I, |b|\leqslant C\} \xrightarrow{p} 0. \quad (5.5.3)$$

In the same way as Theorem 4.5.3 follows from Theorem 4.5.2 and the inequalities (4.5.17) and (4.5.18), the proof of (5.5.3) follows from (5.5.2) and similar inequalities; we leave the details as an exercise (see Problem 5.5.3).

Now, by virtue of Theorem 5.4.2, we are in a position to strengthen (5.5.2) under additional regularity conditions.

THEOREM 5.5.4. *Under* (5.4.14), (5.4.15), *and the hypothesis of Theorem* 5.5.2,

$$\sup_{b\,:\,|b|\leqslant c} \{N^{-1/2}|S_N(N^{-1/2}b) - S_N(0) + N^{1/2}bAI^{1/2}(f)\rho_1^*|\} \to 0$$

$$a.s., \quad as \quad N\to\infty. \quad (5.5.4)$$

Proof. Using the monotonicity property in Theorem 5.5.1 and proceeding as in (4.5.8) through (4.5.10), it follows that for proving (5.5.4), it suffices to show that for every $b \in [-c, c]$,

$$N^{-1/2}\{S_N(N^{-1/2}b) - S_N(0)\} + bA[I(f)]^{1/2}\rho_1^* \to 0 \quad \text{a.s.,} \quad \text{as} \quad N \to \infty. \tag{5.5.5}$$

Let us now define

$$\begin{aligned} H(x) &= H_0(x) = 2F(x) - 1, \qquad x \geq 0, F_b(x) = F(x + bN^{-1/2}),\\ H_b(x) &= F_b(x) - F_b(-x), \qquad x \geq 0,\\ H_N(x) &= F_N(x) - F_N(-x),\\ F_{N,b}(x) &= F_N(x + bN^{-1/2}) \end{aligned}$$

and

$$H_{N,b}(x) = F_{N,b}(x) - F_{N,b}(-x), x \geq 0.$$

Then, by (5.4.13) through (5.4.19), we obtain

$$\begin{aligned} &N^{-1/2}[S_N(N^{-1/2}b) - S_N(0)] + bAI^{1/2}(f)\rho_1^*\\ &= \left[2\sqrt{N}\left\{\int_0^\infty \phi(H_b(x))\,dF_{N,b}(x)\right.\right.\\ &\qquad \left.\left. - \int_0^\infty \phi(H(x))\,dF_N(x)\right\} + bAI^{1/2}(f)\rho_1^*\right]\\ &\quad + 2\sqrt{N}\left\{\int_0^\infty [H_{N,b}(x) - H_b(x)]\phi'(H_b(x))\,dF_b(x)\right.\\ &\qquad \left. - \int_0^\infty [H_N(x) - H(x)]\phi'(H(x))\,dF(x)\right\}\\ &\quad + o(1) \quad \text{a.s., as} \quad N \to \infty. \end{aligned} \tag{5.5.6}$$

We write the first term on the rhs of (5.5.6) as

$$\begin{aligned} &2\sqrt{N}\left\{\int_0^\infty \phi(H_b(x))d[F_{N,b}(x) - F_b(x)]\right.\\ &\qquad \left. - \int_0^\infty \phi(H(x))d[F_N(x) - F(x)]\right\}\\ &\quad + \left\{2\sqrt{N}\left[\int_0^\infty \phi(H_b(x))\,dF_b(x) - \int_0^\infty \phi(H(x))\,dF(x)\right]\right.\\ &\qquad \left. + bAI^{1/2}(f)\rho_1^*\right\}, \end{aligned} \tag{5.5.7}$$

where the second term is nonstochastic and under the hypothesis of Theorem 5.5.2, it converges to 0 as $N \to \infty$. For the first term, by integrating by parts and noting that $\phi(H(0)) = \phi(H_b(0)) = 0$, we obtain that the same is equal to

$$2\left\{\int_0^\infty N^{1/2}[F_b(x) - F_{N,b}(x)]\phi'(H_b(x))\,dH_b(x) - \int_0^\infty N^{1/2}[F(x) - F_N(x)]\phi'(H(x))\,dH(x)\right\}. \quad (5.5.8)$$

We choose a $K_N^*(= K_{\eta,N}^*, \eta > 0)$ such that $[2K^2\int_{K_N^*}^\infty\{H(1-H)\}^{-1+\delta-\epsilon}dH] < \eta(\log N)^{-1}$, where K, δ, and ϵ are defined by (5.4.14) and (2.6.31) with $\delta > \epsilon$ and η is arbitrary. Then, by (2.6.31), as $N\to\infty$, the following holds a.s.

$$\begin{aligned}&\left|2\int_{K_N^*}^\infty \sqrt{N}\,[F_b(x) - F_{N,b}(x)]\phi'(H_b(x))\,dH_b(x)\right|\\ &\quad\leqslant 2K\log N\int_{K_N^*}^\infty [H_b(1-H_b)]^{1/2-\epsilon}K[H_b(1-H_b)]^{-3/2+\delta}dH_b\\ &\quad= 2K^2\log N\int_{K_N^*}^\infty [H_b(1-H_b)]^{-1+\delta-\epsilon}dH_b < \eta \end{aligned} \quad (5.5.9)$$

and a similar inequality holds for the upper tail of the second term in (5.5.8). On the other hand,

$$\begin{aligned}&2\int_0^{K_N^*}\sqrt{N}\,[F_b(x) - F_{N,b}(x)]\phi'(H_b(x))\,dH_b(x)\\ &\quad -2\int_0^{K_N^*}\sqrt{N}\,[F(x) - F_N(x)]\phi'(H(x))\,dH(x)\\ &= 2\int_0^{K_N^*}\sqrt{N}\,\{F_b(x) - F_{N,b}(x) - F(x) + F_N(x)\}\phi'(H_b(x))\,dH_b(x)\\ &\quad +2\int_0^{K_N^*}\sqrt{N}\,[F(x) - F_N(x)][\phi'(H_b(x)) - \phi'(H(x))]\,dH_b(x)\\ &\quad +2\int_0^{K_N^*}\sqrt{N}\,[F(x) - F_N(x)]\phi'(H(x))d[H_b(x) - H(x)]. \end{aligned} \quad (5.5.10)$$

Note that $[1 - H(K_N^*)]^{\delta-\epsilon} \leqslant 0((\log N)^{-1})$, so that by (5.4.14), $|\phi'(H)| \leqslant 0((\log N)^{(3/2-\delta)/(\delta-\epsilon)})$, $\forall x \leqslant K_N^*$ and by direct expansion, $H_b(x) - H(x) = O(b/\sqrt{N})$, so that on using (5.4.38), the first term on the rhs of (5.5.10) is $O(N^{-1/4}(\log N)^{1+(3/2-\delta)/(\delta-\epsilon)}) = o(1)$, with probability 1 as $N\to\infty$. A similar argument holds for the second term where we use (2.6.31) and the fact that $|\phi'(H_b) - \phi'(H)| = [0(b/\sqrt{N})](\log N)^{(5/2-\delta)/(\delta-\epsilon)}$. For the last term, we use the fact that $d[H_b(x) - H(x)] = b^2/2N[f'(x + \theta_1 b/\sqrt{N}) + f'(-x + \theta_2 b/\sqrt{N})]\,dx$ where $0 < \theta_1, \theta_2 < 1$, so that $I(f) < \infty$ implies that $\int|f'| < \infty$, and hence this term is $O(N^{-1}(\log N)^{1+(3/2-\delta)/(\delta-\epsilon)}) = o(1)$, with probability 1, as $N \to \infty$. The treatment of the second term on the rhs of (5.5.6) is very similar to that of (5.5.8), and hence the same proof holds. Q.E.D.

In (5.4.14), if we let $\delta > \frac{1}{3}$ and make use of (5.4.44) and (5.4.45), then, in (5.5..4), we may replace the a.s. statement by a statement involving a probability $\geqslant 1 - 0(N^{-s})$, for some $s > 1$.

5.6. WEAK CONVERGENCE OF ALIGNED SIGNED RANK STATISTICS

We are concerned here with the model

$$F(x) = F_0(x - \theta), \qquad \theta \text{ unknown}, \qquad F_0(x) + F_0(-x) = 1, \qquad \forall x. \tag{5.6.1}$$

Since θ is unknown, we use the aligned rank order statistics

$$\hat{S}_{k,N} = S(X_1 - \hat{\theta}_N, \ldots, X_k - \hat{\theta}_N), \qquad 1 \leqslant k \leqslant N, \quad \hat{S}_{0,N} = 0, \tag{5.6.2}$$

where $S(\cdot)$ is defined by (5.2.1), the original observations X_i, $i \geqslant 1$ are replaced by the aligned observations $X_i - \hat{\theta}_N, i \geqslant 1$ and the estimator $\hat{\theta}_N$ is defined as follows: Let

$$\hat{\theta}_N^{(1)} = \sup\{a : S_N(a) > 0\}, \hat{\theta}_N^{(2)} = \inf\{a : S_N(a) < 0\}; \tag{5.6.3}$$

$$\hat{\theta}_N = \tfrac{1}{2}\left[\hat{\theta}_N^{(1)} + \hat{\theta}_N^{(2)}\right]. \tag{5.6.4}$$

Then $\hat{\theta}_N$ is a robust, translation-invariant, and consistent estimator of θ and we study its properties in greater detail in Chapter 10. Since $|N^{-1/2}S_N(0)| = O_p(1)$ [by (5.2.6)], by (5.5.2), we claim that $N^{1/2}|\hat{\theta}_N - \theta| = O_p(1)$. Finally, on replacing in (5.3.10) $S^0_{N(t)}$ by $\hat{S}_{N(t),N}, t \in I$, we define the corresponding process by $\hat{Y}_N = \{Y_N(t), t \in I\}$ and let $W_0 = \{W_0(t) = W(t) - tW(1), t \in I\}$ be a standard Brownian Bridge. Then, we have the following theorem due to Sen (1977d).

THEOREM 5.6.1. *Under* (5.6.1) *and the conditions of Theorem* 5.5.2, *as* $N \to \infty$,

$$\hat{Y}_N \xrightarrow{\mathfrak{D}} W_0, \qquad \text{in the } J_1\text{-topology on } D[0,1]. \tag{5.6.5}$$

Proof. Note that by (5.6.3), (5.6.4), the fact that $N^{1/2}|\hat{\theta}_N - \theta| = O_p(1)$ and Theorem 5.5.2, we conclude that as $N \to \infty$

$$N^{-1/2}S_N(0) = N^{1/2}(\hat{\theta}_N - \theta)AI^{1/2}(f)\rho_1^* + o_p(1). \tag{5.6.6}$$

$$Y_N(1) = N^{1/2}(\hat{\theta}_N - \theta)I^{1/2}(f)\rho_1^* + o_p(1). \tag{5.6.7}$$

Further, from (5.5.3) and the fact that $N^{1/2}|\hat{\theta}_N - \theta| = O_p(1)$, we obtain that

$$\sup\left\{|\hat{Y}_N(t) - Y_N(t) + \sqrt{N}(\hat{\theta}_N - \theta)tI^{1/2}(f)\rho_1^*| : t \in I\right\} \xrightarrow{p} 0. \quad (5.6.8)$$

Hence, from (5.6.7) and (5.6.8), we obtain that as $N \to \infty$

$$\sup_{t \in I} |\hat{Y}_N(t) - Y_N(t) + tY_N(1)| \xrightarrow{p} 0. \quad (5.6.9)$$

On the other hand, by Theorem 5.3.1 and some direct steps,

$$\{Y_N(t) - tY_N(t), \quad t \in I\} \xrightarrow{\mathfrak{D}} W_0. \quad (5.6.10)$$

Hence (5.6.5) follows from (5.6.9) and (5.6.10). Q.E.D.

Note that in (5.6.2) if some other estimator $\hat{\theta}_N$ is used, (5.6.6) and (5.6.7) may not hold and as a result the theorem may not hold. However, under fairly general regularity conditions, $\hat{Y}_N$ weakly converge to some Gaussian function; see Problem 5.6.2 in this context.

5.7. ALMOST SURE CONVERGENCE OF $N^{-1}S_N$

From Theorems 5.4.1 and 5.4.3, it follows on using the law of iterated logarithm that $(S_N - E(S_N))/g(N) \to 0$ a.s. as $N \to \infty$ whenever $(N \log\log N)^{1/2}/g(N) \to 0$ as $N \to \infty$. However, the a.s. convergence of $N^{-1}S_N$ to $N^{-1}E(S_N)$ (or to $\mu(F,\phi)$), defined by (5.4.12)) can be established under much weaker conditions.

THEOREM 5.7.1 *Suppose that the score function ϕ satisfy both the conditions* (4.6.6) *and* (4.6.7). *Then as $N \to \infty$,*

$$N^{-1}S_N \to \mu(F,\phi) \qquad \text{a.s.} \quad (5.7.1)$$

Proof. By using (5.4.13), it suffices to show that

$$N^{-1}T_N = \int_0^\infty a_N(NH_N(x))\,dF_N(x)$$

$$\to \mu^*(F,\phi) = \int_0^\infty \phi(H(x))\,dF(x), \qquad \text{a.s., as } N \to \infty. \quad (5.7.2)$$

Also, $dF_N \leqslant dH_N$ and $dF \leqslant dH$, so that proceeding as in (4.6.9) through (4.6.12), we may argue that it suffices to consider the case of ϕ having bounded variation over all [0, 1]. Then, note that

$$N^{-1}T_N - \mu^*(F,\phi) = \int_0^1 a_N(1+[Nu])\,dF_N(H_N^{-1}(u)) - \int_0^1 \phi(u)\,dF(H^{-1}(u))$$

$$= \int_0^1 a_N(1+[Nu])d\big[F_N(H_N^{-1}(u)) - F(H^{-1}(u))\big] + \int_0^1 \big[a_N(1+[Nu]) - \phi(u)\big]\,dF(H^{-1}(u)), \tag{5.7.3}$$

where by the fact that $dF(H^{-1}(u)) \leqslant dH(H^{-1}(u)) = du$, the second term is bounded by

$$\int_0^1 |a_N(1+[Nu]) - \phi(u)|du \to 0 \qquad (\text{as } N\to\infty) \quad \text{by (4.6.7).} \tag{5.7.4}$$

For the first term, we note that by the Glivenko-Cantelli theorem,

$$\sup_x |F_N(x) - F(x)| \to 0, \qquad \sup_x |H_N(x) - H(x)| \to 0 \qquad \text{a.s.,} \tag{5.7.5}$$

and hence we may virtually repeat the steps in (4.6.18) and (4.6.19) and complete the proof. Q.E.D.

An extension of this theorem for nonidentically distributed independent r.v.'s is due to Sen (1970a).

EXERCISES

5.2.1. In (5.2.1), let $a_N(i) = i/(N+1)$, $1 \leqslant i \leqslant N$ and denote the corresponding S_N by W_N (the Wilcoxon signed rank statistic). Then show that W_N is a linear combination of two U-statistics:

$$U_N^{(1)} = \binom{N}{2}^{-1} \sum_{1\leqslant i<j\leqslant N} \operatorname{sgn}(X_i + X_j) \qquad \text{and} \qquad U_N^{(2)} = N^{-1}\sum_{i=1}^{N} \operatorname{sgn} X_i.$$

5.2.2. Verify (5.2.7) when (i) $a_N(i) = i/(N+1)$, $1 \leqslant i \leqslant N$ and (ii) $a_N(i) = \Phi^{-1}(\frac{1}{2}(1 + i/(n+1)))$, $i = 1, \ldots, N$, where Φ is the standard normal d.f.

5.3.1. Provide a proof of Theorem 5.3.2.

5.3.2. Show that under (4.3.19) and the conditions of Theorem 5.3.1, both $\{Y_{N_n}\}$ and $\{Z_{N_n}\}$ weakly converge to a standard Brownian motion.

5.3.3. Establish the weak convergence in Theorems 5.3.1 and 5.3.2 in the sup-norm metric d_q in (2.4.26).

5.3.4. If F is absolutely continuous with density f and a finite Fisher information $I(f)$, then under (5.3.8), show that $P_N^{(\theta)}$ is contiguous to $P_N^{(0)}$ for every fixed θ (Hájek and Sidák, 1967).

5.3.5. Provide the proof of (5.3.13).

5.4.1. Show that (5.4.7) holds.

5.4.2. Show that if $E|X_i|^r < \infty$ and the X_i are i.i.d.r.v. then $N^{-1}\max_{1 \leqslant i \leqslant N}|X_i|^r \to 0$ in L_1-norm.

5.4.3. If ϕ is the difference of two nondecreasing and square integrable functions then show that under (5.2.2), the law of iterated logarithm holds for $\{S_N\}$ (Sen and Ghosh, 1973b).

5.4.4. Let $\psi(t), t > 0$ be a nondecreasing positive function on $[0, \infty)$ such that for some to $(\geqslant 0)$, $\gamma(t) = \psi^2(t) - \log\log t$ is nonnegative and nondecreasing in $t(\geqslant t_0)$. Then, for every $\epsilon > 0$, there exists positive numbers K and η, such that

$$P_0\left\{S_m \geqslant m^{1/2}A\left[21+\epsilon\right)\right]^{1/2}\psi(m) \quad \text{for some} \quad m \geqslant n\Big\}$$
$$\leqslant K\left\{\exp\left[-\gamma(n) - \eta\psi^2(n)\right]\right\}, \qquad \forall n \geqslant t_0,$$

and a similar inequality holds for the lower tail (Sen and Ghosh, 1973b).

5.4.5 (5.4.4 continued). If, in addition, $\lim_{t\to\infty}(\log\log t)/\psi^2(t) = 0$ and $\psi(t) = (t^{1/6})$, then

$$\left[\psi(n)\right]^{-2}\log P_0\left\{m^{-1/2}S_m > A\psi(m) \quad \text{for some} \quad m \geqslant n\right\} \to -\tfrac{1}{2}$$

(Sen and Ghosh, 1973b).

5.5.1. Let $\{c_1, \ldots, c_N\}$ be known constants, $\bar{c}_N = (1/N)\sum_{i=1}^N c_i$ and let

$$c_{Ni}^+ = \max\left[0, c_i - \bar{c}_N\right], c_{Ni}^- = \min\left[0, c_i - \bar{c}_N\right], \qquad 1 \leqslant i \leqslant N.$$

Define $S_{\Delta N}^+ = \sum_{i=1}^N \mathrm{sgn}(X_i + \Delta c_{Ni}^+)a_N(R_{Ni}^{\Delta+})$ and $S_{\Delta N}^- = \sum_{i=1}^N \mathrm{sgn}(X_i + \Delta c_{Ni}^-)$ $a_N(R_{Ni}^{\Delta-})$ where $R_{Ni}^{\Delta+}$ = Rank of $|X_i + \Delta c_{Ni}^+|$ among $|X_1 + \Delta c_{N1}^+|, \ldots,$ $|X_N + c_{NN}^+|$, $1 \leqslant i \leqslant N$ and $R_{Ni}^{\Delta-}$ = Rank of $|X_i + \Delta c_{Ni}^-|$ among $|X_1 + \Delta c_{N1}^-|, \ldots, |X_N + \Delta c_{NN}^-|$, $1 \leqslant i \leqslant N$. Then, show that under the assumptions of Theorem 5.5.2, $S_{\Delta N}^+$ (or $S_{\Delta N}^-$) is $\nearrow$ (or $\searrow$) in Δ, with probability 1 (Jurečková, 1971a).

5.5.2 (5.5.1 continued). Let $\{d_1, \ldots, d_N\}$ be known constants such that either (a) $c_i d_i \geqslant 0$ and $(|c_i| - |c_j|)(|d_i| - |d_j|) \geqslant 0$, $\forall i, j \in [1, \ldots, N]$ or (b)

$c_i d_i \geqslant 0$ $(|c_i| - |c_i|)(|d_i| - |d_j|) \geqslant 0$, $\forall i, j \in [1, \ldots, N]$ and define

$$S_N^*(\Delta) = \sum_{i=1}^{N} a_N(R_{Ni}^{+})\operatorname{sgn}(X_i - \Delta c_i),$$

where $a_N(1) \leqslant \cdots \leqslant a_N(N)$ and $R_{Ni}^{\Delta+}$ = Rank of $|X_i - \Delta c_i|$ among $|X_1 - \Delta c_1|, \ldots, |X_N - \Delta c_N|$, $1 \leqslant i \leqslant N$. Then $S_N^*(\Delta)$ is $\nearrow$ (or $\searrow$) in Δ when (a) [or (b)] holds (van Eeden, 1972).

5.5.3. Provide the proof of Theorem 5.5.2.

5.5.4. Provide the proof of Theorem 5.5.3.

5.6.1. Show that $\hat{\theta}_N$, defined by (5.6.4) is a consistent, robust, and translation-invariant estimator of θ (Hodges and Lehmann, 1963; Puri and Sen, 1971, Ch. 6).

5.6.2. Suppose $\tilde{\theta}_N$ is an estimator of θ with the property, that $\sqrt{N}(\tilde{\theta}_N - \theta) = \gamma W_N^*(1) + o_p(1)$ where $\gamma(> \theta)$ is a positive constant and $W_N^* = \{W_N^*(b), t \in I\}$ is an element of $D[0, 1]$, such that $(W_N, W_N^*) \xrightarrow{\mathfrak{D}} (W, W^*)$ where (W, W^*) is a bivariate Gaussian function on $C[0, 1]$. Then for $\tilde{Y}_N$ defined as in $\hat{Y}_N$ with $\hat{\theta}_N$ replaced by $\tilde{\theta}_N$, the weak convergence to a Gaussian function holds (Sen, 1977d).

CHAPTER 6

Invariance Principles for Rank Statistics for Testing Bivariate Independence

6.1. INTRODUCTION

This chapter is devoted to the study of some invariance principles for the usual (pure and mixed) rank order statistics employed for testing the hypothesis of stochastic independence in bivariate distributions. Section 6.2 deals with the preliminary notions. Weak convergence (under the hypothesis of stochastic independence) to Brownian motions is studied in Section 6.3. Embedding of Wiener processes is established in Section 6.4. Almost sure representation of such pure and mixed rank statistics in terms of a sum of independent random variables is considered in Section 6.5. The last section deals with the almost sure convergence of such statistics.

6.2. PRELIMINARY NOTIONS

Let $\{\mathbf{Z}_i = (X_i, Y_i), i \geqslant 1\}$ be a sequence of i.i.d.r.v. (vectors) with each $\mathbf{Z}_i$ having a continuous (bivariate) d.f. $H(x, y)$, $-\infty < x, y < \infty$. Let $F(x) = H(x, \infty)$ and $G(y) = H(\infty, y)$ be the two marginal d.f.'s and we frame the null hypothesis of stochastic independence of (X, Y) as

$$H_0 : H(x, y) = F(x)G(y), \qquad \forall (x, y) \in (-\infty, \infty)^2. \tag{6.2.1}$$

For testing H_0 against suitable alternatives, a general class of nonparametric tests is based on the following type of (pure) rank order statistics:

$$M_N = \sum_{i=1}^{N} a_N(R_{Ni}) b_N(S_{Ni}), \qquad N \geqslant 1, \tag{6.2.2}$$

where $R_{Ni}(=\sum_{j=1}^{N} c(X_i - X_j))$ (and $S_{Ni} = \sum_{j=1}^{N} c(Y_i - Y_j)$) is the rank of X_i (and Y_i) among $X_1, \ldots, X_N$ (and $Y_1, \ldots, Y_N$), for $i = 1, \ldots, N$ and the two sets $\{a_N(1), \ldots, a_N(N)\}$ and $\{b_N(1), \ldots, b_N(N)\}$ of scores are defined as in (4.2.2). Under H_0 in (6.2.1), $\mathbf{R}_N = (R_{N1}, \ldots, R_{NN})$ and $\mathbf{S}_N = (S_{N1}, \ldots, S_{NN})$ are stochastically independent, each one assuming all possible permutations of $(1, \ldots, N)$ with the equal probability $(N!)^{-1}$. Thus, under H_0, M_N is a distribution-free statistic. Note that by the same arguments as in (4.2.3) through (4.2.10), one obtains the following results where E_0 stands for the expectation under H_0:

$$E_0(M_N) = 0, \qquad E_0(M_N^2) = N^2(N-1)^{-1}A_N^2 B_N^2 \tag{6.2.3}$$

where

$$A_N^2 = N^{-1}\sum_{i=1}^{N}\left(a_N(i) - \bar{a}_N\right)^2, \qquad B_N^2 = N^{-1}\sum_{i=1}^{N}\left(b_N(i) - \bar{b}_N\right)^2 \tag{6.2.4}$$

and $\bar{a}_N = N^{-1}\sum_{i=1}^{N} a_N(i)$, $\bar{b}_N = N^{-1}\sum_{i=1}^{N} b_N(i)$. Also, parallel to (4.2.10),

$$N^{-1/2}A_N^{-1}B_N^{-1}M_N \xrightarrow{\mathfrak{D}} \mathfrak{N}(0,1), \qquad \text{as} \quad N \to \infty. \tag{6.2.5}$$

We are primarily interested here in deriving deeper results concerning the sequence $\{M_k; k \leqslant N\}$ and the tail sequence $\{M_k; k \geqslant N\}$ when $N \to \infty$. In passing, we may remark that particular cases of M_N are the Spearman rank correlation, normal score statistics and other ones; see Problems 6.2.1 and 6.2.2. Among the other well-known measures not belonging to this class, Kendall's tau coefficient deserves mention. This is, however, expressible as a U-statistic, and hence the results of Chapter 3 apply.

In the context of nonparametric tests for no regression with stochastic predictors (Ghosh and Sen, 1971), one may consider the following model relating to the sequence $\{\mathbf{Z}_i\}$. Let

$$G(y \mid x) = P\{Y \leqslant y \mid X = x\} = G(y - \beta_0 - \beta a(x)), \qquad \forall (x, y) \in (-\infty, \infty)^2, \tag{6.2.6}$$

where β_0 and β are unknown parameters, G is an unknown (continuous) d.f. and $a(x)$ is a specified (nonconstant) function of x. The hypothesis of no regression is then formulated as

$$H_0^*: \beta = 0 \quad \text{vs.} \quad H_1^*: \beta > 0 \quad (\text{or} \quad < 0 \quad \text{or} \quad \neq 0). \tag{6.2.7}$$

Note that under H_0^*, X and Y are independent, though under H_1^*, here one is more interested in a specific pattern of dependence (regression) of Y on X, and the roles of X and Y are not symmetric in this setup. For such a regression model, as $a(x)$ is of specified form (while G is not), it may be more informative to incorporate the $a(X_i)$ in the formulation of the test statistics (and this is also justified from considerations of local optimality of tests based on the ranks of the Y_i). The following mixed rank statistics may be considered for this purpose:

$$Q_N = \sum_{i=1}^{N} a(X_i)\left[b(S_{Ni}) - \bar{b}_N\right], \qquad N \geqslant 1, \tag{6.2.8}$$

where $\mathbf{S}_N$ and the $b_N(i)$ are defined as in (6.2.2). Let $\mathbf{X}^{(N)} = (X_{N,1} \leqslant \cdots \leqslant X_{N,N})$ be the vector of order statistics corresponding to $X_1, \ldots, X_N$. Then, under H_0^*, $\mathbf{S}_N$ is stochastically independent of $\mathbf{X}^{(N)}$ and the conditional distribution of Q_N, given $\mathbf{X}^{(N)}$, is generated by the $N!$ equally likely permutations of $(1, \ldots, N)$ (for $\mathbf{S}_N$). If we denote this uniform conditional probability measure by $\mathcal{P}_N$, then we have by steps similar to in (4.2.3) through (4.2.10),

$$E_{\mathcal{P}_N} Q_N = 0 \qquad \text{and} \qquad E_{\mathcal{P}_N} Q_N^2 = s_N^2 B_N^2, \tag{6.2.9}$$

where B_N^2 is defined by (6.2.4) and

$$s_N^2 = (N-1)^{-1} \sum_{i=1}^{N} \{a(X_i) - \bar{a}_N\}^2; \qquad \bar{a}_N = N^{-1} \sum_{i=1}^{N} a(X_i). \tag{6.2.10}$$

Further, under quite general conditions (see Ghosh and Sen, 1971),

$$Q_N / s_N B_N \xrightarrow{\mathfrak{D}} \mathfrak{N}(0,1), \qquad \text{when} \quad H_0^* \quad \text{holds}. \tag{6.2.11}$$

We are interested here in the invariance principles for $\{Q_k; k \leqslant N\}$ or $\{Q_k; k \geqslant N\}$ when $N \to \infty$. In the remainder of this section, we present the basic martingale property of $\{M_N\}$ and $\{Q_N\}$, when H_0 and H_0^* hold. As in (4.2.2), we let

$$a_N^0(i) = E\phi_1(U_{Ni}) \qquad \text{and} \quad b_N^0(i) = E\phi_2(U_{Ni}) \qquad \text{for} \quad i = 1, \ldots, N \tag{6.2.12}$$

where ϕ_1 and ϕ_2 are integrable inside (0, 1) and we set

$$\int_0^1 \phi_j(u)\,du = 0 \qquad \text{for} \quad j = 1, 2. \tag{6.2.13}$$

With the scores a_N^0 and b_N^0, the corresponding M_N in (6.2.2) is denoted by M_N^0. Let $\mathcal{F}_N^{(1)}$ (and $\mathcal{F}_N^{(2)}$) be the sigma field generated by $\mathbf{R}_N$ (and $\mathbf{S}_N$) and let $\mathcal{F}_N = \mathcal{F}_N^{(1)} \times \mathcal{F}_N^{(2)}$ for $N \geqslant 1$; $\mathcal{F}_0$ stands for the trivial sigma field. Then $\mathcal{F}_N$ is nondecreasing in N.

THEOREM 6.2.1. *Under* (6.2.1) *and* (6.2.13), $\{M_N^0, \mathcal{F}_N; N \geqslant 1\}$ *is a martingale.*

Proof. By definition,

$$E_0(M_{N+1}^0 \mid \mathfrak{F}_N) = \sum_{i=1}^{N} E_0\{a_{N+1}^0(R_{N+1i})b_{N+1}^0(S_{N+1i}) \mid \mathfrak{F}_N\}$$
$$+ E_0\{a_{N+1}^0(R_{N+1N+1})b_{N+1}^0(S_{N+1N+1}) \mid \mathfrak{F}_N\}. \quad (6.2.14)$$

Given $\mathfrak{F}_N$, R_{N+1N+1}, S_{N+1N+1} are stochastically independent and each can assume the values $1, \ldots, N+1$ with equal (conditional) probability $(N+1)^{-1}$. Hence

$$E_0\{a_{N+1}^0(R_{N+1N+1})b_{N+1}^0(S_{N+1N+1}) \mid \mathfrak{F}_N\}$$
$$= E_0\{a_{N+1}^0(R_{N+1N+1}) \mid \mathfrak{F}_N^{(1)}\} E_0\{b_{N+1}^0(S_{N+1N+1}) \mid \mathfrak{F}_N^{(2)}\}$$
$$= \left\{(N+1)^{-1} \sum_{j=1}^{N+1} a_{N+1}^0(j)\right\}\left\{(N+1)^{-1} \sum_{j=1}^{N+1} b_{N+1}^0(j)\right\}$$
$$= \left\{\int_0^1 \phi_1(u)\,du\right\}\left\{\int_0^1 \phi_2(u)\,du\right\} = 0, \qquad \text{by (6.2.13)}. \quad (6.2.15)$$

Also, given $\mathfrak{F}_N$, R_{N+1i} can only assume the two values R_{Ni} and $R_{Ni}+1$ with respective conditional probabilities $1-(N+1)^{-1}R_{Ni}$ and $(N+1)^{-1}R_{Ni}$, S_{N+1i} can assume the two values S_{Ni} and $S_{Ni}+1$ with respective conditional probabilities $1-(N+1)^{-1}S_{Ni}$ and $(N+1)^{-1}S_{Ni}$, and R_{N+1i}, S_{N+1i} are conditionally stochastically independent. Hence, using (4.2.20),

$$\sum_{i=1}^{N} E_0\{a_{N+1}^0(R_{N+1i})b_{N+1}^0(S_{N+1i}) \mid \mathfrak{F}_N\}$$
$$= \sum_{i=1}^{N} \{(N+1)^{-1}R_{Ni}a_{N+1}^0(R_{Ni}+1) + (1-(N+1)^{-1}R_{Ni})a_{N+1}^0(R_{Ni})\}.$$
$$\{(N+1)^{-1}S_{Ni}b_{N+1}^0(S_{Ni}+1) + (1-(N+1)^{-1}S_{Ni})b_{N+1}^0(S_{Ni})\}$$
$$= \sum_{i=1}^{N} a_N^0(R_{Ni})b_N^0(S_{Ni}) = M_N^0. \quad (6.2.16)$$

Thus, from (6.2.14) through (6.2.16), $E_0(M_{N+1} \mid \mathfrak{F}_N) = M_N^0$, $N \geqslant 1$. Q.E.D.

Remark. Let us define

$$M_N^* = \sum_{i=1}^{N} \phi_1(F(X_i))\phi_2(G(Y_i)), \qquad N \geqslant 1. \quad (6.2.17)$$

Then we have

$$E_0(M_N^* \mid \mathcal{F}_N) = \sum_{i=1}^{N} E_0\{\phi_1(F(X_i))\phi_2(G(Y_i)) \mid \mathcal{F}_n\}$$

$$= \sum_{i=1}^{N} E_0\{\phi_1(F(X_i)) \mid \mathbf{R}_N\} E_0\{\phi_2(G(Y_i)) \mid \mathbf{S}_N\}$$

$$= \sum_{i=1}^{N} a^0(R_{Ni}) b_N^0(S_{Ni}) = M_N^0, \qquad \forall N \geqslant 1. \tag{6.2.18}$$

We shall find this representation very useful in subsequent theorems.

Let now $\mathcal{B}_N^{(1)} = \mathcal{B}(\mathbf{X}^{(N)})$ be the sigma field generated by $\mathbf{X}^{(N)}$ and let $\mathcal{F}_N^* = \mathcal{B}_N^{(1)} \times \mathcal{F}_N^{(2)}$, for $N \geqslant 0$; $\mathcal{F}_0^*$ is then the trivial σ-field and $\mathcal{F}_N^*$ is nondecreasing in N. Also, letting $b_N(i) = b_N^0(i), i = 1, \ldots, N$, we define the corresponding Q_N as Q_N^0.

THEOREM 6.2.2. *Under H_0^* and $Ea(X) < \infty$, $\{Q_N^0, \mathcal{F}_N^*; N \geqslant 1\}$ is a martingale.*

Proof. We may assume without any loss of generality that (6.2.13) holds, so that $\bar{b}_N^0 = 0$ for every $N \geqslant 1$. Then,

$$Q_{N+1}^0 = Q_N^0 + \sum_{i=1}^{N} a(X_i)\{b_{N+1}^0(S_{N+1i}) - b_N^0(S_{Ni})\}$$

$$+ a(X_{N+1}) b_{N+1}^0(S_{N+1N+1}). \tag{6.2.19}$$

As in (6.2.15),

$$E_0\{a(X_{N+1}) b_{N+1}^0(S_{N+1N+1}) \mid \mathcal{F}_N^*\} = \left(\int_{-\infty}^{\infty} a(x)\, dF(x)\right)\bar{b}_{N+1}^0 = 0, \tag{6.2.20}$$

while the $a(X_i), i = 1, \ldots, N$ are held fixed, given $\mathcal{B}_N^{(1)}$, and, as in (6.2.16),

$$E_0\{b_{N+1}^0(S_{N+1i}) - b_N^0(S_{Ni}) \mid \mathcal{F}_N^{(2)}\} = 0 \qquad \text{for every} \quad i = 1, \ldots, N. \tag{6.2.21}$$

Hence $E_0\{Q_{N+1}^0 \mid \mathcal{F}_N^*\} = Q_N^0$, for every $N \geqslant 1$. Q.E.D.

Remark. If we let

$$Q_N^* = \sum_{i=1}^{N} a(X_i)\phi_2(G(Y_i)), \qquad N \geqslant 1, \tag{6.2.22}$$

then

$$E_0(Q_N^* \mid \mathfrak{F}_N^*) = \sum_{i=1}^{N} a(X_i) E_0\{\phi_2(G(Y_i)) \mid \mathfrak{F}_N^{(2)}\} = \sum_{i=1}^{N} a(X_i) b_N^0(S_{Ni})$$
$$= \sum_{i=1}^{N} a(X_i)\{b_N^0(S_{Ni}) - \bar{b}_N^0\} = Q_N^0, \qquad \forall N \geq 1. \quad (6.2.23)$$

Note that though $\{Q_N^*\}$ involves a sum of i.i.d.r.v.'s and hence forms a martingale sequence, $\{Q_N^0\}$ and $\{Q_N^*\}$ are not martingales with respect to a common sequence of sigma fields.

6.3. WEAK CONVERGENCE OF $\{M_N\}$ AND $\{Q_N\}$

We consider first the case of M_N. Assume that ϕ_1 and ϕ_2 are both square integrable inside (0, 1), and, without any loss of generality, we set

$$A_{(j)}^2 = \int_0^1 \phi_j^2(u)\,du = 1, \qquad \text{for} \quad j = 1, 2. \quad (6.3.1)$$

For every $N(\geq 1)$, we define a stochastic process $Y_N^0 = \{Y_N^0(t), t \in [0, 1]\}$ by letting

$$Y_N^0(t) = N^{-1/2} A_N^{-1} B_N^{-1} M_{N(t)}^0;$$
$$N(t) = \max\{k : k/N \leq t\}, \qquad t \in [0, 1], \quad (6.3.2)$$

where A_N and B_N are defined by (6.2.4), but for the scores $a_N^0(i)$ and $b_N^0(i)$. Note that $Y_N(t) = 0$, $\forall t \in [0, 1]$ when $N = 0$ or 1, and Y_N belongs to the $D[0, 1]$ space for every N (≥ 1). The following theorem is adapted from Sen and Ghosh (1974b).

THEOREM 6.3.1. *Under* (6.2.1), (6.2.13), *and* (6.3.1),

$$Y_N^0 \xrightarrow{\mathfrak{D}} W, \qquad \textit{in the } J_1\textit{-topology on} \quad D[0, 1], \quad (6.3.3)$$

where W is a standard Wiener process on $I = [0, 1]$.

Proof. Note that $A_N^2 = N^{-1}\sum_{i=1}^N (a_N^0(i) - \bar{a}_N^0)^2 = N^{-1}\sum_{i=1}^N \{a_N^0(i)\}^2 = N^{-1}\sum_{i=1}^N \{E\phi_1(U_{Ni})\}^2 \leq N^{-1}E\{\sum_{i=1}^N \phi_1^2(U_{Ni})\} = E\phi_1^2(U_1) = A_{(1)}^2 = 1$, $\forall N \geq 1$. Similarly, $B_N^2 \leq 1$, $\forall N \geq 1$. On the other hand, as $N \to \infty$, A_N^2 and B_N^2 both $\to 1$, so that $A_N^2 B_N^2 \to 1$. Thus, under the hypothesis of the theorem, by (6.2.18),

$$E_0(M_N^* - M_N)^2 = E_0(M_N^{*2}) - E_0(M_N^2)$$
$$= N - N^2(N-1)^{-1} A_N^2 B_N^2$$
$$= o(N), \qquad \text{as} \quad N \to \infty, \quad (6.3.4)$$

so that for every (fixed) t (ϵI),

$$N^{-1/2}|M^*_{N(t)} - M^0_{N(t)}| \xrightarrow{p} 0, \quad \text{as} \quad N \to \infty. \tag{6.3.5}$$

On the other hand, if in (6.3.2) we replace $\{M_k^0; k \leqslant N\}$ by $\{M_k^*; k \leqslant N\}$ and denote the corresponding process by Y_N^*, then our Theorem 2.4.3 applies directly to Y_N^*. Thus, by (6.3.5) and the weak convergence of Y_N^* to W, the convergence of the f.d.d.'s of Y_N to those of W follows. Further, by Theorem 6.2.1, Corollary 2.4.4.1, and the convergence of f.d.d.'s of $\{Y_N^0\}$ to those of W, the tightness of $\{Y_N^0\}$ follows readily. Q.E.D.

Since $\{N^{-1}M_N^*\}$ is a reverse martingale sequence motivated by the backward invariance principle, we consider the following. Let $Z_N^0 = \{Z_N^0(t), t \in I\}$ be defined by

$$Z_N^0(t) = A_N^{-1}B_N^{-1}N^{1/2}\{N(t)\}^{-1}M^0_{N(t)}; \qquad N(t) = \min\{k: N/k \leqslant t\}, \qquad t \in I. \tag{6.3.6}$$

THEOREM 6.3.2. *Under the hypothesis of Theorem* 6.3.1, Z_N^0 *converges weakly to* W, *in the* J_1*-topology on* $D[0,1]$.

The proof runs on lines parallel to (4.3.15) through (4.3.18) and hence is left as an exercise (see Problem 6.3.1). Further, both the theorems also hold for stochastic sample sizes $\{N_n\}$, satisfying (4.3.19); toward this, see Problem 6.3.2.

Now, for general scores, we note that

$$|M_k - M_k^0| \leqslant \left|\sum_{i=1}^k a_k(R_{ki})\{b_k(S_{ki}) - b_k^0(S_{ki})\}\right| + \left|\sum_{i=1}^k b_k^0(S_{ki})\{a_k(R_{ki}) - a_k^0(R_{ki})\}\right|. \tag{6.3.8}$$

Thus, applying the Schwarz inequality on each of the terms on the rhs of (6.3.8), we have

$$k^{-1/2}|M_k - M_k^0| \leqslant \left\{\left[k^{-1}\sum_{i=1}^k a_k^2(i)\right]\left[\sum_{i=1}^k \{b_k(i) - b_k^0(i)\}^2\right]\right\}^{1/2} + \left\{\left[k^{-1}\sum_{i=1}^k b_k^{02}(i)\right]\left[\sum_{i=1}^k \{a_k(i) - a_k^0(i)\}^2\right]\right\}^{1/2}. \tag{6.3.9}$$

Thus, if

$$\lim_{N\to\infty} \sum_{i=1}^{N} \{a_N(i) - a_N^0(i)\}^2 = 0 \qquad \text{and} \qquad \lim_{N\to\infty} \sum_{i=1}^{N} \{b_N(i) - b_N^0(i)\}^2 = 0, \tag{6.3.10}$$

then, from (6.3.8) through (6.3.10), we have for $N \to \infty$

$$\max_{1 \leqslant k \leqslant N} N^{-1/2}|M_k - M_k^0| \to 0 \qquad \text{and} \qquad \sup_{k \geqslant N} N^{1/2}k^{-1}|M_k - M_k^0| \to 0, \tag{6.3.11}$$

with probability 1. Thus, in (6.3.2) and (6.3.6), $\{M_k^0\}$ may be replaced by $\{M_k\}$ whenever (6.3.10) holds.

In Chapters 4 and 5, we studied weak convergence for contiguous alternatives relating to shift (location) or regression alternatives. It is also possible to extend Theorems 6.3.1 and 6.3.2 to contiguous alternatives. But, unlike Chapters 4 and 5, there may not be a very simple dependence model that is naturally appealing and ensures contiguity. Among various possibilities, one of the commonly employed models is the following.

We conceive of a sequence $\{(U_i, V_i, W_i), i \geqslant 1\}$ of i.i.d.r.v.'s, where for each i, U_i, V_i and W_i are stochastically independent. Let $\Delta(\geqslant 0)$ be a nonstochastic constant and set

$$X_i = U_i + \Delta W_i \qquad \text{and} \qquad Y_i = V_i \pm \Delta W_i, \qquad i \geqslant 1, \tag{6.3.12}$$

where for a plus (minus) sign of Δ in Y_i, (X_i, Y_i) have a positive- (negative-) type association. The positive (negative) quadrant-association is defined by

$$P\{X \leqslant x, Y \leqslant y\} - P\{X \leqslant x\}P\{Y \leqslant y\} \geqslant (\text{or} \leqslant) 0, \qquad \forall -\infty < x, y < \infty, \tag{6.3.13}$$

with the strict inequality sign on a set of measure nonzero. Keeping this in mind, we introduce (see Behnen, 1971) the following type of alternative hypotheses.

Let $\beta_j(t), t \in I, j = 1,2$ be two real-valued functions such that

$$\int_0^1 \beta_j(t)\,dt = 0 \qquad \text{and} \quad 0 < \sigma_j^2 = \int_0^1 \beta_j^2(t)\,dt < \infty, \qquad j = 1,2, \tag{6.3.14}$$

and let $\gamma_{Nj} = \{\gamma_{Nj}(t), t \in I\}, j = 1,2, N \geqslant 1$ be such that

$$\int_0^u \gamma_{N1}(t)\,dt \int_0^v \gamma_{N2}(s)\,ds \geqslant \quad (\text{or} \leqslant) 0, \qquad \forall 0 < u, v < 1 \tag{6.3.15}$$

with the strict inequality on a set of Lebesgue measure nonzero,

$$\int_0^1 \gamma_{Nj}(t)\,dt = 0, \qquad \int_0^1 \{\gamma_{Nj}(t) - \beta_j(t)\}^2\,dt \to 0 \quad \text{as} \quad N \to \infty, \qquad j = 1, 2, \tag{6.3.16}$$

$$\sup_{u,v \in I} \{N^{-1}(\gamma_{N1}(u)\gamma_{N2}(v))^4\} \to 0 \quad \text{as} \quad N \to \infty. \tag{6.3.17}$$

Finally, let f and g be the p.d.f. corresponding to the marginal d.f.'s F and G, respectively. Then, for every $N(\geqslant N_0)$ and $\Delta(>0)$, consider the alternative hypothesis:

$$K_\Delta^{(N)} : h(x,y) = h_{\Delta N}(x,y) = f(x)g(y)\{1 + \Delta N^{-1/2}\gamma_{N1}(F(x))\gamma_{N2}(G(y))\},$$
$$-\infty < x, y < \infty, \tag{6.3.18}$$

where N_0 is the least positive integer for which the rhs of (6.3.24) is nonnegative (a.e.); by (6.3.17), such an N_0 exists. The (joint) d.f. H under $K_\Delta^{(N)}$ is denoted by $H_{\Delta N}$ and the joint d.f. of $\mathbf{Z}_1, \ldots, \mathbf{Z}_N$ under H_0 and $K_\Delta^{(N)}$ are denoted by P_N^0 and P_N^Δ,

$$\rho_j = \frac{\left(\int_0^1 \phi_j(t)\beta_j(t)\,dt\right)}{\sigma_j} \quad \text{for} \quad j = 1, 2; \tag{6.3.19}$$

$$\mu = \{\mu(t) = \Delta t \rho_1\sigma_1\rho_2\sigma_2, \quad t \in I\}. \tag{6.3.20}$$

THEOREM 6.3.3. *Under* (6.3.14) *through* (6.3.18), *as* $N \to \infty$,

$$Y_N^0 \xrightarrow{\mathcal{D}} W + \mu, \qquad \textit{in the } J_1\textit{-topology on} \quad D[0,1]. \tag{6.3.21}$$

Proof. Problem 6.3.3 is to show that under the hypothesis of the theorem, P_N^Δ, the joint distribution of $Z_1, \ldots, Z_N$ under $K_\Delta^{(N)}$ is contiguous to P_N, the same under H_0. Hence, by Theorems 4.3.4 and 6.3.1, we conclude that $\{Y_N^0\}$ remains tight under $K_\Delta^{(N)}$ as well. Also, the preceding contiguity implies that (6.3.5) remains true under $K_\Delta^{(N)}$. On the other hand, the central limit theorem applies to the $M_{N(t)}^*$ (for finitely many $t \in I$) under $K_\Delta^{(N)}$, and hence the convergence of the f.d.d.'s of Y_N^0 to those of $W + \mu$ follows by some standard steps (see Problem 6.3.4). Q.E.D.

Let us now consider the invariance principles for the mixed-rank statistics. Suppose that in (6.3.2) and (6.3.6), we replace $M_{N(t)}^0$ and A_N by $Q_{N(t)}^0$ and s_N, respectively [which are defined by (6.2.8) and (6.2.10)], and the resulting processes are denoted by $\tilde{Y}_N^0$ and $\tilde{Z}_N^0$, respectively. Then we have the following.

THEOREM 6.3.4. *Under H_0^* in (6.2.7), whenever $0 < V(a(X)) < \infty$ and $0 < \int_0^1 \phi_{(2)}^2(u)\,du < \infty$,*

$$\tilde{Y}_N^0 \xrightarrow{\mathfrak{D}} W, \quad \text{in the } J_1\text{-topology on} \quad D[0,1], \tag{6.3.22}$$

$$\tilde{Z}_N^0 \xrightarrow{\mathfrak{D}} W, \quad \text{in the } J_1\text{-topology on} \quad D[0,1], \tag{6.3.23}$$

where W is a standard Wiener process on I.

The proof is very similar to the ones in Theorems 6.3.1 and 6.3.2. By (6.2.22) and (6.2.23), it follows that $N^{-1}E_0(Q_N^0 - Q_N^*)^2 = V(a(X))\{A_{(2)}^2 - B_N^2\} \to 0$ as $N \to \infty$. Hence, for the convergence of the finite-dimensional distributions of $\tilde{Y}_N^0$ (or $\tilde{Z}_N^0$), we may replace the $Q_{N(t)}^0$ by $Q_{N(t)}^*$ on which Theorem 2.4.3 applies directly. Also, Theorem 6.2.2 and the above, in accordance with Corollary 2.4.4.1, ensure the tightness.

We may also consider the case of the sequence $\{K_\Delta^{(N)}\}$ of alternative hypotheses in (6.3.18) and define σ_1^2, σ_2^2 as in (6.2.14), ρ_2 as in (6.3.19). Let then

$$\tilde{\rho}_1 = \frac{\left(\int_{-\infty}^{\infty} a(x)\beta_1(F(x))\,dF(x)\right)}{\sigma_1}, \tag{6.3.24}$$

$$\tilde{\mu} = \{\tilde{\mu}_1(t) = \Delta t \tilde{\rho}_1 \rho_2 \sigma_1 \sigma_2, \quad t \in I\}. \tag{6.3.25}$$

Then, we have the following theorem whose proof follows on the same line as in the proof of Theorem 6.6.3 (and hence is left as an exercise; see Problem 6.3.5).

THEOREM 6.3.5. *Under (6.3.14) through (6.3.18), as $N \to \infty$,*

$$Y_N^0 \xrightarrow{\mathfrak{D}} W + \tilde{\mu}, \quad \text{in the } J_1\text{-topology on} \quad D[0,1]. \tag{6.3.26}$$

Instead of (6.3.18), one can also consider a sequence of local alternatives where for the model (6.2.6), $\beta = N^{-1/2}\lambda$, λ real and fixed and G has an absolutely continuous density g with a finite Fisher information $I(g)$. This, however, is a special case of (6.3.18) where $\beta_2 \equiv 1$.

6.4. EMBEDDING OF WIENER PROCESSES FOR $\{M_N\}$ AND $\{Q_N\}$

Under the hypothesis of stochastic independence, the martingale property in Theorems 6.2.1 and 6.2.2 enables us to establish these embedding of Wiener processes under fairly general conditions. We consider the case of

$\{M_N^0\}$ first and define

$$V_N = \sum_{k=2}^{N} \operatorname{Var}\{M_k^0 \mid \mathfrak{F}_{k-1}\}, \qquad N \geqslant 2; \qquad V_0 = V_1 = 0; \tag{6.4.1}$$

$$\tilde{M}_{V_N}^0 = M_N^0, \qquad N \geqslant 0 \quad \text{where} \quad M_0^0 = M_1^0 = 0, \tag{6.4.2}$$

and complete the definition of $\tilde{M}^0 = \{\tilde{M}_t^0, t \geqslant 0\}$ by linear interpolation between $(V_N, \tilde{M}_{V_N}^0)$ and $(V_{N+1}, \tilde{M}_{V_{N+1}}^0)$, for $V_N \leqslant t \leqslant V_{N+1}, N \geqslant 0$. Finally, for a function $h = \{h(t), t \in I\}$, we set a condition: for some $r > 2$,

$$n^{-1} \sum_{i=1}^{n} E\left\{h(U_{ni}) - h\left(\frac{i}{n+1}\right)\right\}^2 = o(n^{-2/r}), \tag{6.4.3}$$

where the U_{ni} are defined after (4.2.2). Then, the following theorem is essentially due to Sen and Ghosh (1974b).

THEOREM 6.4.1. *Suppose that the score functions* $\phi_1, \phi_2 \in L^r$, *for some* $r > 2$, *and satisfy* (6.4.3) *and the hypothesis of Theorem* 6.3.1. *Then there exists a standard Wiener process* $W = \{W(t), t \geqslant 0\}$, *such that*

$$\tilde{M}_t^0 = W(t) + o(t^{1/2}) \quad a.s., \quad as \quad t \to \infty. \tag{6.4.4}$$

If, in addition, (6.3.10) *holds, then the result extends to general* $\{M_N\}$ *as well.*

Proof. First, we show that under the hypothesis of the theorem,

$$N^{-1} V_N \to 1 (\Rightarrow V_N \to \infty) \qquad \text{a.s.,} \quad \text{as} \quad N \to \infty. \tag{6.4.5}$$

Let $m_k = M_{k+1}^0 - M_k^0, k \geqslant 1$, so that $V_N = \sum_{k=1}^{N-1} E(m_k^2 \mid \mathfrak{F}_k)$, and,

$$\begin{aligned} m_k &= a_{k+1}^0(R_{k+1k+1}) b_{k+1}^0(S_{k+1k+1}) \\ &\quad + \sum_{i=1}^{k} \{a_{k+1}^0(R_{k+1i}) b_{k+1}^0(S_{k+1i}) - a_k^0(R_{ki}) b_k^0(S_{ki})\} \\ &= m_{k1} + m_{k2}, \text{ say.} \end{aligned} \tag{6.4.6}$$

Note that by the same arguments as in Theorem 6.2.1,

$$\begin{aligned} E(m_{k1}^2 \mid \mathfrak{F}_k) &= \left\{\frac{1}{k+1} \sum_{i=1}^{k+1} a_{k+1}^{02}(i)\right\}\left\{\frac{1}{k+1} \sum_{i=1}^{k+1} b_{k+1}^{02}(i)\right\} \\ &= A_{k+1}^2 B_{k+1}^2 \to 1, \qquad \text{as} \quad k \to \infty. \end{aligned} \tag{6.4.7}$$

Also, note that

$$\Big[\{E(m_{k1}^2 \mid \mathscr{F}_k)\}^{1/2} - \{E(m_{k2}^2 \mid \mathscr{F}_k)\}^{1/2}\Big]^2$$
$$\leqslant E\{(m_{k1} + m_{k2})^2 \mid \mathscr{F}_k\} \leqslant \Big[\{E(m_{k1}^2 \mid \mathscr{F}_k)\}^{1/2} + \{E(m_{k2}^2 \mid \mathscr{F}_k)\}^{1/2}\Big]^2,$$
$$\forall k \geqslant 1, \tag{6.4.8}$$

so that to prove (6.4.5), it suffices to show that

$$E(m_{k2}^2 \mid \mathscr{F}_k) \to 0 \quad \text{a.s.,} \quad \text{as} \quad k \to \infty. \tag{6.4.9}$$

Since, given $\mathscr{F}_k, R_{k+1i}, S_{k+1i}$ are (conditionally) independent, by arguments similar to Theorem 6.2.1, it follows that

$$\begin{aligned}
E(m_{k2}^2 \mid \mathscr{F}_k) &= \sum_{i=1}^{k}\sum_{j=1}^{k} b_k^0(S_{ki}) b_k^0(S_{kj})(k+1)^{-2}(R_{ki} \vee R_{kj})(k+1-R_{ki} \wedge R_{kj}) \\
&\quad \times \big[a_{k+1}^0(R_{ki}+1) - a_{k+1}^0(R_{ki})\big]\big[a_{k+1}^0(R_{kj}+1) - a_{k+1}^0(R_{kj})\big] \\
&\quad + \sum_{i=1}^{k}\sum_{j=1}^{k} a_k^0(R_{ki}) a_k^0(R_{kj})(k+1)^{-2}(S_{ki} \wedge S_{kj})(k+1-S_{ki} \vee S_{kj}) \\
&\quad \times \big[b_{k+1}^0(S_{ki}+1) - b_{k+1}^0(S_{ki})\big]\big[b_{k+1}^0(S_{kj}+1) - b_{k+1}^0(S_{kj})\big] \\
&\quad + \sum_{i=1}^{k}\sum_{j=1}^{k} (k+1)^{-4}(R_{ki} \wedge R_{kj})(k+1-R_{ki} \vee R_{kj}) \\
&\quad \times (S_{ki} \wedge S_{kj})(k+1-S_{ki} \vee S_{kj}) \\
&\quad \times \{a_{k+1}^0(R_{ki}+1) - a_{k+1}^0(R_{ki})\}\{a_{k+1}^0(R_{kj}+1) - a_{k+1}^0(R_{kj})\} \\
&\quad \times \{b_{k+1}^0(S_{ki}+1) - b_{k+1}^0(S_{kj})\}\{b_{k+1}^0(S_{kj}+1) - b_{k+1}^0(S_{kj})\} \\
&= \tau_{k1} + \tau_{k2} + \tau_{k3}, \qquad \text{say.}
\end{aligned} \tag{6.4.10}$$

Now,

$$\begin{aligned}
\tau_{k1} &\leqslant \Big\{\max_{1 \leqslant i \leqslant k} b_k^{02}(i)\Big\}\left\{\sum_{i=1}^{k} \frac{\sqrt{i(k+1-i)}}{k+1} |a_{k+1}^0(i+1) - a_{k+1}^0(i)|\right\}^2 \\
&\leqslant \Big\{\max_{1 \leqslant i \leqslant k} b_k^{02}(i)\Big\}\left\{k^{-1}\sum_{i=1}^{k} E\Big[\phi_1(U_{ki}) - \phi_1\Big(\frac{i}{(k+1)}\Big)\Big]^2\right\} \\
&= \big[o(k^{2/r})\big]\big[o(k^{-2/r})\big] = o(1), \qquad \text{as} \quad k \to \infty,
\end{aligned} \tag{6.4.11}$$

where the second inequality follows from (4.4.21) and the hypothesis of the theorem. Similarly, $\tau_{k2} \to 0$ a.s. as $k \to \infty$, while $\tau_{k3} = o(k^{-4/r}) \to 0$ as $k \to \infty$. Hence (6.4.9) holds.

Let us now choose an η such that $2/r > (1 + \eta)/2$. Then, by direct steps, similar to those in (6.4.10), it follows that

$$\begin{aligned}
|m_{k2}| &\leqslant \sum_{i=1}^{k} \{|a_k^0(R_{ki})|\,|b_{k+1}^0(S_{k+1i}) - b_{k+1}^0(S_{ki})| \\
&\quad + |b_k^0(S_{ki})|\,|a_{k+1}^0(R_{k+1i}) - a_{k+1}^0(R_{ki})| \\
&\quad + |a_{k+1}^0(R_{k+1i}) - a_{k+1}^0(R_{ki})|\,|b_{k+1}^0(S_{k+1i}) - b_{k+1}^0(S_{ki})|\} \\
&= \left[o(k^{1/r})\right]\left\{ \sum_{i=1}^{k} |b_{k+1}^0(S_{ki}+1) - b_{k+1}^0(S_{ki})| \right\} \\
&\quad + \left[o(k^{1/r})\right]\left\{ \sum_{i=1}^{k} |a_{k+1}^0(R_{ki}+1) - a_{k+1}^0(R_{ki})| \right\} + o(k^{1-2/r}) \\
&= o(k^{1-2/r}) = o(k^{(1-\eta)/2}), \qquad \text{as} \quad k \to \infty \quad (\text{where } \eta > 0).
\end{aligned} \tag{6.4.12}$$

Hence, for every $\eta > 0$, there exists an $n_0(\eta)$, such that for $n \geqslant n_0(\eta)$,

$$E\left\{ m_{k2} I\left(m_{k2}^2 > \tfrac{1}{4}k^{1-\eta}\right) \mid \mathfrak{F}_k \right\} = 0 \qquad \text{a.s.} \qquad \forall k \geqslant n. \tag{6.4.13}$$

Further, for every $k \geqslant 1$,

$$\begin{aligned}
E\{|m_{k1}|^r \mid \mathfrak{F}_k\} &= \left\{ \frac{1}{k+1} \sum_{i=1}^{k+1} |a_{k+1}^0(i)|^r \right\} \left\{ \frac{1}{k+1} \sum_{i=1}^{k+1} |b_{k+1}^0(i)|^r \right\} \\
&\leqslant \left(\int_0^1 |\phi_1(u)|^r \, du \right) \left(\int_0^1 |\phi_2(v)|^r \, dv \right) < \infty,
\end{aligned} \tag{6.4.14}$$

so that

$$\begin{aligned}
E\left(m_{k1}^2 I\left(m_{k1}^2 > \tfrac{1}{4}k^{1-\eta}\right) \mid \mathfrak{F}_k\right) &\leqslant \left(\tfrac{1}{4}k^{1-\eta}\right)^{-(r-2)} E(|m_{k1}|^r \mid \mathfrak{F}_k) \\
&= O(k^{-(r-2)(1-\eta)}).
\end{aligned} \tag{6.4.15}$$

From the last two equations, it follows that

$$\sum_{n \geqslant 1} n^{-1+\eta} E\left(m_n^2 I\left(m_n^2 > n^{1-\eta}\right) \mid \mathfrak{F}_n\right) < \infty \qquad \text{a.s.,} \tag{6.4.16}$$

and hence our desired result follows from (2.5.9) and (2.5.10). Q.E.D.

We consider now the case of $\{Q_N^0\}$. We replace in (6.4.1), M_N^0 by Q_N^0 and denote the resulting statistics by V_N^*, $N \geqslant 0$. Also, in (6.4.2), we replace M_N^0 and V_N by Q_N^0 and V_N^*, respectively, and denote the resulting statistic by $\tilde{Q}_{V_N^*}$, $N \geqslant 0$; by linear interpolation we complete then the definition of $\tilde{Q}^0 = \{\tilde{Q}_t^0, t \geqslant 0\}$.

THEOREM 6.4.2. *Suppose that for some* $r > 2$, $E|a(X)|^r < \infty$, $\phi_2 \in L^r$ *and it satisfies* (6.3.4). *Then, under the hypothesis of Theorem* 6.6.4, *there exists a standard Wiener process* $W = \{W(t), t > 0\}$, *such that*

$$\tilde{Q}_t^0 = W(t) + o(t^{1/2}) \qquad a.s., \quad as \quad t \to \infty. \tag{6.4.17}$$

If, in addition, (6.3.10) *holds for* ϕ_2, *then the result extends for general* Q_N.

The proof follows along the same line as in the proof of Theorem 6.4.1 and hence is left as an exercise (see Problem 6.4.1).

6.5. A.S. REPRESENTATION OF M_N AND Q_N IN TERMS OF I.I.D.R.V.

The results of Section 6.4 are extended here (under additional regularity conditions) to the case where the null hypothesis in (6.2.1) or (6.2.7) may not hold. Bhuchongkul (1964) has considered a Chernoff-Savage representation (in probability) for M_N in terms of i.i.d.r.v.'s and this weak representation has been studied further by Ruymgaart (1972, 1973), Ruymgaart, Shorack, and van Zwet (1972), and others. Bönner (1976) has strengthened this representation (in the same spirit as in Theorems 5.4.2 and 5.4.3) in having an a.s. equivalence; some improvements of this result are due to Bönner, Müller-Funk, and Witting (1980). We discuss this result here. In this context, we assume that the d.f.'s H, F, and G are all absolutely continuous with continuous density functions h, f, and g. Also, for the score functions ϕ_1 and ϕ_2, on letting $\phi_j^{(0)} = \phi_j$ and $\phi_j^{(r)}(u) = (d^r/du^r)\phi_j(u), r = 1, 2, j = 1, 2$, we assume that

$$|\phi_j^{(r)}(u)| \leqslant K[u(1-u)]^{-(1/4)-r+\delta}, \qquad \forall 0 < u < 1, r = 0, 1, 2, \tag{6.5.1}$$

where $\delta > 0$ (and, without any loss of generality, we may take $\delta \leqslant \frac{1}{4}$). Finally, let

$$U_i = U_{i,1} + U_{i,2} + U_{i,3}, \qquad i \geqslant 1 \quad \text{and} \quad S_N = \sum_{i=1}^{N} U_i \qquad \text{for} \quad N \geqslant 1, \tag{6.5.2}$$

where

$$U_{i,1} = \phi_1(F(X_i))\phi_2(G(Y_i)), \tag{6.5.3}$$

$$U_{i,2} = \int_{R^2}\int [c(x - X_i) - F(x)]\phi_1^{(1)}(F(x))\phi_2(G(y))\,dH(x, y), \tag{6.5.4}$$

$$U_{i,3} = \int_{R^2}\int [c(y - Y_i) - G(y)]\phi_1(F(x))\phi_2^{(1)}(G(y))\,dH(x, y). \tag{6.5.5}$$

Then we have the following.

THEOREM 6.5.1. *Under* (6.5.1),

$$M_N = S_N + R_N, \tag{6.5.6}$$

where

$$N^{-1/2}|R_N| = o(N^{-\eta}) \qquad a.s., \quad as \quad N \to \infty, \qquad for\ some\ \eta > 0. \tag{6.5.7}$$

Proof. In view of the similarity of the proof to that of Theorem 5.4.2 (or 5.4.3), we present only the salient features. We write

$$\phi_{1N}(t) = a_N(i) \qquad \text{and} \quad \phi_{2N}(t) = b_N(i)$$
$$\text{for} \quad \frac{i-1}{N} < t \leqslant \frac{i}{N}, i = 1, \ldots, N. \tag{6.5.8}$$

Then, by (6.2.2) and (6.5.8), we have

$$N^{-1}M_N = \int_{-\infty}^{\infty}\int_{-\infty}^{\infty} \phi_{1N}(F_N(x))\phi_{2N}(G_N(y))\,dH_N(x, y), \tag{6.5.9}$$

where F_N, G_N, and H_N are the empirical d.f.'s. As in (5.4.19), we may replace in (6.5.9) $\phi_{jN}(i/N)$ by $\phi_j(i/(N+1))$, for $i = 1, \ldots, N$ and $j = 1, 2$. Then, by the Taylor expansion of $\phi_1(NF_N(x)/(N+1))$ [and $\phi_2(NG_N \cdot (y)/(N+1))$] around $\phi_1(F(x))$ [and $\phi_2(G(y))$] along with the decomposition that $dH_N = dH + d[H_N - H]$, we obtain

$$R_N = \sum_{k=1}^{6} R_N^{(k)}, \tag{6.5.10}$$

where, writing $F_N^* = NF_N/(N+1)$ and $G_N^* = NG_N/(N+1)$, we have (for some $0 < \theta < 1$)

$$N^{-1}R_N^{(1)} = \int\int_{R^2}(F_N^* - F)\phi_1^{(1)}(F)\phi_2(G)d(H_N - H), \tag{6.5.11}$$

$$N^{-1}R_N^{(2)} = \int\int_{R^2}(G_N^* - G)\phi_2^{(1)}(G)\phi_1(F)d(H_N - H), \tag{6.5.12}$$

$$N^{-1}R_N^{(3)} = \int\int_{I_N}(F_N^* - F)(G_N^* - G)\phi_1^{(1)}(\theta F_N^* + (1-\theta)F)$$
$$\times \phi_2^{(1)}(\theta G_N^* + (1-\theta)G)\,dH_N, \tag{6.5.13}$$

$$N^{-1}R_N^{(4)} = \int\int_{I_N} \left[\tfrac{1}{2}(F_N^* - F)^2 \right] \phi_1^{(2)}(\theta F_N^* + (1-\theta)F)$$

$$\times \phi_2(\theta G_N^* + (1-\theta)G)\, dH_N, \tag{6.5.14}$$

$$N^{-1}R_N^{(5)} = \int\int_{I_N} \left[\tfrac{1}{2}(G_N^* - G)^2 \right] \phi_2^{(2)}(\theta G_N^* + (1-\theta)G)$$

$$\times \phi_1(\theta F_N^* + (1-\theta)F)\, dH_N, \tag{6.5.15}$$

$$N^{-1}R_N^{(6)} = \int\int_{I_N^C} \left[\phi_1(F)\phi_2(G) + (F_N^* - F)\phi_1^{(1)}(F)\phi_2(G) \right.$$

$$\left. + (G_N^* - G)\phi_1(F)\phi_2^{(1)}(G) \right] dH_N, \tag{6.5.16}$$

where $I_N = \{(x, y) : N^{-1-\delta} \leqslant F(x), G(y) \leqslant 1 - N^{-1-\delta}\}$, δ is defined by (6.5.1) and $I_N^C = R^2 - I_N$ is the complement of I_N. As in the proof of Theorem 5.4.2, we make use of (2.6.31). In this context, we need the following result due to Bönner, Müller-Funk, and Witting (1980). Let $\mathbf{u} = (u_1, u_2)$ and $\mathbf{v} = (v_1, v_2)$ be points in E^2 and let $B_N = \{(\mathbf{u}, \mathbf{v}) : |u_i - v_i| \leqslant n^{-1/2}\log n, i = 1, 2\}$. Then, we have the following.

Lemma 6.5.2. For absolutely continuous H, as $N \to \infty$,

$$\sup_{(u,v)\in B_N} \left| H_N\left(F^{-1}(u_1), G^{-1}(u_2)\right) - H\left(F^{-1}(u_1), G^{-1}(u_2)\right) \right.$$

$$\left. - H_N\left(F^{-1}(v_1), G^{-1}(v_2)\right) + H\left(F^{-1}(v_1), G^{-1}(v_2)\right) \right|$$

$$= o\left(N^{-1}(\log N)^2\right) \quad \text{a.s.} \tag{6.5.17}$$

Proof. In the same way that (5.4.38) follows from Theorem 7.3.2, the proof of (6.5.17) follows along the same line. By reference to the proof of Theorem 7.3.1 (and 7.3.2), we note that the only modification here is to consider the bivariate empirical and the true d.f.'s and to replace the $2b_n$ grid-points by $4b_n^2$ grid-points corresponding to the $2b_n$ possible values of each of the two coordinates in $\mathbf{u}$. Therefore, we leave this proof as an exercise (see Problem 6.5.1).

Returning to the proof of Theorem 6.5.1, we observe that once Lemma 6.5.2 is at our disposal, we may virtually repeat the arguments in (5.4.26) and (5.4.43), replacing (5.4.38) and (6.5.17), and making the necessary adjustments in the treatments of $R_N^{(3)}$ through $R_N^{(5)}$. The treatment of $R_N^{(1)}$, $R_N^{(2)}$, and $R_N^{(6)}$ will be somewhat simpler. To avoid repetitions, we therefore omit these details. Q.E.D.

Looking at the definitions of the $U_{i,j}$, $j = 1, 2, 3$ and $i \geqslant 1$ and the U_i in (6.5.2) through (6.5.5), we gather that the U_i are i.i.d.r.v.'s with finite moments up to the order r for some $r > 2$. Further, $EU_{i,2} = EU_{i,3} = 0$ while, say,

$$EU_{i,1} = \int\int_{R^2} \phi_1(F(x))\phi_2(G(y))\,dH(x, y) = \mu(\phi_1, \phi_2). \quad (6.5.18)$$

Hence we have no problem in applying Theorem 2.5.1 to $S_n - n\mu(\phi_1, \phi_2)$, $n \geqslant 1$. This gives an a.s. invariance principle for the S_N, defined by (6.5.2) through (6.5.5), and, by virtue of Theorem 6.5.1, such a result is applicable to the sequence $\{M_N\}$ as well. We may remark that as δ in (6.5.1) is $\leqslant \frac{1}{4}$, in (6.5.7), instead of the a.s. statement, we may, as in Theorem 5.4.3, make a statement involving an order $N^{-1-\nu}$ (for some $\nu > 0$) for the probability of $N^{-1/2}|R_N|$ being greater than $N^{-\eta}$, $\eta > 0$.

We conclude this section with a parallel result for the mixed rank statistics. Let $S_N^* = \sum_{i=1}^N (U_{i1}^* + U_{i2}^*)$, $N \geqslant 1$, where

$$U_{i1}^* = a(X_i)\phi_2(G(Y_i)), \qquad i \geqslant 1, \quad (6.5.19)$$

$$U_{i2}^* = \int\int_{R^2} \{c(y - Y_i) - G(y)\}\phi_2^{(1)}(G(y))a(x)\,dH(x, y), \qquad i \geqslant 1. \quad (6.5.20)$$

Then, we have the following.

THEOREM 6.5.3. *If* $E|a(X)|^r < \infty$ *for some* $r > 2$ *and* ϕ_2 *satisfies* (6.5.1), *then* $Q_N = S_N^* + R_N^*$, *where*

$$R_N^* = O(N^{(1/2)-\eta}) \quad a.s., \quad as \quad N \to \infty, \quad and\ \eta\ is\ a\ positive\ number. \quad (6.5.21)$$

The proof runs parallel to that of Theorem 6.5.1 and hence is left as an exercise (see Problem 6.5.2).

6.6. ALMOST SURE CONVERGENCE OF M_N AND Q_N

We study the a.s. convergence of the statistics M_N and Q_N when H_0 may not hold. First, let us define $\mu(\phi_1, \phi_2)$ as in (6.5.18). Then, we are interested here in the a.s. convergence of $N^{-1}M_N$ to $\mu(\phi_1, \phi_2)$. For this purpose, we define the $\phi_{jN}(t)$, $j = 1, 2$ as in (6.5.8) and consider the integral representation for $N^{-1}M_N$ in (6.5.9). Further, we assume that as $N \to \infty$,

$$N^{-1}\sum_{i=1}^{N}(a_N(i) - \phi_1(U_{Ni}))^2 \to 0 \quad \text{and}$$
$$N^{-1}\sum_{i=1}^{N}(b_N(i) - \phi_2(U_{Ni}))^2 \to 0 \quad \text{a.s.,} \tag{6.6.1}$$

where the U_{Ni} are defined in (4.2.2).

THEOREM 6.6.1. *If* ϕ_1 *and* ϕ_2 *are square integrable inside* $(0,1)$ *and* (6.6.1) *holds, then*

$$N^{-1}M_N \to \mu(\phi_1,\phi_2) \quad a.s. \quad as \quad N\to\infty. \tag{6.6.2}$$

Proof. By (6.5.9) and (6.5.18), we have

$$N^{-1}M_N - \mu(\phi_1,\phi_2) = I_{N1} + I_{N2} + I_{N3}, \tag{6.6.3}$$

where

$$I_{N1} = \int_{-\infty}^{\infty}\int_{-\infty}^{\infty}\phi_{1N}(F_N(x))\{\phi_{2N}(G_N(y)) - \phi_2(G(y))\}\,dH_N(x,y), \tag{6.6.4}$$

$$I_{N2} = \int_{-\infty}^{\infty}\int_{-\infty}^{\infty}\phi_2(G(y))\{\phi_{1N}(F_N(x)) - \phi_1(F(x))\}\,dH_N(x,y), \tag{6.6.5}$$

$$I_{N3} = \int_{-\infty}^{\infty}\int_{-\infty}^{\infty}\phi_1(F(x))\phi_2(G(y))d\{H_N(x,y) - H(x,y)\}. \tag{6.6.6}$$

Now, by the Schwarz inequality,

$$\begin{aligned} I_{N1}^2 &\leqslant \left\{\int_{-\infty}^{\infty}\phi_{1N}^2(F_N(x))\,dF_N(x)\right\} \\ &\quad\times\left\{\int_{-\infty}^{\infty}\left[\phi_{2N}(G_N(y)) - \phi_2(G(y))\right]^2 dG_N(y)\right\} \\ &= \left\{N^{-1}\sum_{i=1}^{N}a_N^2(i)\right\}\left\{N^{-1}\sum_{i=1}^{N}\left[b_N(i) - \phi_2(U_{Ni})\right]^2\right\} \\ &\to 0 \quad \text{a.s.,} \quad \text{as} \quad N\to\infty, \quad \text{by (6.6.1).} \end{aligned} \tag{6.6.7}$$

Similarly, on noting that (by the Kintchine strong law of large numbers), as $N\to\infty$, $\int_{-\infty}^{\infty}\phi_2^2(G(y))\,dG_N(y)$ a.s. converges to $\int_{-\infty}^{\infty}\phi_2^2(G(y))\,dG(y) < \infty$, it follows that $I_{N2}\to 0$ a.s., as $N\to\infty$. Finally, we rewrite I_{N3} as $N^{-1}\sum_{i=1}^{N}\phi_1(F(X_i))\phi_2(G(Y_i)) - E\phi_1(F(X))\phi_2(G(Y))$, so that, by the Kintchine strong law of large numbers again, $I_{N3}\to 0$ a.s., as $N\to\infty$.

Q.E.D.

Let us now define

$$\mu^*(a,\phi_2) = \int\int_{R^2} a(x)\phi_2(G(y))\,dH(x,y). \tag{6.6.8}$$

Then, we have the following.

THEOREM 6.6.2. *If* $Ea^2(X) < \infty$ *and* ϕ_2 *is square integrable and satisfy* (6.6.1), *then*

$$N^{-1}Q_N \to \mu^*(a,\phi_2) \quad a.s., \quad as \quad N\to\infty. \tag{6.6.9}$$

The proof is very similar to that of the preceding theorem and hence is left as an exercise (see Problem 6.6.1).

We conclude this section with some discussion on (6.6.1). Suppose that (i) ϕ_1 and ϕ_2 are continuous inside every closed interval $[\epsilon, 1-\epsilon]$, $\epsilon > 0$ and (ii) $\{\phi_{jN}^2(t), 0 < t < 1\}$ is uniformly (in N) integrable inside I, for $j = 1,2$. Then, the assumed square integrability of ϕ_1 implies that for every ϵ $(0 < \epsilon < \frac{1}{4})$, there exists an $\eta(>0)$, such that

$$\int_0^{2\epsilon} + \int_{1-2\epsilon}^{1} \phi_1^2(u)\,du < \eta. \tag{6.6.10}$$

Further, by the Glivenko-Cantelli theorem, $F_N^{-1}(\epsilon) \leqslant F^{-1}(2\epsilon)$ a.s. and $F_N^{-1}(1-\epsilon) \geqslant F^{-1}(1-2\epsilon)$ a.s., as $N\to\infty$, and hence

$$N^{-1}\Big\{\sum_{i\leqslant[N\epsilon]} + \sum_{i\geqslant N-[N\epsilon]} \phi_1^2(U_{Ni})\Big\} = \int_{-\infty}^{F_N^{-1}(\epsilon)} + \int_{F_N^{-1}(1-\epsilon)}^{1} \phi_1^2(F(x))\,dF_N(x)$$

$$\leqslant \int_{-\infty}^{F^{-1}(2\epsilon)} + \int_{F^{-1}(1-2\epsilon)}^{1} \phi_1^2(F(x))\,dF_N(x)$$

$$\text{a.s.,} \quad \text{as} \quad N\to\infty. \tag{6.6.11}$$

Since the rhs of (6.6.11) forms a reverse martingale sequence, by (6.6.10) and the reverse martingale convergence theorem, it can be made arbitrarily small (a.s.) as $N\to\infty$. Also, by condition (ii), for every $\epsilon > 0$, there exists an η' (>0), such that

$$N^{-1}\Big\{\sum_{i\leqslant[N\epsilon]} + \sum_{i\geqslant N-[N\epsilon]} a_N^2(i)\Big\} < \eta' \tag{6.6.12}$$

On the other hand, by assumption (i), inside $[\epsilon, 1-\epsilon]$, ϕ_1 is boundedly continuous, and hence, the a.s. convergence of $\max\{|U_{Ni} - i/(N+1)| : 1 \leqslant i \leqslant N\}$ to 0 implies that

$$N^{-1}\left\{\sum_{[N\epsilon]<i<N-[N\epsilon]}\left[a_N(i) - \phi_1(U_{Ni})\right]^2\right\} \to 0 \quad \text{a.s., as} \quad N \to \infty. \tag{6.6.13}$$

Hence (6.6.1) holds. In practice, conditions (i) and (ii) hold quite generally.

EXERCISES

6.2.1. Consider the problems 3.2.4 and 3.2.5. (a) Show that for the Kendall tau statistic, the representation (6.2.2) does not hold but, by virtue of the U-statistic representation, the invariance principles follow from the results of Chapter 3. (b) Express the Spearman rank correlation r_n in the form (6.2.2) as well as a linear function of two U-statistics and derive the invariance principles from either approach.

6.2.2. Let $\Phi(x)$ be the standard normal d.f. and let

$$a_N(i) = b_N(i) = E\{\Phi^{-1}(U_{Ni})\}, \qquad i = 1, \ldots, N;$$

the corresponding M_N in (6.2.2) is termed the normal scores statistics. Show that (6.3.1) holds here.

6.3.1. Provide the proof of Theorem 6.3.2.

6.3.2. Extend Theorems 6.3.1 and 6.3.2 to stochastic sample sizes $\{N_n\}$, satisfying (4.3.19).

6.3.3. Under the hypothesis of Theorem 6.3.3, show that $\{P_N^\Delta\}$ is contiguous to $\{P_N\}$ (Hájek and Šidák, 1967; Behnen, 1971).

6.3.4 (6.6.3 continued). Show that the convergence of the f.d.d.'s of the $\{Y_N^0\}$ to those of W (under H_0) and the contiguity of $\{P_N^\Delta\}$ to $\{P_N\}$ imply that the f.d.d.'s of $\{Y_N^0\}$ converge to those of $W + \mu$, under $\{K^{(N)}\}$.

6.3.5. Provide the proof of Theorem 6.3.5.

6.4.1. Provide the proof of Theorem 6.4.2.

6.5.1. Following the lines of the proof of Theorems 7.3.1 and 7.3.2, provide a proof of Lemma 6.5.2.

6.5.2. Provide a proof of Theorem 6.5.3.

6.6.1. Following the lines of the proof of Theorem 6.6.1, provide a proof of Theorem 6.6.2.

CHAPTER 7

Invariance Principles for Linear Combinations of Functions of Order Statistics

7.1. INTRODUCTION

Linear combinations of functions of order statistics (LCFOS) play a vital role in the theory of statistical inference. For a variety of problems in statistical inference, robust and efficient estimators as well as tests of significance are based on LCFOS; for some detailed accounts, we may refer to Sarhan and Greenberg (1962) and David (1980). These statistics are easily adaptable for censored or truncated data. In the context of life-testing situations (as we shall see in Chapter 11), these LCFOS have special importance. Invariance principles for some LCFOS are studied in this chapter. Section 7.2 deals with the preliminary notions and a basic classification of LCFOS in to two different types. The so-called Bahadur representation of sample quantiles and its impact on invariance principles are considered in Section 7.3. Section 7.4 is devoted to the weak convergence of LCFOS with smooth weight functions to Brownian motions. Embedding of Wiener processes for LCFOS (through an almost sure representation in terms of a sum of i.i.d.r.v.'s) is then established in Section 7.5. The last section deals with the almost sure convergence of certain functions of order statistics having relevance to the main theorems in Section 7.4.

7.2. PRELIMINARY NOTIONS

Let $\{X_i, i \geqslant 1\}$ be a sequence of i.i.d.r.v.'s with a continuous d.f. F, defined on the real line $R = (-\infty, \infty)$. For every $n(\geqslant 1)$, let $X_{n,1}, \dots, X_{n,n}$ be the ordered r.v.'s corresponding to $X_1, \dots, X_n$; by virtue of the assumed continuity of F, ties among the X_i (and hence $X_{n,i}$) may be neglected, with probability 1, so that $X_{n,1} < \cdots < X_{n,n}$. Typically, a LCFOS is defined by

$$T_n = \sum_{i=1}^{n} c_{n,i} g_n(X_{n,i}), \qquad (n \geqslant 1), \tag{7.2.1}$$

where $\{c_{n,i}, 1 \leqslant i \leqslant n; n \geqslant 1\}$ is a triangular array of known constants and $\{g_n(\cdot)\}$ is a sequence of specified functions. We find it convenient to consider the following two types of LCFOS separately:

(i) ***Type I LCFOS.*** Let us define $\xi = \{\xi_p : p \in (0,1)\}$ by letting

$$\xi_p = \inf\{x : F(x) = p\}, \qquad 0 < p < 1. \tag{7.2.2}$$

ξ_p is then the *p-quantile of the d.f. F*. Also, let $F_n = \{F_n(x), x \in R\}$ be the empirical d.f. based on $X_1, \ldots, X_n$ and we define $\xi^{(n)} = \{\xi_p^{(n)} : p \in (0,1)\}$ by letting

$$\xi_p^{(n)} = \inf\{x : F_n(x) \geqslant p\}, \qquad p \in (0,1). \tag{7.2.3}$$

Then, $\xi_p^{(n)}$ is the *sample p-quantile*. Since F_n is a step function, $\xi_p^{(n)} = X_{n,r}$ for some $r : 1 \leqslant r \leqslant n$ (whenever $p \geqslant n^{-1}$). If a and b stand for the lower and upper endpoints of the d.f. F (i.e., $a = \sup\{x : F(x) = 0\}$ and $b = \inf\{x : F(x) = 1\}$ where a may possibly be $-\infty$ and b may be $+\infty$), then we let $\xi_0 = a$, $\xi_1 = b$, and $\xi_p^{(n)} = a$ for every $p < n^{-1}$.

Let $(0 <)p_1 < \cdots < p_k(< 1)$ be $k(\geqslant 1)$ distinct numbers and $\xi_{p_j}^{(n)} = X_{n,r_j}$, for $j = 1, \ldots, k$. Then, a type I LCFOS is defined by

$$T_n = \sum_{j=1}^{k} d_{n,j} g_{nj}(X_{n,r_j}), \tag{7.2.4}$$

where the $d_{n,j}$ are suitable constants and the functions g_{nj} are all specified. It is generally assumed that the $d_{n,j}$ converge to some d_j and the $g_{nj}(\cdot)$ to some smooth $g_j(\cdot)$, for every $j = 1, \ldots, k$. For these statistics, an almost sure representation of $\xi_p^{(n)} - \xi_p$ in terms of an average of i.i.d.r.v.'s (due to Bahadur, 1966) is of considerable importance and this will be considered in the next section.

(ii) ***Type II LCFOS.*** Consider a sequence $\{\phi_n\}$ of *weight functions* $\phi_n = \{\phi_n(t), 0 \leqslant t \leqslant 1\}$ where $\phi_n(t) \to \phi(t), \forall t \in (0,1)$ and $\phi = \{\phi(t), 0 \leqslant t \leqslant 1\}$ is smooth. Suppose that

$$c_{n,i} = n^{-1}\phi_n\left(\frac{i}{(n+1)}\right) = n^{-1}\phi_n(t),$$
$$\text{for} \quad \frac{i-1}{n} < t \leqslant \frac{i}{n}, \quad i = 1, \ldots, n. \tag{7.2.5}$$

Then, T_n, defined by (7.2.1) and (7.2.5), is termed a LCFOS with smooth weight function, or a type II LCFOS. If F_n is the empirical d.f. based on $X_1, \ldots, X_n$, then we may rewrite T_n as

$$T_n = \int_{-\infty}^{\infty} \phi_n(F_n(x)) g_n(x)\, dF_n(x). \qquad (7.2.6)$$

Invariance principles for Type II LCFOS are considered in Sections 7.4 and 7.5.

7.3. BAHADUR REPRESENTATION OF SAMPLE QUANTILES AND INVARIANCE PRINCIPLES FOR TYPE I LCFOS

We start with the Bahadur (1966) representation of sample quantiles and incorporate the same for studying the desired invariance principles. Let $\{U_i, i \geqslant 1\}$ be a sequence of i.i.d.r.v.'s with the uniform (0, 1) d.f., $I_n(u) = n^{-1}\sum_{i=1}^{n} c(u - U_i)$, $u \in (0,1)$ be the empirical d.f., and let

$$G_n(u, p) = n^{1/2}\{[I_n(u) - u] - [I_n(p) - p]\}, \quad 0 \leqslant u, p \leqslant 1. \qquad (7.3.1)$$

Further, let $\{a_n\}$ be a sequence of positive numbers such that for some $\beta : 0 < \beta \leqslant \frac{1}{2}$, as $n \to \infty$,

$$a_n \sim n^{-\beta} \log n. \qquad (7.3.2)$$

Finally, let $J_{n,p} = \{u : |u - p| \leqslant a_n\}$ and

$$G^*_{n,p} = \sup\{|G_n(u, p)| : u \in J_{n,p}\}. \qquad (7.3.3)$$

THEOREM 7.3.1 (Bahadur, 1966). *For every* $0 < p < 1$, *as* $n \to \infty$,

$$G^*_{n,p} = O(n^{-\beta/2} \log n) \qquad a.s. \qquad (7.3.4)$$

Proof. For simplicity of the proof, we let $\beta = \frac{1}{2}$; a similar proof holds for any $\beta \in (0, \frac{1}{2}]$. Let $\{b_n\}$ be a sequence of positive numbers such that as $n \to \infty$, $b_n \sim n^{1/4}$ and let $u_{n,j} = p + j b_n^{-1} a_n$, for $j = 0, \pm 1, \ldots, \pm b_n$. Then, by virtue of the monotonicity of $I_n(u)$ and u ($\in [0,1]$), $G_n(u_{n,j}, p) - n^{1/2}(u_{n,j+1} - u_{n,j}) \leqslant G_n(u, p) \leqslant G_n(u_{n,j+1}, p) + n^{1/2}(u_{n,j+1} - u_{n,j})$, $\forall u \in [u_{n,j}, u_{n,j+1}]$, so that

$$G^*_{n,p} \leqslant \max_{0 \leqslant |j| \leqslant b_n} |G_n(u_{n,j}, p)| + O(n^{-1/4} \log n). \qquad (7.3.5)$$

On the other hand, for each j, $G_n(u_{n,j}, p) = n^{-1}\sum_{i=1}^{n}(Z_{ni}^{(j)} - EZ_{ni}^{(j)})$ where the $Z_{ni}^{(j)}$ are independent $0-1$ valued random variables with $P\{Z_{ni}^{(j)} = 1\}$

$= 1 - P\{Z_{ni}^{(j)} = 0\} = P\{U_i \in (p, u_{n,j})\} = |p - u_{n,j}| = O(n^{-1/2}\log n)$. Since the $Z_{ni}^{(j)}$ are bounded valued r.v.'s, by using the Bernstein inequality, we obtain (by a few standard steps) that for every $s > 0$, there exist positive constants $K_s^{(1)}$, $K_s^{(2)}$ and an integer n_s, such that for $n \geqslant n_s$,

$$P\left\{|G_n(u_{n,j}, p)| > K_s^{(1)} n^{-1/4}\log n\right\} \leqslant K_s^{(2)} n^{-s}. \tag{7.3.6}$$

Hence, choosing $s > \frac{5}{4}$, it follows from (7.3.5), (7.3.6), the Bonferroni inequality, and the Borel-Cantelli lemma that (7.3.4) holds. Q.E.D.

Let us now define

$$\begin{aligned} G_n^* &= \sup\{G_{n,p}^* : 0 < p < 1\} \\ &= \sup\left\{n^{1/2}|I_n(t) - I_n(s) - t + s| : 0 \leqslant s < t \leqslant s + a_n \leqslant 1\right\} \end{aligned} \tag{7.3.7}$$

THEOREM 7.3.2. *As* $n \to \infty$, $G_n^* = O(n^{-\beta/2}\log n)$ *a.s.*

Proof. Here also, we let $\beta = \frac{1}{2}$. Let then $k_n = [n^{3/4}] + 1$ and let $p_{n,j} = jn^{-3/4}$ for $j = 1, \ldots, k_n - 1$. Then proceeding as in the proof of (7.3.5), it can be shown that

$$G_n^* \leqslant 3\left\{\max_{1 \leqslant j \leqslant k_n} G_{n,p_{n,j}}^*\right\} + O(n^{-1/4}). \tag{7.3.8}$$

On the other hand, for each $p_{n,j}$, we may virtually repeat the proof of (7.3.6) where we take $s > 2$ and complete the rest of the proof by using the Bonferroni inequality along with the Borel-Cantelli lemma. Q.E.D.

It is quite apparent from the proof of the preceding two theorems that we have in fact the result that

$$P\left\{G_n^* \geqslant k_1 n^{-\beta/2}\log n\right\} \leqslant k_2 n^{-s}, \qquad \forall n \geqslant n_s, \tag{7.3.9}$$

where $s(>0)$ is any positive number, $\beta > 0$ and the positive constants k_1 and k_2 depend on s and β, and so also the $n_s(>0)$. Consider now the case of d.f.'s not necessarily the uniform one. Let $\mathcal{F}_p = \{F : F$ has a continuous p.d.f. f in some neighborhood of ξ_p where $0 < f(\xi_p) < \infty\}$, $0 < p < 1$ and let $\mathcal{F} = \bigcap_{0<p<1}\mathcal{F}_p = \{F : F$ has a continuous, positive, and bounded density f on its entire support$\}$. Then,

$$\begin{aligned} &\sup\left\{n^{1/2}|F_n(x) - F_n(\xi_p) - F(x) + p| : |x - \xi_p| \leqslant a_n\right\} \\ &\quad = O(n^{-1/2\beta}\log n) \qquad \text{a.s.,} \quad \text{as} \quad n \to \infty, \end{aligned} \tag{7.3.10}$$

and a result parallel to the one in Theorem 7.3.2 holds. Moreover, note that for $\beta = \frac{1}{2}$,

$$P\left\{\xi_p^{(n)} > \xi_p + a_n\right\} = P\left\{F_n(\xi_p + a_n) \leq p\right\}$$
$$= P\left\{F_n\left(\xi_p + n^{-1/2}\log n\right) \leq p\right\} = O(n^{-s}), \qquad s > 1, \tag{7.3.11}$$

and a similar result holds for $p\{\xi_p^{(n)} < \xi_p - a_n\}$, and hence

$$|\xi_p^{(n)} - \xi_p| \leq n^{-1/2}\log n \qquad \text{a.s.,} \quad \text{as} \quad n \to \infty, \tag{7.3.12}$$

and consequently, from (7.3.10) and (7.3.12), we have

$$n^{1/2}\left[F_n\left(\xi_p^{(n)}\right) - F_n(\xi_p)\right] = n^{1/2}\left[F\left(\xi_p^{(n)}\right) - p\right] + O\left(n^{-1/4}\log n\right) \qquad \text{a.s.} \tag{7.3.13}$$

Thus, for $f(\xi_p) > 0$, we have

$$n^{1/2}\left(\xi_p^{(n)} - \xi_p\right) = -n^{1/2}\left[F_n(\xi_p) - p\right]/f(\xi_p) + o(1) \quad \text{a.s.,} \tag{7.3.14}$$

and if f has a bounded first derivative in some neighborhood of ξ_p, then as $n \to \infty$,

$$n^{1/2}\left(\xi_p^{(n)} - \xi_p\right) = -n^{1/2}\left[F_n(\xi_p) - p\right]/f(\xi_p) + O\left(n^{-1/4}\log n\right) \qquad \text{a.s.} \tag{7.3.15}$$

The last result is of considerable importance to our study of the invariance principles. For this purpose, note that

$$F_n(\xi_p) - p = n^{-1}\sum_{i=1}^{n}\{I(X_i \leq \xi_p) - p\} \tag{7.3.16}$$

involves an average over independent and bounded valued random variables, where $EI(X_i \leq \xi_p) = p$ and $V[I(X_i \leq \xi_p)] = p(1-p)$. Thus all the invariance principles studied in Chapter 2 hold for these statistics, and, as a result, by (7.3.14), those invariance principles also hold for the sample quantiles $\xi_p^{(n)}$. We summarize these results (without proof) as follows:

(i) Let $W_n^{(1)} = \{W_n^{(1)}(t) = k_n(t)n^{-1/2}\sigma^{-1}[\xi_p^{(k_n(t))} - \xi_p],\ 0 \leq t \leq 1\}$ where $k_n(t) = \max\{k : k/n \leq t\}$, $0 \leq t \leq 1$ and $\sigma^2 = p(1-p)/f^2(\xi_p)$, and let $W = \{W(t),\ 0 \leq t \leq 1\}$ be a standard Wiener process on $[0,1]$. Then, for $F \in \mathfrak{F}_p$,

$$W_n^{(1)} \xrightarrow{\mathfrak{D}} W, \qquad \text{in the } J_1\text{-topology on} \quad D[0,1]. \tag{7.3.17}$$

(ii) Let $W_n^{(2)} = \{W_n^{(2)}(t) = n^{1/2}\sigma^{-1}(\xi_p^{(k_n^*(t))} - \xi_p), 0 \leq t \leq 1\}$, where

$k_n^*(t) = \min\{k : n/k \leqslant t\}, 0 \leqslant t \leqslant 1$. Then, for $F \in \mathcal{F}_p$,

$$W_n^{(2)} \xrightarrow{\mathfrak{D}} W, \qquad \text{in the } J_1\text{-topology on} \quad D[0,1]. \tag{7.3.18}$$

(iii) There exists a standard Wiener process $\zeta = \{\zeta(t) : 0 < t < \infty\}$ such that on letting

$$\zeta^*(t) = k\sigma^{-1}\left[\xi_p^{(k)} - \xi_p\right], \qquad k \leqslant t < k+1, \qquad k \geqslant 0, \tag{7.3.19}$$

[where $\zeta^*(0) = 0$], for $F \in \mathcal{F}_p$,

$$\zeta^*(t) = \zeta(t) + o(t^{1/2}) \qquad \text{a.s.,} \quad \text{as} \quad t \to \infty. \tag{7.3.20}$$

Let us now consider analogous invariance principles for type I LCFOS defined by (7.2.4). Concerning the $d_{n,j}$, we assume that there exist real and finite numbers $d_1, \ldots, d_k$ such that

$$\lim_{n\to\infty} n^{1/2}|d_{n,j} - d_j| = 0, \qquad \text{for} \quad j = 1, \ldots, k. \tag{7.3.21}$$

For simplicity, we let $g_{nj}(t) = t$ for every $j (= 1, \ldots, k)$ and n, and let then

$$\tau = \sum_{j=1}^{k} d_j \xi_{p_j}. \tag{7.3.22}$$

Then, note that

$$n^{1/2}(T_n - \tau) = \sum_{j=1}^{k} n^{1/2}(d_{n,j} - d_j)\xi_{p_j}^{(n)} + \sum_{j=1}^{k} d_j n^{1/2}\left(\xi_{p_j}^{(n)} - \xi_{p_j}\right), \tag{7.3.23}$$

where by (7.3.12) and (7.3.21), the first term on the rhs of (7.3.23) is $o(1)$ a.s., as $n \to \infty$, and for the second term, we make use of (7.3.14) for each $j = 1, \ldots, k$. Thus

$$n^{1/2}(T_n - \tau) = \sum_{j=1}^{k} \left[\frac{d_j}{f(\xi_{p_j})}\right] n^{1/2}\left[F_n(\xi_{p_j}) - p_j\right] + o(1) \qquad \text{a.s.,} \tag{7.3.24}$$

as $n \to \infty$. Note that

$$\sum_{j=1}^{k} \frac{d_j[F_n(\xi_{p_j}) - p_j]}{f(\xi_{p_j})} = n^{-1} \sum_{i=1}^{n} \left\{ \sum_{j=1}^{k} \left[\frac{d_j}{f(\xi_{p_j})}\right]\left[I(X_i \leqslant \xi_{p_j}) - p_j\right]\right\} \tag{7.3.25}$$

is an average of bounded-valued i.i.d.r.v.'s with mean 0 and variance

$$\sigma^2 = \sum_{j=1}^{k} \sum_{l=1}^{k} \left[\frac{d_j d_l}{f(\xi_{p_j}) f(\xi_{p_l})}\right]\left[p_j \wedge p_l - p_j p_l\right] (< \infty). \tag{7.3.26}$$

Hence the invariance principles of Chapter 2 are all applicable. Thus, if we define

$$W_n^{(1)}(t) = \frac{n^{-1/2}k_n(t)\left[T_{k_n(t)} - \tau\right]}{\sigma}, \qquad t \in [0,1], \tag{7.3.27}$$

$$W_n^{(2)}(t) = \frac{n^{1/2}\left[T_{k_n^*(t)} - \tau\right]}{\sigma}, \qquad t \in [0,1], \tag{7.3.28}$$

$$\zeta^*(t) = \frac{k\left[T_k - \tau\right]}{\sigma}, \qquad k \leqslant t < k+1, \qquad k \geqslant 0, \qquad T_0 = 0, \tag{7.3.29}$$

where $k_n(t)$ and $k_n^*(t)$ are defined before (7.3.17) and (7.3.18), then for the three processes in (7.3.27) through (7.3.29), all of (7.3.17), (7.3.18), and (7.3.20) hold. In (7.3.23) and elsewhere, $\xi_p^{(n)} - \xi_p$ may as well be replaced by $g(\xi_p^{(n)}) - g(\xi_p)$ whenever $g(\,)$ has a continuous derivative in some neighborhood of ξ_p.

There has been a steady flow of research work on the Bahadur representation of sample quantiles. Kiefer (1967) examined this representation in greater depth and continued in this direction (1970, 1972), as have Ghosh (1971) and Eicker (1970), among others. Extensions of this representation for sequences of dependent r.v.'s are due to Sen (1968, 1972a), Dutta and Sen (1971), and others. Since these results are not of direct relevance to our subsequent studies, we shall not enter into their discussions here. However, for more comprehensive reading, we suggest these references.

7.4. WEAK CONVERGENCE OF TYPE II LCFOS

Here, we define T_n by (7.2.6) and we are interested in the case of *smooth weights* where

$$\lim_{n\to\infty} \phi_n(u) = \phi(u) \qquad \text{exists for every} \quad u \in (0,1). \tag{7.4.1}$$

We define T_n as in (7.2.6) and for the sake of simplicity, we let $g_n(x) = g(x) = x$. The results continue to hold for any arbitrary $g(x)$, satisfying some growth-condition and under an additional equicontinuity condition on the $g_n(\cdot)$, they also hold for the general case of T_n in (7.2.6). We shall make more concrete comments on this at a later stage. Let us now define

$$\mu = \int_{-\infty}^{\infty} x\phi(F(x))\,dF(x), \tag{7.4.2}$$

$$\sigma^2 = \int_{-\infty}^{\infty}\int_{-\infty}^{\infty} \left[F(x \wedge y) - F(x)F(y)\right]\phi(F(x))\phi(F(y))\,dx\,dy \tag{7.4.3}$$

and assume that both μ and σ^2 are finite. Before we present the main weak convergence results, we like to study some reverse-martingale approximation for type II LCFOS. For this, we let $c_{n,0} = c_{n,n+1} = 0, \forall n \geqslant 1$, and let

$$c^*_{n+1,i} = [(i-1)/(n+1)]c_{n,i-1} + [(n-i+1)/(n+1)]c_{n,i}, \quad i = 1, \ldots, n+1, \tag{7.4.4}$$

$$d_{n+1,i} = c_{n+1,i} - c^*_{n+1,i}, \qquad i = 1, \ldots, n+1, \tag{7.4.5}$$

$$U_n = n(T_n - T_{n+1}) \quad \text{and} \quad T^*_{n+1} = \sum_{i=1}^{n+1} d_{n+1,i} X_{n+1,i}. \tag{7.4.6}$$

Finally, let $\mathfrak{F}_n = \mathfrak{F}(X_{n,1}, \ldots, X_{n,n}; X_{n+j}, j \geqslant 1)$ be the σ-field generated by $(X_{n,1}, \ldots, X_{n,n})$ and $\{X_{n+j}, j \geqslant 1\}$, for $n \geqslant 1$. Thus $\mathfrak{F}_n$ is nonincreasing in n. Then, we have the following.

Lemma 7.4.1. If for some $n_0 (\geqslant 1)$, $d_{n,i} = 0, \forall 1 \leqslant i \leqslant n, n \geqslant n_0$, then $\{T_n, \mathfrak{F}_n; n \geqslant n_0\}$ is a reverse martingale.

Proof. Given $\mathfrak{F}_{n+1}$, X_{n+1} can assume the values $X_{n+1,1}, \ldots, X_{n+1,n+1}$ with equal conditional probability $(n+1)^{-1}$. Thus the possible realizations of $(X_{n,1}, \ldots, X_{n,n})$, given $\mathfrak{F}_{n+1}$, are $(X_{n+1,1}, \ldots, X_{n+1,k-1}, X_{n+1,k+1}, \ldots, X_{n+1,n+1})$, $k = 1, \ldots, n+1$ and these are conditionally equally likely; here for $k = 1$ (or $n+1$) the vector starts (ends) with $X_{n+1,2}(X_{n+1,n})$. Hence for $n \geqslant 1$,

$$\begin{aligned}
E(T_n \mid \mathfrak{F}_{n+1}) &= \sum_{i=1}^{n} c_{n,i} E[X_{n,i} \mid \mathfrak{F}_{n+1}] \\
&= \sum_{i=1}^{n} c_{n,i} \left\{ \frac{n-i+1}{n+1} X_{n+1,i} + \frac{i}{n+1} X_{n+1,i+1} \right\} \\
&= \sum_{i=1}^{n+1} \left\{ \frac{i-1}{n+1} c_{n,i-1} + \frac{n-i+1}{n+1} c_{n,i} \right\} X_{n+1,i} (c_{n,0} = c_{n,n+1} = 0) \\
&= \sum_{i=1}^{n+1} c^*_{n+1,i} X_{n+1,i} = T_{n+1} - \sum_{i=1}^{n+1} d_{n+1,i} X_{n+1,i}.
\end{aligned} \tag{7.4.7}$$

Since $d_{n,i}$ are all assumed to be equal to 0, the lemma follows from (7.4.7). Q.E.D.

As an illustration, we consider the following.

Example 7.4.1. Let k be a nonnegative integer, $n_0 = 2k+1$ and

$$T_n = \left\{ \sum_{i=1}^{n} \binom{i-1}{k} \binom{n-i}{k} X_{n,i} \right\} \Big/ \binom{n}{2k+1} \qquad \text{for} \quad n \geqslant n_0.$$

Here $(n+1)^{-1}\{(i-1)c_{n,i-1}+(n-i+1)c_{n,i}\}=(n+1)^{-1}\{(i-1)\binom{i-2}{k}\binom{n-i+1}{k}+(n-i+1)\binom{i-1}{k}\binom{n-i}{k}\}\binom{n}{2k+1}^{-1}=\binom{n+1}{2k+1}^{-1}\binom{i-1}{k}\binom{n+1-i}{k}=c_{n+1,i}$, $1\leqslant i\leqslant n+1$. Thus Lemma 3.1 applies here. For $k=0$, T_n is the sample mean while for $k\geqslant 1$, it is a robust competitor of the sample mean; see Sen (1964).

By the same arguments as in the proof of Lemma 7.4.1, it follows that

$$\begin{aligned}\operatorname{Var}(X_{n,i}\mid\mathcal{F}_{n+1})&=\frac{n-i+1}{n+1}X_{n+1,i}^2+\frac{i}{n+1}X_{n+1,i+1}^2\\&\quad-\left\{\frac{n-i+1}{n+1}X_{n+1,i}+\frac{i}{n+1}X_{n+1,i+1}\right\}^2\\&=\frac{i(n+1-i)}{(n+1)^2}\left[X_{n+1,i+1}-X_{n+1,i}\right]^2,\qquad 1\leqslant i\leqslant n\end{aligned}\tag{7.4.8}$$

and, similarly, for $1\leqslant i<j\leqslant n$,

$$\operatorname{Cov}(X_{n,i},X_{n,j}\mid\mathcal{F}_{n+1})=\frac{i(n+1-j)}{(n+1)^2}\left[X_{n+1,i+1}-X_{n+1,i}\right]\left[X_{n+1,j+1}-X_{n+1,j}\right].\tag{7.4.9}$$

Thus, if we define

$$\begin{aligned}\sigma_n^2=\sum_{i=1}^{n}\sum_{j=1}^{n}c_{n,i}c_{n,j}&\frac{\left[(i\wedge j)(n+1)-ij\right]}{(n+1)^2}\\&\times\left[X_{n+1,i+1}-X_{n+1,i}\right]\left[X_{n+1,j+1}-X_{n+1,j}\right],\end{aligned}\tag{7.4.10}$$

we have from (7.2.5), (7.4.8), (7.4.9), and (7.4.10),

$$\sigma_n^2=E\left\{(T_n-T_{n+1})^2\mid\mathcal{F}_{n+1}\right\}.\tag{7.4.11}$$

Let now $I(A)$ be the indicator function of a set A, and we assume that

$$n^2\sigma_n^2\to\sigma^2\qquad\text{a.s.,}\quad\text{as}\quad n\to\infty,\tag{7.4.12}$$

$$E\left[U_n^2I(U_n^2>\epsilon^2n)\mid\mathcal{F}_{n+1}\right]\to 0\qquad\text{a.s.,}\quad\text{as}\quad n\to\infty,\qquad\forall\epsilon>0.\tag{7.4.13}$$

With these notations and assumptions, we define, for every $n(\geqslant 1)$, a stochastic process $W_n=\{W_n(t),t\in[0,1]\}$ by letting

$$W_n(t)=n^{1/2}\sigma^{-1}\left[T_{k_n(t)}-\mu\right],\qquad 0\leqslant t\leqslant 1;\tag{7.4.14}$$

$$k_n(t)=\min\{k:n/k\leqslant t\},\qquad 0\leqslant t\leqslant 1.\tag{7.4.15}$$

In order to define properly $W_n(0)$, we assume that $T_n\to\mu$ a.s., as $n\to\infty$ ($\Rightarrow W_n(0)=0$, in probability). Later on, in Section 7.6, we shall see that this assumption holds under quite general conditions.

THEOREM 7.4.2. *If $d_{n,i} = 0$, $\forall 1 \leqslant i \leqslant n$ and $n \geqslant n_0$, then under* (7.4.12) *and* (7.4.13), $\{W_n\}$ *converges weakly to* W *(a standard Wiener process on* $[0,1]$).

Proof. By (7.4.11) and (7.4.12), we have

$$(n/\sigma^2)\sum_{N \geqslant n} E\{(T_N - T_{N+1})^2 \mid \mathfrak{F}_{N+1}\} \xrightarrow{p} 1 \qquad \text{as} \quad n \to \infty. \tag{7.4.16}$$

Also, by (7.4.13), for every $\epsilon > 0$,

$$\begin{aligned}(n/\sigma^2)&\sum_{N \geqslant n} E\left\{(T_N - T_{N+1})^2 I\left(|T_N - T_{N+1}| > \epsilon\sigma\sqrt{n}\right) \mid \mathfrak{F}_{N+1}\right\}\\ &= (n/\sigma^2)\sum_{N \geqslant n} N^{-2} E\left\{U_N^2 I\left(|U_N|\epsilon\sigma\sqrt{n}\right) \mid \mathfrak{F}_{N+1}\right\}\\ &\to 0 \quad \text{a.s.,} \quad \text{as} \quad n \to \infty.\end{aligned} \tag{7.4.17}$$

Since $\{T_n, \mathfrak{F}_n; n \geqslant n_0\}$ is a reverse martingale (Lemma 7.4.1), by (7.4.16), (7.4.17), and Theorem 2.4.5, the desired result follows. Q.E.D.

In a variety of problems, the asymptotic normality of T_N (or the multinormality of $T_{N_1}, \ldots, T_{N_m}$) can be established by alternative method without requiring explicitly (7.4.12) and (7.4.13). In such a case, we have the following.

THEOREM 7.4.3. *If $d_{n,i} = 0$, $\forall 1 \leqslant i \leqslant n$ and $n \geqslant n_0$, and if*

$$0 < \lim_{n \to \infty} n\operatorname{Var}(T_n) = \sigma^2 < \infty, \tag{7.4.18}$$

the convergence of the f.d.d.'s of $\{W_n\}$ *to those of* W, *ensures the tightness of* $\{W_n\}$.

Proof. Note that for every $0 \leqslant s < s + \delta \leqslant 1$,

$$\sup_{s \leqslant t \leqslant s+\delta} |W_n(t) - W_n(s)| = \max_{k_1 \leqslant q \leqslant k_2} n^{1/2}\sigma^{-1}|T_q - T_{k_2}|, \tag{7.4.19}$$

where

$$k_1^{-1}n \leqslant s + \delta < (k_1 - 1)^{-1}n \qquad \text{and} \quad k_2^{-1}n \leqslant s < (k_2 - 1)^{-1}n. \tag{7.4.20}$$

Since $\{|T_q - T_{k_2})|, \mathfrak{F}_q; q \leqslant k_2\}$ has the reverse submartingale property, by reversing the order of the index set and using (2.2.9), we obtain that for every $\epsilon > 0$,

$$\begin{aligned}P&\left\{\max_{k_1 \leqslant q \leqslant k_2} |T_q - T_{k_2}| > \sigma\epsilon n^{-1/2}\right\}\\ &\leqslant (2/\epsilon\sigma)\left(E\left\{n[T_{k_1} - T_{k_2}]^2\right\} P\left\{n^{1/2}|T_{k_1} - T_{k_2}| > \epsilon\sigma/2\right\}\right)^{1/2},\end{aligned} \tag{7.4.21}$$

where by (7.4.18) and (7.4.20) as $n \to \infty$, $E\{n[T_{k_2} - T_{k_1}]^2\} \to \delta\sigma^2$ and also the convergence of f.d.d.'s of $\{W_n\}$ to those of W implies that as $n \to \infty$

$$n^{1/2}(T_{k_1} - T_{k_2})/\sigma \xrightarrow{\mathcal{L}} \mathfrak{N}(0, \delta). \tag{7.4.22}$$

Finally, if $\Phi(x)$ be the standard normal d.f., then for every $x \geqslant 1$, $2\{1 - \Phi(x)\} \leqslant (2/\pi)^{1/2}x^{-1}\exp(-\frac{1}{2}x^2)$, and hence, for every $\epsilon > 0$, we can choose $\delta(> 0)$, so small that the rhs of (7.4.21) is less than $\eta\delta$, for some arbitrary $\eta > 0$. Hence $\{W_n\}$ is tight. Q.E.D.

We may note that for both Theorems 7.4.2 and 7.4.3, the assumption that the $d_{n,i}$ are all equal to 0 is very crucial. In many cases, the $d_{n,i}$ may not be exactly equal to 0, though they are very nearly so. Toward this, we present the following.

THEOREM 7.4.4. *If $N^{3/2}T_N^* \to 0$ a.s., as $N \to \infty$, then $W_n \xrightarrow{\mathfrak{D}} W$ under the rest of assumptions of Theorems* 7.4.2 *or* 7.4.3.

Proof. Let

$$\tilde{U}_k = T_k - E(T_k \mid \mathfrak{F}_{k+1}) = (T_k - T_{k+1}) + T_{k+1}^*, \qquad \text{by (7.4.6)} \tag{7.4.23}$$

Therefore, for every $q \geqslant n$, by the first condition of Theorem 7.4.4,

$$n^{1/2}\Big(T_q - \sum_{N \geqslant q} \tilde{U}_N\Big) = n^{1/2}\Big(\sum_{N \geqslant q+1} T_N^*\Big) \to 0 \quad \text{a.s.,} \quad \text{as} \quad n \to \infty. \tag{7.4.24}$$

Since $E[\tilde{U}_k \mid \mathfrak{F}_{k+1}] = 0$ for every $k \geqslant n_0$, $\{U_n, \mathfrak{F}_n; n \geqslant n_0\}$ is a reverse martingale difference sequence. Further, $E\{\tilde{U}_k^2\} = E\{E[\tilde{U}_k^2 \mid \mathfrak{F}_{k+1}]\} \leqslant E\{E[(T_k - T_{k+1})^2 \mid \mathfrak{F}_{k+1}]\} = E(T_k - T_{k+1})^2$, $\forall k \geqslant n_0$, and hence

$$E\Big(\sum_{k_1 \leqslant k \leqslant k_2} \tilde{U}_k\Big)^2 \leqslant E(T_{k_2} - T_{k_1})^2, \qquad \forall n_0 \leqslant k_1 < k_2 < \infty. \tag{7.4.25}$$

Also, (7.4.24) and the convergence of the f.d.d.'s of $\{W_n\}$ to those of W ensures the same for the process where T_N is replaced by $\sum_{k \geqslant N} \tilde{U}_k$, $N \geqslant n$. Hence the proof of Theorem 7.4.3 readily extends. For Theorem 7.4.2, we note that (7.4.25) and the first condition of Theorem 7.4.4 imply that

$$\sigma_N^2 - O(N^{-3}) = E\{\tilde{U}_N^2 \mid \mathfrak{F}_{N+1}\} \leqslant \sigma_N^2, \qquad \forall N \geqslant n_0, \tag{7.4.26}$$

Thus, by (7.4.12) and (7.4.26), here also (7.4.16) holds with $T_N - T_{N+1}$

being replaced by $\tilde{U}_N, N \geqslant n$. Also, for every $N \geqslant n$,

$$\begin{aligned}
&E\big[\tilde{U}_N^2 I(|\tilde{U}_N| > \epsilon\sigma n^{-1/2}) \mid \mathcal{F}_{N+1}\big] \\
&\quad = E\big[(T_N - T_{N+1})^2 I(|\tilde{U}_N| > \epsilon\sigma n^{-1/2}) \mid \mathcal{F}_{n+1}\big] \\
&\qquad - E\left\{T_{N+1}^{*2} I(|\tilde{U}_N| > \epsilon\sigma n^{-1/2})\right\} \\
&\quad \leqslant E\big[(T_N - T_{N+1})^2 I(|T_N - T_{N+1} + T_{N+1}^*| > \epsilon\sigma n^{-1/2}) \mid \mathcal{F}_{N+1}\big] \\
&\quad \leqslant E\left\{(T_N - T_{N+1})^2 I(|T_N - T_{N+1}| > \tfrac{1}{2}\epsilon\sigma n^{-1/2}) \mid \mathcal{F}_{n+1}\right\} \\
&\qquad + \sigma_N^2 E\left\{I(|T_{N+1}^*| > \tfrac{1}{2}\epsilon\sigma n^{-1/2}) \mid \mathcal{F}_{n+1}\right\} \\
&\quad = E\left\{(T_N - T_{N+1})^2 I(|T_N - T_{N+1}| > \tfrac{1}{2}\epsilon\sigma n^{-1/2}) \mid \mathcal{F}_{n+1}\right\} \\
&\qquad + o(N^{-2}) \qquad \text{a.s.,}
\end{aligned} \tag{7.4.27}$$

by (7.4.12) and the fact that $N^{-3/2}T_N^* \to 0$ a.s., as $N \to \infty$. Hence (7.4.13) holds with $T_N - T_{N+1}$ being replaced by $\tilde{U}_N, N \geqslant n$. Therefore, the proof of Theorem 7.4.2 readily extends to that of Theorem 7.4.4. Q.E.D.

THEOREM 7.4.5. *Suppose that it is possible to write*

$$T_n = T_n^0 + \tilde{T}_n^*, \qquad \forall n \geqslant n_0, \tag{7.4.28}$$

where $\{T_n^0, \mathcal{F}_n; n \geqslant n_0\}$ *is a reverse martingale, satisfying the hypothesis of Theorem* 7.4.2 *or* 7.4.3 *and*

$$n^{1/2}\Big\{\sup_{N \geqslant n} |\tilde{T}_N^*|\Big\} \xrightarrow{p} 0, \qquad \text{as} \quad n \to \infty. \tag{7.4.29}$$

Then, $\{W_n\}$ *converges weakly to* W.

The proof follows by decomposing W_n in to two parts, applying Theorem 7.2.2 or 7.2.3 on the first part and showing that by (7.4.29), the other part is negligible, in probability. The details are left as exercise (see Problem 7.4.7).

For the rest of this section, we express $x = F^{-1}(F(x)) = b(F(x))$, where $b(u)$ is a nondecreasing and continuous function of u $(0 < u < 1)$. We then have

$$T_n = \int_0^1 b(u)\phi_n(G_n(u))\,dG_n(u), \tag{7.4.30}$$

where $G_n(u) = F_n(F^{-1}(u)), u \in [0, 1]$ is the empirical d.f. of the $U_i(= F(X_i)), i = 1, \ldots, n$ [which are i.i.d.r.v.'s with the uniform (0, 1) d.f.].

Further, we have

$$\mu = \int_0^1 b(u)\phi(u)\,du. \tag{7.4.31}$$

Let us now write

$$T_n^{(1)} = \int_0^1 b(u)\phi(u)d[G_n(u) - u] = n^{-1}\sum_{i=1}^{n}[b(U_i)\phi(U_i) - \mu], \tag{7.4.32}$$

$$T_n^{(2)} = \frac{\sum_{1\leqslant i\neq j\leqslant n}[c(U_i - U_j) - U_i]\phi'(U_i)b(U_i)}{n(n-1)}, \tag{7.4.33}$$

$$T_n^{(3)} = \frac{1}{n^2}\sum_{i=1}^{n}(1 - U_i)b(U_i)\phi'(U_i), \tag{7.4.34}$$

$$R_n = \int_0^1 [\phi_n(G_n(u)) - \phi(u) - (G_n(u) - u)\phi'(u)]b(u)\,dG_n(u). \tag{7.4.35}$$

Then, from (7.4.30) through (7.4.35), we obtain

$$T_n - \mu = T_n^0 + \tilde{T}_n^*, \tag{7.4.36}$$

where

$$T_n^0 = T_n^{(1)} + T_n^{(2)}, \qquad \tilde{T}_n^* = T_n^{(3)} - n^{-1}T_n^{(2)} + R_n. \tag{7.4.37}$$

Note that by (7.4.32), $\{T_n^{(1)}\}$ is a reverse martingale sequence, whenever μ exists. Also, if we define a (symmetric) kernel (of degree 2)

$$g(x, y) = \{[c(y - x) - y]\phi'(y)b(y) + [c(x - y) - x]\phi'(x)b(x)\}/2, \tag{7.4.38}$$

then, by (7.4.33) and (7.4.38),

$$T_n^{(2)} = \binom{n}{2}^{-1}\sum_{1\leqslant i<j\leqslant n} g(U_i, U_j)$$

is a U-statistic, so that by Theorem 3.2.1, $\{T_n^{(2)}\}$ is a reverse martingale; both $\{T_n^{(1)}\}$ and $\{T_n^{(2)}\}$ are reverse martingales with respect to a common sequence of sigma fields. Hence $\{T_n^0\}$ is also a reverse martingale sequence. In fact, $T_n^{(1)} + T_n^{(2)}$, being a sum of two U-statistics, is also a U-statistic, so that Theorem 3.3.4 may be called on to verify the weak convergence of the tail sequence $\{T_k^0, k \geqslant n\}$. Thus, to apply Theorem 7.4.5, we need to verify only (7.4.29). We therefore introduce the following regularity conditions on $b(\cdot)$ and $\phi(\cdot)$.

We allow $b(\cdot)$ to be quite arbitrary and assume that for every $\theta \in (0, \frac{1}{2})$, $b(u)$ is of bounded variation on $(\theta, 1 - \theta)$, and, for some real α and finite positive K,

$$|b(u)| \leqslant K[u(1 - u)]^{-\alpha}, \qquad \forall u \in (0, 1). \tag{7.4.39}$$

Also, we take, as in earlier places, $\phi_n(u) = \phi(i/(n+1))$, for $(i-1)/n < u \leqslant i/n$, $i = 1, \ldots, n$, and, we assume that for some real β,

$$|\phi(u)| \leqslant K[u(1-u)]^{-\beta}, \qquad \forall u \in (0,1), \tag{7.4.40}$$

where

$$\alpha + \beta = \tfrac{1}{2} - \delta, \qquad \text{for some} \quad \delta > 0. \tag{7.4.41}$$

Further, we assume that $\phi(u)$ has a continuous first order derivative $\phi'(u)$ inside $(0,1)$, where

$$|\phi'(u)| \leqslant K[u(1-u)]^{-1-\beta}, \qquad \forall u \in (0,1). \tag{7.4.42}$$

Then, we have the following.

THEOREM 7.4.6. *Under* (7.4.39) *through* (7.4.42), *for* $\tilde{T}_n^*$, *defined by* (7.4.36),

$$n^{1/2} \sup_{N \geqslant n} |\tilde{T}_N^*| \xrightarrow{p} 0 \qquad \text{and} \qquad \max_{k \leqslant n} kn^{-1/2}|\tilde{T}_k^*| \xrightarrow{p} 0, \qquad \text{as} \quad n \to \infty. \tag{7.4.43}$$

Proof. We prove only the first part of (7.4.43); the second part follows on parallel lines. Since $T_n^{(2)}$ has been shown to be a U-statistic, by Theorem 3.2.1, $T_n^{(2)}$ a.s. converges to $ET_n^{(2)}$ $(=0)$, as $n \to \infty$. Hence $\sup_{N \geqslant n} \{N^{-1/2}|T_N^{(2)}|\} \xrightarrow{p} 0$, as $n \to \infty$. Further, by (7.4.34), (7.4.39), (7.4.41), and (7.4.42),

$$\begin{aligned} n^{1/2} \sup_{N \geqslant n} |T_N^{(3)}| &= n^{1/2} \sup_{N \geqslant n} |N^{-2} \sum_{i=1}^{N} (1 - U_i) b(U_i) \phi'(U_i)| \\ &\leqslant \sup_{N \geqslant n} \left\{ N^{-3/2} \sum_{i=1}^{N} K^2 [U_i(1-U_i)]^{-3/2+\delta} \right\} \\ &\leqslant \sup_{N \geqslant n} \left[\left\{ K^2 N^{-1/2} \max_{1 \leqslant i \leqslant N} [U_i(1-U_i)]^{-(1-\delta)/2} \right\} \right. \\ &\qquad \left. \times \left\{ N^{-1} \sum_{i=1}^{N} [U_i(1-U_i)]^{-1+\delta/2} \right\} \right]. \end{aligned} \tag{7.4.44}$$

Now, by the Kintchine strong law of large numbers, as $n \to \infty$,

$$n^{-1} \sum_{i=1}^{n} [U_i(1-U_i)]^{-1+\delta/2} \to \int_0^1 [u(1-u)]^{-1+\delta/2} du (< \infty) \qquad \text{a.s.} \tag{7.4.45}$$

Also, the U_i have the uniform $(0,1)$ d.f., and hence, as $n \to \infty$,

$$\max_{1 \leqslant i \leqslant n} [U_i(1-U_i)]^{-1} = o(n(\log n)^2) \qquad \text{a.s.} \tag{7.4.46}$$

Since, by (7.4.41), $\delta > 0$, it follows from the last three equations that as $n \to \infty$, $n^{1/2} \sup_{N \geqslant n} |T_N^{(3)}| \xrightarrow{p} 0$. Thus it remains to show that as $n \to \infty$,

$$n^{1/2} \sup_{N \geqslant n} |R_N| \xrightarrow{p} 0. \tag{7.4.47}$$

For this, we shall make use of (2.6.32) and, as there, we let $\epsilon_n = (\log n)^4/n$. We break up the integral in (7.4.35) in to three domains $(0, \epsilon_n)$, $(\epsilon_n, 1-\epsilon_n)$ and $(1-\epsilon_n, 1)$ and denote these components by R_{n1}, R_{n2}, and R_{n3}, respectively. Then, we have

$$\begin{aligned}|R_{n1}| &\leqslant \int_0^{\epsilon_n} |\phi_n(G_n(u))b(u)|\,dG_n(u) + \int_0^{\epsilon_n} |\phi(u)b(u)|\,dG_n(u) \\ &\quad + \int_0^{\epsilon_n} |(G_n(u) - u)\phi'(u)b(u)|\,dG_n(u) \\ &= R_{n11} + R_{n12} + R_{n13}, \qquad \text{say,}\end{aligned} \tag{7.4.48}$$

where, by (7.4.39), (7.4.40), (7.4.41), and (7.4.46), we have for $n \to \infty$,

$$\begin{aligned}R_{n11} &\leqslant K^2 \left\{ n^{-1} \sum_{i=1}^{(\log n)^4} \left[i(n-i+1)/(n+1)^2 \right]^{-\beta} \right\} \\ &\quad \times \left\{ \left[n(\log n)^2 \right]^{\alpha} \right\} \qquad \text{(a.s.)} \\ &= O\left(n^{-1+\alpha+\beta}(\log n)^{2\alpha+4(1-\beta)}\right) \qquad \text{(a.s.)} \\ &= O\left(n^{-(1/2)-\eta}\right) \qquad \text{a.s., where } \eta > 0.\end{aligned} \tag{7.4.49}$$

A similar treatment holds for R_{n12} and R_{n13}. Also, the case of R_{n3} is very similar to that of R_{n1}. Hence we obtain that as $n \to \infty$,

$$n^{1/2} \sup_{N \geqslant n} |R_{n1} + R_{n3}| \xrightarrow{p} 0. \tag{7.4.50}$$

The treatment of R_{n2} is little bit more delicate. First, we note that by (2.6.32)

$$\sup_{\epsilon_n \leqslant u \leqslant 1-\epsilon_n} \left| \frac{G_n(u)}{u} - 1 \right| \to 0 \qquad \text{a.s., as } n \to \infty. \tag{7.4.51}$$

Second, we may write

$$\begin{aligned}&\left[\phi_n(G_n(u)) - \phi(u) - (G_n(u) - u)\phi'(u) \right] b(u) \\ &\quad = b(u)(G_n(u) - u)\left[\{\phi_n(G_n(u)) - \phi(u)\}/(G_n(u) - u) - \phi'(u) \right].\end{aligned} \tag{7.4.52}$$

Next, we note that $\{\{G_n(u) - u, 0 \leqslant u \leqslant 1\}, n \geqslant 1\}$ is a reverse martingale (process), so that for every $\epsilon > 0$,

$$\Big\{ \sup_{0<u<1} |G_n(u) - u| / \{u(1-u)\}^{(1/2)-\epsilon}, n \geqslant 1 \Big\} \quad \text{is a reverse submartingale.} \tag{7.4.53}$$

Therefore, using the Kolmogorov inequality in (2.2.6), we obtain that for every $C > 0$,

$$P\Big\{ \sup_{N \geqslant n} \sup_{0<u<1} |G_N(u) - u| / \{u(1-u)\}^{(1/2)-\epsilon} > Cn^{-1/2} \Big\}$$
$$\leqslant C^{-1} E\Big\{ \sup_{0<u<1} n^{1/2} |G_n(u) - u| / \{u(1-u)\}^{(1/2)-\epsilon} \Big\}. \tag{7.4.54}$$

Now, for every (fixed) $n(\geqslant 1)$, $\{G_n(u)/u - 1, 0 \leqslant u < 1\}$ is a reverse martingale, while $\{u(1-u)\}^{-1+2\epsilon}$ is a U-shaped, nonnegative, and integrable function inside (0, 1). Therefore, by breaking up the interval (0, 1) into $(0, \frac{1}{2}]$ and $(\frac{1}{2}, 1)$ and using the Birnbaum-Marshall inequality in (2.2.21) [on $(G_n(u) - u)^2$], we obtain that for every $\lambda > 0$,

$$P\left\{ \sup_{0<u<1} \frac{n^{1/2}|G_n(u) - u|}{\{u(1-u)\}^{1/2-\epsilon}} > \lambda \right\}$$
$$= P\left\{ \sup_{0<u<1} \frac{n(G_n(u) - u)^2}{\{u(1-u)\}^{1-2\epsilon}} > \lambda^2 \right\} \leqslant K\lambda^{-2}, \tag{7.4.55}$$

where K is a positive constant, independent of λ. (7.4.55) ensures the finiteness of the expectation on the rhs of (7.4.54), so that we have for every $\epsilon > 0$,

$$n^{1/2} \sup_{N \geqslant n} \sup_{0<u<1} |G_N(u) - u| / \{u(1-u)\}^{(1/2)-\epsilon} = O_p(1). \tag{7.4.56}$$

For an arbitrary $\eta > 0$ and n so large that $0 < \epsilon_n < \eta$, by using (7.4.39), (7.4.40), (7.4.41), (7.4.42), (7.4.51), and (7.4.56), we obtain that as $n \to \infty$,

$$\sup_{N \geqslant n} \Big| \int_{\epsilon_n}^{\eta} + \int_{1-\eta}^{1-\epsilon_n} n^{1/2} [\phi_N(G_N(u)) - \phi(u) - (G_n(u) - u)\phi'(u)]$$
$$\times b(u)\, dG_N(u) \Big|$$
$$\leqslant [O_p(1)] \sup_{N \geqslant n} \Big\{ \int_0^{\eta} + \int_{1-\eta}^{1} \{u(1-u)\}^{-1+\delta-\epsilon} dG_N(u) \Big\}, \tag{7.4.57}$$

where, in (7.4.56), we choose ϵ to be less than δ, and hence the integral inside $\{\ \ldots\ \}$ in (7.4.57) exists and is a nonnegative reverse submartingale.

Thus, by the convergence theorem on nonnegative reverse submartingales, we conclude that the rhs of (7.4.57) converges (in probability) to a number bounded from above by

$$K\left\{\int_0^\eta + \int_{1-\eta}^1 \{u(1-u)\}^{-1+\delta-\epsilon}\,du\right\}, \tag{7.4.58}$$

where K is a positive constant, independent of η. Since the integral in (7.4.58) can be made arbitrarily small by choosing η small, we conclude that (7.4.56) can be made $o_p(1)$ by choosing η appropriately small. Finally, for $u \in [\eta, 1-\eta]$, $\phi'(u)$ is bounded and continuous, so that by making use of (7.4.39) through (7.4.42), (7.4.52), and (7.4.56), we conclude that as $n \to \infty$,

$$\sup_{N \geqslant n}\left|\int_\eta^{1-\eta} n^{1/2}\left[\phi_N(G_N(u) - \phi(u)) - (G_N(u) - u)\phi'(u)\right]\right.$$
$$\left.\times\, b(u)\,dG_N(u)\right| = o_p(1). \tag{7.4.59}$$

This completes the proof of (7.4.47) and hence of (7.4.43). Q.E.D.

We may note that for the $\{T_n^0\}$, we may virtually repeat the proof of Theorem 3.3.4 and thereby verify that both (7.4.12) and (7.4.13) hold under the assumptions in (7.4.39) through (7.4.42). Thus, by Theorems 7.4.2, 7.4.5, and 7.4.6, we conclude that under (7.4.39) through (7.4.42), $\{W_n\}$, defined by (7.4.14) and (7.4.15), converges weakly to a standard Wiener process. Similarly, for the $\{T_n^0\}$, we can also adapt the proof of Theorem 3.3.1 and claim that the Donsker-type weak convergence result holds for the partial sequence $\{k(T_k^0 - \mu)/(\sigma n^{1/2}), k \leqslant n\}$, where σ^2 is defined by (7.4.3). Further, the second part of (7.4.43) makes the residual $o_p(1)$. Hence we arrive at the following.

THEOREM 7.4.7. *Under* (7.4.39) *through* (7.4.42), *the weak invariance principle holds for both the sequences* $\{n^{-1/2}k(T_k - \mu)/\sigma, k \leqslant n\}$ *and* $\{n^{1/2}(T_k - \mu)/\sigma, k \geqslant n\}$ *where* T_n *and* μ *are defined by* (7.4.30) *and* (7.4.31), *while* σ^2 *is defined as in* (7.4.3), *with* $dx\,dy$ *being replaced by* $db(F(x))\,db\,F(y))$.

The theorems in this section are formulated in the light of the reverse martingale approximations considered in Sen (1978b). Asymptotic normality results were studied by different workers [e.g., Shorack (1969, 1972), Wellner (1974)] by invoking the weak convergence of the empirical process in the ρ_q-metric [defined by (2.6.25) and (2.6.33)]; we prefer to refer to the current proof in detail because it is in line with the martingale approach underlying the proofs in the preceding chapters. Both the forward and backward invariance principles in Theorem 7.4.7 also follow from the a.s. representation of $T_n - \mu$ in terms of an average of i.i.d.r.v.'s, which is considered in the next section under additional regularity conditions.

7.5. EMBEDDING OF WIENER PROCESSES FOR TYPE II LCFOS

Here, we are interested in representing $T_n - \mu$ as an average of i.i.d.r.v.'s plus a remainder term which converges a.s. to 0 at a faster rate and then in using the Skorokhod-Strassen representation on the principal term. Specifically, we consider the representation

$$T_n - \mu = n^{-1} \sum_{i=1}^{n} Z_i + R_n, \tag{7.5.1}$$

where

$$Z_i = \int_{-\infty}^{\infty} \left[c(x - X_i) - F(x) \right] \phi(F(x))\, dx, \qquad i \geqslant 1, \tag{7.5.2}$$

and under suitable regularity conditions, we intend to show that as $n \to \infty$,

$$R_n = O(n^{-(1/2)-\eta}) \qquad \text{a.s.,} \quad \text{where} \quad \eta > 0. \tag{7.5.3}$$

Note that for (7.5.1), the principal term $n^{-1}\sum_{i=1}^{n} Z_i$ is different from $T_n^{(1)}$ in (7.4.32), but differs from T_n^0 by a quantity that is asymptotically negligible. Basically, we intend to incorporate the Skorokhod-Strassen embedding of the Wiener process for the Z_i and to use (7.5.3) to provide a parallel result for T_n. Note that under (7.4.39), (7.4.40), and (7.4.41), $E|Z_i|^k < \infty$ for some $k > 2$, so that by Theorem 2.5.1, there exists a standard Wiener process $W = \{W(t), t \geqslant 0\}$, defined possibly on a new probability space, such that

$$\sigma^{-1}\left(\sum_{i=1}^{n} Z_i \right) = W(n) + O(n^{(1/2)-\eta}) \qquad \text{a.s.,} \quad \text{as} \quad n \to \infty, \tag{7.5.4}$$

where $\eta > 0$. Thus (7.5.1) through (7.5.4) enable us to transmit such a stronger result for the T_n as well. For this purpose, we keep in mind the regularity conditions in (7.4.39) through (7.4.42) and assume that

$$E|X|^r < \infty, \qquad \text{for some positive } r \text{ (not necessarily} \geqslant 1). \tag{7.5.5}$$

Note that (7.5.5) ensures that

$$|x|^r F(x)\left[1 - F(x)\right] \leqslant K(<\infty), \qquad \forall x \in (-\infty, \infty) \tag{7.5.6}$$

and further, it converges to 0 as $x \to \pm\infty$. Second, we assume that $\phi(u), \phi^{(1)}(u)$ and $\phi^{(2)}(u)$ exist for every $u \in (0,1)$, and, with r, defined by (7.5.5),

$$|\phi^{(s)}(u)| \leqslant K\left[u(1-u)\right]^{-s-(1/2)+\delta+r^{-1}}, \qquad \forall u \in (0,1), \qquad s = 0, 1, 2, \tag{7.5.7}$$

where K is a positive constant. Note that by (7.5.5) and (7.5.7) there exists a positive $q(>2)$ such that

$$E|X_i\phi(F(X_i))|^q = \int_{-\infty}^{\infty} |x\phi(F(x))|^q \, dF(x) < \infty. \tag{7.5.8}$$

Then, we have the following.

THEOREM 7.5.1. *Under* (7.5.5) *and* (7.5.7), *both* (7.5.3) *and* (7.5.4) *hold.* [For simplicity of the proof, in the sequel, we let r in (7.5.5) be some finite positive number. If the X_i are bounded with probability 1, then we may take r to be arbitrarily large, so that in (7.5.7), we may even drop the term r^{-1} in the exponent. In such a case, the proof becomes simpler, and hence we consider only the case where the X_i are not bounded r.v.'s. Further, as in the later part of Section 7.4, we may also replace x by an arbitrary function of x [of bounded variation inside (0, 1)] satisfying (7.4.39); the modifications in the proof in this case are straightforward and hence are omitted.]

Proof. The $c_{n,i}$ are defined as in (7.2.5), but, for $\phi_n = \phi$. Then, we have

$$T_n = \int_{-\infty}^{\infty} x\phi\left(\frac{nF_n(x)}{(n+1)}\right) dF_n(x), \qquad n \geqslant 1. \tag{7.5.9}$$

From (7.2.6), (7.4.2), (7.5.1), and (7.5.9), we obtain by some standard steps

$$R_n = R_{1n} + R_{2n} + R_{3n} + R_{4n} + R_{5n} \tag{7.5.10}$$

where

$$R_{1n} = \int_{-\infty}^{a_n'} + \int_{a_n''}^{\infty} x\left[\phi\left(\frac{n}{n+1}F_n(x)\right) - \phi(F(x))\right] dF_n(x), \tag{7.5.11}$$

$$R_{2n} = \int_{a_n'}^{a_n''} x\left[\phi\left(\frac{n}{n+1}F(x)\right) - \phi(F(x)) - \left[\frac{n}{n+1}F_n(x) - F(x)\right]\phi^{(1)}(F(x))\right] dF_n(x), \tag{7.5.12}$$

$$R_{3n} = -(n+1)^{-1}\int_{a_n'}^{a_n''} xF(x)\phi^{(1)}(F(x))\, dF_n(x), \tag{7.5.13}$$

$$R_{4n} = \int_{a_n'}^{a_n''} x[F_n(x) - F(x)]\phi^{(1)}(F(x))d[F_n(x) - F(x)], \tag{7.5.14}$$

$$R_{5n} = -\left(\int_{-\infty}^{a_n'} + \int_{a_n''}^{\infty}[F_n(x) - F(x)]\phi^{(1)}(F(x))x\, dF(x), \tag{7.5.15}$$

and where

$$F(a_n') = 1 - F(a_n'') = n^{-1+\delta'} \qquad \text{with} \quad \delta' = \begin{cases} \delta, & \text{if} \quad r \geqslant 1 \\ r\delta, & \text{if} \quad 0 < r < 1, \end{cases} \tag{7.5.16}$$

with r and $\delta(>0)$ being defined by (7.5.5) and (7.5.7). Also, in (7.5.7), we may without any loss of generality assume that $0 < \delta < \frac{1}{2}$; otherwise, one can always substitute a smaller value of δ in (7.5.7) and carry out the analysis. Further, we noted in (2.6.31) that for every $\epsilon > 0$, as $n \to \infty$,

$$\sup_{-\infty < x < \infty} \{F(x)[1 - F(x)]\}^{-(1/2)+\epsilon} |F_n(x) - F(x)|$$
$$= 0\left((N^{-1}\log\log n)^{1/2}\right) \qquad \text{a.s.} \tag{7.5.17}$$

Note that by definition in (7.5.11),

$$|R_{1n}| \leqslant \int_{-\infty}^{a_n'} + \int_{a_n''}^{\infty} |x\phi(F(x))|\, dF_n(x)$$
$$+ \int_{-\infty}^{a_n'} \int_{a_n''}^{\infty} |x\phi(nF_n(x)/(n+1))|\, dF_n(x). \tag{7.5.18}$$

Now, by (7.5.5) and (7.5.7),

$$|x\phi(F(x))| \leqslant K_1 K_2 \{F(x)[1 - F(x)]\}^{-(1/2)+\delta}, \qquad \forall x \in (-\infty, \infty), \tag{7.5.19}$$

and hence, by the same techniques as in (5.4.26) through (5.4.28), it follows that the first term on the rhs of (7.5.18) is a.s. $O(n^{-(1/2)-\eta})$ where $0 < \eta < \frac{1}{2}\delta(1-\delta)$. For the second term on the rhs of (7.5.18) we consider the two cases (a) $r > 1$ and (b) $0 < r \leqslant 1$ separately. In case (a), apply the Hölder inequality and obtain an upper bound

$$\left\{\int_{-\infty}^{a_n'} + \int_{a_n''}^{\infty} |x|^r\, dF_n(x)\right\}^{1/r} \left\{\int_{-\infty}^{a_n'} + \int_{a_n''}^{\infty} |\phi\left(\frac{n}{n+1}F_n(x)\right)|^q\, dF_n(x)\right\}^{1/q} \tag{7.5.20}$$

where $r^{-1} + q^{-1} = 1$. Now

$$\left\{\int_{-\infty}^{a_n'} + \int_{a_n''}^{\infty} |x|^r\, dF_n(x); \qquad n \geqslant 1\right\}$$

is a reverse submartingale sequence and it converges to 0 a.s., as $n \to \infty$.

Also, on denoting by $n_1 = nF_n(a_n')$ and $n_2 = n - nF_n(a_n'')$, we have by (7.5.7),

$$\left[\int_{-\infty}^{a_n'} + \int_{a_n''}^{\infty} \left|\phi\left(\frac{n}{n+1} F_n(x)\right)\right|^q dF_n(x)\right]^{1/q}$$

$$\leqslant K_2^{1/q}\left\{\frac{1}{n}\sum_{i=1}^{n_1}\left[\frac{i(n+1-i)}{(n+1)^2}\right]^{q(\delta+r^{-1}-(1/2))}\right.$$

$$\left.+\frac{1}{n}\sum_{i=1}^{n_2}\left[\frac{i(n+1-i)}{(n+1)^2}\right]^{q(\delta+r^{-1}-(1/2))}\right\}^{1/q} \qquad (7.5.21)$$

where by (7.5.16) and (7.5.17), both n_1/n and n_2/n are a.s. (as $n \to \infty$) bounded by

$$n^{-1+\delta'} + O\left\{(n^{-1/2}\log n)(n^{-1+\delta'})^{(1/2)-\epsilon}\right\} = n^{-1+\delta'}\{1 + o(1)\}, \qquad (7.5.22)$$

where we let $\epsilon = \frac{1}{2}\delta = \frac{1}{2}\delta'$. Thus the rhs of (7.5.21), for large n, is a.s. bounded by

$$O\left(\left[n^{-1(1-\delta')(q(\delta+r^{-1}-(1/2))+1)}\right]^{1/q}\right) = O(n^{-(1/2)-\delta((1/2)-\delta)}). \qquad (7.5.23)$$

In case (b), the second term on the rhs of (7.5.18) is bounded by

$$\left\{\max_{1\leqslant i\leqslant n}|X_i|\right\}^{1-r}\left\{\max_{1\leqslant i\leqslant n_1\vee n_2} K_2\left[\frac{i(n+1-i)}{(n+1)^2}\right]^{-(1/2)+(1/r)+\delta}\right\} \qquad (7.5.24)$$

$$\left\{\int_{-\infty}^{a_n'} + \int_{a_n''}^{\infty}|x|^r dF_n(x)\right\},$$

where, as before, the last factor of (7.5.24) converges almost surely to 0 as $n \to \infty$. Using the a.s. upper bound (7.5.22) for both n_1 and n_2, the second factor of (7.5.24) is a.s. bounded by

$$O(n^{-(1-\delta')(\delta+r^{-1}-(1/2))}) = O(n^{-(1/r)+(1/2)-\eta}),$$

$$0 < \eta < \delta r(\tfrac{1}{2} - \delta), \qquad \delta' = \delta r. \qquad (7.5.25)$$

Finally, $E|X|^r < \infty \Rightarrow \max_{1\leqslant i\leqslant n}|X_i|^r = o(n)$ a.s., as $n \to \infty$, and hence (7.5.24) is a.s. (as $n \to \infty$), $o(n^{-(1/2)-\eta})$. Thus in either case (7.5.18) is a.s. (as $n \to \infty$), $O(n^{-(1/2)-\eta})$ for some $\eta > 0$.

From (7.5.5) and (7.5.7), we have

$$|x\phi^{(2)}(F(x))| \leqslant K_1K_2\{F(x)[1 - F(x)]\}^{-(5/2)+\delta}, \qquad \forall x \in (-\infty, \infty), \qquad (7.5.26)$$

and hence, as in (5.4.35) through (5.4.37), it follows by some standard steps that $R_{2n} = O(n^{-(1/2)-\eta})$ a.s., as $n \to \infty$, for some $\eta > 0$. Again, by (7.5.5) and (7.5.7), $|x\phi^{(1)}(F(x))| \leqslant K_1 K_2 \{F(x)[1 - F(x)]\}^{-(3/2)+\delta}$, $\forall x \in (-\infty, \infty)$, and hence, the treatment of R_{3n}, R_{4n}, and R_{5n} follows on the same line as in (5.4.32) and (5.4.43), with obvious modificications. Hence the details are omitted. Q.E.D.

We conclude this section with the remark that if in (7.5.3), we seek the order of the remainder term as simply $o(n^{-1/2}(\log\log n)^{1/2})$, then, we do not need the second derivative condition of $\phi(\cdot)$ in (7.5.7). The result follows along the lines of the proof of Theorem 7.4.6, where, instead of (7.4.56), we need to use (2.6.32); the proof is left as an exercise (see Problem 7.5.1).

7.6. A.S. CONVERGENCE OF LCFOS AND SOME RELATED FUNCTIONALS

For type I LCFOS, a.s. convergence directly follows from (7.3.24). Hence we are mainly interested in type II LCFOS. First, consider the a.s. convergence of an arbitrary LCFOS

$$V_n = \sum_{i=1}^{n} h_{n,i} X_{n,i} = \int_{-\infty}^{\infty} \psi_n(F_n(x)) x \, dF_n(x) \tag{7.6.1}$$

where $h_{n,i} = n^{-1}\psi_n(t)$ for $(i-1)/n < t \leqslant i/n, i = 1, \ldots, n$, and we let

$$\nu = \int_{-\infty}^{\infty} \psi(F(x)) x \, dF(x), \tag{7.6.2}$$

where we assume that $\psi_n(u) \to \psi(u)$ as $n \to \infty$ for $\forall 0 < u < 1$ and the integral in (7.6.2) exists. Our concern is to study suitable regularity conditions under which

$$V_n \to \nu \qquad \text{a.s.,} \quad \text{as} \quad n \to \infty. \tag{7.6.3}$$

Note that whenever $\{V_n, \mathfrak{F}_n; n \geqslant n_0\}$ is a reverse martingale, we may use the reverse martingale convergence theorem to prove (7.6.3). Hence, in the sequel, we only consider the case where the above reverse martingale property may not hold. We assume that

$$\psi(t) = \psi_1(t) - \psi_2(t), \qquad 0 < t < 1 \quad \text{where} \quad \psi_j(t) \quad \text{is} \nearrow \text{in } t$$
$$\text{and is continuous inside} \quad I, \qquad j = 1, 2. \tag{7.6.4}$$

Then, the following theorems relate to (7.6.3) under different sets of regularity conditions. A more recent result of van Zwet (1980) also deserves mention in this context.

THEOREM 7.6.1 (Sen, 1978b). *Let r and s be two positive numbers satisfying $r^{-1} + s^{-1} = 1$ and such that*

$$E|X|^r < \infty \quad \textit{and} \quad \limsup_n n^{-1} \sum_{i=1}^{n} \left|\psi_n\left(\frac{i}{n}\right)\right|^s < \infty. \tag{7.6.5}$$

Then, under (7.6.4), (7.6.3) *holds.*

Proof. We may rewrite $V_n - \nu$ as

$$V_n - \nu = \int_{-\infty}^{\infty} [\psi_n(F_n(x)) - \psi(F(x))] x \, dF_n(x) + \int_{-\infty}^{\infty} x\psi(F(x)) d[F_n(x) - F(x)]. \tag{7.6.6}$$

By the Kintchine strong low of large numbers, the second term on the rhs $\to 0$ a.s. as $n \to \infty$. The first term is written as

$$\int_{-\infty}^{-C} + \int_{-C}^{C} + \int_{C}^{\infty} [\psi_n(F_n(x)) - \psi(F(x))] x \, dF_n(x) = I_{n1} + I_{n2} + I_{n3}, \quad \text{say}, \tag{7.6.7}$$

where $C(0 < C < \infty)$ is so chosen that for given $\epsilon > 0$ and $\eta > 0$,

$$\int_{-\infty}^{-C} + \int_{C}^{\infty} |x\psi(F(x))| \, dF(x) < \tfrac{1}{2}\epsilon \quad \text{and} \quad \int_{-\infty}^{-C} + \int_{C}^{\infty} |x|^r \, dF(x) < \eta. \tag{7.6.8}$$

For I_{n2}, $|x|$ is bounded, so that by (7.6.4) and the Clivenko-Cantelli theorem, $\sup\{x|\psi_n(F_n(x)) - \psi(F(x))| : |x| \leqslant C\} \to 0$ a.s., as $n \to \infty$, and hence, $I_{n2} \to 0$ a.s. For $I_{n1} + I_{n3}$, note that $\{\int_{-\infty}^{-C} + \int_{C}^{\infty} \psi(F(x)) x \, dF_n(x), \mathcal{F}_n; n \geqslant 1\}$ is a reverse martingale (uniformly integrable), so that by (7.6.8) $|\int_{-\infty}^{-C} + \int_{C}^{\infty} x\psi(F(x)) \, dF_n(x)| < \frac{1}{2}\epsilon$ a.s., as $n \to \infty$. Further, by the Hölder-inequality

$$\left|\int_{-\infty}^{-C} + \int_{C}^{\infty} \psi_n(F_n(x)) x \, dF_n(x)\right| \leqslant \left[\int_{-\infty}^{-C} + \int_{C}^{\infty} |x|^r \, dF_n(x)\right]^{r^{-1}} \left[\int_{-\infty}^{-C} + \int_{C}^{\infty} |\psi_n(F_n(x))|^s \, dF_n(x)\right]^{s^{-1}}, \tag{7.6.9}$$

where $\{\int_{-\infty}^{-C} + \int_{C}^{\infty} |x|^r \, dF_n(x), \mathcal{F}_n; n \geqslant 1\}$ is a reverse martingale (uniformly integrable), so that by (7.6.8), $\int_{-\infty}^{-C} + \int_{C}^{\infty} |x|^r \, dF_n(x) < \eta$ a.s., as $n \to \infty$. Finally, by (7.6.5)

$$\int_{-\infty}^{-C} + \int_{C}^{\infty} |\psi_n(F_n(x))|^s \, dF_n(x) \leqslant n^{-1} \sum_{i=1}^{n} \left|\psi_n\left(\frac{i}{n+1}\right)\right|^s < \infty. \tag{7.6.10}$$

Thus, by proper choice of $\eta(>0)$, the rhs of (7.6.9) can be made $\leqslant \frac{1}{2}\epsilon$ a.s., as $n\to\infty$. Thus $I_{n1} + I_{n3} \to 0$ a.s., as $n\to\infty$. Q.E.D.

In the preceding theorem, we need that $E|X|^r < \infty$ for some $r > 1$. This may be relaxed (at the cost of additional restrictions on ψ_n) as follows.

THEOREM 7.6.2. *Suppose that both $b(u)$ and $\psi(u)$ are continuous inside* $(0, 1)$ *and*

$$|b(u)| \leqslant K\left[u(1-u)\right]^{-\alpha}, \qquad |\psi(u)| \leqslant K\left[u(1-u)\right]^{-\beta}, \qquad \forall u \in (0,1), \tag{7.6.11}$$

where K is a positive constant and $\alpha + \beta = 1 - \delta$, for some $\delta > 0$. Then (7.6.3) *holds.*

Proof. By (7.6.6) and (7.6.7), here also, we need to show only that $I_{n1} + I_{n3} \to 0$ a.s., as $n\to\infty$, and, it suffices to show that by choosing C adequately large,

$$\left|\int_{-\infty}^{-C} + \int_{C}^{\infty} \psi(F(x))x\,dF_n(x)\right| \to 0 \qquad \text{a.s., as } n\to\infty, \tag{7.6.12}$$

$$\left|\int_{-\infty}^{-C} + \int_{C}^{\infty} \psi_n(F_n(x))x\,dF_n(x)\right| \to 0 \qquad \text{a.s., as } n\to\infty. \tag{7.6.13}$$

By virtue of (7.6.11) and the reverse submartingale property of the integral in (7.6.12), its a.s. convergence follows readily. To prove (7.6.13), we note that

$$P\left\{nF(X_{n,1}) < (\log n)^{-2} \quad \text{or} \quad n\left[1 - F(X_{n,n})\right] < (\log n)^{-2} \right.$$
$$\left. \text{for some } n \geqslant N\right\} \to 0 \text{ as } N\to\infty. \tag{7.6.14}$$

We break up the domain $(-\infty, -C)$ into two parts: $(-\infty, F^{-1}((\log n)^4/n))$ and $(F^{-1}((\log n)^4/n) - C)$. By (7.6.11) and (7.6.14), the contribution of the integral on the first domain is

$$O\left(n^{-1+\alpha+\beta}(\log n)^{4+2\alpha-\beta}\right) \qquad \text{a.s., as } n\to\infty, \tag{7.6.15}$$

and hence, noting that $\alpha + \beta < 1$, we conclude that (7.6.15) is $o(1)$ a.s., as $n\to\infty$. For the integral over the second domain, we make use of (2.6.32) and (7.4.51) and thereby we bound it from above by

$$K^* \int_{-\infty}^{-C} \left\{F(x)\left[1 - F(x)\right]\right\}^{-1+\delta'} dF_n(x), \tag{7.6.16}$$

where K^* is a finite positive number and $\delta' > 0$. Since (7.6.16) is a nonnegative submartingale, the same proof as in the case of (7.6.12) holds for its a.s. convergence to 0. A similar case holds for the domain (C, ∞). Q.E.D.

Let us now examine the regularity conditions pertaining to (7.4.12). For simplicity, we let $c_{n,i} = n^{-1}\phi(i/(n+1))$, $1 \leqslant i \leqslant n$ and assume that ϕ satisfy (7.6.4). Then, we may rewrite σ_n^2 in (7.4.12) as

$$n^2\sigma_n^2 = \int_{-\infty}^{\infty}\int_{-\infty}^{\infty}\left[F_{n+1}(x \wedge y) - F_{n+1}(x)F_{n+1}(y)\right] \phi(F_{n+1}(x))\phi(F_{n+1}(y))\,dx\,dy. \tag{7.6.17}$$

Note that under (7.6.11)

$$L = \int_{-\infty}^{\infty} |\phi(F(x))|\{F(x)[1 - F(x)]\}^{1/2}\,dx < \infty, \tag{7.6.18}$$

which ensures that $\sigma^2 \leqslant L^2 < \infty$. Also let $E = \{(x, y): -\infty < x, y < \infty\}$ and for $C > 0$, $E_C = \{(x, y): |x| < C, |y| < C\}$, $E_C^* = E - E_C$. Further, we choose $C(< \infty)$ such that for some given $\eta > 0$,

$$\int_{-\infty}^{-C} + \int_{C}^{\infty} |\phi(F(x))|\{F(x)[1 - F(x)]\}^{1/2}\,dx < \eta/4L. \tag{7.6.19}$$

Then, from (7.4.3) and (7.6.19), we have

$$\left|\sigma^2 - \int\int_{E_C}[F(x \wedge y) - F(x)F(y)]\phi(F(x))\phi(F(y))\,dx\,dy\right| < \eta/2. \tag{7.6.20}$$

By (7.6.4) (for $\phi = \psi$) and the Glivenko-Cantelli theorem as $n \to \infty$,

$$\int\int_{E_C}\{F_{n+1}(x \wedge y) - F_{n+1}(x)F_{n+1}(y)\}\phi(F_{n+1}(x))\phi(F_{n+1}(y))\,dx\,dy$$
$$\to \int\int_{E_C}[F(x \wedge y) - F(x)F(y)]\phi(F(x))\phi(F(y))\,dx\,dy \qquad \text{a.s.} \tag{7.6.21}$$

Finally,

$$\left|\int\int_{E_C^*}[F_{n+1}(xy) - F_{n+1}(x)F_{n+1}(y)]\phi(F_{n+1}(x))\phi(F_{n+1}(y))\,dx\,dy\right|$$
$$\leqslant 2\left(\int_{-\infty}^{\infty}\{F_{n+1}(x)[1 - F_{n+1}(x)]\}^{1/2}|\phi(F_{n+1}(x))|\,dx\right)$$
$$\times\left(\int_{-\infty}^{-C} + \int_{C}^{\infty}\{F_{n+1}(y)[1 - F_{n+1}(y)]\}^{1/2}|\phi(F_{n+1}(y))|\,dy\right). \tag{7.6.22}$$

As such, if we let

$$\psi_n^*\left(\frac{i}{n+1}\right) = \left\{\left[i(n+1-i)\right]^{1/2}/(n+1)\right\}\left|\phi\left(\frac{i}{n+1}\right)\right|, \qquad 1 \leqslant i \leqslant n;$$

$$\psi_n^*(0) = \psi_n^*(1) = 0, \tag{7.6.23}$$

and note that by (7.6.19), $\psi_n^*(i/(n+1))$ is bounded for every $1 \leqslant i \leqslant n$ and it tends to 0 as $i/(n+1) \to 0$ or 1, we obtain on letting

$$\psi_n\left(\frac{i}{n+1}\right) = n\left[\psi_n^*\left(\frac{i-1}{n+1}\right) - \psi_n^*\left(\frac{i}{n+1}\right)\right], \qquad 1 \leqslant i \leqslant n, \tag{7.6.24}$$

and integrating by parts

$$\begin{aligned}
&\int_{-\infty}^{\infty} \{F_{n+1}(x)[1 - F_{n+1}(x)]\}^{1/2} |\phi(F_{n+1}(x))|\, dx \\
&\quad = \int_{-\infty}^{\infty} \psi_n^*(F_{n+1}(x))\, dx = -\int_{-\infty}^{\infty} x\, d\psi_n^*(F_{n+1}(x)) \\
&\quad = \int_{-\infty}^{\infty} \psi_n(F_{n+1}(x)) x\, dF_{n+1}(x).
\end{aligned} \tag{7.6.25}$$

Thus, if ϕ_n, defined by (7.6.24), satisfies the conditions of Theorem 7.6.1 or 7.6.2, the rhs of (7.6.25) converges a.s. and similarly, the second factor on the rhs of (7.6.22) can be made arbitrarily small (a.s.) as $n \to \infty$. This leads us to the following.

THEOREM 7.6.3. *If* (7.6.18) *holds, for* $c_{n,i} = n^{-1}\phi(i/(n+1))$, $1 \leqslant i \leqslant n$, ϕ *satisfies* (7.6.4) *and* ψ_n, *defined by* (7.6.24), *satisfies the hypothesis of Theorem* 7.6.1 *or* 7.6.2, *then* (7.4.12) *holds.*

We proceed on to (7.4.13). Note that if $X_{n+1} = X_{n+1,k} (1 \leqslant k \leqslant n+1)$, then under the condition that $d_{n,i} = 0$, $\forall 1 \leqslant i \leqslant n$, $n \geqslant n_0$, by (7.4.6),

$$\begin{aligned}
U_n &= n\left\{\sum_{i=1}^{k-1} c_{n,i} X_{n+1,i} + \sum_{i=k+1}^{n+1} c_{n,i-1} X_{n+1,i} - \sum_{i=1}^{n+1} c_{n+1,i} X_{n+1,i}\right\} \\
&= n\left\{\sum_{i=1}^{k-1}\left[\frac{i}{n+1} c_{n,i} - \frac{i-1}{n+1} c_{n,i-1}\right] X_{n+1,i} - c_{n,k} X_{n+1,k}\right. \\
&\qquad \left. - \sum_{i=k+1}^{n+1}\left[\frac{n-i+1}{n+1} c_{n,i} - \frac{n-i+2}{n+1} c_{n,i-1}\right] X_{n+1,i}\right\} \\
&= \int_{-\infty}^{\infty} [I(x \geqslant X_{n+1,k}) - F_{n+1}(x)]\phi(F_{n+1}(x))\, dx \\
&= U_{n+1,k}, \qquad \text{say, for } k = 1, \ldots, n+1.
\end{aligned} \tag{7.6.26}$$

Hence, for every $\lambda > 0$,

$$E\left[U_n^2 I(|U_n| > \lambda) | \mathcal{F}_{n+1} \right] = \frac{1}{n+1} \sum_{k=1}^{n+1} U_{n+1,k}^2 I(|U_{n+1,k}| > \lambda). \quad (7.6.27)$$

Thus it suffices to show that as $n \to \infty$,

$$n^{-1/2} \left| \int_{-\infty}^{\infty} \left[I(x \geqslant X_{n+1}) - F_{n+1}(x) \right] \phi(F_{n+1}(x))\, dx \right| \to 0 \quad \text{a.s.} \quad (7.6.28)$$

For this purpose, we define

$$Z_n = \int_{-\infty}^{\infty} \left[I(x \geqslant X_n) - F(x) \right] \phi(F(x))\, dx, \qquad n \geqslant 1, \quad (7.6.29)$$

$$Z_n^* = \int_{-\infty}^{\infty} \left[I(x \geqslant X_n) - F_n(x) \right] \phi(F_n(x))\, dx, \qquad n \geqslant 1, \quad (7.6.30)$$

and decomposing the range $(-\infty, \infty)$ as $(-\infty, -C)$, $[-C, C]$, (C, ∞), $C > 0$, we denote the corresponding components of Z_n (or Z_n^*) as $Z_n^{(1)}(C)$, $Z_n^{(2)}(C)$ and $Z_n^{(3)}(C)$ [or $Z_n^{*(1)}(C)$, $Z_n^{*(2)}(C)$, and $Z_n^{*(3)}(C)$], respectively. Since $\{Z_n, n \geqslant 1\}$ are i.i.d.r.v. with $EZ_n = 0$ and $EZ_n^2 = \sigma^2$, defined by (7.4.3), it follows that $n^{-1/2}|Z_n| \to 0$ a.s., as $n \to \infty$. In a similar manner it follows that

$$\max_{1 \leqslant l \leqslant 3} n^{-1/2} |Z_n^{(l)}(C)| \to 0 \quad \text{a.s., as } n \to \infty. \quad (7.6.31)$$

Also, if ϕ satisfies (7.6.4), then every $C > 0$,

$$\begin{aligned} Z_n^{*(2)}(C) - Z_n^{(2)}(C) &= \int_{-C}^{C} \left[F(x) - F_n(x) \right] \phi(F_n(x))\, dx \\ &\quad + \int_{-C}^{C} \left[I(x \geqslant X_n) - F(x) \right] \\ &\quad \times \left[\phi(F_n(x)) - \phi(F(x)) \right] dx \\ &\to 0 \quad \text{a.s., as } n \to \infty. \end{aligned} \quad (7.6.32)$$

Further, for every $C > 0$,

$$\begin{aligned} & n^{-1/2} |Z_n^{*(1)}(C) + Z_n^{*(3)}(C)| \\ &\leqslant \int_{-\infty}^{-C} + \int_{C}^{\infty} n^{-1/2} |I(x \geqslant X_n) - F_n(x)| \phi(F_n(x))|\, dx \\ &\leqslant \int_{-\infty}^{-C} + \int_{C}^{\infty} \left\{ F_n(x)\left[1 - F_n(x)\right] \right\}^{1/2} |\phi(F_n(x))|\, dx. \end{aligned} \quad (7.6.33)$$

As such, as in (7.6.22), by choosing C adequately large, the rhs of (7.6.33) can be made $\leqslant \frac{1}{2}\eta$ a.s., as $n \to \infty$. Hence, from (7.6.31) through (7.6.33) we obtain the following.

THEOREM 7.6.4. *Under the assumptions of Theorem* 7.6.3, (7.4.13) *holds*.

It may be remarked that for both Theorems 7.6.3 and 7.6.4, the assumptions in (7.4.39) through (7.4.41) suffice; it is not necessary to impose (7.4.42). This latter condition has been used in dealing with the convergence of the remainder term in (7.4.43).

In Theorem 7.6.3 [see also (7.4.12)], we have noticed that a strongly consistent estimator of the variance σ^2 [defined by (7.4.3)] is

$$\hat{\sigma}_n^2 = \int_{-\infty}^{\infty}\int_{-\infty}^{\infty}\left[F_n(x\wedge y) - F_n(x)F_n(y)\right]\phi_n(F_n(x))\phi_n(F_n(y))\,dx\,dy. \tag{7.6.34}$$

Note that $\hat{\sigma}_n^2$ is a quadratic function of the order statistics $X_{n,i}$ and the constants $c_{n,i}$. So a natural question arises whether the asymptotic theory, studied earlier for the linear functions of order statistics, extends to such quadratic functionals as well. An answer to this question has been provided by Gardiner and Sen (1979). We quote below one of their main results and leave the proof as an exercise (see Problem 7.6.2).

THEOREM 7.6.5. *Under* (7.4.39), (7.4.40), (7.4.42), *and the condition that* $\alpha+\beta<\frac{1}{4}$, *as* $n\to\infty$,

$$\frac{n^{1/2}(\hat{\sigma}_n^2-\sigma^2)}{\gamma} \xrightarrow[\mathfrak{D}]{} \mathfrak{N}(0,1), \tag{7.6.35}$$

where

$$\begin{aligned}\gamma^2 &= \int_0^1\int_0^1 (s\wedge t - st)L_0(s)L_0(t)\,db(s)\,db(t)\\ &= \int_0^1 G_0^2(t)\,dt - \left(\int_0^1 G_0(t)\,dt\right)^2,\end{aligned} \tag{7.6.36}$$

$$G_0(t) = \int_0^t L_0(s)\,db(s), \qquad 0<t<1, \tag{7.6.37}$$

$$L_0(t) = L_1(t)\{\phi(t)+t\phi'(t)\} + L_2(t)\{(1-t)\phi'(t)-\phi(t)\}, \tag{7.6.38}$$

$$L_1(t) = 2\int_t^1 (1-s)\phi(s)\,db(s), \tag{7.6.39}$$

$$L_2(t) = 2\int_0^t s\phi(s)\,db(s). \tag{7.6.40}$$

The result can be extended to a weak invariance principle too and hence remains valid even when n, the sample size, is a positive integer valued r.v.

satisfying a condition similar to (3.3.19). We pose this also as an exercise (see Problem 7.6.3). These results are useful in the study of the sequential procedures based on LCFOS.

It may be remarked that as regards the a.s. representation in (7.5.1) is concerned, one can also consider a linear combination of both type I and type II LCFOS and get a similar representation. For type I LCFOS, one needs to use the Bahadur-representation considered in Section 7.2 and then combine the principal term there with the one in (7.5.1). This feature remains true for the a.s. convergence results considered in this section too.

We conclude this chapter with some references to the diverse methods for the study of the asymptotic normality of LCFOS with smooth weight functions. Chernoff, Gastwirth, and Johns (1967) have employed an elegant representation of type II LCFOS in terms of r.v.'s having simple exponential d.f.'s and have utilized the properties of the order statistics from exponential d.f.'s to derive the desired asymptotic normality. Bickel (1967) and Shorack (1969, 1972a, b) have utilized the weak convergence of the uniform empirical distribution process (and the related quantile process) for the study of the asymptotic normality of type II LCFOS; Wellner (1974) has considered an extension of the Shorack-approach to the sequential case. Stigler (1969) has employed the Hájek (1968)-projection technique to approximate a type II LCFOS in terms of a linear function of independent r.v.'s and has utilized the central limit theorem on the latter for the asymptotic normality of the former. In this respect, the contribution of the Hungarian school (Komlós, Major, and Tusnády, 1975; Csörgo and Revesz, 1975c) deserves special mention. Though most of these studies are concentrated to the asymptotic normality of LCFOS, their techniques can be extended (under additional regularity conditions) to cover the weak invariance principles considered in Section 7.4. A few problems relate to these possibilities. Ghosh (1972) has studied the a.s. representation in Section 7.5 under more restrictive regularity conditions and the theorems considered here relax some of these conditions. Extensions of the techniques in the other papers referred to earlier to cover this a.s. invariance principle may involve more elaborate analysis. The martingale (or reverse martingale) approach considered in this chapter, besides being in line with the general motivations underlying the earlier chapters, provides a simple approach to the study of the invariance principles for LCFOS.

EXERCISES

7.3.1. Let $k = [np] + 1$, $0 < p < 1$ and let $F(\xi_p) = p$. Show that $X_{n,k}$ is a consistent estimator of ξ_p whenever ξ_p is uniquely defined. Is it strongly consistent too?

7.3.2. Let $M_{n,k} = (X_{n,k} + X_{n,n-k+1})/2$ and $W_{n,k} = (X_{n,n-k+1} - X_{n,k})$ for $k = 1, \ldots, [(n+1)/2]$. Let F be (a) normal, (b) double exponential, (c) Cauchy d.f. with location parameter θ and scale parameter $\gamma(>0)$. Let $k = [np] + 1$, $0 < p < 1$. Find an optimal value of p for which $M_{n,k}$ (or $W_{n,k}$) provides an estimator of θ (or γ) having the smallest asymptotic variance. Compare this optimal variance with the Cramér-Rao bound for the same parameter.

7.3.3. Let F be a simple exponential d.f. with scale parameter θ, $k_j = [np_j] + 1$, for $j = 1, \ldots, m(\geqslant 1)$ and $0 \leqslant p_1 < \cdots < p_m < 1$. Obtain an estimator of θ based on a linear combination of $X_{n,k_1}, \ldots, X_{n,k_m}$ having the smallest asymptotic variance with (i) $p_1, \ldots, p_m$ are given and (ii) only m is specified while $p_1, \ldots, p_m$ are to be determined in an optimal way.

7.3.4. In each of the foregoing problems, identify the estimators as type I LCFOS and derive the invariance principles considered in Section 7.3.

7.4.1. Let $f(x)$, the p.d.f. corresponding to the d.f. F, be of the form $f_0((x-\mu)/\gamma)/\gamma$, where μ and γ are the location and scale parameters and f_0 has a continuous first order derivative f_0' a.e. Let $\theta = \alpha_1\mu + \alpha_2\gamma$ where α_1 and α_2 are given constants. Consider a type II LCFOS of the form (7.2.6) where ϕ_n converges to a smooth ϕ. Also, let $g_1(x) = -f_0'(x)/f_0(x)$ and $g_2(x) = -xf_0'(x)/f_0(x) - 1, \forall x \in R$, $\boldsymbol{\alpha} = (\alpha_1, \alpha_2)'$, $\mathbf{a} = (a_1, a_2)' = \boldsymbol{\alpha}'\mathbf{D}^{-1}$ where

$$\mathbf{D} = ((d_{ij})) = \left(\left(\int_{-\infty}^{\infty} g_i(x) g_j(x) f_0(x)\, dx\right)\right)_{i,j=1,2}$$

and, finally, let

$$\phi_0(u) = -a_1 g_1'\big(F^{-1}(u)\big) - a_2 g_2'\big(F^{-1}(u)\big), \qquad \forall u \in (0,1).$$

Show that T_n, defined by (7.2.6) with ϕ_n converging to ϕ_0, is asymptotically optimal for θ (Jung, 1962: Chapter 4 of Sarhan and Greenberg, 1962).

7.4.2. Based on the set $(X_{n,1}, \ldots, X_{n,m})$ where $m = [n\alpha] + 1$ and $0 < \alpha < 1$ find an asymptotically optimal estimator of μ in the problem 7.4.1 and show that it is a linear combination of type I and type II LCFOS. For this statistic, obtain the invariance principles considered in Sections 7.3 through 7.5.

7.4.3. Suppose that F is an exponential d.f. with parameter $\theta(>0)$. Suppose that you want to test for $H_0: \theta = \theta_0$ vs $H_1: \theta = \theta_1 > \theta_0$. Based on the set $X_{n,1}, \ldots, X_{n,m}$, consider the usual likelihood ratio statistic, express it in terms of an LCFOS and apply the invariance principles in Section 7.3 and 7.4 to study possible sequential versions of your tests.

7.4.4. Rewrite $X_{n,i} = F^{-1}(U_{n,i})$, $i = 1, \ldots, n$ where the $U_{n,i}$ are the order statistics from a uniform $(0, 1)$ d.f. Let then $V_{nj} = -\log(1 - U_{nj})$ for $j = 1, \ldots, n$. Show that the V_{nj} may be represented by

$$V_{nj} = Z_1/n + Z_2/(n-1) + \cdots + Z_j/(n-j+1), \qquad j = 1, \ldots, n$$

where the Z_j are i.i.d.r.v.'s with the simple exponential d.f. (Chernoff, Gastwirth, and Johns, 1967).

7.4.5 (7.4.4 continued). Use the representation in the Problem 7.4.4 to express $T_n = \mu_n + Q_n + R_n$ where μ_n is a nonstochastic constant, $Q_n = n^{-1}\sum_{j=1}^{n}\alpha_{nj}(Z_j - 1)$ and the remainder term R_n is asymptotically negligible. Hence, or otherwise, specify conditions under which Q_n (and hence $T_n - \mu_n$)) is asymptotically normal. Compare to Theorems 7.4.2 and 7.4.3 (Chernoff, Gastwirth, and Johns, 1967).

7.4.6. If ϕ is continuous on $[0, 1]$ except for jump discontinuities at $a_1, \ldots, a_m$ and ϕ' is continuous and of bounded variation on $[0, 1] - \{a_1, \ldots, a_m\}$, then the asymptotic normality in Theorem 7.4.3 holds whenever $E|X| < \infty$ (Moore, 1968).

7.4.7. Consider the projection $T_n^* = \sum_{i=1}^{n} E(T_n \mid X_i) - (n-1)ET_n$ and study conditions (on the $c_{n,i}$ and F) under which T_n and T_n^* are asymptotically equivalent in the quadratic mean. Express T_n^* as a sum of i.i.d.r.v.'s and apply the central limit theorem to study the asymptotic normality of T_n^* (and hence of T_n) (Stigler, 1969).

7.4.8. Use the weak convergence of $\{T_n^0\}$ along with (7.4.28) and (7.4.29) to prove the same for $\{T_n\}$. (Hint. Use Theorem 2.3.7 along with Theorem 7.4.2.)

7.5.1. Use the law of iterated logarithm for the empirical process in (2.6.32) and show that under (7.4.36) through (7.4.42), the same holds for T_n (Wellner, 1977a, b).

7.5.2. For the proof of Theorem 7.5.1, in (7.5.10), show that for each of the terms R_{3n}, R_{4n}, and R_{5n}, (7.5.3) holds (Sen, 1977c).

7.6.1. Verify (7.6.15).

Hints. Use the monotonicity of $X_{n,1}$ and $X_{n,n}$ (in n) and follow the classical proof of the Kolmogorov law of large numbers to replace the tail-sequence of events by a countable number of subsequences and apply the Borel-Cantelli lemma.

7.6.2. Provide a proof of Theorem 7.6.5 (Gardiner and Sen, 1979).

Hints. Define G_n as in after (7.4.30) and let $\xi_{n,i} = F(X_{n,i}), i = 1, \ldots, n$. Let then $L_{n,1}(t) = 2\int_t^{\xi_{n,n}}(1 - G_n(u))\phi_n(G_n(u))\,db(u)$ and $L_{n,2}(t) = 2\int_{\xi_{n,1}}^{t} G_n(u)$

$\phi_n(G_n(u))\,db(u)$ for $t \in (\xi_{n,1}, \xi_{n,n})$ and set both of them equal to 0, otherwise. Then, $n^{1/2}(\hat{\sigma}_n^2 - \sigma^2) = \frac{1}{2}(S_{n,1} + S_{n,2} + R_n)$, where $S_{n,1} = \int_0^1 n^{1/2}(L_{n,1}(t) - L_1(t))\,dL_2(t)$, $S_{n,2} = \int_0^1 L_1(t)\,d\{n^{1/2}(L_{n,2}(t) - L_2(t))\}$ and $R_n = n^{1/2}\int_0^1 (L_{n,1}(t) - L_1(t))\,d(L_{n,2}(t) - L_2(t))$. Use the weak convergence of the two processes $n^{1/2}\{L_{n,j}(t) - L_j(t), t \in [0,1]\}$, $j = 1, 2$ to prove the desired result.

7.6.3. Extend the result in Theorem 7.6.5 to the case of random sample sizes satisfying (3.3.19).

Additional Problems

Induced order statistics

7.*.1. Let (X_i, Y_i), $i = 1, \ldots, n$ be i.i.d.r.v.'s with a bivariate d.f. F, defined on R^2. Let $X_{n,1} < \cdots < X_{n,n}$ be the order statistics corresponding to the X_i. Then, the induced order statistics (or concomitants of order statistics) $Y_{n1}, \ldots, Y_{nn}$ are defined by $Y_{nk} = Y_j$ if $X_{n,k} = X_j$ for $j, k = 1, \ldots, n$. Let $m(x) = E(Y_1 | X_1 = x)$ and set $S_{nk} = n^{-1/2}\sum_{j=1}^{k}\{Y_{nj} - m(X_{n,j})\}$, $k = 1, \ldots, n$; $S_{n0} = 0$. Establish a Donsker-type (weak) invariance principle for the partial sequence $\{S_{nk}, k \leqslant n\}$ (Bhattacharya, 1974, 1976; Sen, 1976e).

7.*.2. Study the relationship between the induced order statistics and the mixed rank statistics of Chapter 6 (Sen, 1981e).

Rank-discounted partial sums

7.*.3. Let $X_i, i \geqslant 1$ be a sequence of i.i.d.r.v.'s with a continuous d.f. F, defined on R. Let R_{ni} be the rank of X_i among $X_1, \ldots, X_n$, for $i = 1, \ldots, n$, and let $g(\cdot)$ be a measurable function. A rank-discounted partial sum is defined by $T_n = \sum_{i=1}^{n} a_i(R_{ii})g(X_i), n \geqslant 1$ and $T_0 = 0$. Show that T_n can be expressed in either of the following two forms: $\sum_{k=1}^{n} a_k(R_{kk})\, g(X_{k,R_{kk}})$ or $\sum_{k=1}^{n} a_k(R_{kk})g(X_{n,R_{nk}})$ and, in either form, it differs from an LCFOS in the sense studied in Section 7.2. Nevertheless, the invariance principles hold for the T_n under parallel conditions (Sen, 1978c).

Capture, mark, and recapture methods

7.*.4. From a population containing an unknown number N of individuals, the individuals are drawn one by one, marked, and released before the next drawing is made. Let M_k be the number of marked individuals in the population just before the kth drawal, for $k \geqslant 1$. Show that for the sequential sampling tagging, at the nth stage ($n \geqslant 2$), the (partial) likelihood function is

$$L_n(N) = N^{-(n-1)} \prod_{k=2}^{n} \{ M_k^{X_k}(N - M_k)^{1-X_k} \},$$

where X_k is equal to 1 or 0 according as the kth drawal yields a marked individual or not, for $k \geqslant 2$. Let $Z_n(N) = N^{1/2}(\partial/\partial N) \log L_n(N)$, $n \geqslant 2$. Show that the $Z_n(N), n \geqslant 2$ form a martingale sequence (though the increments are not independent or homogeneous). Hence verify the regularity conditions of Theorem 2.4.7 and obtain the weak convergence of the related likelihood ratio process. Utilize this result for the asymptotic normality of the maximum likelihood estimator of N when the sample size n is itself a positive integer valued random variable (Samuel, 1968; Sen, 1981g).

7.*.5 (7.*.4 continued). An urn contains an unknown number N of white balls and no others. We repeatedly draw a ball at random, observe its color, and replace it by a black ball, so that before each draw, there are N balls in the urn. Let W_n be the number of white balls observed in the first n draws, $n \geqslant 1$. For every $c > 0$, consider the stopping variable $t_c = \inf\{n : n \geqslant (c+1)W_n\}$. Show that the W_n can be expressed in terms of a linear combination of some martingale-differences. Hence obtain a renewal theorem for the $t_c, c > 0$ (Sen, 1982a).

[The last two problems have nothing to do with LCFOS. But, they are based on some martingale theory very similar to the one employed in this chapter, and hence are presented here.]

CHAPTER 8

Invariance Principles for M-Estimators and Extrema of Certain Sample Functions

8.1. INTRODUCTION

The *U-statistics*, considered in Chapter 3, play a fundamental role in the theory of optimal (unbiased) nonparametric estimation of general parameters. In parametric models, the *maximum likelihood estimators* (*MLE*) possess certain (asymptotic) optimality properties. Quite often, the MLE are expressible as functions of U-statistics, though, in general, they need to be solved as the roots of certain implicit functions. The MLE are, in general, not so *robust* against *outliers* or *gross-errors*. From considerations of robustness and validity for a wider class of underlying distributions, often, some alternative estimators are advocated. These include the so-called *L-estimators*, *R-estimators*, and *M-estimators*, among others. The L-estimators are based on linear combinations of functions of order statistics, and hence the theory (as well as the invariance principles) developed in Chapter 7 applies to these estimators. The R-estimators rest on linear and signed linear rank statistics; the monotonicity and asymptotic linearity (in the regression or location parameter) of the rank statistics studied in Sections 4.5 and 5.5 (along with the invariance principles presented there) provide the necessary tools for studying invariance principles for these R-estimators. The M-estimators, introduced and popularized by Huber (1964, 1973), are MLE-type, but robust in character. Some invariance principles for such M-estimators are studied in this chapter. These M-estimators include the classical *least squares estimators* (*LSE*) as special cases, and hence these invariance principles apply to these estimators as well.

The M-estimators, being the MLE-type, are related to the extrema of certain sample functions. In some other context too, one encounters some other statistics that are expressible as the extrema of certain sample functions. A notable example of this type of statistics is the *bundle strength of filaments*. We intend to study invariance principles for this type of statistics too.

Section 8.2 includes a general introduction to M-estimators of location and regression along with some of their (asymptotic) properties. Some invariance principles relating to these M-estimators are studied in Section 8.3. Embedding of Wiener processes for M-estimators is considered (under additional regularity conditions) in Section 8.4. Section 8.5 is devoted to a comparative study of the asymptotic properties of M-, L-, and R-estimators. Invariance principles for the bundle-strength type of statistics are presented in Sections 8.6 and 8.7. Some general remarks are made in the concluding section.

8.2. M-ESTIMATORS OF LOCATION AND REGRESSION

Let $X_1, \ldots, X_N$ be N independent r.v. with continuous d.f. $F_1, \ldots, F_N$, respectively, all defined on the real line $E = (-\infty, \infty)$, where

$$F_i(x) = F(x - \Delta c_i), \qquad x \in E, \qquad i = 1, \ldots, N, \tag{8.2.1}$$

Δ is an unknown *parameter*, the c_i are specified constants (not all equal to 0) and F is an unknown d.f. belonging to some class $\mathfrak{F}$ of d.f.'s. If the c_i are all equal to 1, (8.2.1) reduces to the location model (where Δ is the location parameter), while, for the c_i not all equal, we have a simple regression model (where Δ is the regression slope). More general models with vector $\boldsymbol{\Delta}$ and $\mathbf{c}_i$ may also be conceived and will be considered later on.

In the classical least squares estimation theory, one minimizes

$$\sum_{i=1}^{N} (X_i - \Delta c_i)^2 \tag{8.2.2}$$

with respect to Δ and obtains the LSE

$$\tilde{\Delta}_N = \sum_{i=1}^{N} c_i X_i \Big/ \left(\sum_{i=1}^{N} c_i^2 \right). \tag{8.2.3}$$

If F is assumed to be a normal d.f., then $\tilde{\Delta}_N$ is also the MLE of Δ. In general, the MLE of Δ, for a specified form of f, is obtained as a solution of the equation

$$\sum_{i=1}^{N} c_i f'(X_i - \Delta c_i) \Big/ f(X_i - \Delta c_i) = 0. \tag{8.2.4}$$

Keeping (8.2.3) and (8.2.4) in mind, we conceive of some function $\psi(=\rho')$ and set to minimize

$$\sum_{i=1}^{N} \rho(X_i - \Delta c_i) \qquad \text{(with respect to } \Delta) \tag{8.2.5}$$

which leads us to the implicit equation

$$S_N(t) = \sum_{i=1}^{N} c_i \psi(X_i - tc_i) = 0 \tag{8.2.6}$$

whose solution is taken as the estimator of Δ. If we assume that $\psi = \{\psi(y), y \in E\}$ is *nondecreasing and nonconstant*, then, by (8.2.6), $S_N(t)$ is $\searrow$ in $t \in E$, so that we may formally define an M-estimator $\hat{\Delta}_N$ of Δ by

$$\hat{\Delta}_N = \tfrac{1}{2}(\hat{\Delta}_{N1} + \hat{\Delta}_{N2}); \tag{8.2.7}$$

$$\hat{\Delta}_{N1} = \sup\{t : S_N(t) > 0\} \qquad \text{and} \quad \hat{\Delta}_{N2} = \inf\{t : S_N(t) < 0\}. \tag{8.2.8}$$

In particular, for $\psi(x) \equiv x$ [i.e., $\rho(x) \equiv \frac{1}{2}x^2$], $\hat{\Delta}_N$ reduces to $\tilde{\Delta}_N$ in (8.2.3), and for $\psi(x) \equiv -f'(x)/f(x)$, it yields the MLE of Δ. Among other possibilities, if we take $\rho(x) \equiv |x|$, so that $\psi(x)$ is $+1$ or -1 according as x is positive or negative, then, $\hat{\Delta}_N$ in (8.2.7) reduces to a *median type* estimator. Often in actual practice, ψ is taken to be a bounded function. For example, Huber (1964) has suggested the use of

$$\psi(x) = \begin{cases} x, & \text{if } \ |x| \leqslant k, \\ k \operatorname{sgn} x, & \text{if } \ |x| > k, \end{cases} \tag{8.2.9}$$

where k is some specified positive number. We assume that

$$\psi(x) = \psi_1(x) + \psi_2(x), \qquad x \in E, \tag{8.2.10}$$

where both ψ_1 and ψ_2 are nondecreasing and skew-symmetric functions, ψ_1 is absolutely continuous on any bounded interval in E and ψ_2 is a step-function having finitely many jumps. Let $E_j = (a_j, a_{j+1})$, $j = 0, \ldots, p$ (for some $p \geqslant 0$) where $-\infty = a_0 < a_1 < \cdots < a_p < a_{p+1} = +\infty$ and we assume that

$$\psi_2(x) = \beta_j, \qquad \text{for all } \ x \in E_j, \qquad 0 \leqslant j \leqslant p, \tag{8.2.11}$$

where the β_j are real and finite numbers, not all equal. We let, conventionally,

$$\psi_2(x) = \tfrac{1}{2}(\beta_{j-1} + \beta_j), \qquad \text{for } j = 1, \ldots, p \qquad (\text{whenever } p \geqslant 1). \tag{8.2.12}$$

Further, we assume that

$$0 < \sigma_0^2 = \int_{-\infty}^{\infty} \psi^2(x)\, dF(x) < \infty. \tag{8.2.13}$$

The d.f. F is assumed to be symmetric about 0 ($\Rightarrow F(x)+F(-x)=1$, $\forall x \in E$). Since ψ is assumed to be skew-symmetric, this implies that

$$\bar{\psi}_F = \int_{-\infty}^{\infty} \psi(x)\,dF(x) = 0. \tag{8.2.14}$$

Thus, by (8.2.6), (8.2.13), and (8.2.14), we obtain

$$E_\Delta S_N(\Delta) = 0 \qquad \text{and} \quad E_\Delta S_N^2(\Delta) = C_N^2 \sigma_0^2, \tag{8.2.15}$$

where E_Δ denotes the expectation when Δ holds [in (8.2.1)] and

$$C_N^2 = \sum_{i=1}^{N} c_i^2. \tag{8.2.16}$$

Note that by (8.2.7), (8.2.8), and the assumptions on F and ψ made above, $\hat{\Delta}_N$ has a distribution symmetric about Δ, and hence $\hat{\Delta}_N$ is a *midian-unbiased* estimator of Δ. The *translation-invariance* of $\hat{\Delta}_N$ also follows from (8.2.6) to (8.2.8). To study the asymptotic properties of $\hat{\Delta}_N$, we further assume that

$$C_N^{-2}\left\{\max_{1 \leqslant i \leqslant N} c_i^2\right\} \to 0 \qquad \text{as} \quad N \to \infty \tag{8.2.17}$$

and the d.f. F possesses an absolutely continuous p.d.f. f with a finite Fisher information $I(f)$. Let

$$\gamma = \int_{-\infty}^{\infty} \psi(x)\left\{\frac{-f'(x)}{f(x)}\right\} dF(x). \tag{8.2.18}$$

Then, $I(f) < \infty$ and $\sigma_0 < \infty \Rightarrow \gamma^2 < \infty$. On the other hand, by (8.2.10), (8.2.11), and (8.2.18), whenever $f(a_j) > 0$, for $j = 1, \ldots, p$,

$$\gamma = \int_{-\infty}^{\infty} \psi_1^{(1)}(x)\,dF(x) + \sum_{j=1}^{p} (\beta_j - \beta_{j-1}) f(a_j) > 0, \tag{8.2.19}$$

where the second term vanishes when $p = 0$.

Now, by (8.2.15), (8.2.17), and the central limit theorem, we obtain, when Δ holds,

$$S_N(\Delta)/(C_N\sigma_0) \xrightarrow{\mathfrak{D}} \mathfrak{N}(0,1). \tag{8.2.20}$$

Let us consider then a null hypothesis $H_0 : \Delta = 0$ against a sequence $\{K_N\}$ of alternative hypotheses, where under K_N, (8.2.1) holds for $\Delta = -t/C_N$, for some real (fixed) t. Then, under Δ, $S_N(\Delta + t/C_N)$ has the same distribution as $S_N(0)$ under K_N. Let P_N be the probability measure (for $X_1, \ldots, X_N$) under H_0 and Q_N be the corresponding one under K_N. Then, as in the proof of Theorem 4.3.6, $\{Q_N\}$ is contiguous to $\{P_N\}$. Therefore, by an appeal to (4.3.28) and (4.3.29) along with (8.2.20), we conclude that under $\{K_N\}$, $S_N(0)/(C_N\sigma_0)$ is asymptotically normal with mean $-t\gamma/\sigma_0$ and unit variance. Thus, under (2.17) and $I(f) < \infty$, when Δ holds, for every fixed $t(\in E)$,

$$S_N(\Delta + t/C_N)/(C_N\sigma_0) \xrightarrow{\mathfrak{D}} \mathfrak{N}(-t\gamma/\sigma_0, 1). \tag{8.2.21}$$

Further, we note that by (8.2.7) and (8.2.8), for every (fixed) $t(\in E)$,

$$\begin{aligned} &\lim_{N\to\infty} P_\Delta\{C_N(\hat{\Delta}_N - \Delta) \leqslant t\} \\ &\quad = \lim_{N\to\infty} P_\Delta\{S_N(\Delta + t/C_N) \leqslant 0\} \\ &\quad = \lim_{N\to\infty} P_0\{S_N(t/C_N)/(C_N\sigma_0) \leqslant 0\}. \end{aligned} \tag{8.2.22}$$

From (8.2.21) and (8.2.22), we conclude that for every (fixed) $t(\in E)$,

$$\lim_{N\to\infty} P_\Delta\{C_N(\hat{\Delta}_N - \Delta) \leqslant t\} = \Phi(t\gamma/\sigma_0) = \Phi(t/\nu), \tag{8.2.23}$$

where Φ is the standard normal d.f. and

$$\nu^2 = \sigma_0^2/\gamma^2. \tag{8.2.24}$$

Note that by (8.2.13), (8.2.18), and the Cauchy-Schwarz inequality,

$$\nu^2 \geqslant [I(f)]^{-1}, \tag{8.2.25}$$

where the equality sign holds iff $\psi(x)$ is proportional to $\{-f'(x)/f(x)\}$. Thus, from the asymptotic efficiency point of view, the M-estimator attains its full efficiency when it agrees with the MLE. However, from considerations of robustness (against outliers or for the error-contamination model), the MLE may not be very desirable and some other M-estimators may have some *minimax properties* (see Huber, 1981). We intend to study some invariance principles for M-estimators and incorporate these in subsequent chapters in the study of the properties of some sequential procedures based on these M-estimators.

8.3. WEAK INVARIANCE PRINCIPLES FOR M-ESTIMATORS

For every $N(\geqslant 1)$, $k(1 \leqslant k \leqslant N)$ and $t \in A = [-a, a]$, $a > 0$, let

$$M_{Nk}(t) = (\sigma_0 C_N)^{-1}\left\{\sum_{i\leqslant k} c_i[\psi(X_i - tc_i/C_N) - \psi(X_i)] + t\gamma C_k^2/C_N\right\}, \tag{8.3.1}$$

and, conventionally, we let $M_{N0}(t) = 0$, $\forall t \in A$. Further, we asume that

$$\lim_{t\to 0} E_0\{\psi(X_i + t) - \psi(X_i) - E_0\psi(X_i + t)\}^2 = 0, \qquad \forall i \geqslant 1, \tag{8.3.2}$$

where E_0 stands for the expectation under $H_0 : \Delta = 0$. Then, as a basis for the subsequent results, first, we consider the following.

THEOREM 8.3.1. *Under* $H_0 : \Delta = 0$, (8.2.10) *through* (8.2.13), (8.2.17), (8.2.18), *and* (8.3.2), *for every (finite)* $a(>0)$, *as* $N \to \infty$,

$$\max_{k \leqslant N} \sup_{t \in A} |M_{Nk}(t)| \to 0, \qquad \textit{in probability}. \tag{8.3.3}$$

Proof. Note that by (8.3.1), $\{M_{Nk+1}(t) - M_{Nk}(t),\ t \in A\}$ is stochastically independent of $\{M_{Nk}(t),\ t \in A\}$, for every $k \geqslant 0$. Hence $\{\sup_{t \in A} |M_{Nk}(t) - E_0 M_{Nk}(t)|;\ 0 \leqslant k \leqslant N\}$ is a nonnegative submartingale and by the Kolmogorov inequality, for every $\epsilon > 0$,

$$P\Big\{\max_{0 \leqslant k \leqslant N} \sup_{t \in A} |M_{Nk}(t) - E_0 M_{Nk}(t)| > \epsilon | H_0\Big\}$$
$$\leqslant \epsilon^{-1} E_0\Big\{\sup_{t \in A} |M_{NN}(t) - E_0 M_{NN}(t)|\Big\}. \tag{8.3.4}$$

Also, by (8.2.6) and (8.3.1), $M_{NN}(t) = (\sigma_0 C_N)^{-1}\{S_N(t/C_N) - S_N(0) + t\gamma C_N\}$, $t \in A$, where $S_N(b)$ is $\searrow$ in b $\in E$. Hence, for every $\epsilon > 0$, there exist an $m\ (= m_\epsilon)$ and $t_1, \ldots, t_m$ (all $\in A$), such that

$$\epsilon^{-1} E_0\Big\{\sup_{t \in A} |M_{NN}(t) - E_0 M_{NN}(t)|\Big\}$$
$$\leqslant \epsilon^{-1} E_0\Big\{\max_{1 \leqslant j \leqslant m} |M_{NN}(t_j) - E_0 M_{NN}(t_j)|\Big\} + \epsilon$$
$$\leqslant \epsilon^{-1} \sum_{j=1}^{m} E_0\{|M_{NN}(t_j) - E_0 M_{NN}(t_j)|\} + \epsilon. \tag{8.3.5}$$

Thus, to prove (8.3.3), it suffices to show that as $N \to \infty$,

$$E_0 M_{NN}(t) \to 0 \quad \text{and} \quad E_0\{|M_{NN}(t) - E_0 M_{NN}(t)|\} \to 0,$$
$$\text{uniformly in } t \in A. \tag{8.3.6}$$

Since $(\sigma_0 C_N)^{-1} E_0 S_N(t/C_N) = (\sigma_0 C_N)^{-1} \sum_{i=1}^{N} c_i \int_{-\infty}^{\infty} \psi(x - tc_i/C_N)\, dF(x)$ $= (\sigma_0 C_N)^{-1} \sum_{i=1}^{N} c_i \int_{-\infty}^{\infty} \psi(x)\, d[F(x + tc_i/C_N) - F(x)]$ [as by (8.2.14), $\int \psi\, dF = 0$], by (8.2.17), (8.2.18), and (8.3.1), it follows that $\lim_{N \to \infty} E_0 M_{NN}(t) = 0$, $\forall t \in A$. On the other hand, by (8.3.2), (8.2.13), and (8.2.17), for every $t \in A$, $[E_0|M_{NN}(t) - E_0 M_{NN}(t)|]^2 \leqslant E_0\{M_{NN}(t) - E_0 M_{NN}(t)\}^2 \to 0$ as $N \to \infty$. Q.E.D.

For every $N(\geqslant 1)$, we consider now a stochastic process $W_N^0 = \{W_N^0(s) = S_{N(s)}(0)/(\sigma_0 C_N),\ s \in [0, 1]\}$, by letting

$$N(s) = \max\{k : C_k^2 \leqslant s C_N^2\}, \qquad s \in [0, 1]. \tag{8.3.7}$$

Since $S_N(0)$ has independent summands, by an appeal to Theorem 2.4.4, we arrive at the following (the proof is left as an exercise; see Problem 8.3.1).

THEOREM 8.3.2. *Under* (8.2.13), (8.2.14), (8.2.17) *and* $H_0 : \Delta = 0$:

$$W_N^0 \xrightarrow{\mathfrak{D}} W, \qquad \text{in the } J_1\text{-topology on} \quad D[0,1], \tag{8.3.8}$$

where W *is a standard Wiener process on* $[0, 1]$.

Note that (8.3.8) ensures that under $H_0 : \Delta = 0$,

$$\sup_{0 \leqslant t \leqslant 1} |W_n^0(t)| = \max_{k \leqslant N} |S_k(0)|/(\sigma_0 C_N) = O_p(1). \tag{8.3.9}$$

By (8.2.6) through (8.2.8), Theorem 8.3.1, and (8.3.9), we conclude that for every ϵ $(0 < \epsilon < 1)$, defining $N(\epsilon)$ as in (8.3.6),

$$\max_{N(\epsilon) \leqslant k \leqslant N} C_N|\hat{\Delta}_k - \Delta| = O_p(1), \tag{8.3.10}$$

so that by (8.3.3) and (8.3.8) through (8.3.10), we have

$$\max_{N(\epsilon) \leqslant k \leqslant N} |C_n(\hat{\Delta}_k - \Delta)\gamma C_k^2/C_N^2 - C_N^{-1} S_k(\Delta)| \xrightarrow{P} 0. \tag{8.3.11}$$

Thus, if we consider a stochastic process ${}_\epsilon W_N = \{W_N(x) = C_N(\hat{\Delta}_{N(s)} - \Delta)\gamma/\sigma_0, \epsilon \leqslant s \leqslant 1\}$, where $N(s)$ is defined by (8.3.6), and let ${}_\epsilon W^* = \{t^{-1}W(t), \epsilon \leqslant t \leqslant 1\}$ where $\{W(t), t \geqslant 0\}$ is a standard Wiener process, then from Theorem 8.3.2 and (8.3.11), we arrive at the following.

THEOREM 8.3.3. *Under* (8.2.10) *through* (8.2.13), (8.2.17), (8.2.18), *and* (8.3.2), *for every* $\epsilon : 0 < \epsilon < 1$,

$${}_\epsilon W_N \xrightarrow{\mathfrak{D}} {}_\epsilon W^*, \qquad \text{in the } J_1 \text{ topology on} \quad D[0,1]. \tag{8.3.12}$$

Note that neither $t^{-1}W(t)$ nor $W_N(t)$ behaves smoothly as $t \downarrow 0$ and the small exclusion in (8.3.12) eliminates this problem.

For drawing statistical inference based on M-estimators, mostly, we need to estimate σ_0 as well as γ, defined by (8.2.13) and (8.2.18), respectively. We estimate σ_0 by

$$s_N^{02} = N^{-1} \sum_{i=1}^{N} \psi^2(X_i - \hat{\Delta}_N c_i) = \int_{-\infty}^{\infty} \psi^2(x)\, d\hat{F}_N(x), \tag{8.3.13}$$

where $\hat{\Delta}_N$ is defined by (8.2.7) and (8.2.8) and $\hat{F}_N(x) = N^{-1}\sum_{i=1}^{N} I(X_i - \hat{\Delta}_N c_i \leqslant x)$ is the sample d.f. of the residuals $X_i - \hat{\Delta}_N c_i$, $i = 1, \ldots, N$. If, we let side by side

$$s_N^2 = N^{-1} \sum_{i=1}^{N} \psi^2(X_i - \Delta c_i), \tag{8.3.14}$$

then, under (8.2.1), s_N^2 is an average over i.i.d.r.v.'s with finite mean [by (8.2.13)], so that by the Kintchine strong law of large numbers,

$$s_N^2 \to \sigma_0^2 \quad \text{a.s.,} \quad \text{as} \quad N \to \infty. \tag{8.3.15}$$

Further, if we assume that

$$\int_{-\infty}^{\infty} \psi^4(x)\, dF(x) < \infty, \tag{8.3.16}$$

then, by Theorem 2.4.3, the weak invariance principle holds for the partial sequence $\{kN^{-1/2}(s_k^2 - \sigma_0^2),\ 0 \leqslant k \leqslant N\}$. Thus, for our purpose, it suffices to study the asymptotic equivalence of $\{s_k\}$ and $\{s_k^0\}$.

Parallel to (8.3.1), we consider the process $\{M_{Nk}^*(t) = N^{-1/2}\sum_{i \leqslant k}[\psi^2(X_i - \Delta c_i) - \psi^2(X_i - (\Delta + tC_N^{-1})c_i)],\ k \leqslant N,\ t \in A\}$. Note that $\psi^2(x)$ is symmetric about 0 and is expressible as the sum of two nonnegative functions, one of which is ↗ and the other one is ↘ in $x \in E$. Further $\int \psi(x)\psi'(x)\, dF(x) = 0$. As such, we may virtually repeat the proof of Theorem 8.3.1 and obtain that as $N \to \infty$,

$$\max_{k \leqslant N} \sup_{t \in A} |M_{Nk}^*(t)| \to 0, \quad \text{in probability.} \tag{8.3.17}$$

As such, by (8.3.10) and (8.3.17), we conclude that for every $0 < \epsilon < 1$,

$$\max\left\{ N^{1/2}|s_k^0 - s_k| : N(\epsilon) \leqslant k \leqslant N \right\} \xrightarrow{P} 0, \quad \text{as} \quad N \to \infty. \tag{8.3.18}$$

This provides the desired result.

Next, to estimate γ, we note that $S_N(t)$ is ↘ in t, and therefore keeping in mind (8.2.20), (8.3.15), and (8.3.18), we let

$$\hat{\Delta}_{NL} = \sup\left\{ t : S_N(t) > C_N s_N^0 \tau_{\alpha/2} \right\}, \quad \hat{\Delta}_{NU} = \inf\left\{ t : S_N(t) < -C_N s_N^0 \tau_{\alpha/2} \right\}; \tag{8.3.19}$$

$$L_N = \hat{\Delta}_{NU} - \hat{\Delta}_{NL} (\geqslant 0), \tag{8.3.20}$$

where $\Phi(\tau_\alpha) = 1 - \alpha (0 < \alpha < 1)$ and Φ is the standard normal d.f. Note that by Theorems 8.3.1 and 8.3.2 and by (8.3.19), parallel to (8.3.10), we have for every ϵ: $0 < \epsilon < 1$,

$$\max_{N(\epsilon) \leqslant k \leqslant N} C_N |\hat{\Delta}_{NU} - \Delta| = O_p(1) = \max_{N(\epsilon) \leqslant k \leqslant N} C_N |\hat{\Delta}_{NL} - \Delta|. \tag{8.3.21}$$

Thus, if we define

$$\hat{\gamma}_N = \left(2 s_N^0 \tau_{\alpha/2}\right) / (C_N L_N), \tag{8.3.22}$$

then, by (8.3.20), (8.3.21), (8.3.22), and Theorem 8.3.1, we obtain for every

ϵ: $0 < \epsilon < 1$, as $N \to \infty$,

$$\max\{|\hat{\gamma}_k - \gamma| : N(\epsilon) \leqslant k \leqslant N\} \xrightarrow{p} 0. \tag{8.3.23}$$

For invariance principles relating to the partial sequence $\{kN^{-1/2}(\hat{\gamma}_k - \gamma), k \leqslant N\}$ (or the asymptotic normality of the standardized form of these estimates), one needs some extra regularity conditions. Also, in this context, the normalizing constants depend explicitly on the function ψ, that is, whether it is purely absolutely continuous or it possesses jump-discontinuities. In the latter case, a slower rate of convergence holds. These results are due to Jurečková and Sen (1981a, b) and we pose some of them as problems at the end of this chapter.

8.4. A.S. REPRESENTATION OF M-ESTIMATORS

Under additional regularity conditions, (8.3.11) will be strengthened here to an a.s. equivalence. Further, for $\{S_N(\Delta)\}$, when Δ holds, the Skorokhod-Strassen embedding holds under fairly general conditions, and hence, this a.s. equivalence enables us to study the embedding of Wiener processes for the M-estimators as well.

Let us now assume that there exists an $\lambda > 0$, such that

$$\sup_{a : |a| \leqslant \lambda} \left\{ E_0\left[\psi(X_1 + a) - E_0\psi(X_1 + a)\right]^6 \right\} < \infty \tag{8.4.1}$$

and

$$\sup_{a : 0 < |a| \leqslant \lambda} \left\{ E_0\left[a^{-1}\{\psi(X_1 + a) - \psi(X_1) - E_0\psi(X_1 + a)\}\right]^4 \right\} < \infty. \tag{8.4.2}$$

Also, we strengthen (8.2.17) to

$$C_N^{-2}\left\{\max_{1 \leqslant i \leqslant N} c_i^2\right\} = O\left(N^{-1}(\log N)^q\right), \qquad \text{for some q} \geqslant 0. \tag{8.4.3}$$

Then, we have the following.

THEOREM 8.4.1. *Under* (8.2.10) *through* (8.2.13), (8.2.18), (8.4.1), (8.4.2), *and* (8.4.3), *for every* $\epsilon > 0$, *there exists a constant* K $(= K_\epsilon)$ *and a sample size* N_ϵ, *such that for every* $N \geqslant N_\epsilon$,

$$P_\Delta\{|\gamma C_N(\hat{\Delta}_N - \Delta) - C_N^{-1}S_N(\Delta)| > \epsilon\} \leqslant K_\epsilon N^{-s}, \tag{8.4.4}$$

where $s > 1$, *and hence*

$$|\gamma C_N(\hat{\Delta}_N - \Delta) - C_N^{-1}S_N(\Delta)| \to 0 \qquad a.s., \quad as \quad N \to \infty. \tag{8.4.5}$$

Proof. Note that for every $a > 0$, by the Markov inequality

$$
\begin{aligned}
P_\Delta\{|\hat{\Delta}_N - \Delta| \geqslant a\} &= 2P_\Delta\{\hat{\Delta}_N \geqslant \Delta + a\} \\
&\leqslant 2P_\Delta\{S_N(\Delta + a) \geqslant 0\} = 2P_0\{S_N(a) \geqslant 0\} \\
&= 2P_0\{S_N(a) - E_0S_N(a) \geqslant -E_0S_N(a)\} \\
&\leqslant 2\{E_0(S_N(a) - E_0S_N(a))^6/(-E_0S_N(a))^6\} \qquad (8.4.6)
\end{aligned}
$$

where for any $a > 0$, $E_0S_N(a) < 0$. Let us choose then

$$a = a_N = C_N^{-1}N^r \qquad \text{where} \quad \tfrac{1}{6} < r < \tfrac{1}{5}. \qquad (8.4.7)$$

Note that by (8.4.3) and (8.4.7), $a_N(\max_{1 \leqslant i \leqslant N}|c_i|) = O(N^{r-1/2}(\log N)^{q/2})$ $\to 0$ as $N \to \infty$. Hence, proceeding as in the discussion following (8.3.6), it follows that as N increases

$$-E_0S_N(a_N) \sim C_N\gamma N^r, \qquad \text{where} \quad \gamma > 0. \qquad (8.4.8)$$

On the other hand, since the $S_N(b)$ has independent summands, we obtain by some routine steps that under (8.4.1) and (8.4.3)

$$C_N^{-6}E_0[S_N(a_N) - E_0S_N(a_N)]^6 \to 15\sigma_0^6 \qquad \text{as} \quad N \to \infty; \qquad (8.4.9)$$

the proof is left as an exercise (see Problem 8.4.1). Thus, from (8.4.6) through (8.4.9), we obtain for N adequately large,

$$P_\Delta\{C_N|\hat{\Delta}_N - \Delta| \geqslant N^r\} \leqslant KN^{-6r}, \qquad \text{where by (8.4.7)}, \quad 6r > 1. \quad (8.4.10)$$

The Borel-Cantelli lemma on (8.4.10) yields that

$$|C_N(\hat{\Delta}_N - \Delta)| \leqslant N^r \qquad \text{a.s., as} \quad N \to \infty. \qquad (8.4.11)$$

Now, proceeding as in (8.3.5), but using $M_N = N^r m_\epsilon$ grid-points, we conclude that

$$
\begin{aligned}
&\sup_{|t| \leqslant N^r} |M_{NN}(t) - E_0M_{NN}(t)| \\
&\quad \leqslant \max_{1 \leqslant j \leqslant M_N} |M_{NN}(t_j) - E_0M_{NN}(t_j)| + \epsilon/2, \qquad (8.4.12)
\end{aligned}
$$

where, for every $\epsilon > 0$,

$$
\begin{aligned}
&P_0\Big\{\max_{1 \leqslant j \leqslant M_N} |M_{NN}(t_j) - E_0M_{NN}(t_j)| \geqslant \epsilon/2\Big\} \\
&\quad \leqslant \sum_{j \leqslant M_N} P_0\{|M_{NN}(t_j) - E_0M_{NN}(t_j)| > \epsilon/2\} \\
&\quad \leqslant (16M_N\epsilon^{-4})\Big\{\sup_{|t| \leqslant N^r} E_0[M_{NN}(t) - E_0M_{NN}(t)]^4\Big\}. \qquad (8.4.13)
\end{aligned}
$$

Finally, by (8.4.2) and (8.4.3), for every $t : |t| \leqslant N^r$,

$$E_0\big[M_{NN}(t) - E_0 M_{NN}(t)\big]^4 = O\Big(t^4\big(N^{-1}(\log N)^q\big)^2\Big)$$

$$= O\big(N^{-2+4r}(\log N)^{2q}\big), \qquad (8.4.14)$$

so that the rhs of (8.4.13) is $O(N^{-2+5r}(\log N)^{2q})$, and, as by (8.4.7), $5r < 1$, we conclude from the above that for every $\epsilon > 0$, there exist a constant K_ϵ^* and a sample size N_ϵ, such that for every $N \geqslant N_\epsilon$,

$$P_0\Big\{\sup_{|t| \leqslant N^r} |M_{NN}(t) - E_0 M_{NN}(t)| \geqslant \epsilon\Big\} \leqslant K_\epsilon^* N^{-s}, \qquad \text{where} \quad s > 1. \tag{8.4.15}$$

(8.2.7), (8.2.8), (8.4.10), and (8.4.16) ensure (8.4.4), and (8.4.5) follows from (8.4.4) and the Borel-Cantelli lemma. Q.E.D.

Remarks. It may be noted that (8.4.3) holds generally, while (8.4.1) is also not very restrictive. On the other hand, (8.4.2) may not hold when the score function (ψ) possesses jump-discontinuities [as in (8.2.10) through (8.2.12)]. However, the theorem remains valid in such a case with very few modifications. In such a case, the proof of (8.4.11) poses no problem, but we need to provide a different proof for (8.4.15) as (8.4.14) may not hold without (8.4.2). As in (8.2.10) through (8.2.12), we express $\psi = \psi_1 + \psi_2$ and write $M_{NN}(t) = M_{NN}^{(1)}(t) + M_{NN}^{(2)}(t)$, where $M_{NN}^{(1)}$ rests on an absolutely continuous score function and $M_{NN}^{(2)}$ on the jump-function ψ_2. For absolutely continuous and differentiable ψ_1, (8.4.2) holds under quite general conditions, and hence, for the proof of (8.4.15) for $M_{NN}^{(1)}$, we may proceed as in (8.4.13) through (8.4.15). On the other hand, by (8.2.11) and (8.2.12), for $M_{NN}^{(2)}$, we have

$$M_{NN}^{(2)}(t) = (\sigma_0 C_N)^{-1} \sum_{i \leqslant N} c_i \left[\psi_2\left(X_i - \frac{tc_i}{C_N}\right) - \psi_2(X_i)\right] + \frac{t\gamma_2}{\sigma_0}, \tag{8.4.16}$$

where

$$\gamma_2 = \sum_{j=1}^{p} (\beta_j - \beta_{j-1}) f(a_j) > 0 \tag{8.4.17}$$

and

$$\psi_2\left(X_i - \frac{tc_i}{C_N}\right) - \psi_2(X_i)$$

$$= \sum_{j=1}^{p} (\beta_j - \beta_{j-1}) \left\{ I(tc_i < 0) I\left(a_j + \frac{tc_i}{C_N} \leqslant X_i \leqslant a_j\right) - I(tc_i > 0) I\left(a_j \leqslant X_i \leqslant a_j + \frac{tc_i}{C_N}\right)\right\} \tag{8.4.18}$$

($i = 1, \ldots, N$) are all bounded valued r.v.'s. Thus instead of the Markov inequality in (8.4.14), we may first evaluate the moment generating function of $M_{NN}^{(2)}(t) - E_0 M_{NN}^{(2)}(t)$ and then use the Bernstein inequality to verify (8.4.15). Actually, here we will end up with an exponential rate instead of the power rate N^{-s}. The details are left in the form of exercises (see Problems 8.4.2 and 8.4.3). We may also remark that if the score function ψ itself has a finite moment generating function (or is bounded with probability one), then, instead of (8.4.1), we may again use the moment generating function and in (8.4.11) will end up with $\log N$ instead of N^r.

Let us now proceed on to the case of s_N^{02}, defined by (8.3.13). Here also, under the assumptions of Theorem 8.4.1, we may proceed as for (8.4.11) through (8.4.15) and strengthen (8.3.17) as follows: for every $\epsilon > 0$, there exists a positive constant K_ϵ and a sample size N_ϵ, such that for r defined by (8.4.7),

$$P\Big\{ \sup_{t:|t| \leqslant N^r} |M_{NN}^*(t)| \geqslant \epsilon \Big\} \leqslant K_\epsilon N^{-s}, \qquad \forall N \geqslant N_\epsilon, \tag{8.4.19}$$

where $s > 1$. By (8.4.11) and (8.4.19), we conclude that for every $\epsilon > 0$, there exists a positive constant K_ϵ^0 and a sample size N_ϵ such that

$$P\big\{ N^{1/2} |s_N^{02} - s_N^2| \geqslant \epsilon \big\} \leqslant K_\epsilon^0 N^{-s}, \qquad \text{for every} \quad N \geqslant N_\epsilon, \tag{8.4.20}$$

where $s > 1$, so that, by (8.4.20) and the Borel-Cantelli Lemma,

$$N^{1/2} |s_N^{02} - s_N^2| \to 0 \qquad \text{a.s.,} \quad \text{as} \quad N \to \infty. \tag{8.4.21}$$

For either $\{S_N(\Delta)\}$ or $\{s_N^2\}$, the embedding of Wiener process in Theorem 2.5.1 holds and, as a result, by (8.4.5) and (8.4.21), such a.s. representations remain true for the M-estimators and their variance estimators. We conclude this section with the remark that by virtue of Theorem 8.4.1 and (8.4.20), we are also in a position to show that under the hypothesis of Theorem 8.4.1, (8.3.23) can be strengthened to the following: for every $\epsilon > 0$, there exist a positive constant K_ϵ and a sample size N_ϵ, such that for every $N \geqslant N_\epsilon$,

$$P\{ |\hat{\gamma}_N - \gamma| \geqslant \epsilon \} \leqslant K_\epsilon N^{-s}, \text{ where } s > 1, \tag{8.4.22}$$

so that by the Borel-Cantelli lemma,

$$\hat{\gamma}_N \to \gamma \qquad \text{a.s.,} \quad \text{as} \quad N \to \infty; \tag{8.4.23}$$

the proof of (8.4.22) is left as an exercise (see Problem 8.4.4).

8.5. ASYMPTOTIC EQUIVALENCE OF M-, R-, AND L-ESTIMATORS

Consider the basic model (8.2.1). For Δ, the LSE and MLE are given by (8.2.3) and (8.2.4), respectively, while a general definition of the M-estimator is given in (8.2.6) through (8.2.8). In this context, the equivalence result in (8.3.11) [or the stronger form in (8.4.5)] is of prime interest. We denote the M-estimator by $\hat{\Delta}_N^M$ and using (8.2.6), rewrite (8.3.11) as $N \to \infty$

$$\max_{N(\epsilon) \leqslant k \leqslant N} \left| \frac{\gamma C_k^2(\hat{\Delta}_k^M - \Delta)}{C_N} - \sum_{i=1}^{k} c_i \frac{\psi(X_i - \Delta c_i)}{C_N} \right| = o_p(1), \qquad (8.5.1)$$

for every $0 < \epsilon < 1$, where $N(\epsilon)$ and γ are defined by (8.2.18) and (8.3.7), respectively and ψ is the score function, defined by (8.2.10) through (8.2.12).

Let us next consider the case of R-estimators, defined by (4.5.20); we denote the same by $\hat{\Delta}_N^R$. Also, we write

$$\gamma^* = AI^{1/2}(f)\rho_1 = -\int_{-\infty}^{\infty} \phi(F(x))f'(x)\,dx, \qquad (8.5.2)$$

where ϕ is the score function for the rank statistics and A, $I(f)$, and ρ_1 are defined by (4.3.1), (4.3.37), and (4.3.40), respectively. Further, using the tightness of both the partial sequence $\{C_N^{-1}L_k^0,\ k \leqslant N\}$ and $\{C_N^{-1}L_k^*,\ k \leqslant N\}$ [ensured by the martingale property of each sequence when H_0 in (4.2.3) holds] along with the convergence equivalence in (4.3.10), we conclude that under H_0 in (4.2.3),

$$\max_{1 \leqslant k \leqslant N} \{C_N^{-1}|L_k^0 - L_k^*|\} \xrightarrow{p} 0, \qquad \text{as} \quad N \to \infty. \qquad (8.5.3)$$

As such, by (4.5.24), (8.5.2), (8.5.3), and some standard arguments, we conclude that for every ϵ: $0 < \epsilon < 1$, defining $N(\epsilon)$ as in (8.3.7),

$$\max_{N(\epsilon) \leqslant k \leqslant N} C_N^{-1}|\gamma^* C_k^2(\hat{\Delta}_k^R - \Delta) - \sum_{i \leqslant k} c_i(\phi(F(X_i - \Delta c_i)) - \bar{\phi})| = o_p(1), \qquad (8.5.4)$$

as $N \to \infty$, where $\bar{\phi} = \int_0^1 \phi(u)\,du$. From (8.5.1) and (8.5.4), it is quite clear that if for some real $a (\neq 0)$ and b,

$$\psi(y) = \phi(F(y))a + b, \quad \text{for all} \quad y \in E, \qquad (8.5.5)$$

then, for every ϵ: $0 < \epsilon < 1$,

$$\max_{N(\epsilon) \leqslant k \leqslant N} \{C_N^{-1}|C_k^2(\hat{\Delta}_k^R - \hat{\Delta}_k^M)|\} \xrightarrow{p} 0, \qquad \text{as} \quad N \to \infty; \qquad (8.5.6)$$

and conversely, if (8.5.6) holds, then (8.5.5) also holds. Actually, if we let

$$\zeta = \int_{-\infty}^{\infty} \psi(y)\left[\phi(F(y)) - \bar{\phi}\right] dF(y), \qquad (8.5.7)$$

$$\sigma^{*2} = \sigma_0^2/\gamma^2 + A^2/\gamma^{*2} - 2\zeta/(\gamma\gamma^*), \qquad (8.5.8)$$

$$\tilde{W}_N(s) = C_N^{-1} C_{N(s)}^2 \left(\hat{\Delta}_{N(s)}^R - \hat{\Delta}_{N(s)}^M\right)/\sigma^*, \qquad \text{for} \quad s \in (0, 1], \quad (8.5.9)$$

where $N(s)$ is defined by (8.3.7) and use the weak convergence result in Theorem 2.4.3 for the partial sequence $\{C_N^{-1}\sum_{i \leqslant k} c_i[\gamma^{*-1}(\phi(F(X_i - \Delta c_i)) -$

$\bar{\phi}) - \gamma^{-1}\psi(X_i - \Delta c_i)], k \leqslant N\}$ (involving independent summands), then, from (8.5.1), (8.5.4), and (8.5.7) through (8.5.9), we conclude as in Jurečková (1977) that for every ϵ: $0 < \epsilon < 1$ and $\sigma^* > 0$,

$$_\epsilon\tilde{W}_N = \{\tilde{W}_N(s), \epsilon \leqslant s \leqslant 1\} \xrightarrow{\mathfrak{D}} {}_\epsilon W, \quad \text{in the } J_1\text{-topology on} \quad D[\epsilon, 1], \tag{8.5.10}$$

where $_\epsilon W = \{W(t), \epsilon \leqslant t \leqslant 1\}$ is a standard Wiener process on $[\epsilon, 1]$. For R-estimators of location, we need to use the results of Section 5.5, but (8.5.10) remains true with $C_N^{-1}C_k^2$ being replaced by $N^{-1/2}k$, $k \leqslant N$.

We may note that for the rank estimator of regression, we do not need the d.f. to be symmetric about 0 or the score function ϕ to be skew-symmetric, but these constitute a part of the assumptions underlying the M-estimators. For the location problem, for rank estimators, the symmetry of F is needed, except in the simplest case of the median estimator. Also, the R-estimators being based on the ranks can accommodate unbounded scores in a more flexible manner and therefore remain robust. In particular, the score function ψ for the M-estimator may not be scale-invariant [i.e., $\psi(x)$ may not be equal to $a\psi(x/a)$ a.e., for every $a > 0$], and this can cause some concern for the choice of an optimal ψ when the scale of the underlying d.f. is of unknown magnitude (as is generally the case in practice). Finally, the rank statistics have completely specified distributions under suitable null hypotheses, so that distribution-free confidence intervals for location or regression based on them can be obtained under very general conditions (see Chapter 10 where this has been displayed in detail). On the other hand, the distribution of $S_N(0)$, even under H_0: $\Delta = 0$, is generally dependent on the underlying d.f. and is only asymptotically normal; in any case, the variance of this distribution is unknown and as in (8.3.19), one needs to use the estimator s_N to derive confidence intervals for Δ, which may not be genuinely distribution-free. As a general rule, the M-estimators can be justified mostly on the ground of *local optimality* (viz., minimax property for small amount of error-contamination with some specified d.f.) whereas the rank estimators are justifiable on the ground of *global optimality*.

Let us now consider the case of L-estimators. We confine ourselves to the location model and consider the estimator

$$\hat{\Delta}_N^L = \sum_{i=1}^{N} c_{Ni} X_{N,i}, \tag{8.5.11}$$

where $X_{N,1} \leqslant \cdots \leqslant X_{N,N}$ are the order statistics corresponding to $X_1, \ldots, X_N, c_{Ni} = N^{-1}J_N(i/(N+1))$, $i = 1, \ldots, N$ are suitable constants where $J_N(\cdot)$ converges to some smooth function $J(\cdot)$ as $N \to \infty$. If we impose the regularity conditions in Section 7.4 or 7.5, we have the invariance principles for these estimators as well. In fact, as in (7.5.1) through

(7.5.3), we have

$$\hat{\Delta}_N^L - \Delta = N^{-1}\sum_{i=1}^{N}\int_{-\infty}^{\infty}\left[c(x-X_i)-F(x)\right]J(F(x))\,dx + o(N^{-1/2})$$
$$\text{a.s.,} \quad \text{as} \quad N\to\infty. \tag{8.5.12}$$

Thus, if we define $J^* = \{J^*(F(x)), x\in E\}$ by letting

$$\left(\frac{d}{dx}\right)J^*(F(x)) = J(F(x)), \qquad x\in E, \tag{8.5.13}$$

then, by integrating by parts [in (8.5.12)], we obtain that

$$\hat{\Delta}_N^L = N^{-1}\sum_{i=1}^{N} J^*(F(X_i)) + o(N^{-1/2}) \qquad \text{a.s.,} \quad \text{as} \quad N\to\infty. \tag{8.5.14}$$

In (8.5.12) and (8.5.14), actually, $F(X_i)$ should read as $F(X_i-\Delta)$ and $c(x-X_i)$ as $c(x-X_i+\Delta)$; however, because of the translation-invariance, we may take $\Delta=0$.

By virtue of (8.5.14), we claim that (8.5.6) (for $C_N^2 = N$, for all N) holds for the R-estimator being replaced by the L-estimator iff (8.5.5) holds with $\phi(F(x))$ being replaced by $J^*(F(x))$. Thus, if $\psi(x)$, $\phi(F(x))$, and $J^*(F(x))$ are all the same, then all the three estimators are asymptotically equivalent in a strong sense. A result parallel to (8.5.10) also holds for the L-estimators and is posed as an exercise (see Problem 8.5.1). Like the M-estimators, the L-estimators may not provide a genuinely distribution-free confidence interval for Δ.

8.6. EXTREMA OF CERTAIN SAMPLE FUNCTIONS

Daniels (1945) in the context of the distribution theory of the strength of a bundle of parallel filaments (see Problem 8.6.1), considered the statistic

$$D_N = \max_{1\leqslant i\leqslant N}\{(N-i+1)X_{N,i}\}, \tag{8.6.1}$$

where, as before, the $X_{N,i}$ are the order statistics of a sample of size N drawn from a continuous d.f. F. Note that D_N is not an LCFOS, so that the theory developed in Chapter 7 may not be properly applicable here; the technique employed by Daniels for the study of the asymptotic normality of D_N involves tedious analysis. We may note that if F_N stands for the empirical d.f., then, by (8.6.1),

$$\begin{aligned} Z_N = N^{-1}D_N &= \max\left\{\left(1-N^{-1}(i-1)\right)X_{N,i}\colon 1\leqslant i\leqslant N\right\} \\ &= \max\{(1-F_N(X_{N,i}-))X_{N,i}\colon 1\leqslant i\leqslant N\} \\ &= \sup\left\{x\left[1-F_N(x)\right]\colon x\in E\right\}. \end{aligned} \tag{8.6.2}$$

As such, the asymptotic behavior of the empirical process F_N should provide us with a comparatively easier approach to the asymptotic behavior of D_N or Z_N. With this motivation, we consider a sequence $\{X_i, i \geqslant 1\}$ of i.i.d.r.v.'s with a continuous d.f. $F(x)$, $x \in R^p$, the $p(\geqslant 1)$-dimensional Euclidean space and we consider the statistic

$$Z_N = \sup\{\psi(x, F_N(x)) \colon x \in A\}$$
$$= \max\{\psi(X_i, F_N(X_i))I(X_i \in A) \colon \quad 1 \leqslant i \leqslant N\} \qquad (8.6.3)$$

where I stands for the indicator function, $A \subset R^p$ and ψ is nonnegative and satisfies certain regularity conditions. We may note that for $\psi(x, y) = x[1-y]$, $0 \leqslant x < \infty$, $0 \leqslant y \leqslant 1$, (8.6.2) becomes a special case of (8.6.3). We are primarily concerned with the asymptotic behavior of $\{Z_N\}$ and with the various invariance principles relating to this sequence.

Let us assume that there exists a unique $x_0(\in A)$ such that

$$\theta = \sup\{\psi(x, F(x)) : x \in A\} = \psi(x_0, F(x_0)), \qquad (8.6.4)$$

and for every $\epsilon > 0$ $(0 < \epsilon < \theta)$, there exists a $\delta > 0$ such that

$$\psi(x, F(x)) < \theta - \epsilon \forall x : \|x - x_0\| > \delta, \qquad (8.6.5)$$

where $\|u\|$ is the Euclidean norm of $u \in R^p$. Further, there exists a metric $\rho(x, x')$; $x, x' \in R^p$ such that for $\delta(>0)$ sufficiently small and $\|x - x_0\| < \delta$,

$$\psi(x, F(x) = \psi(x_0, F(x_0)) - \rho(x, x_0); \qquad (8.6.6)$$

$$c_2\|x - x_0\|^{k_2} \leqslant \rho(x, x_0) \leqslant c_1\|x - x_0\|^{k_1} \qquad (8.6.7)$$

where $(0<)c_1, c_2(<\infty)$ and $(1 \leqslant)k_1, k_2(<\infty)$ are positive constants. Also, assume that for all x: $\|x - x_0\| < \delta$, $F(x)$ has a continuous (and finite) density function $f(x)$ where

$$0 < \pi_0 = F(x_0) < 1 \quad \text{and} \quad 0 < f(x_0) < \infty. \qquad (8.6.8)$$

Moreover, $\psi(x, y)$ is well defined on $A^* = \{x \in A,\ 0 \leqslant y \leqslant 1\}$ and has continuous first-order partial derivatives with respect to x and y, and we define

$$\psi_{01}(x, y) = (\partial/\partial y)\psi(x, y), \qquad \xi = \psi_{01}(x_0, \pi_0). \qquad (8.6.9)$$

Finally, we assume that for $\delta(>0)$, sufficiently small and $A_\delta^* = \{(x, y) : x \in A, |y - F(x)| \leqslant \delta,\ 0 < y < 1\}$,

$$|\psi_{01}(x, y)| \leqslant g(x), \qquad \forall (x, y) \in A_\delta^*, \qquad (8.6.10)$$

$$0 < \int_{R^p} g^2(x)\, dF(x) = \lambda^2 < \infty. \qquad (8.6.11)$$

Another condition, which will be needed in Section 8.7, is that for $\delta_1(>0)$, $\delta_2(>0)$ sufficiently small, there exist positive (and finite) constants

K_1, K_2 and d_1, d_2 such that

$$|\psi_{01}(x, y) - \psi_{01}(x_0, \pi_0)| \leqslant K_1 \|x - x_0\|^{d_1} + K_2 |y - \pi_0|^{d_2}, \quad (8.6.12)$$

for all $(x, y) : \|x - x_0\| \leqslant \delta_1$ and $|y - y_0| \leqslant \delta_2$.

For the special case of (8.6.2), all these conditions are easily verifiable. The stochastic convergence of Z_N to θ is intuitively suggested by (8.6.3), (8.6.4), and the Glivenko-Cantelli Theorem. First, we consider the following.

THEOREM 8.6.1. *Under* (8.6.4) *through* (8.6.11): *as* $N \to \infty$,

$$N^{1/2}(Z_N - \theta) \xrightarrow{\mathfrak{D}} \mathfrak{N}\left(0, \xi^2 \pi_0 (1 - \pi_0)\right). \quad (8.6.13)$$

Proof. Choose a $\delta(> 0)$, satisfying (8.6.6), (8.6.7), (8.6.10), and (8.6.11). Then, whenever $\sup\{|F_N(x) - F(x)| : x \in A\} \leqslant \delta$,

$$\begin{aligned}
&\max_{1 \leqslant i \leqslant N} \{\psi(X_i, F_N(X_i)) : \|X_i - x_0\| > \delta\} \\
&\quad \leqslant \max_{1 \leqslant i \leqslant N} \{\psi(X_i, F(X_i)) : \|X_i - x_0\| > \delta\} \\
&\quad\quad + \max_{1 \leqslant i \leqslant N} \{ g(X_i) | F_N(X_i) - F(X_i)| : \|X_i - x_0\| > \delta\} \\
&\quad \leqslant \theta - \epsilon + \left\{ \max_{1 \leqslant i \leqslant N} g(X_i) \right\} \left\{ \sup_x |F_N(x) - F(x)| \right\}. \quad (8.6.14)
\end{aligned}$$

Note that

$$\sup\left\{ \sqrt{N} \, |F_N(x) - F(x)| : x \in R^p \right\} = O_p(1), \quad (8.6.15)$$

so that for every $\delta > 0$, as $N \to \infty$, $P\{\sup\{|F_N(x) - F(x)| : x \in R^p\} > \delta\} \to 0$. Further, by (8.6.11), $g(X_i)$, $1 \leqslant i \leqslant N$ are i.i.d.r.v. with a bounded second moment, so that

$$N^{-1/2}\left\{ \max_{1 \leqslant i \leqslant N} g(X_i) \right\} \to 0 \quad \text{a.s.,} \quad \text{as} \quad n \to \infty \quad (8.6.16)$$

(the proof is left as an exercise; see Problem 8.6.4). Consequently, the rhs of (8.6.14) can be made $\leqslant \theta - \eta$, $\eta > 0$, in probability, as $N \to \infty$.

With the definition of k_2 in (8.6.7), we let $\beta = 1/(2k_2)$ and

$$J_N(\beta) = \left\{ x : \|x - x_0\| < N^{-\beta} \log N \right\}, \quad (8.6.17)$$

$$J_N^*(\beta) = \left\{ x : \|x - x_0\| \leqslant \delta \quad \text{but} \quad x \notin J_N(\beta) \right\}. \quad (8.6.18)$$

Then, by (8.6.6), (8.6.7), (8.6.17), and (8.6.18),

$$\begin{aligned}
\sup\{\psi(x, F(x)) : x \in J_N^*(\beta)\} &\leqslant \theta - c_2 (N^{-\beta} \log N)^{k_2} \\
&= \theta - c_2 N^{-1/2} (\log N)^{k_2}, \quad k_2 \geqslant 1. \quad (8.6.19)
\end{aligned}$$

On the other hand, by (8.6.10)

$$\sup\{|\psi(x,F_N(x))-\psi(x,F(x))|:x\in J_N^*(\beta)\}$$
$$\leqslant \sup\{g(x)|F_N(x)-F(x)|:x\in J_N^*(\beta)\}=O_p(N^{-1/2}). \quad (8.6.20)$$

Since $(\log N)^k \to \infty$ as $N\to\infty$, $k \geqslant 1$, by (8.6.19) and (8.6.20), we have

$$\sup\{\psi(x,F_N(x)):x\in J_N^*(\beta)\}\leqslant \theta - c_2^* N^{-1/2}(\log N)^{k_2}, \quad (8.6.21)$$

in probability, as $N\to\infty$, where $0<c_2^*<c_2$. Finally, for $x\in J_N(\beta)$, we write

$$\psi(x,F_N(x))=\psi(x,F(x))+\left[F_N(x)-F(x)\right]\psi_{01}(x,\nu_N(x));$$
$$\nu_N(x)=hF(x)+(1-h)F_N(x) \qquad \text{for some} \quad 0<h<1. \quad (8.6.22)$$

Therefore, letting $V_N=\psi_{01}(x_0,\pi_0)[F_N(x_0)-F(x_0)]$, we obtain, by (8.6.21), (8.6.22), and the continuity of $\psi_{01}(x, y)$ [in the neighborhood of (x_0,π_0)] as $N\to\infty$,

$$\sup\{|N^{1/2}\{\psi(x,F_N(x))-\psi(x,F(x))-V_N\}|:x\in J_N(\beta)\}$$
$$\leqslant \sup\{|N^{1/2}\left[F_N(x)-F(x)-F_N(x_0)+\pi_0\right]\psi_{01}(x,\nu_N(x))|$$
$$+|N^{1/2}\left[F_N(x_0)-\pi_0\right]\left[\psi_{01}(x,\nu_N(x))-\psi_{01}(x_0,\pi_0)\right]|$$
$$:x\in J_N(\beta)\}, \quad (8.6.23)$$

where the first term on the rhs of (8.6.23) is $o_p(1)$, while by continuity of $\psi_{01}(x, y)$, the fact that $\sup_x|F_N(x)-F(x)|\overset{p}{\to}0$ as $N\to\infty$ and the fact that $N^{1/2}|F_N(x_0)-\pi_0|=O_p(1)$, the second term is $o_p(1)$ and $N\to\infty$. Consequently, on noting that $\sup\{\psi(x,F(x)):x\in J_N(\beta)\}=\theta$, we obtain from (8.6.23) and a few standard steps as $N\to\infty$

$$\left[\sup\{|\psi(x,F_N(x))-\theta-V_N|:x\in J_N(\beta)\}\right]=o_p(N^{-1/2}). \quad (8.6.24)$$

Since, by (8.6.15), $|V_N|=O_p(1)$, from (8.6.24), we also have

$$\sup\{\psi(x,F_N(x)):x\in J_N(\beta)\}\geqslant \theta-\epsilon_N \quad \text{where} \quad \epsilon_N=O_p(N^{-1/2}). \quad (8.6.25)$$

Consequently, by (8.6.14), (8.6.21), and (8.6.22), as $N\to\infty$,

$$\sup\{\psi(x,F_N(x)):x\in R^p\}=\sup\{\psi(x,F_N(x)):x\in J_N(\beta)\}, \quad (8.6.26)$$

in probability, and hence from (8.6.3), (8.6.24), and (8.6.26), we have for $N\to\infty$,

$$N^{1/2}\left[(Z_N-\theta)-V_N\right]=o_p(1). \quad (8.6.27)$$

The proof of the theorem follows then by using (8.6.27), the definition of $N^{1/2}V_N$ and the fact that $V_N \overset{\mathfrak{D}}{\sim} \mathfrak{N}(0, \xi^2\pi_0(1-\pi_0))$. Q.E.D.

We may remark that essentially the same proof goes through when $\{X_i, i \geqslant 1\}$ form a dependent sequence, not necessarily stationary. We may refer to Sen, Bhattacharyya, and Suh (1973) and Sen and Bhattacharyya (1976) for some related studies.

8.7. INVARIANCE PRINCIPLES FOR $\{Z_N\}$

We assume in addition to the regularity conditions of Section 8.6 that

$$\sup\{g^2(x)F(x)[1-F(x)] : x \in A\} = (g^*)^2 < \infty. \tag{8.7.1}$$

If in (8.6.10), $g(x)$ is nondecreasing (nonincreasing) in each of its p arguments [while the remaining $(p-1)$ are held fixed], then (8.7.1) follows from (8.6.11). We first prove the following.

THEOREM 8.7.1. *Under* (8.6.4) *through* (8.6.12) *and* (8.7.1), *as* $N \to \infty$,

$$N^{1/2}|Z_N - \theta - V_N| \to 0 \qquad a.s. \tag{8.7.2}$$

Proof. As in (8.6.13),

$$\begin{aligned}
&\max_{1 \leqslant i \leqslant N}\{\psi(X_i, F_N(X_i)) : \|X_i - x_0\| > \delta\} \\
&\quad \leqslant \theta - \epsilon + \max_{1 \leqslant i \leqslant N}\{g(X_i)|F_N(X_i) - F(X_i)| : \|X_i - x_0\| > \delta\} \\
&\quad \leqslant \theta - \epsilon + \Big\{\max_{1 \leqslant i \leqslant N}(F(X_i)[1-F(X_i)])^{\gamma} g(X_i)\Big\} \\
&\quad \Big\{\sup_{x \in R^p}(F(x)[1-F(x)])^{-\gamma}|F_N(x) - F(x)|\Big\}
\end{aligned} \tag{8.7.3}$$

where we let $0 < \gamma < \frac{1}{4}$. Then, by (8.7.1),

$$\begin{aligned}
&\max_{1 \leqslant i \leqslant N}\{(F(X_i)[1-F(X_i)])^{\gamma} g(X_i)\} \\
&\quad = \max_{1 \leqslant i \leqslant N}\{(F(X_i)[1-F(X_i)]g^2(X_i))^{\gamma} \cdot [g(X_i)]^{1-2\gamma}\} \\
&\quad \leqslant (g^*)^2\Big[\max_{1 \leqslant i \leqslant N} g(X_i)\Big]^{1-2\gamma} = O(N^{1/2-\gamma}) \qquad \text{a.s.}
\end{aligned} \tag{8.7.4}$$

On the other hand, by (2.6.31), as $N \to \infty$,

$$\begin{aligned}
&\sup_{x \in R^p}\{(F(x)[1-F(x)])^{-\gamma}|F_N(x) - F(x)|\} \\
&\quad = O(N^{-1/2}\log\log N) \qquad \text{a.s.,}
\end{aligned} \tag{8.7.5}$$

and hence, from (8.7.3) through (8.7.5), we get that as $N \to \infty$.

$$\max_{1 \leqslant i \leqslant N} \{\psi(X_i, F_N(X_i)) : \|X_i - x_0\| > \delta\} \leqslant \theta - \tfrac{1}{2}\epsilon \quad \text{a.s.} \quad (8.7.6)$$

Also, using (8.7.5) on the rhs of (8.6.19), we obtain as in (8.6.18) and (8.6.19) that as $N \to \infty$, (8.6.20) holds a.s. Finally, on looking at the rhs of (8.6.23), we note that (i) $N^{1/2}(\log N)^{-1/2}|F_N(x_0) - F(x_0)| = O(1)$ a.s., (ii) by (8.6.12) and (i), $\sup\{|\psi_{01}(x, \nu_N(x)) - \psi_{01}(x_0, \pi_0)| : x \in J_N(\beta)\} = O(N^{-\beta d_1}(\log N)^{d_1}) + O(N^{-1/2}\log N) + O(N^{-\beta d_1}(\log N)^{d_1}) = O(N^{-q})$ a.s., for some $0 < q < \frac{1}{2}$, and (iii) by (8.6.16) and (7.3.4), $\sup\{N^{1/2}|F_N(x) - F(x) - F_N(x_0) + \pi_0| : x \in J_N(\beta)\} = O(N^{-\beta/2}\log N)$ a.s., as $N \to \infty$. Hence the rhs of (8.6.23) $\to 0$ a.s., as $N \to \infty$. As a result, by (8.6.7) and the above discussions, we have for $N \to \infty$,

$$\max_{1 \leqslant i \leqslant N} \{\psi(X_i, F_N(X_i)) : X_i \in J_N(\beta)\} \geqslant \theta - O\left(N^{-1/2}(\log N)^{1/2}\right) \quad \text{a.s.,} \quad (8.7.7)$$

$$|\sup\{\psi(x, F_N(x)) : x \in J_N(\beta)\} - \theta - V_N| = o(N^{-1/2}) \quad \text{a.s.,} \quad (8.7.8)$$

and the rest of the proof of (8.7.2), follows for (8.7.6), (8.7.7), (8.7.8), and the strengthened version of (8.6.20). Q.E.D.

Recall that $NV_N = N\psi_{01}(x_0, \pi_0)[F_N(x_0) - \pi_0]$ can be written as

$$NV_N = \sum_{i=1}^{N} \xi[c(x - X_i) - \pi_0] = \sum_{i=1}^{N} U_i, \quad \text{say,} \quad (8.7.9)$$

where the U_i are i.i.d. (bounded-valued) r.v. with mean 0 and variance $\xi^2\pi_0(1 - \pi_0) = \sigma^2$, say. Since the invariance principles of Chapter 2 directly apply to $\{NV_N\}$, we have the following theorems (whose proofs are omitted).

(i) Let $W_N^{(1)} = \{W_N^{(1)}(t) = k_N(t)N^{-1/2}\sigma^{-1}[Z_{k_N(t)} - \theta],\ 0 \leqslant t \leqslant 1\}$ where

$$k_N(t) = \max\left\{k : \frac{k}{N} \leqslant t\right\}, \quad 0 \leqslant t \leqslant 1; \quad Z_0 = 0. \quad (8.7.10)$$

THEOREM 8.7.2. *Under the hypothesis of Theorem* 8.4.1, *as* $N \to \infty$,

$$W_N^{(1)} \xrightarrow{\mathfrak{D}} W, \quad \textit{in the } J_1\textit{-topology on} \quad D[0,1] \quad (8.7.11)$$

where W is a standard Brownian motion on [0, 1].

(ii) Let $W_N^{(2)} = \{W_N^{(2)}(b) = N^{1/2}\sigma^{-1}[Z_{k_N^*(t)} - \theta], 0 \leqslant t \leqslant 1\}$ where

$$k_N^*(t) = \min\left\{k : \frac{N}{k} \leqslant t\right\}, \quad 0 \leqslant t \leqslant 1. \quad (8.7.12)$$

THEOREM 8.7.3. *Under the hypothesis of Theorem* 8.4.1, *as* $N \to \infty$,

$$W_N^{(2)} \xrightarrow{\mathcal{D}} W, \quad \text{in the } J_1\text{-topology on} \quad D[0,1]. \tag{8.7.13}$$

Both the preceding theorems remain valid for stochastic sample sizes.

(iii) There exists a standard Wiener process $W = \{W(t),\ 0 < t < \infty\}$ such that on letting $Z_0 = 0$, $V_0 = 0$, and

$$W^*(t) = k\sigma^{-1}[Z_k - \theta] \qquad \text{for} \quad k \leqslant t < k+1, k \geqslant 0, \tag{8.7.14}$$

$$W^0(t) = kV_k \qquad \text{for} \quad k \leqslant t < k+1, k \geqslant 0, \tag{8.7.15}$$

we have

$$W^0(t) = W(t) + O(t^{1/4}\log t) \quad \text{a.s.,} \quad \text{as} \quad t \to \infty. \tag{8.7.16}$$

$$|W^*(t) - W^0(t)| = o(t^{1/2}) \quad \text{a.s.,} \quad \text{as} \quad t \to \infty. \tag{8.7.17}$$

We consider now some a.s. convergence results of some related statistics. Note that by (8.7.9), $V_N \to 0$ a.s., as $N \to \infty$ and hence, by (8.7.2), under the hypothesis of Theorem 8.7.1,

$$Z_N \to \theta \qquad \text{a.s.,} \quad \text{as} \quad N \to \infty. \tag{8.7.18}$$

In fact, for (8.7.18), we may not require all the assumptions of Theorem 8.7.1. Note that (8.7.6) holds under (8.6.6), (8.6.7), (8.6.10), (8.6.11), and (8.7.1). Also, $\theta = \sup\{\psi(x, F(x)) : \|x - x_0\| \leqslant \delta\}$, and hence

$$\begin{aligned}
&|\sup\{\psi(x, F_N(x)) : \|x - x_0\| \leqslant \delta\} - \theta| \\
&\quad \leqslant \sup\{|\psi(x, F_N(x)) - \psi(x, F(x))| : \|x - x_0\| \leqslant \delta\} \\
&\quad \leqslant [\sup\{g(x) : \|x - x_0\| \leqslant \delta\}]\Big[\sup_{\|x - x_0\| \leqslant \delta} |F_N(x) - F(x)|\Big] \\
&\quad = O(N^{-1/2}\log N) \qquad \text{a.s.,} \quad \text{as} \quad N \to \infty.
\end{aligned} \tag{8.7.19}$$

Hence $Z_N = \sup\{\psi(x, F_N(x)) : \|x - x_0\| \leqslant \delta\}$ a.s., as $N \to \infty$, and by (8.7.18) $Z_N \to \theta$ a.s. Further, it can be shown that under (8.6.6), (8.6.7), (8.6.10), (8.6.11), and (8.7.1), for every $\epsilon > 0$, there exists a $\rho(\epsilon) : 0 < \rho(\epsilon) < 1$ and an N_0, such that for $N \geqslant N_0$,

$$P\{|Z_N - \theta| > \epsilon\} \leqslant [\rho(\epsilon)]^N; \tag{8.7.20}$$

see Problem 8.7.2. Note that by (8.6.3),

$$Z_N = \psi(X_{N,r}, F_N(X_{N,r})), \tag{8.7.21}$$

where $r = r_N$ is a positive integer valued random variable ($1 \leqslant r \leqslant N$). Also, recall that $\xi = \psi_{01}(x_0, \pi_0)$ and $\sigma^2 = \xi^2\pi_0(1 - \pi_0)$. We consider the following estimator of σ^2:

$$\hat{\sigma}_N^2 = \psi_{01}^2(X_{N,r}, F_N(X_{N,r}))F_N(X_{N,r})[1 - F_N(X_{N,r})]. \tag{8.7.22}$$

We observe that $Z_N = \sup\{\psi(x, F_N(x)) : \|x - x_0\| \leqslant \delta\}$ a.s. and (8.7.21) that $\|X_{N,r} - x_0\| \leqslant \delta$ a.s., as $N \to \infty$, and hence, by (8.7.5) and the continuity of $F(x)$, we conclude that

$$F_N(X_{N,r}) \overset{\text{a.s.}}{\sim} F(X_{N,r}) \to \pi_0 \quad \text{a.s., as } N \to \infty. \tag{8.7.23}$$

Finally, the above and the continuity of $\psi(x, y)$ (in x and y) in the neighborhood of (x_0, π_0) implies that $\psi_{01}(X_{N,r}, F_N(X_{N,r})) \to \xi$ a.s., as $N \to \infty$. Hence we conclude that

$$\hat{\sigma}_N^2 \to \sigma^2 \quad \text{a.s., as } N \to \infty. \tag{8.7.24}$$

In fact, parallel to (8.7.20), we have

$$P\left\{\left|\frac{\hat{\sigma}_N^2}{\sigma^2} - 1\right| > \epsilon\right\} \leqslant [\rho(\epsilon)]^N, \qquad \forall N \geqslant N_0; \tag{8.7.25}$$

see Problem 8.7.3.

All these results are useful for sequential tests and estimates for θ, to be considered in Chapters 9 and 10.

8.8. SOME GENERAL REMARKS ON SOME OTHER ESTIMATORS

In addition to the M-estimators (including the LSE and MLE) and Z_N, considered in earlier sections, there are other estimators which are also derived by minimizing some sample functions. Among these, we like to mention the following.

(i) ***Blackman (1955) Estimator of Location.*** Let $X_1, \ldots, X_N$ be N i.i.d.r.v.'s with an absolutely continuous d.f. F, defined on the real line E. We set $F(x) = F_0(x - \theta)$ where F_0 is a completely specified d.f. and θ is the location (unknown) parameter. Let F_N be the sample d.f. and consider the functional

$$\begin{aligned} M_N(t) &= \int_{-\infty}^{\infty} [F(x - t) - F_N(x)]^2 \, dF(x) \\ &= \int_{-\infty}^{\infty} [F(x) - F_N(x + t)]^2 \, dF(x), \qquad t \in E. \end{aligned} \tag{8.8.1}$$

Then the Blackman estimator $\hat{\theta}_N^B$ of θ is given by a solution of

$$M(\hat{\theta}_N^B) = \min_{t \in E} M_N(t). \tag{8.8.2}$$

(ii) *Schuster and Narvarte (1973) Estimator of Location.* With the same notations as in (i) but assuming that the d.f. F is symmetric about its (unknown) median θ, consider the statistic

$$L_N(t) = \max_x |F_N(x) + F_N((2t - x)^-) - 1|; \tag{8.8.3}$$

the statistic $L_N(0)$ has been used for testing the symmetry of F around 0 by Butler (1969) and Chatterjee and Sen (1973b), among others. The estimator of θ is then given by a solution of

$$L(\hat{\theta}_N^S) = \min_{t \in E} L_N(t). \tag{8.8.4}$$

(iii) *Rao, Schuster, and Littell (1975) Estimator of the Difference of Location Parameters in the Two-Sample Case.* Let $X_1, \ldots, X_{N_1}$ be i.i.d.r.v. with a continuous d.f. F and $Y_1, \ldots, Y_{n_2}$ be a second sample of i.i.d.r.v. from a d.f. G, where both F and G are defined on the real line E. Set $G(x) = F(x - \Delta)$, so that Δ is the difference of the locations of the two d.f.'s. Let F_{n_1} and G_{n_2} be respectively the first and second sample d.f. and consider the functional

$$D_{n_1,n_2}(t) = \max_x |F_{n_1}(x) - G_{n_2}(x + t)|, \qquad t \in E. \tag{8.8.5}$$

The statistic $D_{n_1,n_2}(0)$ is the classical two-sample Kolmogorov-Smirnov statistic for testing the identity of F and G. Then the estimator of Δ is given by a solution of

$$D_{n_1,n_2}(\hat{\Delta}_N^S) = \min_{t \in E} D_{n_1,n_2}(t), \qquad \text{where} \quad N = n_1 + n_2. \tag{8.8.6}$$

For the study of the asymptotic behavior of the Blackman estimator, following Pyke (1970), we may proceed as follows. We define $w_N = \{w_N(x), x \in E\}$ by letting

$$w_N(x) = N^{1/2}\{F_N(x) - F(x)\}, x \in E. \tag{8.8.7}$$

By the results in Section 2.6,

$$\sup_{x \in E} |w_N(x)| = O_p(1) \qquad \text{and} \qquad \max_{k \leqslant N} \sup_{x \in E} \left(\frac{k}{N}\right)^{1/2} |w_k(x)| = O_p(1), \tag{8.8.8}$$

and hence, by (8.8.1), (8.8.7), and (8.8.8), asymptotically, for every fixed t ($\neq 0$), $M_N(t)$ is dominated by the first term $\int_{-\infty}^{\infty}(F(x - t) - F(x))^2 \, dF(x)$ (where we may set without any loss of generality that $\theta = 0$). On the other hand, if we set $t = N^{-1/2}\lambda$ for some real λ, then, by (8.8.1), (8.8.7), and

(8.8.8), we obtain

$$M_N(\lambda/N^{1/2}) = \lambda^2 \int_{-\infty}^{\infty} f^2(x)\,dF(x) - \int_{-\infty}^{\infty} w_N^2(x)\,dF(x)$$
$$- 2\lambda \int_{-\infty}^{\infty} w_N(x) f(x)\,dx + o_p(1), \tag{8.8.9}$$

where we assume that the d.f. F possesses a p.d.f. f which is uniformly continuous. The first term on the rhs is nonstochastic, while both the second and third terms are $O_p(1)$. Hence a local minimum is attained at

$$\lambda = \lambda_N = \frac{\int_{-\infty}^{\infty} f(x) w_N(x)\,dF(x)}{\int_{-\infty}^{\infty} f^2(x)\,dF(x)} + o_p(1). \tag{8.8.10}$$

Thus we obtain that asymptotically

$$N^{1/2}(\hat{\theta}_N^B - \theta) \sim \frac{\int_{-\infty}^{\infty} f(x) w_N(x)\,dF(x)}{\int_{-\infty}^{\infty} f^2(x)\,dF(x)}, \tag{8.8.11}$$

so that using the weak convergence of w_N, we conclude from (8.8.11) that $N^{1/2}(\hat{\theta}_N^B - \theta)/\sigma_0$ is asymptotically normal with 0 mean and unit variance, where

$$\sigma_0^2 = \frac{\left(2\int\int_{x \leqslant y} f^2(x) f^2(y) F(x)\left[1 - F(y)\right] dx\,dy\right)}{\left(\int_{-\infty}^{\infty} f^2(x)\,dF(x)\right)^2}. \tag{8.8.12}$$

In fact, using the second part of (8.8.2), defining $N(s) = \max\{k : k/N \leqslant s\}$ and incorporating the weak convergence of $\{(k/N)^{1/2} w_k(x) : k \leqslant N, x \in E\}$ (to a tied-down Brownian sheet), we conclude that for every ϵ: $0 < \epsilon < 1$,

$$N^{-1/2}\left\{N(s)\left(\hat{\theta}_{N(s)}^B - \theta\right)/\sigma_0,\ s \in [\epsilon, 1]\right\} \xrightarrow{\mathfrak{D}} {}_{\epsilon}W, \tag{8.8.13}$$

where ${}_{\epsilon}W = \{W(t), t \in [\epsilon, 1]\}$ is a standard Wiener process on $[\epsilon, 1]$.

In a similar manner, it follows from (8.8.3), (8.8.8), and the assumed symmetry of the d.f. F that

$$N^{1/2}|\hat{\theta}_N^S - \theta| = O_P(1) \tag{8.8.14}$$

and the same result holds for the two-sample case. However, in either case, even for $t = N^{-1/2}\lambda$, $L_N(N^{-1/2}\lambda)$ [or $D_{n_1,n_2}(N^{-1/2}\lambda)$] does not behave like a polynomial or some other simple function of λ (with the coefficients random in nature), and hence, the minimization in (8.8.4) [or (8.8.6)] may not lead to any simple expression. In fact, compared to (8.8.13), the limiting distribution here is not normal and the weak convergence holds for some other limiting processes. See Rao, Schuster, and Littell (1975) for some of these details.

EXERCISES

8.3.1. For W_N^0, defined by (8.3.7), verify the conditions (2.4.10) and (2.4.11) and hence, use Theorem 2.4.4 to prove Theorem 8.3.2.

8.3.2. Provide a proof of (8.3.17).

8.3.3. Let $\{c_{ni},\ i \geqslant 0,\ n \geqslant 0\}$ and $\{d_{ni},\ i \geqslant 0,\ n \geqslant 0\}$ be two triangular arrays of real constants, such that if $\{k_n\}$ be any sequence of positive integers for which $k_n \uparrow \infty$ as $n \to \infty$, then $\lim_{n\to\infty}\{\max_{i \leqslant k_n}(c_{ni}^r d_{ni}^s)/\sum_{j \leqslant k_n}(c_{nj}^r d_{nj}^s)\} = 0$, for $(r,s) = (2,0)$, $(0,2)$ and $(2,2)$, and $D_n^2 = \sum_{j \leqslant k_n} d_{nj}^2 \leqslant D^{*2} < \infty$, $\forall n \geqslant 1$. Conventionally, let $c_{n0} = d_{n0} = 0$, $\forall n \geqslant 0$ and let $A_{nk}^2 = \sum_{j<k} c_{nj}^2 d_{nj}^2$, $k \geqslant 0$, $n \geqslant 0$. Let then $M_{nk}(t) = \sum_{i \leqslant k} c_{ni}\{\psi(X_i - td_{ni}) - \psi(X_i)\}$, $k \geqslant 1$, $n \geqslant 1$, $t \in (-k^*, k^*)$, for some finite and positive k^*, where ψ is assumed to be absolutely continuous on any bounded interval in R. Let $\psi^{(1)}$ be the derivative of ψ and assume that the X_i are i.i.d.r.v.'s with an absolutely continuous p.d.f. f having a finite Fisher information $I(f)$. Define γ by (8.2.18) and assume that $[\psi^{(1)}(X_i)]$ has a finite and positive variance σ_1^2 and (8.3.2) holds. Consider then a two-dimensional time parameter stochastic process $W_n^* = \{W_n^*(t,s) = (\sigma_1 A_{nn})^{-1}\{M_{n(t)n(t)}(s) - E_0 M_{n(t)n(t)}(s)\},\ 0 \leqslant t \leqslant 1,\ s \in (-k^*, k^*)\}$, where $n(t) = \max\{k : A_{nk}^2 \leqslant A_{nn}^2\}$, $t \in [0,1]$. Show that if for every $(t,t') : 0 \leqslant t \leqslant t' \leqslant 1$,

$$\lim_{n\to\infty} A_{nn}^{-2} \sum_{i=0}^{n(t)} c_{n(t)i} c_{n(s)i} d_{n(t)i} d_{n(s)i} = g(t,t')$$

exists and is a continuous function of (t,t') and further, there exists a positive number M^*, such that uniformly in $0 \leqslant q < k \leqslant k_n$, $n \geqslant 1$,

$$\left\{ \sum_{i=0}^{q} (c_{ki}d_{ki} - c_{qi}d_{qi}) + \sum_{i=q+1}^{k} c_{ki}^2 d_{ki}^2 \right\} \Big/ (A_{nk}^2 - A_{nq}^2) \leqslant M^*,$$

then, $\{W_n^*\}$ converges weakly to a Gaussian function W^*, where $W^* = \{W^*(t,s), (t,s) \in [0,1] \times [-k^*, k^*]\}$ has mean zero and

$$EW^*(t,s)W^*(t',s') = ss'g(t,t'),\ \forall (t,s),(t',s') \in [0,1] \times [-k^*,k^*].$$

In particular, if $c_{ni} = d_{ni}$, $i \geqslant 1$, $n \geqslant 1$, then, show that the last two conditions hold (Jurečková and Sen, 1981a).

Hints. Consider a sequence of processes $\{W_n^{*0}\}$, where $W_n^{*0}(t,s) = s(\sum_{i \leqslant n(t)} c_{n(t)i} d_{n(t)i}\{\psi^{(1)}(X_i) - \gamma\}/(A_{nn}\sigma_1)$, for $0 \leqslant t \leqslant 1$, $-k^* \leqslant s \leqslant k^*$. Then show that under the assumed conditions, $\sup\{|W_n^*(t,s) - W_n^{*0}(t,s)| : 0 \leqslant t \leqslant 1, -k^* \leqslant s \leqslant k^*\} \xrightarrow{p} 0$, as $n \to \infty$. Also, the process W_n^{*0} involves independent summands; hence, the invariance principles developed in Section 2.4 apply directly. Thus the desired result follows.

8.3.4. Using (8.3.18), (8.3.20), (8.3.21), (8.3.22), and the weak convergence result of the preceding exercise, show that for every ϵ: $0 < \epsilon < 1, \{N(t) \cdot N^{-1/2}(\hat{\gamma}_{N(t)} - \gamma), t \in [\epsilon, 1]\}$ converges weakly to a Gaussian function, where $N(t), t \in [\epsilon, 1]$ is defined by (8.3.7) and the c_i are all equal to 1 (location model).

8.3.5. Suppose now that in the set up of Exercise 8.3.3, ψ is a step function for which (8.2.11) through (8.2.14) hold. In addition, $\gamma^{*2} = \sum_{j=1}^{p}(\beta_j - \beta_{j-1})^2 f(a_j)$ is positive. Let then $B_{nk}^2 = \sum_{i \leqslant k} c_{ni}^2 |d_{ni}|$, $k \geqslant 1$, $n^*(s) = \max\{k : B_{nk}^2 \leqslant sB_{nn}^2\}$, $s \in [0, 1]$, and, let $W_n(t, s) = \{M_{n,n^*(t)}(s) - E_0 M_{n,n^*(t)}(s)\} /(B_{nn}\gamma^*)$, for $0 \leqslant t \leqslant 1$, $-k^* \leqslant s \leqslant k^*$, where the $M_{nk}(s)$ are defined as in Problem 8.3.3. Show that in this case, W_n weakly converges to a Gaussian function with 0 drift and covariance function $(t \wedge t')g(s, s')$, where $g(s, s') = |s| \wedge |s'|$, if $s \cdot s' > 0$ and is 0, otherwise. (Jurečková and Sen, 1981b).

Hint. In this case, express $M_{n,k}(s)$ as $\sum_{i \leqslant k} \sum_{i=1}^{p} \{I(X_i < a_j < X_i - sd_{ni}) - I(X_i - sd_{ni} < a_j < X_i)\}(\beta_j - \beta_{j-1})c_{ni}$ and use Theorem 2.3.3 to verify the tightness condition. For the finite dimensional laws, apply the central limit theorem.

8.3.6. In the setup of Problem 8.3.3, consider now the case where the score function $\psi = \psi_1 + \psi_2$, ψ_1 is absolutely continuous on any bounded interval in R and ψ_2 is a step function. Suppose that ψ_1 satisfies the regularity conditions imposed in Problem 8.3.3 while ψ_2 does so as in Problem 8.3.5. Show that $B_{nn}^2 / A_{nn}^2 \to 0$ as $n \to \infty$, and hence, using the results of Problems 8.3.3 and 8.3.5, show that the invariance principle in the Problem 8.3.5 holds in this case as well. Note that in this case, as in the case of a step function, the rate of convergence is slower compared to the case of absolutely continuous ψ (Jurečková and Sen, 1981b).

8.3.7. In the setup of Problem 8.3.5 or 8.3.6, show that for the location model, for every ϵ: $0 < \epsilon < 1$, $\{N^{1/4}(\hat{\gamma}_{N^*(t)} - \gamma), t \in [\epsilon, 1]\}$ (where $N^*(t)$ is defined as in Problem 8.3.5) converges weakly to a Gaussian function. Extend this result to the regression case, where the normalizing factor $N^{1/4}$ has to be replaced by a different function of the c_i (Jurečková, 1980; Jurečková and Sen, 1981b).

8.4.1. Verify (8.4.9).

8.4.2. Note that the r.v.'s in (8.4.18) are independent and bounded. Hence show that uniformly in t: $|t| < N^r$, for some $r > 0$, there exists a $T(> 0)$, such that

$$\phi_N(\theta, t) = E_0\left(\exp\left\{\theta\left[M_{NN}^{(2)}(t) - E_0 M_{NN}^{(2)}(t)\right]\right\}\right)$$
$$\leqslant \phi_N(\theta) < \infty, \qquad 0 < \theta < T.$$

8.4.3. Use the result in Problem 8.4.2, replace the last step in (8.4.13) by $M_N(\sup_{|t|<N'}\inf_{\theta>0}\exp(-\frac{1}{2}\theta\epsilon)\phi_N(t,\theta))$, and hence show that the power rate in (8.4.15) may be replaced by an exponential one.

8.4.4. Show that (8.4.11) holds for the confidence limits $\hat{\Delta}_{NU}$ and $\hat{\Delta}_{NL}$, defined in (8.3.19). Also, show that for $(s_N^2 - \sigma_0^2)$, an inequality similar to (8.4.22) holds. Hence, using (8.3.22), (8.4.15), (8.4.20), and the above, show that (8.4.22) holds.

8.5.1. In (8.5.9), replace the M-estimators by L-estimators and establish an invariance principle parallel to the one in (8.5.10).

8.6.1. Consider a bundle of n parallel filaments of equal length and let the nonnegative r.v.'s $X_1, \ldots, X_n$ denote the strength of individual filaments while $X_{n,1} < \cdots < X_{n,n}$ stand for the associated order statistics. If a bundle breaks under load L, then all the inequalities $L/n \geqslant X_{n,1}, L/(n-1) \geqslant X_{n,2}, \ldots, L/1 \geqslant X_{n:n}$ are simultaneously satisfied. Consequently, the bundle strength D_n [viz. (8.6.1)] can be represented as $D_n = \max\{nX_{n,1}, (n-1)X_{n,2}, \ldots, X_{n,n}\}$. Let $Z_n = n^{-1}B_n$, $n \geqslant 1$. Assume that the X_i have a common d.f. F, defined on R^+. Show that

(a) $Z_n \leqslant \bar{X}_n = n^{-1}\sum_{i=1}^n X_i$ for every $n \geqslant 1$,

(b) $\{Z_n\}$ is a reverse semimartingale sequence, and

(c) $Z_n \to \theta = \sup\{x[1-F(x)] : x \in R^+\}$ both a.s. and in the first mean (Suh, Bhattacharyya, and Grandage, 1970).

8.6.2. Define $\psi(x, y)$ as in the discussion following (8.6.3) and assume that

(a) $\psi(x, y)$ is convex in $y(\in E)$, that is, $\psi(x, y) \leqslant (1-y)\psi(x,0) + y\psi(x, y)$ for every $y \in E$, where $\psi(x,0)$ and $\psi(x,1)$ are nonnegative,

(b) $\psi(x,0)$ is $\uparrow$ in x and $\psi(x,1)$ is $\downarrow$ in x $(x \in R^+)$, and

(c) $\int_{R^+}\psi^2(x,\delta)\,dF(x) \leqslant \mu_2(\delta) < \infty$, for $\delta = 0, 1$.

Under **(a)**, **(b)**, **(c)**, and the hypothesis of Theorem 8.6.1, Z_n converges in L_1 norm to θ as $n \to \infty$ (Sen, Bhattacharyya, and Suh, 1973).

8.6.3. For the Problem 8.6.1, assume in addition that $x[1-F(x)]$ assumes a unique maximum θ at $x = x_0$ and for some $x_1(> x_0)$, $x[1-F(x)]$ is $\downarrow$ in $x(> x_1)$. Then, $\lim_{n\to\infty}\{n^{1/2}|E(Z_n) - \theta|\} = 0$ (Sen, Bhattacharyya, and Suh, 1973).

8.6.4. Provide the proof of (8.6.16).

8.7.1. Show that for Problem 8.6.1, if F admits of a finite first moment, for every $\epsilon > 0$, there exist positive constants $C(< \infty), \rho(\epsilon)$: $0 < \rho(\epsilon) < 1$

and a positive number $n_0(\epsilon)$, such that for $n \geqslant n_0(\epsilon)$, $P\{\sup_x[x|F_n(x) - F(x)| : x \in R^+] > \epsilon\} \leqslant C[\rho(\epsilon)]^n$, and hence $\sup[x(1 - F_n(x)) : x \in R^+] \to \sup[x(1 - F(x)) : x \in R^+]$ a.s., as $n \to \infty$ (Sen, 1973a; though Sen assumed the existence of the second moment for other reasons, his proof goes through under the first moment condition as well.)

8.7.2. Since $Z_n = \sup\{\psi(x, F_n(x)) : x \in R\}$ and $\theta = \sup\{\psi(x, F(x)) : x \in R\}$, use the result in the preceding problem and the regularity conditions in Section 8.7 to show that (8.7.20) holds.

8.7.3. For both (8.7.22) and (8.7.23), use the exponential rate of convergence of the sample quantiles to verify (8.7.25).

PART 2

Statistical Inference

CHAPTER 9

Asymptotic Theory of Nonparametric Sequential Tests

9.1. INTRODUCTION

The invariance principles relating to various (nonparametric) statistics, studied in Part 1, are incorporated in this chapter in the formulation and study (of the asymptotic properties) of different types of sequential tests based on these statistics. Basically, the following four types of sequential tests are considered here:

(i) The Darling-Robbins type tests with power one.
(ii) Nonparametric repeated significance tests.
(iii) Analogues of the classical SPRT/SLRT.
(iv) Life testing or time-sequential tests.

The basic features of nonparametric tests with power 1 are presented in Section 9.2. Section 9.3 is devoted to the study of the formulation and properties of nonparametric repeated significance tests. The main body of this chapter is concerned with sequential rank tests along with other allied ones, and these are studied in Section 9.4 through 9.7. Nonparametric time sequential tests are studied separately in detail in Chapter 11.

9.2. SOME NONPARAMETRIC SEQUENTIAL TESTS WITH POWER 1

In a class of statistical decision problems, the implications of the null and the alternative hypotheses may have different physical importance. For example, in a clinical study involving an existing and a newly introduced drug, so long as the performance of the new one is not significantly better than the existing one, there may not be enough grounds to introduce this new drug in the market. On the other hand, if its performance is clearly superior, it should be detected as early as possible so that for the benefit of the patients (under the clinical trial or elsewhere) this drug can be adapted as soon as possible. Similarly, if the new drug is not better than the existing one, it may be eliminated from the market. In such a case or in many other situations involving a continuous production process, one may be interested in testing suitable null hypotheses against suitable alternatives in such a way that the test has a small type I error and its power is equal to 1 if the null hypothesis is not true. Obviously, fixed sample size procedures do not satisfy these two requirements and we need to turn to suitable sequential procedures.

In a series of papers dealing with various specific statistics, Darling and Robbins (1967a, b, c; 1968a, b; see also Robbins and Siegmund, 1968, 1969, 1970, 1974) have developed the theory of sequential tests having the properties that (a) by choosing an appropriately large initial sample size, the type I error can be made arbitrarily small, and (b) for every alternative hypothesis belonging to a suitable family, the power is equal to one. As we shall see shortly that the theory rests on the a.s. convergence of suitable statistics and on the invariance principles relating to them.

Let $\mathbf{X} = \{X_1, X_2, \ldots\}$ be a sequence of random elements defined on a probability space $(\Omega, \mathcal{Q}, P)$, where we assume that the probability measure P belongs to a family $\mathcal{P}$, and the null hypothesis to be tested, states that $H_0 : P \in \mathcal{P}_0 \subset \mathcal{P}$, while we are interested in the set of alternative hypotheses $H_1 : P \in \mathcal{P}_1 \subset \mathcal{P} \backslash \mathcal{P}_0$.

In the classical Neyman-Pearsonian setup (nonsequential case), for some given $n(\geqslant 1)$, based on $\mathbf{X}^{(n)} = (X_1, \ldots, X_n)$ (with the corresponding probability measure $P^{(n)}$), we construct a test with the following properties: (a) for some preassigned α $(0 < \alpha < 1)$, the type I error,

$$P\{H_0 \text{ is rejected when it is actually true}\} \leqslant \alpha \tag{9.2.1}$$

and (b) within the class of size-α tests,

$$\begin{aligned} \text{power} &= 1\text{-type II error} \\ &= P\{H_0 \text{ is rejected when actually } H_1 \text{ holds}\} \end{aligned} \tag{9.2.2}$$

is a maximum for the desired one. Of course, if H_0 and H_1 are not simple hypotheses, other restrictions, such as similarity, unbiasedness, and local alternatives, may be needed to reduce the domain of competing tests and choose an appropriate one within that class. The choice of an appropriate sample size to ensure a good power of the test depends on H_1 and, in many cases, may not be done without additional restrictions. In the Wald sequential setup, with a view to controlling both the type I and type II errors, the sample size is treated as a random variable and the procedure is chosen in such a way that subject to given bounds on both these errors, the expected sample size of the desired test is a minimum.

In a variety of statistical problems, particularly, relating to continuous production processes, one is not interested in unnecessarily stopping the process if the null hypothesis is true (i.e., the process outcomes conform to the specifications), while any change of the specifications due to extraneous causes (i.e., deviations from the null hypothesis) has to be detected with a view to adjusting the production line to meet the set specifications. For this reason, we consider (following Darling and Robbins) some sequential procedures that have power one for suitable alternative hypotheses and have small type I error.

In what follows, we shall be concerned with a sequence $\{T_n = T(\mathbf{X}^{(n)});\ n \geqslant 1\}$ of real valued statistics, and for every $n \geqslant 1$, let $G_n(t) = P\{T_n \leqslant t\}$, $-\infty < t < \infty$; $G = \{G_n; n \geqslant 1\}$ and let

$$\mathcal{G}_0 = \{G : P \in \mathcal{P}_0\} \qquad \text{and} \quad \mathcal{G}_1 = \{G : P \in \mathcal{P}_1\}. \tag{9.2.3}$$

Concerning $\{T_n\}$ and G, we assume that the following conditions hold:

(i) For every $P \in \mathcal{P}$, there exists a real valued $\tau(P)$ such that

$$T_n \to \tau(P) \qquad \text{a.s.,} \quad \text{as} \quad n \to \infty, \text{ when } P \text{ holds,} \tag{9.2.4}$$

(ii) There exists a sequence $J = \{J_n;\ n \geqslant 1\}$ of closed, half-open or open intervals on the real line, such that

$$\text{under } \mathcal{P}_0,\ T_n \in J_n \qquad \text{a.s.,} \quad \text{as} \quad n \to \infty \tag{9.2.5}$$

(where J_n does not depend on the specific $G \in \mathcal{G}_0$), and

(iii) For every $P \in \mathcal{P}_1$, there exists a positive integer $n_0 = n_0(P)$,

$$\tau(P) \notin J_n, \qquad \forall n \geqslant n_0. \tag{9.2.6}$$

Then, to test for H_0 versus H_1, based on $\{T_n\}$, we conceive of an initial sample of size n_0 and define a *stopping variable*

$$\begin{aligned} N &= \text{least positive integer} \quad n\,(\geqslant n_0) \text{ such that} \quad T_n \notin J_n \\ &= \infty, \qquad \text{if no such } n \text{ exists.} \end{aligned} \tag{9.2.7}$$

Thus we continue drawing observations, until, for the first time, for some N $(\geqslant n_0)$, $T_N \notin J_N$. If $N < \infty$, we stop at the Nth stage along with the rejection of H_0, while for $N = \infty$, H_0 is not rejected. Note that by (9.2.4), (9.2.5), and (9.2.7),

$$\begin{aligned} &P\{H_0 \quad \text{is rejected when it is really true}\} \\ &\qquad = P\{T_n \notin J_n \quad \text{for some} \quad n \geqslant n_0 \mid \mathcal{P}_0\} \to 0 \qquad \text{as} \quad n_0 \to \infty, \end{aligned} \tag{9.2.8}$$

so that, the lhs of (9.2.8) can be made arbitrarily small by choosing n_0 appropriately large. Also, by (9.2.4), (9.2.6), and (9.2.7), we have

$$\begin{aligned} P\{H_0 \text{ is rejected when } H_1 \text{ holds}\} \\ = P\{T_n \notin J_n \quad \text{for some} \quad n \geqslant n_0 \mid \mathcal{P}_1\} = 1. \end{aligned} \tag{9.2.9}$$

Thus the sequential test based on $\{T_n;\ n \geqslant n_0\}$ and the stopping rule in (9.2.7) meet both the requirements (a) and (b). It is quite clear that one can generate a class of tests by choosing first a class of $\{T_n\}$ and second, for each specific $\{T_n\}$, a class of $\{J_n\}$, each satisfying (9.2.4) through (9.2.6). A useful tool for discriminating among the various rival tests of this type is the *average sample number* (ASN)

$$E(N) = \sum_{n \geqslant n_0} nP\{N = n\} = n_0 + \sum_{n \geqslant n_0} P\{N > n\}. \tag{9.2.10}$$

Minimization of $E(N \mid \mathcal{P}_1)$ is a natural criterion of choosing an optimal test. However, the choice (in accordance to this criterion) may depend on the particular $P(\in \mathcal{P}_1)$ and hence may not be uniform for all P and $\mathcal{P}_1$. Moreover, in the majority of cases, exact evaluation of $E(N \mid \mathcal{P}_1)$ becomes prohibitively laborious, if not impracticable. In such cases, suitable bounds for $E(N \mid \mathcal{P}_1)$ frequently are used. Note that by (9.2.7), for every $n > n_0$,

$$P\{N > n\} = P\{T_m \in J_m, \quad \forall n_0 \leqslant m \leqslant n\}. \tag{9.2.11}$$

Thus, if $\tau(P) = 0$, $\forall P \in \mathcal{P}_0$ and if $J_n = \{x : |x| \leqslant c_n\}$ where $\{c_n\}$ is a suitable sequence of positive numbers, we have

$$P\{N > n\} = P\{|T_m| < c_m, \quad \forall n_0 \leqslant m \leqslant n\}. \tag{9.2.12}$$

Then, if $\{|T_n|\}$ is a forward (reverse) submartingale sequence (under $\mathcal{P}_0$) and $\{c_n\}$ is nondecreasing (nonincreasing), one can use the Hájek-Rényi inequality to obtain a suitable bound for (9.2.12). A cruder bound would be to replace the rhs of (9.2.12) by

$$P\{N > n\} \leqslant P\{T_n \in J_n\} \qquad \forall n \geqslant n_0. \tag{9.2.13}$$

Hence, to verify that $E(N \mid \mathcal{P}_1) < \infty$, it suffices to show that

$$P\{T_n \in J_n \mid \mathcal{P}_1\} = 0(n^{-1-\delta}), \qquad \delta > 0, \qquad \forall n \geqslant n^*, \tag{9.2.14}$$

where $n^* \geqslant n_0$. In the rest of this section, we consider various specific problems of special interest and illustrate the basic feature of the Darling-Robbins type tests for these problems.

9.2.1. One-Sample Location Problem

Let $\{X_i,\ i \geqslant 1\}$ be a sequence of i.i.d.r.v. with a d.f. F, defined on the real line $(-\infty, \infty)$. We may formulate the problem and the corresponding test procedure under different sets of regularity conditions on F.

(a) ***Test for the population mean: variance known.*** We assume that both $\mu = EX = \int x\,dF(x)$ and $\sigma^2 = \int x^2\,dF(x) - \mu^2$ exist and $0 < \sigma < \infty$. Our problem is to test for

$$H_0 : \mu = 0 \quad \text{versus} \quad H_1 : \mu > 0 \quad \text{or} \quad H_2 : \mu \neq 0, \qquad (9.2.15)$$

where σ is assumed to be given. Let $\bar{X}_n = n^{-1}\sum_{i=1}^{n} X_i$, $n \geqslant 1$, and set

$$T_n = \bar{X}_n / \sigma, \qquad n \geqslant 1. \qquad (9.2.16)$$

By the Kintchine strong law of large numbers, it follows that $\bar{X}_n \to \mu$ a.s., as $n \to \infty$. Hence $T_n \to \delta = \mu/\sigma$ a.s., as $n \to \infty$. Also, by the law of iterated logarithm, we have with probability 1

$$\limsup_{n\to\infty} \left\{ n^{1/2} |\bar{X}_n - \mu| / \sigma (2 \log\log n)^{1/2} \right\} = 1. \qquad (9.2.17)$$

Thus if we set

$$J_n = \left\{ x : |x| \leqslant n^{-1/2} (2 \log\log n)^{1/2} \right\}, \qquad n \geqslant 3, \qquad (9.2.18)$$

it follows from the above results that (9.2.4) through (9.2.6) all hold here. When testing H_0 against H_1, we set $J_n = \{x : x < n^{-1/2}\sigma(2 \log\log n)^{1/2}\}$, the rest remaining the same. Further, note that $\{\bar{X}_n - \mu\}$ is a reverse martingale sequence, and hence, as in (9.2.12), for testing H_0 versus H_1, for $\mu \neq 0$,

$$\begin{aligned} P\{N > n \mid \mu\} &= P\left\{ \bar{X}_m \leqslant \sigma m^{-1/2} (2 \log\log m) \quad \forall n_0 \leqslant m \leqslant n \mid \mu \right\} \\ &= P\left\{ \bar{X}_m - \mu \leqslant \sigma n^{-1/2} (2 \log\log m)^{1/2} - \mu, \quad \forall n_0 \leqslant m \leqslant n \mid \mu \right\} \\ &= P\left\{ T_m - \delta \leqslant -\delta + m^{-1/2} (2 \log\log m)^{1/2}, \quad \forall n_0 \leqslant m \leqslant n \mid \mu \right\}. \end{aligned} \qquad (9.2.19)$$

The form (9.2.19) is not very convenient for manipulations. However, we note that for any $\delta >$ (or $\neq$) 0, $-\delta + n^{-1/2}(2 \log\log n)^{1/2} \to -\delta$ as $n \to \infty$. Thus for every $\delta > 0$, there exists an n^0, such that for $n \geqslant n^0$, $\delta - n^{-1/2}(2 \log\log n)^{1/2} > \delta/2$, and hence

$$\begin{aligned} P\{N > n \mid \delta > 0\} &\leqslant P\{|T_m - \delta| > \delta/2, \quad n^0 \leqslant m \leqslant n \mid \delta\} \\ &\leqslant P\{|T_n - \delta| > \delta/2 \mid \delta\}. \end{aligned} \qquad (9.2.20)$$

Thus if we assume that

$$E|X|^r < \infty \qquad \text{for some } r > 2 \left(\Rightarrow E|\bar{X}_n - \mu|^r = O(n^{-r/2}) \right), \qquad (9.2.21)$$

by the Markov inequality, the rhs of (9.2.20) is $O(n^{-r/2})$, and hence $E[N \mid \delta > 0] < \infty$. A similar case holds for the two-sided test.

We may remark that in this problem, (9.2.5) holds by virtue of the law of iterated logarithm for $\{\bar{X}_n\}$. However, this does not provide us with suitable expressions for the type I error; some asymptotic expressions for this can be obtained when we restrict ourselves to wider $\{J_n\}$. For example, if we let $J_n = \{x : x < n^{-1/2}(c \log n)^{1/2}\}$ (and replace x by $|x|$ for the two-sided case), where $c(>0)$ is a suitable constant, then (9.2.20) and (9.2.21) still hold, while

$$P\{T_n \notin J_n \text{ for some } n \geqslant n_0 \mid H_0\}$$
$$= P\left\{T_n > n^{-1/2}(c \log n)^{1/2} \text{ for some } n \geqslant n_0 \mid \delta = 0\right\}, \quad (9.2.22)$$

so that using Theorem 2.5.1 [and letting $f(t) = t(\log t)^{-2}$, $t > 1$], we may approximate (9.2.22) (for large n_0) by

$$P\left\{W(t) > (ct \log t)^{1/2} \quad \text{for some} \quad t \geqslant n_0\right\}$$
$$\simeq \left[1 - \Phi\left((c \log n_0)^{1/2}\right)\right]\left[\frac{1+c}{c} + \log n_0\right], \; \left[\text{by } (2.7.20)\right]$$
$$(9.2.23)$$

where Φ is the standard normal d.f. A similar case holds for the two-sided test.

(b) ***Test for the population mean: variance unknown.*** Here, we are interested in (9.2.15), but σ is not known. Let $s_n^2 = (n-1)^{-1}\sum_{i=1}^{n}(X_i - \overline{X}_n)^2$, $n \geqslant 2$ and set

$$T_n = \overline{X}_n / s_n \qquad \text{for} \quad n \geqslant 2. \quad (9.2.24)$$

Note that (by the Kintchine strong law of large numbers)

$$s_n^2 = \frac{n}{n-1}\left\{\frac{1}{n}\sum_{i=1}^{n}(X_i - \mu)^2 - \left(\overline{X}_n - \mu\right)^2\right\} \to \sigma^2 \qquad \text{a.s.} \quad \text{as} \quad n \to \infty$$

so that on replacing σ by s_n in (9.2.16) and (9.2.17), we can proceed as in the case of σ being known. To show that $E(N \mid \delta \neq 0) < \infty$, we need to show that $\sum p\{|s_n^2/\sigma^2 - 1| > \epsilon\}$ $(\epsilon > 0)$ converges, and this is possible if we strengthen (9.2.21) to $r > 4$.

(c) ***Test for the population quantile (median): F not specified.*** We denote the p-quantile $(0 < p < 1)$ of F by $\xi_p(\Rightarrow F(\xi_p) = p)$ and assume that ξ_p is uniquely defined. We desire to test

$$H_0 : \xi_p = \xi_p^0 \qquad \text{versus} \quad H_1 : \xi_p > \xi_p^0 \qquad \text{or} \quad H_2 : \xi_p \neq \xi_p^0, \quad (9.2.25)$$

where ξ_p^0 is specified. (Typically, we take $p = \frac{1}{2}$, so that ξ_p is the population median of F.) Define

$$Y_i = \begin{cases} 1, & X_i \leqslant \xi_p^0, \\ 0, & \text{otherwise}; \quad i \geqslant 1, \end{cases} \quad (9.2.26)$$

$$\overline{Y}_n = n^{-1}\sum_{i=1}^{n} Y_i \qquad \text{and} \quad \sigma^2 = p(1-p)\;(\leqslant \tfrac{1}{4}). \quad (9.2.27)$$

Note that the Y_i are i.i.d.r.v. and under H_0, Y_1 has mean $F(\xi_p^0) = p$ and variance $p(1-p) = \sigma^2$. Also, $T_n = (\overline{Y}_n - p)/\sigma \to [F(\xi_p^0) - p]/\sqrt{p(1-p)}$ a.s. as $n \to \infty$. Hence we can proceed as in case (a) with $T_n = (\overline{Y}_n - p)/[p(1-p)]^{1/2}$ and $J_n = \{x : |x| \leqslant (2n^{-1}\log\log n)^{1/2}\}$, $n \geqslant 3$. Moreover, in this case, the Y_i are bounded r.v.'s, and hence by the Bernstein inequality, we have for every $\epsilon > 0$, there exists an n_0 such that

$$P\left\{|\overline{Y}_n - F(\xi_p^0)| > \epsilon\right\} \leqslant 2[\rho(\epsilon)]^n, \qquad \forall n \geqslant n_0, \tag{9.2.28}$$

so that as in (9.2.20), $P\{|T_n - (F(\xi_p^0) - p)/\sqrt{p(1-p)}\,| > \epsilon\}$ converges to zero at an exponential rate, ensuring that $E(N \mid H_1) < \infty$ (or $E(N \mid H_2) < \infty$).

(d) ***Rank tests for the median of a symmetric d.f.*** We assume that $F(x)$ is symmetric about its median θ and we desire to test

$$H_0 : \theta = 0 \qquad \text{versus} \quad H_1 : \theta > 0 \quad \text{or} \quad H_2 : \theta \neq 0; \tag{9.2.29}$$

we could have taken $\theta = \theta^0$ for any specified θ^0, but subtracting θ^0 from each observation, we can always reduce it to the case in (9.2.29). We base our test on the class of one-sample rank order statistics, studied in detail in Chapter 5. Let us define $\{S_n, n \geqslant 1\}$ as in (5.2.1) and note that under H_0, S_n has a distribution symmetric about 0 (with mean 0) and variance nA_n^2 with A_n^2 defined by (5.2.5). Here, for testing against the two-sided alternative H_2, we choose

$$J_n = \left\{x : |x| \leqslant A_n(2n\log\log n)^{1/2}\right\}, \qquad n \geqslant 3, \tag{9.2.30}$$

and for the one-sided case in (9.2.30) we replace $|x|$ by x. Then, by Theorem 5.4.1, under H_0,

$$S_n \in J_n \qquad \text{a.s.,} \quad \text{as} \quad n \to \infty. \tag{9.2.31}$$

Let us define $\mu(\phi, F)$ as in (5.4.12) and note that if ϕ is nondecreasing and skew-symmetric and if F is symmetric about θ, then

$$\mu(\phi, F) \text{ is } \gtreqless 0 \qquad \text{according as} \quad \theta \gtreqless 0. \tag{9.2.32}$$

Hence by theorem 5.7.1,

$$P\{S_m \notin J_m \text{ for some } m \geqslant n_0 \mid H_0 \text{ is not true}\} = 1, \tag{9.2.33}$$

so that the test has power 1. Further, by the same method of proof as in Theorem 5.7.1, it can be shown (see Problem 9.2.2) that if $\phi \in L_r$ for some $r > 2$ and ϕ is $\nearrow$ and continuous, then for every $\epsilon > 0$, there exist a $K (0 < K < \infty)$ and an $n_0 = n_0(\epsilon)$, such that for $n \geqslant n_0$,

$$P\left\{|n^{-1}S_n - \mu(\phi, F)| > \epsilon\right\} \leqslant Kn^{-r/2}. \tag{9.2.34}$$

Thus, as in (9.2.20), $P\{N > n \mid \theta \neq 0\} = O(n^{-r/2})$, $\forall n \geqslant n_0$, and hence $E(N \mid \theta \neq 0) \leqslant n_0 + O(n_0^{-r/2+1}) = n_0 + O(1) < \infty$. Further, the results in Exercises 5.4.4 and 5.4.5 may be used to verify that (9.2.22) and (9.2.23) hold in this case.

9.2.2. General Estimable Parameter I

Here also, let $\{X_i, i \geqslant 1\}$ be i.i.d.r.v. with a d.f. F, and, in the setup of Chapter 3, we conceive of an estimable parameter $\theta = \theta(F) = E\phi(X_1, \ldots, X_m)$ where ϕ is a symmetric kernel of degree $m(\geqslant 1)$. As in (3.2.3) and (3.2.4), we consider the U-statistic and von Mises functional corresponding to θ and denote these by U_n and $\theta(F_n)$, respectively. Suppose that we want to test

$$H_0 : \theta = \theta_0 \quad \text{(specified)} \quad \text{versus} \quad H_1 : \theta > \theta_0 \quad \text{or} \quad H_2 : \theta \neq \theta_0. \tag{9.2.35}$$

Let us define s_n^2 as in (3.7.5) and let

$$T_n = \frac{U_n - \theta_0}{ms_n} \quad \text{or} \quad \frac{\theta(F_n) - \theta_0}{ms_n}, \tag{9.2.36}$$

$$J_n = \left\{x : |x| \leqslant (2(1+\epsilon)(\log\log n)/n)^{1/2}\right\}, \epsilon > 0; \tag{9.2.37}$$

for the one-sided case, replace $|x|$ by x in (9.2.37). Then, by Theorem 3.2.1, (3.2.10), and (3.7.7), under H_0,

$$T_n \in J_n \quad \text{a.s.,} \quad \text{as} \quad n \to \infty. \tag{9.2.38}$$

On the other hand, by Theorem 3.4.1, one can show that (9.2.6) holds. Finally, (3.7.21) ensures that $E[N \mid \theta \neq \theta_0] < \infty$. Of course, if the variance of U_n [or $\theta(F_n)$] is known, one need not use the estimator s_n^2 and, in that case, $E[N \mid \theta \neq \theta_0] < \infty$, even when $E|\phi|^r < \infty$ for some $r > 2$.

9.2.3. General Estimable Parameter, II

In Chapter 7, in the context of the invariance principles for linear combinations of order statistics, we have considered functionals of the d.f. F, defined by (7.3.22) or (7.4.2). The results of Sections 7.3, 7.5, and 7.6 ensure that for such parameters, tests having the properties (9.2.8) and (9.2.9) can be based on LCFOS where $E(N) < \infty$ when H_0 does not hold. Similarly, in Chapter 8, we considered suitable functionals of the empirical distributions that stochastically converge to appropriate functionals of the true d.f. The results of Sections 8.4 and 8.7 ensure that sequential tests based on such functionals have also the desired properties (9.2.8), (9.2.9), and $E(N) < \infty$ when H_0 does not hold.

9.2.4. Bivariate Independence Problem

As in Chapter 6, let $\{Z_i = (X_i, Y_i),\ i \geqslant 1\}$ be a sequence of (bivariate) i.i.d.r.v. with a continuous d.f. $F(x, y)$ and consider the null hypothesis (6.2.1) against either positive (negative) type of association or nonzero association. Define $\{M_n,\ n \geqslant 1\}$ as in (6.2.2) and $\{A_n^2, B_n^2\colon\ n \geqslant 1\}$ as in (6.2.4). Let then $T_n = M_n / A_n B_n$, $n \geqslant 2$. For the two-sided alternatives, we set here

$$J_n = \left\{x : |x| \leqslant (2n \log\log n)^{1/2}\right\}, \qquad n \geqslant 3, \tag{9.2.39}$$

and for the one-sided case, we replace in (9.2.39) $|x|$ by x. Then, by Theorem 6.4.1, under H_0 in (6.2.1), $M_n \in J_n$ a.s., as $n \to \infty$, while defining $\mu(\phi_1, \phi_2)$ by (6.6.1), we obtain by Theorem 6.6.1 and the fact that $\mu(\phi_1, \phi_2)$ is $>$ or < 0 according as there is a positive or negative association between the two variates that $P\{M_n \notin J_n \text{ for some } n \geqslant n^0 \mid H_0 \text{ is not true}\} = 1$. Finally, the results of Section 6.5 ensure that $E(N) < \infty$ when H_0 does not hold. Theorem 6.4.1 ensures (9.2.22) and (9.2.23).

9.2.5. Two-Sample Location and Scale Problem

Let $\{X_i,\ i \geqslant 1\}$ be a sequence of independent r.v.'s with continuous d.f.'s $\{F_i,\ i \geqslant 1\}$, all defined on the real line $(-\infty, \infty)$. For simplicity, suppose that

$$F_{2i-1} = F \qquad \text{and} \qquad F_{2i} = G \quad \text{for every} \quad i \geqslant 1, \tag{9.2.40}$$

and our null hypothesis relates to

$$H_0 : F = G (\Rightarrow F_i = F \qquad \text{for all} \quad i \geqslant 1). \tag{9.2.41}$$

We are interested in location and scale alternatives, where we set

$$H_L : G(x) = F(x - \Delta) \qquad \text{and} \quad H_S : G(x) = F\left(\frac{x}{1 + \Delta}\right), \tag{9.2.42}$$

so that H_0 relates to $\Delta = 0$, while, under H_L (or H_S), Δ is different from 0 (Δ being > -1 for H_S). For this testing problem, we define the L_N by (4.2.1) where we note that $c_{2i} = 1$ and $c_{2i-1} = 0$ for every $i \geqslant 1$, so that by (4.2.9), $C_{2n}^2 = n/2$ for every $n \geqslant 1$. We choose J_n as in (9.2.30). Then, by Theorem 4.4.1, under H_0, $L_n \in J_n$ a.s., as $n \to \infty$. Also, as a corollary to Theorem 4.6.1, we have

$$n^{-1} L_n \to \lambda(F, G) \qquad \text{a.s.,} \quad \text{as} \quad n \to \infty, \tag{9.2.43}$$

where $\lambda(F, G) = \int_{-\infty}^{\infty} \phi(\frac{1}{2}(F(x) + G(x))\, dG(x)$ and under H_L or H_S with $\Delta \neq 0$, $\lambda(F, G)$ is also nonzero, having the same sign as of Δ. Hence here also (9.2.8) and (9.2.9) hold. Further, if the score function ϕ satisfies

$\int_0^1 |\phi(u)|^r \, du < \infty$ for some $r > 2$, then along with same lines as in the proof of Theorem 4.6.1, it can be shown (see Problem 9.2.3) that for every $\epsilon > 0$

$$P\left\{|n^{-1}L_n - \lambda(F,G)| > \epsilon\right\} \leqslant K_\epsilon n^{-r/2}, \qquad \forall n \geqslant n_0, \qquad (9.2.44)$$

where both $K_\epsilon(<\infty)$ and n_0 may depend on ϵ. Thus, as in (9.2.20), $E[N \mid \Delta \neq 0] < \infty$ for both the location and scale models. Finally, Theorem 4.4.1 ensures that (9.2.22) and (9.2.23) hold in this case, too [when in (9.2.30), we replace $2\log\log n$ by $c \log n$].

9.2.6. Simple Regression Problem

Instead of (9.2.40), we may consider the general regression model:

$$F_i(x) = F(x - \beta_0 - \beta c_i), \qquad i \geqslant 1, \qquad (9.2.45)$$

where (β_0, β) are unknown parameters and the c_i are known regression constants. We desire to test

$$H_0 : \beta = 0 \quad \text{versus} \quad H_1 : \beta > 0 \quad \text{or} \quad H_2 : \beta \neq 0. \qquad (9.2.46)$$

Thus, under H_0, all the F_i are identical. Here also, we use the L_n defined by (4.2.1) and proceed in the same manner as in the special case of (9.2.40). We need to use Theorem 4.4.1 to establish (9.2.8) as well as (9.2.22) and (9.2.23), where in the later case, in (9.2.30), we replace $(2\log\log n)$ by $(c \log n)$. Theorem 4.6.1 along with the assumed existence of $\underline{\lim}|\lambda_N|$ (when H_0 does not hold) ensures (9.2.9). The proof of $E[N \mid \beta \neq 0] < \infty$ is somewhat involved and demands extra regularity conditions; we pose this as Problem 9.2.4.

9.2.7. Goodness of Fit Tests

For the one- or two-sample goodness of fit problem, one may also use the classical Komogorov-Smirnov type test statistics:

$$D_n^+ = \sup_x \left[F_n(x) - F(x)\right], \qquad D_n = \sup_x |F_n(x) - F(x)|, \qquad (9.2.47)$$

$$D_{n_1 n_2}^+ = \sup_x \left[F_{n_1}(x) - G_{n_2}(x)\right] \quad \text{and} \quad D_{n_1 n_2} = \sup_x |F_{n_1}(x) - G_{n_2}(x)|, \qquad (9.2.48)$$

where F_n (F_{n_1}) and G_{n_2} are the sample d.f.'s and F is the hypothetical d.f. Results of Section 2.6 ensure that the law of iterated logarithm holds for these statistics, and hence they enjoy the properties (9.2.8) and (9.2.9). Moreover (see Problem 9.2.5), it is easy to verify that for every $\epsilon > 0$, if F_0 is the true d.f.,

$$P\left\{\sup_x |F_n(x) - F_0(x)| > \epsilon\right\} \leqslant K\left[\rho(\epsilon)\right]^n, \qquad \forall n \geqslant n_0, \qquad (9.2.49)$$

where $\rho(\epsilon) \in (0, 1)$ and n_0 and K are positive numbers, depending on ϵ. A similar exponential bound holds for the two-sample case, where we note that

$$\sup_x |F_{n_1}(x) - G_{n_2}(x)| \leqslant \sup_x |F_{n_1}(x) - F_0(x)| + \sup_x |G_{n_2}(x) - G_0(x)| + \sup_x |F_0(x) - G_0(x)|,$$

and (9.2.49) applies to the first two terms on the rhs. Hence in either case $E(N) < \infty$ when H_0 does not hold. The results remain true even when the r.v. are vector-valued, as for the multivariate case, exponential bounds for the empirical processes also are available.

We conclude this section with the remark that nonparametric tests with power 1 can also be worked out easily for the analysis of variance problems based on rank statistics.

9.3. NONPARAMETRIC REPEATED SIGNIFICANCE TESTS

An experimenter frequently faces the situation where the individuals under study may not all enter the scheme at the same time. For example, in a clinical trial involving the application of two or more drugs (possibly one standard and the rest newly introduced), the patients undergoing teatment may not enter into the clinic simultaneously. To make statistical inferences reliable and sensitive in such a case, one needs to fix in advance a target sample size (usually moderately or very large) on which to base the statistical procedures. This, in turn, may need a considerable waiting time to accommodate the desired number of individuals, and the statistical procedures are then employed following the termination of the experiment. From time and cost considerations, it may be advisable to monitor the experiment from the beginning with the objective of updating the statistical evidence as observations increasingly become available so that if, at an early or intermediate stage, a clear statistical decision is feasible, the experimentation may be stopped, resulting in savings of time and cost, among other things. In medical trials or clinical experiments, such repeated significance tests (RST) are popular; a detailed account of some of these RST procedures in the parametric case (based on the sample proportions and means or other standard test statistics) is given in Armitage (1975). Though his treatment is mostly nonmathematical, it provides nice statistical interpretations and usages of the RSTs. For the nonparametric case, some treatment of the RSTs is due to Sen (1978a).

In this section we develop some nonparametric RST procedures. The invariance principles studied in detail in Chapters 2 through 8 play a fundamental role here.

As in Section 9.2, we conceive of a sequence $X = \{X_i,\ i \geqslant 1\}$ of r.v.'s, defined on a probability space $(\Omega, \mathscr{Q}, P)$ and we denote by $\mathbf{X}^{(n)} = (X_1, \ldots, X_n)$ for $n \geqslant 1$. We fix in advance a target sample size N (usually large). Let $\{T_n = T(\mathbf{X}^{(n)});\ n \geqslant 1\}$ be a sequence of real valued statistics, so that we are concerned primarily with the partial sequence

$$\{T_n : n_0 \leqslant n \leqslant N\} \tag{9.3.1}$$

where $n_0(\geqslant 1)$ designate the initial sample size with which the statistical screening/monitoring procedure initiates on the experiment. As in Section 9.2, we frame the null hypothesis $H_0 : P \in \mathscr{P}_0 \subset \mathscr{P}$ against the alternative $H_1 : P \in \mathscr{P}_1 \subset \mathscr{P} \backslash \mathscr{P}_0$. Our problem is to formulate a test procedure, based on the set (9.3.1), which allows a statistical decision in favor of one of the hypotheses at an early stage if the accumulated evidence up to that stage so indicates. We shall find it convenient to classify such tests as type A, genuinely distribution-free, and type B, asymptotically distribution-free. For rank statistics, studied in detail in Chapters 4, 5, and 6, type A tests hold, whereas for others studied in Chapters 3, 7, and 8, the second type holds.

9.3.1. Type A Tests

We assume that under H_0, the joint d.f. of $\{T_n : n_0 \leqslant n \leqslant N\}$ does not depend on the underlying d.f. of the X_i, so that a suitable test based on this set of statistics will also be distribution-free under H_0. We conceive of a real-valued function

$$h_N = h(T_n : n_0 \leqslant n \leqslant N) \tag{9.3.2}$$

as a test statistic, and, as h_N is distribution-free under H_0, for every $\alpha : 0 < \alpha < 1$ and (n_0, N), there exist an α_N and h_N^0, such that

$$0 \leqslant \alpha_N = P\{h_n > h_N^0 \mid H_0\} < \alpha \leqslant P\{h_N \geqslant h_N^0 \mid H_0\}. \tag{9.3.3}$$

Typically, we may have the following choice of h_N:

$$h_N = \max_{n_0 \leqslant k \leqslant N} T_k \quad \text{or} \quad \max_{n_0 \leqslant k \leqslant N} |T_k|. \tag{9.3.4}$$

Corresponding to the *level of significance* α and the critical value h_N^0, we may associate a r.v.

$$M = M_N = \begin{cases} \min k(n_0 \leqslant k \leqslant N) : T_k \quad (\text{or} \quad |T_k|) > h_N^0, \\ N, \qquad \text{if no such } k \text{ exists.} \end{cases} \tag{9.3.5}$$

We designate M as the *stopping number*. Then, operationally, the test procedure consists in computing T_k for successive values of k $(\geqslant n_0)$. If, for the first time, for some $k = M(\leqslant N)$, T_M (or $|T_M|$) exceeds h_N^0, the experiment is terminated at that stage along with the rejection of H_0; if no such M ($\leqslant N$) exists, the null hypothesis H_0 is accepted along with the termination of the experiment at the target sample size N. By (9.3.3) and (9.3.4), the test has overall level of significance α.

In the conventional case, a test for H_0 based solely on the target sample of size N and corresponding to the same level of significance α, will involve a critical value smaller than h_N^0. Alternatively, for the given h_N^0 in (9.3.3), this target sample size test will have a level of significance smaller than α. Thus, operationally, in an RST procedure, we make a test at each sample size attained, and for this reason, the overall significance level is higher than the conventional level for the target sample size procedure. On the other hand, in an RST procedure M may be smaller than N, resulting in savings of time and cost of sampling. Because of the dependence of the successive elements of the set $\{T_k : n_0 \leqslant k \leqslant N\}$ in an RST procedure, the crucial point is the determination of h_N^0 in (9.3.3). Though marginally the T_n may have some nice d.f. (under H_0), their joint d.f. or the distribution of their maximum is in general far too complicated. For small N, one may venture to enumerate the same, but the task becomes prohibitively laborious as N increases. However, in this context the invariance principles studied in Chapters 4, 5, and 6 may be incorporated to yield some good approximations for the desired distributions.

We standardize the T_k by setting

$$E(T_n \mid H_0) = 0 \qquad \text{and} \quad \sigma_n^2 = E(T_n^2 \mid H_0) > 0, \qquad n \geqslant n_0 \quad (9.3.6)$$

where, by definition, the σ_n are known and we assume them to be finite. Also, let $\{k_N(t),\ t \in [0,1]\}$ be a sequence of nondecreasing, right-continuous, and nonnegative integers, such that $k_N(1) = N$ and $k_N(0) = 0$. For every N, define a stochastic process $Y_N = \{Y_N(t),\ t \in [0,1]\}$ by letting

$$Y_N(t) = \sigma_N^{-1} T_{k_N(t)}, \qquad \text{for} \quad 0 \leqslant t \leqslant 1. \quad (9.3.7)$$

Further, for the initial sample size n_0, we define t_N^0 by $k_N(t_N^0) = n_0$ and assume that as $N \to \infty$, t_N^0 converges to some t_0 ($0 \leqslant t_0 \leqslant 1$). Also, let $Y = \{Y(t),\ t \in [0,1]\}$ be a suitable random function on $[0,1]$ and set

$$D_{t_0}^+ = \sup_{t_0 \leqslant t \leqslant 1} Y(t) \qquad \text{and} \quad D_{t_0} = \sup_{t_0 \leqslant t \leqslant 1} |Y(t)|. \quad (9.3.8)$$

We denote the upper $100\alpha\%$ points of the d.f. of $D_{t_0}^+$ and D_{t_0} by $D_{t_0}^+(\alpha)$ and $D_{t_0}(\alpha)$, respectively. Finally, let

$$Y_N^0 = \{Y_N(t), t_N^0 \leqslant t \leqslant 1\} \qquad \text{and} \quad Y^0 = \{Y(t), t_0 \leqslant t \leqslant 1\}. \quad (9.3.9)$$

THEOREM 9.3.1. *If, under H_0, $Y_N^0 \xrightarrow{\mathfrak{D}} Y^0$, then, for (9.3.3) and (9.3.4),*

$$\sigma_N^{-1} h_N^0 \to D_{t_0}^+(\alpha) \qquad (\text{or } D_{t_0}(\alpha)), \quad \text{as} \quad N \to \infty. \quad (9.3.10)$$

The proof is straightforward and hence is left as an exercise (see Problem 9.3.1).

We proceed now to discuss some specific RST procedures and make use of Theorem 9.3.1. Consider first the case of a population quantile. We frame the null hypothesis H_0 and the alternatives H_1 and H_2 as in (9.2.25). Also, we define the Y_i as in (9.2.26) and let $r_k = \sum_{i=1}^{k} Y_i$, for $k \geqslant 1$. Note that for every n ($\geqslant 1$), r_n has the binomial distribution with parameters n and $p = F(\xi_p^0)$. Among other possibilities, we consider here the following choices for h_N in (9.3.4):

(i) We let $T_n = (r_n - np)$, $n \geqslant 1$, so that under H_0, $ET_n = 0$ and $\sigma_n^2 = ET_n^2 = np(1-p)$. Thus by (9.3.4) we have $h_N = \max\{(r_k - kp) : 1 \leqslant k \leqslant N\}$ (in the one-sided case) and $\max\{|r_k - kp| : 1 \leqslant k \leqslant N\}$ (in the two-sided case). Since the Y_i are i.i.d.r.v., the invariance principles in Chapter 2 applies to the T_n; hence, under H_0, $h_N/\{Np(1-p)\}^{1/2}$ converges in law to $\sup\{W(t) : 0 \leqslant t \leqslant 1\}$ (in the one-sided case) and $\sup\{|W(t)| : 0 \leqslant t \leqslant 1\}$ (in the two-sided case), where $W = \{W(t), 0 \leqslant t \leqslant 1\}$ is a standard Wiener process on $[0, 1]$. The percentile points of the distribution of these statistics are known (see Section 2.7 and the Tables in Section 5 of the Appendix).

(ii) In this case, we take for T_n the standardized form $(r_n - np)/\{np(1-p)\}^{1/2}$, $n \geqslant 1$, so that in (9.3.4), we have for h_N, $\max\{(r_k - kp)/\{kp(1-p)\}^{1/2} : n_0 \leqslant k \leqslant N$ (in the one-sided case) and $\max\{|r_k - kp|/\{kp(1-p)\}^{1/2} : n_0 \leqslant k \leqslant N\}$ (in the two-sided case). Unlike case (i), here we need to choose n_0 in such a way that $N^{-1}n_0$ converges to some $t_0 > 0$, that is, we initiate the RST only after a specified proportion of the observations are available. Note that here, h_N converges in law to $\sup\{t^{-1/2}W(t) : t_0 \leqslant t \leqslant 1\}$ (in the one-sided case) or $\sup\{t^{-1/2}|W(t)| : t_0 \leqslant t \leqslant 1\}$ (in the two-sided case), where, for the above weak convergence to hold, we need to set $t_0 > 0$. Further, as $t \downarrow 0$, $t^{-1/2}|W(t)|$ bounces indefinitely and, by the law of iterated logarithm, its supremum as $t \downarrow 0$ goes to $+\infty$. For specific values of t_0, the critical points can be determined by reference to the tables by De Long (1981): See the tables in Section 5 of the Appendix.

Consider next repeated rank order tests for location. Frame the null (H_0) and the alternative (H_1 or H_2) hypotheses as in (9.2.29) and S_n and A_n^2 as in (5.2.1) and (5.2.5), respectively. Here also, we may have two cases:

(i) $T_n = S_n$, so that $h_N = \max\{S_n : 1 \leqslant n \leqslant N\}$ (in the one-sided case) or $\max\{|S_n| : 1 \leqslant n \leqslant N\}$ (in the two-sided case), and by the invariance principles developed in Chapter 5, under H_0, $h_N/(N^{1/2}A_N)$ converges in law to $\sup\{W(t) : 0 \leqslant t \leqslant 1\}$ (in the one-sided case) or $\sup\{|W(t)| : 0 \leqslant t \leqslant 1\}$ (in the two-sided case).

(ii) $T_n = S_n/(n^{1/2}A_n)$, $n \geqslant n_0$, where $n_0/N \to t_0 > 0$. Then, h_N is $\max\{n^{-1/2}S_n/A_n : n_0 \leqslant n \leqslant N\}$ (in the one-sided case) and $\max\{n^{-1/2}|S_n|/A_n : n_0 \leqslant n \leqslant N\}$ (in the two-sided case).

Then h_N converges in law in both cases to $\sup\{t^{-1/2}W(t) : t_0 \leqslant t \leqslant 1\}$ and $\sup\{t^{-1/2}|W(t)| : t_0 \leqslant t \leqslant 1\}$, respectively, and we may proceed then as in the case of the sample quantiles.

Next we consider RST for the two-sample location/scale problem and the regression problem. We frame the null hypothesis H_0 and the alternatives H_1, H_2 as in (9.2.41), (9.2.42), and (9.2.46). In each of these cases, we may work with appropriate linear rank statistics, as defined in Section 4.2, and, with the aid of the invariance principles developed in Section 4.3, we can verify Theorem 9.3.1. As in the case of the signed rank statistics, we may consider the two choices of h_N based on the L_n in (4.2.1) or their standardized forms [i.e., $L_n/(A_n C_n)$].

We also consider here the RST for the bivariate independence problem. We frame the null hypothesis H_0 and the alternatives as in Section 9.2.4 and work with the statistics M_n (or Q_n), defined by (6.2.2) and (6.2.8), respectively, and in each case, we may use either the sequence itself or the corresponding normalized ones and either the one-sided or the two-sided case. The invariance principles in Section 6.3 provide the necessary tools for the applicability of Theorem 9.3.1.

As a final example, we consider the multiple regression problem where $X_1, \ldots, X_N$ are independent r.v.'s with d.f.'s $F_1, \ldots, F_N$, respectively, and $F_i(x) = F(x - \beta_0 - \boldsymbol{\beta}'\mathbf{c}_i)$, $i = 1, \ldots, N$. We want to test for the hypothesis $H_0 : \boldsymbol{\beta} = \mathbf{0}$ against $H_1 : \boldsymbol{\beta} \neq \mathbf{0}$. In this case, we consider the sequence of q-vectors $\mathbf{L}_n$, $n \geqslant 1$, defined as in Problem 4.6.2, with $\mathbf{C}_N$ and A_N as defined there. Then, for T_n, we have (as in Problem 4.6.2) $A_n^{-1}(\mathbf{L}_n'\mathbf{C}_n^{-1}\mathbf{L}_n)^{1/2}$, $n \geqslant n_0$, so that if we choose n_0 such that $N^{-1}n_0 \to t_0 > 0$, then for h_N, defined by $\max\{T_n : n_0 \leqslant n \leqslant N\}$, we have under H_0,

$$h_N \xrightarrow{\mathfrak{D}} \sup\left\{t^{-1/2}B(t) : t_0 \leqslant t \leqslant 1\right\}, \tag{9.3.11}$$

where $B = \{B(t),\ t \geqslant 0\}$ is a q-variate Bessel process. As such, Theorem 9.3.1 may be used along with the results in Section 2.7 to provide the asymptotic critical values. See Section 5 of the Appendix.

We can also study the asymptotic power properties of the type A tests for local alternatives under fairly general regularity conditions. For simplicity, we consider the specific case of linear rank statistics and consider the case of alternative hypotheses $\{K_N\}$ contiguous to H_0 in (4.2.3). We assume that $\{K_N\}$ satisfies the hypothesis of Theorem 4.3.5, so that under K_N, in the one-sided case, we have the following:

(i) For $h_N = \max\{L_k/(A_N C_N): 1 \leqslant k \leqslant N\}$, the asymptotic power of the RST is given by

$$P\{W(t) + \mu^*(t) \geqslant D_0^+(\alpha), \quad \text{for some} \quad 0 \leqslant t \leqslant 1\}, \tag{9.3.12}$$

where $D_0^+(\alpha)$ is defined by $P\{W(t) < D_0^+(\alpha) \ \forall t \in [0,1]\} = 1 - \alpha$ and $\mu^* = \{\mu^*(t), t \in [0,1]\}$ is defined as in Theorem 4.3.5.

(ii) For $h_N = \max\{L_k/(A_k C_k): n_0 \leqslant k \leqslant N\}$ (with $n_0/N \to t_0 > 0$), the asymptotic power of the RST is given by

$$P\{t^{-1/2}\{W(t) + \mu^*(t)\} \geqslant \tilde{D}_{t_0}^+(\alpha), \quad \text{for some} \quad t \in [t_0, 1]\}, \tag{9.3.13}$$

where $\tilde{D}_{t_0}^+(\alpha)$ is defined by $P\{t^{-1/2}W(t) \leqslant \tilde{D}_{t_0}^+(\alpha), t \in [t_0, 1]\} = 1 - \alpha$.

Similar expressions are available in the case of the two-sided alternatives. Now, under (4.3.36) through (4.3.41), we conclude that $\mu^*(t)$ is maximized (for every fixed t) when ρ_1, defined by (4.3.40), is equal to 1, that is, the score functions ϕ and ψ agree everywhere, so that by (9.3.12) or (9.3.13) the asymptotic power of the test is also maximized (over the choice of ϕ) for the same choice $\phi = \psi$. A comparison of (9.3.12) and (3.3.13) also reveals the relative efficiency of the two types of test. Unfortunately, though the drift of the Wiener process is the same in both the cases, in one case we have a fixed boundary, whereas in the other we have a square root boundary. Hence the relative picture depends not only on the choice of t_0 but also on the actual form of μ^*. A single numerical measure of efficiency (like the Pitman efficiency) is not applicable in this context. However, some simulation studies made with different μ^* indicate that there is not much difference in the power functions of the two tests when t_0 is small. On the other hand, the RST based on the nonstandardized sequence has the advantage of starting with $n_0 = 1$ and therefore eliminates the need of choosing some arbitrary $t_0 > 0$. Hence we recommend the use of this.

Note that by definition, for the RST, with M $(= M_N)$, defined by (9.3.5), and for the one-sided case of h_N in (9.3.4), for every $m \in [n_0, N-1], [N > m] \equiv [T_k \leqslant h_N^0 \ \forall k: n_0 \leqslant k \leqslant m]$, while $M = N$ when $T_k \leqslant h_N^0$ for every $k: n_0 \leqslant k \leqslant N-1$. Further,

$$\begin{aligned} N^{-1}EM_N &= N^{-1}P\{M = N\} + N^{-1}\sum_{k=n_0}^{N-1} P\{M = k\} \\ &= N^{-1}n_0 + N^{-1}P\{M = N\} + N^{-1}\sum_{k=n_0}^{N-2} P\{M > k\}. \end{aligned} \tag{9.3.14}$$

Thus if $N \to \infty$ and $n_0/N \to 0$, then under $\{K_N\}$, (9.3.14), for the first case (with $h_N = \max\{L_k/(A_N C_N): 1 \leqslant k \leqslant N\}$), we have

$$\lim_{N\to\infty} N^{-1}EM_N = \int_0^1 P\{W(s) + \mu^*(s) \leqslant D_0^+(\alpha),\ \forall 0 \leqslant s \leqslant t\}\, dt. \tag{9.3.15}$$

For the second case with the standardized L_n and with $n_0/N \to t_0 > 0$, it is given by

$$\lim_{N\to\infty} N^{-1}EM_N = t_0 + \int_{t_0}^1 P\left\{s^{-1/2}\{W(s) + \mu^*(s)\} \leqslant \tilde{D}_{t_0}^+(\alpha),\ \forall t_0 \leqslant s \leqslant t\right\} dt. \tag{9.3.16}$$

Now, under (4.3.36) through (4.3.41), for every fixed $t \in [0,1]$, $P\{W(s) + \mu^*(s) \leqslant D_0^+(\alpha),\ \forall 0 \leqslant s \leqslant t\}$ or $P\{s^{-1/2}\{W(s) + \mu^*(s)\} \leqslant \tilde{D}_{t_0}^+(\alpha),\ \forall t_0 \leqslant s \leqslant t\}$ is minimized when $\phi = \psi$ so that, by (9.3.14), (9.3.15), and the above, we conclude that in the light of the smallness of the expected stopping time (relative to N), the asymptotically optimal score function is $\phi = \psi$. Under H_0, $\mu^* = 0$, and $\forall x > 0$, $P\{W(s) \leqslant x,\ \forall 0 \leqslant s \leqslant t\} = 1 - (2/\pi t)^{1/2} \int_x^\infty \exp(-u^2/2t)\, du$; hence (9.3.14) reduces to

$$\lim_{N\to\infty} N^{-1}E_0 M_N = 1 - (2/\pi)^{1/2} \int_{D_0^+(\alpha)}^\infty \int_0^1 t^{-1/2} \exp(-u^2/2t)\, du\, dt. \tag{9.3.17}$$

The parallel expression for (9.3.16) is more complicated and only an infinite series form is available.

Similar results hold for the two-sided alternative case. Also, for the RST procedures for the population quantile, median, location parameter, and the bivariate independence problem, the asymptotic power can be studied in a similar fashion. We pose some of these results as exercises.

9.3.2. Type B Tests

In some situations, the T_n appearing in (9.3.2) are not genuinely distribution-free, so that h_n is also not so. However, for large N, asymptotically distribution-free RST procedures can be constructed. Note that σ_N^2, defined by (9.3.6), may not be specified if the T_N are not distribution-free. We assume that there exists a sequence $\{\tilde{s}_n^2,\ n \geqslant n_0\}$ of estimators such that

$$\tilde{s}_n/\sigma_n \to 1 \quad \text{a.s.,} \quad as \quad n \to \infty. \tag{9.3.18}$$

In fact, to be more flexible, we assume a bit more. We assume that

$$\sigma_n^2 = a(n)\sigma^2, \qquad \tilde{s}_n^2 = a(n)s_n^2, \qquad n \geqslant n_0,$$
$$s_n^2 \to \sigma^2 \quad \text{a.s.,} \quad \text{as} \quad n \to \infty, \tag{9.3.19}$$

where $a(n)$ is a specified function of n. Then in (9.3.7) we define a process $\tilde{Y}_N = \{\tilde{Y}_N(t) = T_{k_N(t)}/(a(N)s^2_{k_N(t)})^{1/2},\ t \in [0, 1]\}$ where the $k_n(t)$ is defined in the equation. Then let

$$\tilde{h}_N = \max_{n_0 \leqslant k \leqslant N} T_k/\left(s_k^2 a(N)\right)^{1/2} \quad \text{or} \quad \max_{n_0 \leqslant k \leqslant N} |T_k|/\left(s_k^2 a(N)\right)^{1/2}. \quad (9.3.20)$$

By virtue of (9.3.18) and (9.3.19), whenever n_0 is so chosen that $N^{-1}n_0 \to t_0 > 0$, the weak convergence of Y_N^0 to Y^0 in (9.3.8) ensures the weak convergence of $\tilde{Y}_N^0 = \{\tilde{Y}_N(t),\ t_0 \leqslant t \leqslant 1\}$ to Y^0, so that Theorem 9.3.1 remains applicable to $\tilde{h}_N$ as well. Thus we may proceed as in Section 9.3.1 with h_N being replaced by $\tilde{h}_N$. For contiguous alternatives, results parallel to (9.3.12) through (9.3.17) hold for these tests as well.

Let us now elaborate this RST procedure for some typical tests based on some statistics studied in Chapters 3, 7, and 8. First, consider the estimable parameter $\theta(F)$, defined by (3.2.1). Suppose that we want to test here for

$$H_0 : \theta(F) = \theta_0 \quad \text{against} \quad H_1 : \theta(F) > \theta_0 \quad \text{or} \quad H_2 : \theta(F) \neq \theta_0. \quad (9.3.21)$$

Let us define $\theta(F_n)$ and U_n as in (3.2.3) and (3.2.4) and let

$$T_k = k^{1/2}(U_k - \theta_0)/m \quad \text{or} \quad k^{1/2}(\theta(F_k) - \theta_0)/m, \qquad k \geqslant m, \quad (9.3.22)$$

where m is the degree of the kernel. For this case, $a(N)$, defined by (9.3.19), is $1 + O(N^{-1})$ and $\sigma^2 = \zeta_1(F)$ is defined by (3.2.12). Further, s_N^2, defined by (3.7.5), converges a.s. to $\zeta_1(F)$ [see (3.7.7)]. Hence type B RST procedures can be based on U-statistics and von Mises functionals. Theorem 3.3.1 and 3.3.2 provide the justifications for the validity of the asymptotic theory for both the null hypothesis and local alternatives situations. The theory applies, in particular, to tests based on cumulative sums of the sample observations where $U_n = \theta(F_n) = n^{-1}\sum_{i=1}^n X_i$ is a special case of U-statistics with $m = 1$, and, further, here $s_n^2 = (n-1)^{-1}\sum_{i=1}^n (X_i - U_n)^2$.

In Chapter 7, we have studied invariance principles for the LCFOS. If we frame suitable null hypotheses and alternatives in terms of the parameters these LCFOS actually estimate, then the invariance principles studied in Sections 7.3 and 7.4, along with the a.s. convergence results on the estimated variance in Section 7.6, provide the justification for adapting Theorem 9.3.1. Hence we may proceed as in the case of U-statistics. Finally, in Chapter 8, we studied invariance principles for a class of M-estimators and some extrema of some sample functions. Here also, the invariance principles studied in Chapter 8 and the results on the a.s. convergence of the estimated variances enable us to adapt Theorem 9.3.1 for the asymptotic simplification of the critical values of $\tilde{h}_N$. Further, for local alternatives, the asymptotic power and expected stopping times can be studied as in (9.3.12) through (9.3.17).

We conclude this section with the following remarks on the relative efficiency of some competing RST procedures. If two or more RST procedures are considered for a common testing problem (e.g., test for the median of a symmetric d.f. where the median test, signed rank test, and tests based on M-estimators of location or LCFOS can be applied in an RST setup), for each of them, the asymptotic power function (for a common sequence of local alternatives) will resemble (9.3.12) or (9.3.13) with some μ^* depending on the specific procedure. A little careful analysis reveals that these μ^* for different tests are really proportional to each other, where the proportionality factor is directly related to the Pitman efficiency of the terminal tests based on the corresponding T_N. For example, consider the problem of testing for the location of a symmetric d.f. and base the RST procedures on (i) $\bar{X}_n = n^{-1}\sum_{i=1}^{n} X_i$ and $s_n^2 = (n-1)^{-1}\sum_{i=1}^{n}(X_i - \bar{X}_n)^2$, $2 \leqslant n \leqslant N$ and (ii) the signed rank statistics S_n, $1 \leqslant n \leqslant N$, defined by (5.2.1). Suppose that in each case, we use the one-sided statistic h_N in (9.3.4) or (9.3.20) and for testing the null hypothesis of the location parameter $\theta = 0$, consider a sequence $\{K_N\}$ of alternative hypotheses, where under K_N, $\theta = N^{-1/2}\lambda$ for some fixed $\lambda \in E$. Next consider two sequences $\{N_1\}$ and $\{N_2\}$ of sample sizes, where

$$N_i/N \to k_i^2, \quad \text{as} \quad N \to \infty, \quad \text{for} \quad i = 1,2, \tag{9.3.23}$$

and base the RST procedure for the parametric case on the target sample size N_1 and the nonparametric case on N_2. Then for the first procedure in (9.3.12) for the drift we have

$$\mu_1^*(t) \frac{k_1 t\lambda}{\sigma}, \quad \text{for} \quad t \in [0,1]; \quad \sigma^2 = \text{Var}(X_1), \tag{9.3.24}$$

while for the signed rank procedure, we have

$$\mu_2^*(t) = k_2 t\lambda \frac{B(F)}{A}, \quad \text{for} \quad t \in [0,1], \tag{9.3.25}$$

where, for the underlying score function $\phi(u) = \phi^*(\frac{1}{2}(1+u))$, $0 < u < 1$,

$$B(F) = \int_{-\infty}^{\infty} (d/dx)\phi^*(F(x))\,dF(x) \quad \text{and} \quad A^2 = \int_0^1 \phi^2(u)\,du. \tag{9.3.26}$$

Thus, looking at (9.3.12), (9.3.24), and (9.3.25), we conclude that the two RST procedures based on the target sample sizes N_1 and N_2 will have the same asymptotic power (for the common sequence of alternatives $\{K_N\}$) if

$$\frac{k_1}{\sigma} = \frac{k_2 B(F)}{A} \quad \text{or} \quad \frac{k_1}{k_2} = \frac{\sigma B(F)}{A}. \tag{9.3.27}$$

In other words, if

$$\lim_{N\to\infty}\left(\frac{N_1}{N_2}\right) = \frac{\sigma^2[B(F)]^2}{A^2}, \tag{9.3.28}$$

then the two procedures based on the target sample sizes N_1 and N_2 have the same asymptotic power function. The same conclusion holds if we use the second type of statistics leading to the asymptotic power in (9.3.13). Further, by virtue of (9.3.15) or (9.3.16) and (9.3.24) through (9.3.28), we conclude that (9.3.28) also ensures that the expected stopping times for these two procedures will be the same when (9.3.28) holds. Now (9.3.28) agrees with the classical Pitman ARE (asymptotic relative efficiency) in the nonsequential case. Therefore, the definition of Pitman ARE is adoptable in the case of RST procedures. However, in equating the asymptotic power functions in (9.3.12) and (9.3.13), the Pitman ARE measure does not serve any useful purpose. Also, if we want to compare an RST procedure with that of a fixed sample size procedure based on the target sample size, the Pitman ARE is not adoptable. In fact, in such a case the two asymptotic power functions cannot be made equal (for all values of the parameter under the alternative) by simply adjusting the relative target sample sizes. Among other possibilities, the adoption of the Bahadur efficiency (in a limiting sense) is possible. Some of these alternative measures are discussed in Sen (1978a) and we pose some of these as exercises. Basically, under fairly general conditions, the limiting Bahadur ARE of an RST procedure with respect to a parallel fixed sample size procedure based on the target sample size is equal to 1, so that by taking into account (9.3.15) or (9.3.16), we may conclude that the RST is asymptotically as efficient as the fixed sample size procedure, but it needs on an average a smaller sample size. However, we need not overemphasize the insensitiveness of the limiting Bahadur ARE in most nonstandard situations. For some related results on likelihood ratio tests, we may refer to So and Sen (1981).

9.4. SEQUENTIAL RANK ORDER AND RELATED TESTS

Let $\{X_i,\ i \geqslant 1\}$ be a sequence of independent r.v.'s, defined on a common probability space $(\Omega, \mathcal{A}, P)$ and as in Section 9.2, we frame $H_0 : P \in \mathcal{P}_0$ versus $H_1 : P \in \mathcal{P}_1 \subset \mathcal{P} \backslash \mathcal{P}_0$. When both H_0 and H_1 are simple hypotheses, we define the sequence of likelihood ratio statistics by

$$\lambda_n = \frac{p_{1n}(\mathbf{X}^{(n)})}{p_{0n}(\mathbf{X}^{(n)})}, \qquad \mathbf{X}^{(n)} = (X_1, \ldots, X_n), \qquad n \geqslant 1, \tag{9.4.1}$$

where p_{1n} and p_{0n} are the densities of $\mathbf{X}^{(n)}$ (with respect to a common measure) under H_1 and H_0, respectively. Then the Wald (1947) SPRT (sequential probability ratio test) is based on $\{\lambda_n\}$ and depends on two constants $a = \log A$ and $b = \log B$ (where $-\infty < b < 0 < a < \infty$), such that one continues drawing observations one by one as long as

$$b < \log \lambda_n < a \qquad (\text{or} \quad B < \lambda_n < A), \qquad n \geqslant 1; \tag{9.4.2}$$

If, for the first time, for $n = N$, $\lambda_N \notin (B, A)$, the experiment is terminated and

$$\begin{cases} H_0 & \text{is accepted if} \quad \lambda_N \leqslant B, \\ H_1 & \text{is accepted if} \quad \lambda_N \geqslant A; \end{cases} \tag{9.4.3}$$

if no such N exists, we let $N = \infty$. N is called a *stopping variable* and is an integer valued r.v. The constants A, B are so chosen that

$$P\{H_0 \quad \text{is rejected} \quad |H_0\} \leqslant \alpha, P\{H_0 \quad \text{is accepted} \quad |H_1\} \leqslant \beta \tag{9.4.4}$$

where $0 < \alpha < 1$ and $0 < \beta < 1$ (we take $0 < \alpha + \beta < 1$) are the type I and II errors of the test. In fact, it is known (see Wald, 1947) that

$$A \leqslant \frac{1-\beta}{\alpha} \quad \text{and} \quad B \geqslant \frac{\beta}{1-\alpha}, \tag{9.4.5}$$

with the approximate equality signs holding under fairly general condtions. If we have any other test for H_0 versus H_1 with type I and II errors α' and β', respectively, and if $\alpha' \leqslant \alpha$, $\beta' \leqslant \beta$, then the Wald-Wolfowitz (1948) optimality of the SPRT asserts that the (expected) sample size for the other test cannot be smaller than that of the SPRT (under both hypotheses). Usually the SPRT results in a considerable reduction of EN over the fixed sample size test of the same strength (α, β).

When H_0 and H_1 are not simple hypotheses, the situation becomes somewhat complicated. Some of these sequential tests are discussed by Wald (1947) and B. K. Ghosh (1970).

Bartlett (1946) initiated a line of attack that was later explored by Cox (1963); their procedure is termed the (asymptotic) sequential likelihood ratio test (SLRT). Basically, the procedure consists in replacing λ_n in (9.4.1) by

$$\lambda_n^* = \frac{p_{1n}^*}{p_{0n}^*}, \qquad n \geqslant n_0, \tag{9.4.6}$$

[where p_{in}^* is the maximum value of $p_{in}(\mathbf{X}^{(n)})$, obtained by substituting the maximum likelihood estimates of the parameters involved, under H_i, for $i = 0$, 1 and n_0 is a moderately large positive integer], and continuing as in (9.4.2) and (9.4.3) with λ_N being replaced by λ_N^*, but starting with the initial sample of size n_0. Essentially, the Wiener process approximations for likelihood ratio processes enable one to show that the properties of the SPRT hold approximately for the SLRT too, when H_0 and H_1 relate to two close alternatives. To clarify this, we consider the following.

Suppose that the X_i are i.i.d.r.v. with a normal d.f. with mean μ and variance $\sigma^2 : 0 < \sigma < \infty$. Let H_0: $\mu = 0$ and $H_1 : \mu = \Delta > 0$. If σ is known, then by (9.4.1)

$$\log \lambda_n = \sigma^{-2}\Delta\left(\sum_{i=1}^{n} X_i - \tfrac{1}{2}n\Delta\right), \qquad n \geqslant 1. \tag{9.4.7}$$

If Δ (specified) is small, $\{\Delta\sum_{i=1}^{n}(X_i - \mu),\ n \geqslant n_0\}$ can be approximated (in law) by a Wiener process, so that in (9.4.2) the boundary crossing probabilities for a (drifted) Wiener process enable us to determine a and b so that (9.4.4) holds. In this setup, if σ is not known, we estimate σ^2 by $s_n^2 = (n-1)^{-1}\sum_{i=1}^{n}(X_i - \bar{X}_n)^2$, $\bar{X}_n = n^{-1}\sum_{i=1}^{n}X_i$ and replace (9.4.7) by

$$\frac{n\Delta\left(\bar{X}_n - \frac{1}{2}\Delta\right)}{s_n^2}, \qquad n \geqslant n_0(\geqslant 2) \tag{9.4.8}$$

where again the a.s. convergence of s_n^2 to σ^2 and the Wiener process approximation for $\{n\Delta(\bar{X}_n - \mu),\ n \geqslant n_0\}$ (for small Δ) provide us with the parallel results. In fact, if one starts with the sequence (9.4.8), the assumption of normality of the underlying d.f. F may be replaced by a less restrictive one that F belongs to the class of all d.f.'s with finite second moments. In Chapter 3 we saw that a general class of U-statistics and von Mises functionals have similar invariance principles and that these may be employed to formulate suitable sequential tests based on these statistics. The statistics studied in Chapters 7 and 8 can be similarly employed. This leads us to formulate the first type of sequential tests applicable to a general class of statistics and valid for a broad class of underlying distributions.

Type C sequential tests. We conceive of a real-valued parameter θ (a functional of the underlying d.f.) and a sequence $\{T_n\}$ of real-valued statistics, where T_n is based on $\mathbf{X}^{(n)} = (X_1, \ldots, X_n)$ and is an estimator of θ, for every $n \geqslant n_0$ ($\geqslant 1$). We frame

$$H_0 : \theta = \theta_0 \qquad \text{versus} \quad H_1 : \theta = \theta_1 = \theta_0 + \Delta, \qquad \Delta > 0, \tag{9.4.9}$$

where θ_0 and Δ are specified. Suppose that T_n has an asymptotic mean square $\sigma_\theta^2/\psi(n)$, where $\psi(n)$ is $\nearrow$ in n with $\lim_{n\to\infty}\psi(n) = \infty$ and σ_θ^2 is a continuous function of θ in some neighborhood of θ_0; ψ is known but σ_θ^2 is unknown. Finally, we consider a sequence $\{s_n^2\}$ of consistent estimators of σ_θ^2, where s_n^2 is based on $\mathbf{X}^{(n)}$, for $n \geqslant n_0$. Define

$$Z_n = \frac{\Delta\psi(n)\left[T_n - \frac{1}{2}(\theta_0 + \theta_1)\right]}{s_n^2}, \qquad n \geqslant n_0. \tag{9.4.10}$$

Then, starting sampling with the initial sample of size n_0, we proceed as in the case of SPRT, where in (9.4.2) and (9.4.3) we replace $\log\lambda_N$ by Z_n. We will show that under fairly general regularity conditions these type C tests possess (for small Δ) the essential properties of the corresponding SPRTs. The invariance principles studied in Part 1 play a vital role in this context.

Type D sequential tests. In the context of rank statistics, studied in Chapters 4, 5, and 6, we have observed that under suitable hypotheses of invariance, these statistics are distribution-free. However, under alternative hypotheses, they are not, and their (asymptotic or exact) means depend on some unknown functionals of the underlying d.f.'s [see (5.4.12) and (5.7.1)]. Keeping this in mind we formulate the following type of sequential tests.

Suppose we have some real-valued parameter (functional of the underlying d.f.) θ and let

$$H_0:\theta=\theta_0 \qquad \text{versus} \quad H_1:\theta=\theta_1=\theta_0+\Delta, \qquad \Delta\,(>0) \text{ is small}, \tag{9.4.11}$$

where, as before, θ_0 and Δ are specified. We conceive of a sequence $\{U_n=U(\mathbf{X}^{(n)});\ n\geqslant n_0\}$ of real-valued statistics possessing the following properties: (a) under H_0, U_n has a known d.f. with mean (or location) 0 (always can be made so by proper translation) and a known variance $\sigma^2/\psi(n)$ where $\psi(n)$ is $\nearrow$ in n, (b) under local alternatives, that is, for the true parameter θ close to θ_0, the asymptotic mean of U_n is $(\theta-\theta_0)\gamma(F)$ where $\gamma(F)$ $(\neq 0)$ is a (possibly unknown) functional of the underlying d.f. F, (c) for every n $(\geqslant n_0)$ and $\theta(\in[\theta_0,\theta_1])$, there exists a transformation $\mathbf{X}^{(n)}\to\mathbf{X}_\theta^{(n)}$, such that when θ is the true parameter, $\mathbf{X}_\theta^{(n)}$ has the same joint d.f. as of $\mathbf{X}^{(n)}$ when $\theta=\theta_0$ $(\Rightarrow U_n(\theta)=U(\mathbf{X}_\theta^{(n)})$ is distribution-free, under $\theta)$, and (d) there exists a sequence $\{D_n\}$ [where $D_n=D(\mathbf{X}^{(n)})$] of consistent estimators of $\gamma(F)$. Let us then define

$$\bar\theta=\tfrac{1}{2}(\theta_0+\theta_1)=\theta_0+\tfrac{1}{2}\Delta \qquad \text{and} \quad \bar U_n=U_n(\bar\theta\,)=U\big(\mathbf{X}_{\bar\theta}^{(n)}\big), \qquad n\geqslant n_0. \tag{9.4.12}$$

Keeping (9.4.8) in mind, we conceive of the following sequence

$$Z_n=\frac{\Delta D_n\psi(n)\bar U_n}{\sigma^2}, \qquad n\geqslant n_0. \tag{9.4.13}$$

Then starting with an initial sample of size n_0, we proceed as in the case of SPRT, replacing $\log\lambda_n$ in (9.4.2) and (9.4.3) by Z_n from (9.4.13). Such sequential procedures are termed type D tests and it will be seen later that they possess the essential properties of SPRTs.

For both type C and D tests, the stopping variable is denoted by N. In the remaining part of this section, we show that $\lim_{n\to\infty} P\{N > n\} = 0$, ensuring the termination of the procedures with probability 1.

THEOREM 9.4.1. *Suppose that when θ holds* (i) $\psi^{1/2}(n)[T_n - \theta]/\sigma_\theta$ *has a limiting* (*nondegenerate*) *continuous d.f.*, (ii) $s_n^2 \xrightarrow{P} \sigma_\theta^2$ *and*, (iii) $\psi(n) \to \infty$ *as* $n \to \infty$. *Then, for every* (*fixed*) θ *and* Δ,

$$P\{N > n \mid \theta, \Delta\} \to 0 \quad \text{as} \quad n \to \infty. \tag{9.4.14}$$

Proof. For every $n \geqslant n_0$ and every θ, $\Delta > 0$,

$$P\{N > n \mid \theta, \Delta\} = P\left\{b < \frac{\Delta\psi(m)\left[T_m - \bar{\theta}\right]}{s_m^2} < a, \qquad \forall n_0 \leqslant m \leqslant n \mid \theta, \Delta\right\}$$

$$\leqslant P\left\{\frac{bs_n^2}{\sigma_\theta \psi^{1/2}(n)} < \frac{\Delta\psi^{1/2}(n)\left[T_n - \bar{\theta}\right]}{\sigma_\theta} < \frac{as_n^2}{\sigma_\theta \psi^{1/2}(n)} \,\middle|\, \theta, \Delta\right\}. \tag{9.4.15}$$

First, for $\theta \neq \bar{\theta}, \psi^{1/2}(n)[T_n - \theta]/\sigma_\theta$ has a limiting d.f., while (a) $\Delta\psi^{1/2}(n) \cdot |\theta - \bar{\theta}|\sigma_\theta \to \infty$ and (b) both $bs_n^2/\sigma_\theta\psi^{1/2}(n)$ and $as_n^2/\sigma_\theta\psi^{1/2}(n)$ converge in probability to 0, so that the rhs of (9.4.15) converges to 0 as $n \to \infty$. Next, for $\theta = \bar{\theta}$, $\psi^{1/2}(n)\,[T_n - \bar{\theta}]/\sigma_\theta$ has a limiting continuous and nondegenerate d.f. while (b) holds. Hence again the rhs of (9.4.15) converges to 0 as $n \to \infty$. Q.E.D.

For type D tests, we may note that when $\theta \neq \bar{\theta}$, $\psi^{1/2}(n)\bar{U}_n/\sigma$ may not have a limiting d.f. We assume that for every θ, there exists a sequence $\{h_n(\theta)\}$ of real numbers such that (a) when $\theta = \bar{\theta}$ holds, $\psi^{1/2}(n)\bar{U}_n/\sigma$ has a nondegenerate and continuous distribution and (b) when $\theta \neq \bar{\theta}$, $\bar{U}_n - h_n(\theta) \xrightarrow{P} 0$ as $n \to \infty$, where

$$\lim|h_n(\theta)| > 0, \qquad \theta \neq \bar{\theta}. \tag{9.4.16}$$

Note that the stochastic convergence of $\bar{U}_n - h_n(\theta)$ (to 0) is implied by the relatively stringent condition that under θ, $\psi^{1/2}(n)[\bar{U}_n - h_n(\theta)]$ has a limiting d.f. where $\psi(n) \to \infty$ as $n \to \infty$; we do not need the later for the following.

THEOREM 9.4.2. *If* (i) $D_n \xrightarrow{P} \gamma(F)$ *and* (ii) $\psi(n) \to \infty$ *as* $n \to \infty$, *then under the conditions* (a) *and* (b) *stated above, for every* θ *and* $\Delta > 0$, (9.4.14) *holds, that is, the process terminates with probability* 1.

Proof. For $\theta = \bar{\theta}$, the proof is analogous to that of the preceding theorem. For $\theta \neq \bar{\theta}$, we have as in (9.4.25)

$$P\{N > n \mid \theta, \Delta\} \leqslant P\left\{ \frac{b\sigma^2}{\Delta D_n \psi(n)} < \bar{U}_n < \frac{a\sigma^2}{\Delta D_n \psi(n)} \middle| \theta \right\} \quad (9.4.17)$$

where both $b\sigma^2/\Delta D_n \psi(n)$ and $a\sigma^2/\Delta D_n \psi(n) \xrightarrow{P} 0$ as $n \to \infty$, while by (9.4.16), $U_n \overset{P}{\sim} h_n(\theta)$ does not converge to 0. Hence the rhs of (9.4.17) converges to 0 as $n \to \infty$. Q.E.D.

9.5. OC FUNCTION OF SROT AND RELATED TESTS

In this section, we are interested in studying the probability of acceptance of the null hypothesis $H_0 : \theta = \theta_0$ when θ is the true parameter value; this is termed the operating characteristic (OC) function of the test. Unlike the case of (9.4.14), the expression for the OC function is generally complicated and depends on the behavior of the sequence $\{Z_n;\ n \geqslant n_0\}$, defined by (9.4.10) or (9.4.13). We are able to make some simplifications, when, in either case, Δ (> 0) is small, so that we are in a position to make use of the invariance principles developed in Part 1. For this reason, for the sake of theoretical justification, in the sequel, we let $\Delta \to 0$; the asymptotic results derived in this way should work out well, in practice, whenever the stipulated Δ is small.

Now, for every fixed θ ($> \theta_0$), as we let $\Delta \to 0$ [in (9.4.9) or (9.4.11)], θ eventually becomes larger than θ_1. On the other hand, for every fixed n, $\Delta\psi(n) \to 0$ as $\Delta \to 0$, so that in (9.4.10) and (9.4.23), Z_n has insignificant role unless n is large. But for large n, by using the stochastic convergence of T_n (or $\bar{U}_n$), it can be shown that for every fixed $\theta > \theta_0$ the OC function converges to 0 as $\Delta \to 0$, and, similarly, for $\theta < \theta_0$ it converges to 1 as $\Delta \to 0$. To avoid this limiting degeneracy of the OC function, we assume that our interest is confined to

$$\theta = \theta_0 + \lambda\Delta \qquad \text{where} \quad \lambda \in J = \{\lambda : |\lambda| \leqslant K\}, \quad (9.5.1)$$

for some $K : 1 < K < \infty$. This setup is quite conceivable: as we make Δ small, we confine our range of alternative hypotheses to a comparable interval of θ around θ_0. OC functions will be studied here only for such local alternatives. Moreover as we let $\Delta \to 0$, for every fixed n, $\Delta\psi(n) \to 0$, and hence it is very unlikely that a final decision will be reached at an early stage. Thus we also allow our initial sample size n_0 to depend on Δ [i.e., we let $n_0 = n_0(\Delta)$] and assume that

$$\lim_{\Delta \to 0} n_0(\Delta) = \infty \qquad \text{but} \quad \lim_{\Delta \to 0} \Delta^2 n_0(\Delta) = 0; \quad (9.5.2)$$

the last assumption is made on the ground that the average sample number (ASN) for our procedures is $O(\Delta^{-2})$, as $\Delta \to 0$ (as will be seen in Section 9.6), and hence, allowing $\Delta^2 n_0(\Delta) \to 0$ as $\Delta \to 0$, we are not affecting the asymptotic form of the ASN.

Let us denote the OC function for θ specified by (9.5.1) by $L(\lambda, \Delta)$. First consider the type C tests. We assume that

$$s_n^2 \to \sigma_\theta^2 \quad \text{a.s.,} \quad \text{as} \quad n \to \infty; \qquad \lim_{\theta \to \theta_0} \sigma_\theta^2 = \sigma^2 = \sigma_{\theta_0}^2 < \infty. \tag{9.5.3}$$

Also, for every $\Delta > 0$, we introduce a sequence or right-continuous, nondecreasing and integer-valued functions $k_\Delta(t) = \max\{k : \Delta^2 \psi(k) \leqslant t\}$, $t \geqslant 0$ and define a stochastic process $W_\Delta = \{W_\Delta(t) = \Delta\psi(k_\Delta(t))[T_{k_\Delta(t)} - \theta]/\sigma, t \geqslant 0\}$. Then we assume that when (9.5.1) holds, for every finite T $(0 < T < \infty)$, as $\Delta \to 0$,

$$\{W_\Delta(t), t \in [0, T]\} \xrightarrow{\mathfrak{D}} \{\xi(t), t \in [0, T]\}, \tag{9.5.4}$$

where $\xi = \{\xi(t), t \geqslant 0\}$ is a standard Wiener process on $[0, \infty)$.

THEOREM 9.5.1. *For type C sequential tests, under* (9.5.1) *through* (9.5.4), *as* $\Delta \to 0$, *for every* $\lambda \in J$,

$$L(\lambda, \Delta) \to L(\lambda) = \begin{cases} (A^{1-2\lambda} - 1)/(A^{1-2\lambda} - B^{1-2\lambda}), & \lambda \neq \frac{1}{2}, \\ a/(a-b), & \lambda = \frac{1}{2}, \end{cases} \tag{9.5.5}$$

where a $(= \log A)$ *and* b $(= \log B)$ *are defined in* (9.4.2) *and* (9.4.5).

Remarks. First note that $L(\lambda)$ depends only on λ, a and b (but not on the underlying d.f.), and hence $L(\lambda, \Delta)$ is asymptotically (as $\Delta \to 0$) distribution-free. Second, suppose that in (9.4.5) we replace the inequality signs by equality signs, so that $A = (1-\beta)/\alpha$ and $B = \beta/(1-\alpha)$. Then

$$L(0) = \frac{A-1}{A-B} = 1 - \alpha \qquad \text{and} \quad L(1) = \frac{B(A-1)}{A-B} = \beta,$$

and hence *asymptotically the test is consistent with strength* (α, β).

Proof of the Theorem. Let us introduce the two events

$$E_1(\Delta) = \{s_m^{-2}\Delta\psi(m)[T_m - \bar{\theta}] \leqslant b \quad \text{for some} \quad m\ (\geqslant n_0(\Delta)) \quad \text{before being} \quad \geqslant a \quad \text{for a smaller} \quad m\} \tag{9.5.6}$$

$$E_2(\Delta, \eta) = \left\{\left|\frac{s_m^2}{\sigma^2} - 1\right| \leqslant \eta \quad \text{for all} \quad m \geqslant n_0(\Delta)\right\}, \tag{9.5.7}$$

where η (> 0) is arbitrary, and let $E_1^c(\Delta)$ and $E_2^c(\Delta, \eta)$ be the complementary events. Further, let P_λ stand for the probability under the setup

$\theta = \theta_0 + \lambda\Delta$. Then, by definition,

$$L(\lambda,\Delta) = P_\lambda\{E_1(\Delta)\}$$
$$= P_\lambda\{E_1(\Delta)E_2(\Delta,\eta)\} + P_\lambda\{E_1(\Delta)E_2^c(\Delta,\eta)\} \quad (9.5.8)$$

where by (9.5.3) and (9.5.7),

$$P_\lambda\{E_1(\Delta)E_2^c(\Delta,\eta)\} \leqslant P_\lambda\{E_2^c(\Delta,\eta)\} \to 0 \quad \text{as} \quad \Delta \to 0. \quad (9.5.9)$$

We designate the stopping variable associated with the type C test based on $\{Z_n\}$ in (9.4.10) by $N(\Delta)$, that is,

$$N(\Delta) = \min\{n \geqslant n_0(\Delta) : Z_n \notin (b,a)\}. \quad (9.5.10)$$

Let us also denote by $Z_n^* = \Delta\psi(n)[T_n - \bar{\theta}]/\sigma^2$, $n \geqslant n_0(s)$, and let

$$N^{(ij)}(\Delta) = \min\left\{n \geqslant n_0(\Delta) : Z_n^* \notin \left(\left(1 + (-1)^i\eta\right)b, \left(1 + (-1)^j\eta\right)a\right)\right\}, \quad (9.5.11)$$

for $i,j = 1,2$ and let $L_{ij}^*(\lambda,\Delta)$ be the OC function of a parallel sequential test based on $\{Z_n^*, n \geqslant n_0(\Delta)\}$ and the boundaries $\{(1 + (-1)^i\eta)b, (1 + (-1)^j\eta)a\}$, $i,j = 1,2$. Then, by (9.5.8) through (9.5.11), we conclude that for every $\epsilon > 0$ and $\eta > 0$, there exists a $\Delta_0(> 0)$, such that for all $0 < \Delta < \Delta_0$,

$$L_{21}^*(\lambda,\Delta) - \epsilon \leqslant L(\lambda,\Delta) \leqslant L_{12}^*(\lambda,\Delta) + \epsilon, \qquad \forall\lambda \in J. \quad (9.5.12)$$

Since ϵ and η are arbitrary, to prove (9.5.5) it suffices to show that

$$\lim_{\eta\to 0}\left\{\lim_{\Delta\to 0} L_{ij}^*(\lambda,\Delta)\right\} = L(\lambda), \qquad \text{for} \quad i,j = 1,2. \quad (9.5.13)$$

Defining the standard Wiener process ξ as in the discussion following (9.5.4), we note that by (2.7.16) and (2.7.17), for $c_1 < 0 < c_2$,

$$\lim_{T\to\infty} P\{\xi(t), t \in [0,T] \text{ first crosses the line } \gamma^{-1}c_1 + \gamma t c_0 \text{ before crossing the line } \gamma^{-1}c_2 + \gamma c_y t\}$$
$$= \begin{cases} (e^{2c_0c_2} - 1)/(e^{2c_0c_2} - e^{2c_0c_1}), & c_0 \neq 0 \\ c_2/(c_2 - c_1), & c_0 = 0. \end{cases} \quad (9.5.14)$$

Note that by the definitions of $k_\Delta(t)$ and Z_n^*,

$$Z_{k_\Delta(t)}^* = \frac{\Delta\psi(k_\Delta(t))\left[T_{k_\Delta(t)} - \theta_0 - \frac{1}{2}\Delta\right]}{\sigma^2}$$
$$= \frac{\Delta\psi(k_\Delta(t))\left[T_{k_\Delta(t)} - (\theta_0 + \lambda\Delta)\right]}{\sigma^2} + \frac{(\lambda - \frac{1}{2})\Delta^2\psi(k_\Delta(t))}{\sigma^2}$$
$$= \frac{1}{\sigma}W_\Delta(t) + \frac{(\lambda - \frac{1}{2})t}{\sigma^2} + o(1), \qquad \forall t \in [0,T]. \quad (9.5.15)$$

Thus, letting $\gamma = \sigma^{-1}$, $c_0 = (\lambda - \frac{1}{2})$, $c_1 = [1 + (-1)^i\eta]b(<0)$ and $c_2 = [1 + (-1)^j\eta]a(>0)$ (where we let $0 < \eta < 1$), we obtain from (9.5.2), (9.5.4), (9.5.14), (9.5.15), and the definition of $L_{ij}^*(\lambda, \Delta)$ that as $\Delta \to 0$, $L_{ij}^*(\lambda, \Delta) \to L_{ij}^*(\lambda)$, given by the rhs of (9.5.14) with the specific values of c_0, c_1, c_2 as mentioned above. Since $\eta(>0)$ is arbitrary, (9.5.13) follows readily from the above by letting $\eta \to 0$. Q.E.D.

Let us now consider the case of type D sequential tests. We have the sequence $\{D_n\}$ of consistent estimators of $\gamma(F)$ and we define $\{Z_n\}$ as in (9.4.13). Parallel to (9.5.4), we define $k_\Delta(t) = \max\{k : \Delta^2\psi(k) \leqslant t\}$, $t \geqslant 0$, and $W_\Delta = \{W_\Delta(t) = \Delta\psi(k_\Delta(t))\sigma^{-1}\overline{U}_{k_\Delta(t)},\ t \geqslant 0\}$. Then we assume that for every finite $T(0 < T < \infty)$, under (9.5.1), as $\Delta \to 0$,

$$\{W_\Delta(t), t \in [0, T]\} \xrightarrow{\mathfrak{D}} \left\{\xi(t) + \frac{(\lambda - \frac{1}{2})t}{\nu}, t \in [0, T]\right\}, \qquad (9.5.16)$$

where $\xi = \{\xi(t), t \geqslant 0\}$ is a standard Wiener process on $[0, \infty)$ and

$$\nu = \sigma/\gamma(F). \qquad (9.5.17)$$

Since σ is known and $\gamma(F)$ is unknown, we assume that

$$\max\{|D_N/\gamma(F) - 1| : \epsilon \leqslant \Delta^2\psi(N) \leqslant T\} \xrightarrow{p} 0, \qquad \text{as} \quad \Delta \to 0, \qquad (9.5.18)$$

where $\epsilon > 0$ and T is defined in (9.5.16).

THEOREM 9.5.2. *For type D sequential tests, under* (9.5.1), (9.5.2), (9.5.16), *and* (9.5.18), *for every* $\lambda \in J$,

$$\lim_{\Delta \to 0} L(\lambda, \Delta) = L(\lambda), \qquad \textit{defined by } (9.5.5). \qquad (9.5.19)$$

The proof is similar to that of Theorem 9.5.1; we need to change only $E_2(\Delta, \eta)$ by $\{|D_m/\gamma(F) - 1| \leqslant \eta, \forall n_0(\Delta) \leqslant m \leqslant n^0(\Delta)\}$ (where $\Delta^2 n^0(\Delta) \geqslant T > \Delta^2[n^0(\Delta) - 1]$), and Z_n^* by $\Delta\sigma^{-2}\gamma(F)\psi(k_\Delta(t))\overline{U}_{k_\Delta(t)}$. Hence the details are omitted.

In the rest of this section, we consider various specific problems of special interest, and in each case examine the applicability of the theorems in Sections 9.4 and 9.5.

(a) ***Sequential test for the population mean: σ known.*** Consider the null hypothesis H_0 (vs. H_1) in (9.2.15) where σ is assumed to be known. Then T_n is defined as in (9.2.16). In this special case, $\psi(n) = n$ and $n^{1/2}(T_n - \mu)/\sigma \sim \mathfrak{N}(0, 1)$ so that Theorem 9.4.1 holds. Also, we do not need (9.5.3), while (9.5.4) holds by the Donsker Theorem, studied in Chapter 2. Hence Theorem 9.5.1 also holds. Both theorems apply to the entire class of d.f.'s for which $0 < \sigma < \infty$.

(b) *Sequential test for the population mean: σ unknown.* In the same problem as in (a), suppose now that σ is unknown. Then s_n^2, defined after (9.2.24), satisfies (9.5.3) and hence both Theorems 9.4.1 and 9.5.1 hold whenever the underlying d.f. has a finite second-order moment.

(c) *Sequential test for a population quantile (median).* Consider the setup of (9.2.25) through (9.2.28). The $E(Y_i \mid H_0) = p_0$ and $V(Y_i \mid H_0) = p_0(1 - p_0)$. If the alternative H_1 is framed in terms of $p = p_1 = p_0 + \Delta$, we may proceed as in (a). On the other hand, if $H_1 : \xi_p = \xi_p^0 + \Delta$, $\Delta > 0$, then we need to assume that F has a continuous p.d.f.f. in some neighborhood of ξ_p^0 and need to estimate $f(\xi_p^0)$. This case will be discussed in detail in (f) where we deal with rank statistics, and precisely, a similar treatment holds.

(d) *Sequential test for a regular functional of a distribution function.* Consider the model of Section 9.2.2 and define T_n as in (9.2.36). Note that s_n^2, defined by (3.7.5), converges a.s. to $\sigma^2 = m^2\zeta_1$ where ζ_1 is defined by (3.7.12), so that (9.5.3) holds. Here also, we have $\psi(n) = n$, so that Theorems 3.3.1 and 3.3.2 ensure that both Theorems 9.4.1 and 9.5.1 hold in this case. Whereas the SPRT is applicable for testing a simple H_0 versus simple H_1 when F is specified, these tests do not demand the form of F to be specified nor that H_0 and H_1 be simple. A broad class of parameters (variance, moments, functions of moments, etc.) can be tested in this way. For some details refer to Sen (1973a).

(e) *Sequential tests for other functionals of distribution functions.* In Chapter 7 [see (7.3.22) and (7.4.2)], in the context of linear functions of order statistics, we have considered some functionals of the d.f. F that may be appropriate for framing suitable hypotheses (e.g., concerning location or scale parameters of a d.f.). In such a case, our sequential tests are based on linear functions of order statistics. By virtue of (7.3.17), Theorems 7.4.2 through 7.4.5, and the proof of (7.4.12) (provided in Section 7.6), we conclude that here also Theorems 9.4.1 and 9.5.1 hold. Similarly, in Chapter 8, we have another class of functionals, expressible as the extrema of some sample functions. Theorems 8.7.1 and 8.7.2 and (8.7.18) ensure that Theorems 9.4.1 and 9.5.1 hold in this case too.

(f) *Sequential rank tests for location.* Let $\{X_i,\ i \geqslant 1\}$ be a sequence of i.i.d.r.v. with an absolutely continuous d.f. $F_\theta(x)$, defined on the real line $(-\infty, \infty)$. It is assumed that $F_\theta(x) = F(x - \theta)$ where F is symmetric about 0, and the null and alternative hypotheses (H_0, H_1) are framed as in (9.4.11). For this sequential testing problem, we intend to employ the signed rank statistics studied in detail in Chapter 5. Since, under H_1, the drift function of the signed rank statistics is not necessarily equal to Δ (but is proportional to Δ for small Δ), we return to type D sequential tests, formulated in Section 9.4.

Let us define $\{S_N\}$ as in (5.2.1) and assume that the underlying score function $\phi(u)$ is $\nearrow$ in $u \in (0, 1)$. Let

$$\gamma(F) = \int_{-\infty}^{\infty} \left(\frac{d}{dx} \right) \phi(F(x))\, dF(x). \tag{9.5.20}$$

Then, note that under $H_1 : \theta = \Delta$, asymptotically, $N^{-1}S_N$ has mean $\Delta\gamma(F)$. Let us define $\{S_N(b)\}$ as in (5.5.1) and note that for nondecreasing ϕ, $S_N(b)$ is $\searrow$ in $b : -\infty < b < \infty$. Also, under H_0, $N^{-1/2}S_N$ has mean 0 and variance $A_N^2 = N^{-1}\sum_{i=1}^N a_N^2(i) \to A^2 = \int_0^1 \phi^2(u)\, du$ (as $N \to \infty$) and $N^{-1/2}A^{-1}S_n \sim \mathfrak{N}(0, 1)$. Thus for every $\alpha \in (0, 1)$ and N there exist an $\alpha_N \leqslant \alpha$ and an $S_{N,\alpha} (> 0)$ such that

$$P\{-S_{N,\alpha} \leqslant S_N \leqslant S_{N,\alpha} \mid H_0\} = 1 - \alpha_N \geqslant 1 - \alpha, \tag{9.5.21}$$

and as $N \to \infty$, $\alpha_N \to \alpha$ and $N^{-1/2}A^{-1}S_{N,\alpha} \to \tau_{\alpha/2}$, where τ_ϵ is the upper $100\epsilon\%$ point of the standard normal d.f. Let

$$\hat{\theta}_{L,N} = \sup\{b : S_N(b) > S_{N,\alpha}\}, \tag{9.5.22}$$

$$\hat{\theta}_{U,N} = \inf\{b : S_N(b) > -S_{N,\alpha}\}. \tag{9.5.23}$$

Then, from (9.5.21) through (9.5.23), we have

$$P\{\hat{\theta}_{L,N} \leqslant \theta \leqslant \hat{\theta}_{U,N} \mid \theta\} = 1 - \alpha_N \geqslant 1 - \alpha, \qquad \forall N \geqslant 1. \tag{9.5.24}$$

By using Theorem 5.5.2 and (9.5.22) and (9.5.23), it follows that both $N^{1/2}(\hat{\theta}_{L,N} - \theta)$ and $N^{1/2}(\hat{\theta}_{U,N} - \theta)$ are $O_p(1)$. Hence if we let

$$D_N = \frac{2S_{N,\alpha}}{N(\hat{\theta}_{U,N} - \hat{\theta}_{L,N})}, \tag{9.5.25}$$

then, by Theorem 5.5.2, $D_N \to \gamma(F)$, in probability, as $N \to \infty$. Hence as in (9.4.13), we define

$$Z_n = \frac{\Delta D_n S_n(\frac{1}{2}\Delta)}{A^2}, \qquad n \geqslant n_0(\Delta), \tag{9.5.26}$$

and the test is based on the sequence $\{Z_n, n \geqslant n_0(\Delta)\}$ with the stopping rule in (9.4.2) and (9.4.3).

Note that Theorem 5.7.1 ensures the a.s. convergence of $n^{-1}S_n(\frac{1}{2}\Delta)$, and hence the stochastic convergence of D_n to $\gamma(F)$ as well as the asymptotic normality of $n^{-1/2}S_n/A$ ensure that Theorem 9.4.2 holds.

For the asymptotic OC function, we note that Theorem 5.3.4 ensures (9.5.16). Also, if we let for $\epsilon > 0$, $0 < T < \infty$, $n_0(\Delta) = [\epsilon\Delta^{-2}]$ and $n^0(\Delta) = [T\Delta^{-2}]$, then, by using Theorem 5.5.3, it can be shown that

$$\max_{n_0(\Delta) \leqslant n \leqslant n^0(\Delta)} \left\{ \max\left[|n^{1/2}(\hat{\theta}_{L,n} - \theta)|, |n^{1/2}(\hat{\theta}_{U,n} - \theta)| \right] \right\} = O_p(1),$$

$$\text{as } \Delta \to 0, \tag{9.5.27}$$

so that (9.5.18) follows by using Theorem 5.5.3. Hence Theorem 9.5.2 holds here.

The problem of difference of locations can be treated similarly. Let $\{X_i, i \geqslant 1\}$ be a sequence of i.i.d.r.v. with an absolutely continuous d.f. F_1 and let $\{Y_i, i \geqslant 1\}$ be an independent sequence of i.i.d.r.v. with an absolutely continuous d.f. F_2, both defined on the real line $(-\infty, \infty)$. It is assumed that $F_2(x) = F_1(x - \theta)$, where θ is the difference of locations. If we let $d_i = Y_i - X_i$, $i \geqslant 1$, then the d_i are i.i.d.r.v. with an absolutely continuous d.f. symmetric about θ. Hence we can construct sequential rank tests based on the d_i as in (9.5.26). Alternatively, we may proceed as in (h), treating this model as a special case of the simple regression model.

(g) ***Sequential M-tests for location.*** Consider the same model as in (f) with the same hypotheses H_0 and H_1. Here we employ the M-estimators of location and the derived estimator of the variance of the M-estimators to construct suitable sequential tests. In (8.6.2), we take the c_i all equal to 1 [so that in (8.2.1) we have the simple location model] and the resulting solution in (8.2.7) and (8.2.8) (the M-estimator) is denoted by $\tilde{\theta}_N$, for $N \geqslant 1$. Also, in (8.3.13), replacing $\hat{\Delta}_N$ by $\tilde{\theta}_N$ and all the c_i by 1, we denote the resulting estimator by s_N^{02}, $N \geqslant 2$. Further, in (8.3.19) and (8.3.20), we replace C_N by $N^{1/2}$ and define $\tilde{\theta}_{NL}$, $\tilde{\theta}_{NU}$, and $L_N = \tilde{\theta}_{NU} - \tilde{\theta}_{NL}$ in the same manner. Let

$$\hat{\nu}_n = \frac{n^{1/2} L_n}{2\tau_{\alpha/2}} \quad \text{and} \quad \nu = \frac{\sigma_0}{\gamma} \tag{9.5.28}$$

be defined by (8.2.24). By (8.3.15), (8.3.18), (8.3.22), and (8.3.23), we have for every $\epsilon > 0$,

$$\max\{|\hat{\nu}_k - \nu| : [n\epsilon] \leqslant k \leqslant n\} \xrightarrow{P} 0, \qquad \text{as} \quad n \to \infty. \tag{9.5.29}$$

Thus, if we define $S_N(t)$ as in (8.2.6), then we may define

$$Z_n = \frac{\Delta S_n\left(\frac{1}{2}(\theta_0 + \theta_1)\right)}{s_n^0 \hat{\nu}_n}, \qquad n \geqslant n_0, \tag{9.5.30}$$

and proceed as in the discussion that follows (9.3.10). Equation (8.2.23) and (9.5.29) ensure that the process terminates with probability 1, while Theorems 8.3.1, 8.3.2, and 8.3.3 and (9.5.29) imply (9.5.4), and Theorem 9.5.1 holds in this case. Note that the class of M-estimators studied in Chapter 8 is a broad one and a class of sequential tests can be generated by using specific members from this class. The sample median and the MLE (for some assumed d.f.) both belong to this class.

(h) ***Sequential rank tests for regression.*** Let $\{X_i, i \geqslant 1\}$ be a sequence of independent r.v.'s and we assume that

$$X_i = \beta_0 + \beta c_i + \epsilon_i, \qquad i \geqslant 1; \qquad \boldsymbol{\beta} = (\beta_0, \beta)', \tag{9.5.31}$$

where the c_i are known regression constants, $\boldsymbol{\beta}$ is an unknown parameter vector and the ϵ_i are i.i.d.r.v. with an absolutely continuous d.f. F, defined on the real line $(-\infty, \infty)$. Our problem is to test

$$H_0 : \beta = 0 \quad \text{versus} \quad H_1 : \beta = \Delta \quad \text{where} \quad \Delta(>0) \quad \text{is specified,} \tag{9.5.32}$$

and for this purpose we like to employ the linear rank statistics, studied in Chapter 4. Here also we develop some type D sequential tests.

Let us define L_N as in (4.2.1), $\gamma(F)$ as in (9.5.20) and assume that ϕ is nondecreasing. We further define the $L_N(b)$ as in (4.5.1), so that by Theorem 4.5.1, $L_N(b)$ is $\searrow$ in $b: -\infty < b < \infty$. Note that under H_0, L_N is a distribution-free statistic with mean 0, variance $A_N^2 C_N^2$, defined by (4.2.6) through (4.2.9), and $A_N^{-1} C_N^{-1} L_N \sim \mathfrak{N}(0,1)$. We define A^2 by (4.3.12), so that $A_N^2 \to A^2$ as $N \to \infty$. Hence for every $N(\geqslant 2)$ and $\alpha \in (0,1)$, there exist an $\alpha_N \leqslant \alpha$ and $(L_{N,\alpha}^{(1)}, L_{N,\alpha}^{(2)})$ such that

$$P\left\{ L_{N,\alpha}^{(1)} \leqslant L_N \leqslant L_{N,\alpha}^{(2)} \mid H_0 \right\} = 1 - \alpha_N \geqslant 1 - \alpha, \tag{9.5.33}$$

where for $N \to \infty$, $\alpha_N \to \alpha$ and $A_N^{-1} C_N^{-1} L_{N,\alpha}^{(i)} \to (-1)^i \tau_{\alpha/2}$, $i = 1, 2$. Let

$$\hat{\beta}_{L,N} = \sup\left\{ b : L_N(b) > L_{N,\alpha}^{(2)} \right\}, \tag{9.5.34}$$

$$\hat{\beta}_{U,N} = \inf\left\{ b : L_N(b) < L_{N,\alpha}^{(1)} \right\}. \tag{9.5.35}$$

Then, from (9.5.33) through (9.5.35), we have

$$P\{ \hat{\beta}_{L,N} \leqslant \beta \leqslant \hat{\beta}_{U,N} \mid \beta \} = 1 - \alpha_N \geqslant 1 - \alpha. \tag{9.5.36}$$

Since, under H_0, $A_N^{-1} C_N^{-1} L_n(0) \sim \mathfrak{N}(0,1)$, by Theorem 4.5.2 and (9.5.34) and (9.5.35), both $N^{1/2} |\hat{\beta}_{L,N} - \beta|$ and $N^{1/2} |\hat{\beta}_{U,N} - \beta|$ are $O_p(1)$, and hence

$$D_N^* = \frac{L_{N,\alpha}^{(2)} - L_{N,\alpha}^{(1)}}{\left\{ C_N^2 (\hat{\beta}_{U,N} - \hat{\beta}_{L,N}) \right\}} \xrightarrow{p} \gamma(F), \quad \text{as} \quad N \to \infty. \tag{9.5.37}$$

Here, for (9.4.13), we take

$$Z_n = \frac{\Delta D_n^* L_n(\frac{1}{2}\Delta)}{A^2}, \qquad n \geqslant n_0, \tag{9.5.38}$$

and proceed as in (9.5.2) and (9.5.3).

Note that when β is the true parameter, $L_N(\frac{1}{2}\Delta)$ has the same distribution as of $L_N(\frac{1}{2}\Delta - \beta)$ under H_0. Thus if $\beta = \frac{1}{2}\Delta$, (9.5.37) and the asymptotic normality of $C_N^{-1} L_N(0)$ ensure the applicability of Theorem 9.4.2. For $\beta > \frac{1}{2}\Delta$, that is, $\beta - \frac{1}{2}\Delta < 0$, as $C_N \to \infty$ with $N \to \infty$, for every $K(<\infty)$, there exists an $N_0 = N_0(k)$ such that $\theta = \beta - \frac{1}{2}\Delta \leqslant -K/C_N$ for every $N \geqslant N_0$, so that by the monotonicity (in b) of $L_N(b), L_N(\theta) \geqslant L_N(-$

$C_N^{-1}K)$, $\forall N \geqslant N_0$. On the other hand, by the asymptotic normality of $L_N(0)$ and Theorem 4.5.2, under $\beta = 0$, $C_N^{-1}L_N(-C_N^{-1}K) \sim \mathfrak{N}(K\gamma(F), A^2)$. Thus, as in (9.4.17), $P\{N > n \mid \beta < \frac{1}{2}\Delta\} \leqslant P\{b[\Delta D_n^* C_n]^{-1} < C_n^{-1}L_n(-C_n^{-1}K) < a[\Delta D_n^* C_n]^{-1} \mid H_0: \beta = 0\} \leqslant P\{C_n^{-1}L_n(-C_n^{-1}K) < a[\Delta D_n^* C_n]^{-1} \mid H_0: \beta = 0\} \to 0$, as $n \to \infty$. A similar case holds for $\theta > 0$. Hence Theorem 9.4.2 holds.

For the asymptotic OC function, we note that (9.5.16) holds under the conditions of Theorem 4.3.5. If we proceed as in the case of sequential rank tests for location, it similarly follows that (9.5.18) holds under the conditions of Theorem 4.5.3. Hence Theorem 9.5.2 holds for the sequential rank tests for regression.

In passing, we may remark that for both the sequential rank tests for location and regression, we have framed the alternative hypotheses H_1 in terms of the natural parameter θ and β. This, in turn, requires the estimation of $\gamma(F)$ (by D_n or D_n^*) at each stage of sampling. One could have avoided this estimation problem by expressing the alternative hypotheses as $H_1: \theta\gamma(F) = \Delta$ (for the location test) and $\beta\gamma(F) = \Delta$ (for the regression case). But $\gamma(F)$ is an unknown functional of the unspecified d.f. F and its inclusion in the specification of H_1 is somewhat awkward and is not practically justifiable. For some sequential rank tests of this type, we may refer to Hall and Loynes (1977). Also, the asymptotic theory developed for rank statistics by Lai (1975a) remains applicable in the case of such sequential tests but needs more refinements in the usual case of tests for location or scale or regression based on rank statistics.

(i) ***Sequential rank tests for independence.*** Let $\{X_i = (X_{1i}, X_{2i}),\ i \geqslant 1\}$ be a sequence of i.i.d.r.v. with a continuous (bivariate) d.f. $F(x, y) - \infty < x, y < \infty$. We wish to test $H_0: F(x, y) = F(X, \infty)F(\infty, y) = F_1(x)F_2(y)$ for every $(x, y) \in$ real plane (i.e., X_{1i} and X_{2i} are stochastically independent). In this context we intend to employ the rank statistics $\{M_N\}$, introduced in (6.2.2). With the definitions of the score functions ϕ_1, ϕ_2 as in (6.3.1) and (6.3.2), we define now

$$\theta(F) = \int_{-\infty}^{\infty} \int_{-\infty}^{\infty} \phi_1(F_1(x))\phi_2(F_2(y))\, dF(x, y). \qquad (9.5.39)$$

Since $\int_0^1 \phi_j(u)\, du = 0$ and $\int_0^1 \phi_j^2(u)\, du = 1$ (by assumption), we have $-1 \leqslant \theta(F) \leqslant +1$, and under $H_0: \theta(F) = 0$. In fact, for monotonic scores, $\theta(F)$ may be termed a generalized grade correlation (see Chapter 8 of Puri and Sen, 1971) and $\theta(F) = 0$ relates to the case of uncorrelation (implied by independence). In testing H_0, we may be interested in an alternative $H_1: \theta(F) = \Delta(>0)$, where Δ is specified. In this setup, we can construct (a type C) sequential test based on $\{M_n,\ n \geqslant n_0\}$, where in (9.4.10) we take $Z_n = \Delta[M_n - \frac{1}{2}n\Delta]/A_n^2 B_n^2$, $n \geqslant n_0$ and the A_n^2, B_n^2 are defined by (6.2.4). In some cases, we may set some structure on the d.f. F, namely,

$$F(x, y) = F_1(x)F_2(y)\{1 + \Delta L(F_1(x), F_2(y))\} \tag{9.5.40}$$

where $L(u, v)$ is defined on the unit square I^2, so that $\Delta = 0$ corresponds to H_0. If we let $L(u, v) = \beta_1(u)\beta_2(v)$ where the β_j are defined as in (6.3.14), (6.3.16), and (6.3.18), then we may frame the alternative hypothesis in terms of Δ. Mixed-rank statistics, considered in Chapter 6, can also be employed in this context.

(j) *Sequential M-tests for regression.* For the regression problem, we may also consider the M-estimators of regression [as in (8.2.7) and (8.2.8)] and base sequential tests on these estimators and their estimated variances. For this, we consider the estimator s_N^{02} in (8.3.13), L_N in (8.3.19) and (8.3.20), and in (9.5.28) we replace $n^{1/2}$ by C_n and define the estimator $\hat{\nu}_n$. Similarly, in (9.5.30), we replace θ_0 and θ_1 by 0 and Δ, respectively, and then proceed as in the text following (9.3.10). Theorems 8.3.1 through 8.3.3 and (8.3.15), (8.3.18), and (8.3.23) ensure the applicability of Theorems 9.4.1 and 9.5.1 for such tests.

9.6. ASN OF SROT AND RELATED TESTS

We intend to study in this section the asymptotic form of the ASN (average sample number) for both types C and D sequential tests. Consider the hypotheses in (9.4.9) or (9.4.11), and for a specified $\Delta(> 0)$, denote the *stopping variable* by $N(\Delta)$, so that

$$N(\Delta) = \inf\{n \geqslant n_0(\Delta) : Z_n \notin (b, a)\}, \tag{9.6.1}$$

where $\{Z_n\}$ is defined by (9.4.10) or (9.4.13). As in Section 9.5, we consider here the asymptotic situation where we define θ by (9.5.1), allow λ to lie in the finite interval J, and let $\Delta \to 0$. We also denote by $E_\theta\{\ldots\}$ or $\tilde{E}_\lambda\{\ldots\}$ and $P_\theta\{\ldots\}$ or $\tilde{P}_\lambda\{\ldots\}$, the expectation and probability under $\theta = \theta_0 + \lambda\Delta$, respectively. Then it can be shown that under fairly general regularity conditions that for every $\lambda \in J$,

$$\lim_{\Delta \to 0} \tilde{E}_\lambda\{N(\Delta)\} = \infty \qquad \text{but} \quad \lim_{\Delta \to 0} \{\Delta^2 \tilde{E}_\lambda(N(\Delta))\} < \infty. \tag{9.6.2}$$

Our interest centers around the existence of the limit of $\Delta^2 E_\lambda\{N(\Delta)\}$ (as $\Delta \to 0$) and we incorporate this limit in the study of the asymptotic relative efficiency (ARE) results of competing tests for a common hypothesis.

Note that the "convergence of the first moment" in (9.6.2), being stronger than the "weak convergence" results of Section 9.5, usually demands comparatively stronger regularity conditions. Also, in (9.4.10) and (9.4.13), we have considered nondecreasing $\psi(n)$ with $\lim_{n\to\infty}\psi(n) = \infty$. In that setup, we will end up with $\lim_{\Delta\to 0}\Delta^2 E_\lambda\{\psi(N(\Delta))\}$, which may be different from $\lim_{\Delta\to 0}\Delta^2 E_\lambda\{N(\Delta)\}$, unless $\psi(x)$ is linear in x (for large x, at least). In fact, if $\psi(x)$ is convex in x, we have $\tilde{E}_\lambda\{\psi(N(\Delta))\} \geqslant \psi(\tilde{E}_\lambda\{N(\Delta)\})$, and we may need additional assumptions to claim that $\lim_{\Delta\to 0}\Delta^2 \cdot \tilde{E}_\lambda\{\psi(N(\Delta))\} = \xi$ ensures a similar limit for $\Delta^2 E_\lambda\{N(\Delta)\}$, when $\psi(n)$ is not linear in n. Further, we have seen in Section 9.5 that in the majority of the cases, $\psi(n) = n$. Hence for the sake of simplicity of presentation, we consider the case of $\psi(n) = n$; the general case of $\psi(n)$ will be treated briefly at the end.

Consider first the case of type C tests. We strengthen here (9.5.3) to the following. We assume that there exists a $\delta > 0$, such that for every $\eta > 0$, there exists a positive constant K and an integer $n^* = n^*(\delta, \eta, K)$, for which

$$P_\theta\left\{|s_n^2/\sigma_\theta^2 - 1| > \eta\right\} \leqslant Kn^{-1-\delta}, \qquad \forall n \geqslant n^*, \tag{9.6.3}$$

where σ_θ^2 and s_n^2 are defined after (9.4.9). Also, keeping in mind the invariance principles considered in Part 1, we assume that when θ obtains, there exists a martingale sequence $\{T_{n,\theta}^* = T_n^*,\ n \geqslant n_0\}$, such that for every θ lying in some neighborhood of θ_0,

$$P_\theta\left\{|n(T_n - \theta) - T_{n,\theta}^*| > e_n\right\} < Kn^{-1-\delta}, \qquad n \geqslant n^*, \tag{9.6.4}$$

where K and δ are defined as in (9.6.3) and

$$n^{-1/2}e_n \to 0 \qquad \text{as} \quad n\to\infty. \tag{9.6.5}$$

For the martingale $\{T_{n,\theta}^*\}$, we assume that $E_\theta(T_{n,\theta}^*) = 0$

$$E_\theta\left(\frac{T_{n,\theta}^{*2}}{n}\right) \to \sigma_\theta^2 \qquad \text{as} \quad n\to\infty; \qquad \lim_{\theta\to\theta_0}\sigma_\theta^2 = \sigma_{\theta_0}^2 = \sigma^2; \tag{9.6.6}$$

for some $r > 2$,

$$n^{-r/2}E|T_{n,\theta}^*|^r \leqslant c_r < \infty, \qquad \forall n \geqslant n_0, \tag{9.6.7}$$

and further,

$$n^{-1/2}\left\{\max_{n_0\leqslant k\leqslant n}|T_{k,\theta}^* - T_{k-1,\theta}^*|\right\} \to 0, \qquad \text{in} \quad L_2\text{-norm}. \tag{9.6.8}$$

Finally, let us define for every $\lambda \in J$,

$$\zeta(\lambda,\sigma^2) = \begin{cases} \left[bL(\lambda) + a\{1 - L(\lambda)\}\right]\sigma^2/(\lambda - \frac{1}{2}), & \lambda \neq \frac{1}{2}, \\ -ab\sigma^2, & \lambda = \frac{1}{2} \end{cases} \tag{9.6.9}$$

where $L(\lambda)$ is defined by (9.5.5), (a,b) by (9.4.2) through (9.4.5), and $\sigma^2 = \sigma_{\theta_0}^2$.

THEOREM 9.6.1. *Under* (9.5.1), (9.5.2), (9.5.4), *and* (9.6.3) *through* (9.6.9), *for type C sequential tests, when* $\psi(n) = n$,

$$\lim_{\Delta\to 0}\left[\Delta^2\tilde{E}_\lambda\{N(\Delta)\}\right] = \zeta(\lambda,\sigma^2), \qquad \forall\lambda \in J. \tag{9.6.10}$$

Proof. First, we consider the case of $\lambda \neq \frac{1}{2}$. For an arbitrary $\epsilon > 0$, we define

$$n_1(\Delta) = \max\{n : \Delta^2 n \leqslant \epsilon\} \qquad \text{and} \quad n_2(\Delta) = \min\{n : \Delta^2 n > \epsilon^{-1}\}. \quad (9.6.11)$$

Then,

$$\Delta^2 \tilde{E}_\lambda N(\Delta) = \Delta^2 \left[\sum_{n \leqslant n_1(\Delta)} + \sum_{n_1(\Delta) < n \leqslant n_2(\Delta)} + \sum_{n > n_2(\Delta)} n\tilde{P}_\lambda\{N(\Delta) = n\} \right], \tag{9.6.12}$$

where by (9.6.11), $\Delta^2 \sum_{n \leqslant n_1(\Delta)} n\tilde{P}_\lambda\{N(\Delta) = n\} \leqslant \epsilon \tilde{P}_\lambda\{N(\Delta) \leqslant n_1(\Delta)\} \leqslant \epsilon$. Note also that

$$\sum_{n > k} n\tilde{P}_\lambda\{N(\Delta) = n\} = k\tilde{P}_\lambda\{N(\Delta) \geqslant k\} + \sum_{n > k} \tilde{P}_\lambda\{N(\Delta) > n\} \qquad \forall k \geqslant n_0(\Delta). \tag{9.6.13}$$

Further, by (9.6.3) and the continuity of σ_θ^2 around $\theta = \theta_0$, for Δ sufficiently small [so that $n_1(\Delta) \geqslant n^*$ and $|\sigma_\theta^2/\sigma^2 - 1| < \eta \; \forall \lambda \in J$],

$$\begin{aligned} \tilde{P}_\lambda\{N(\Delta) > n\} &\leqslant \tilde{P}_\lambda\left\{bs_n^2 < \Delta n\left[T_n - \tfrac{1}{2}(\theta_0 + \theta_1)\right] < as_n^2\right\} \\ &\leqslant \tilde{P}_\lambda\left\{b\sigma^2(1 + 2\eta) < \Delta n\left[T_n - \theta_0 - \lambda\Delta + (\lambda - \tfrac{1}{2})\Delta\right] < a\sigma^2(1 + 2\eta)\right\} \\ &\quad + O(n^{-1-\delta}). \end{aligned} \tag{9.6.14}$$

Since $\lambda \neq \frac{1}{2}$, both $b\sigma^2(1 + 2\eta)/\Delta n^{1/2}$ and $a\sigma^2(1 + 2\eta)/\Delta n^{1/2} \to 0$ as $n \to \infty$, $\Delta n^{1/2}|\lambda - \frac{1}{2}| \geqslant \epsilon^{-1/2}|\lambda - \frac{1}{2}| > 0$, $\forall n \geqslant n_2(\Delta)$, and, by (9.6.7), for every $c > 0$ and $n \geqslant n_2(\Delta)$,

$$\begin{aligned} P_\theta\{|T_{n,\theta}^*| > c\Delta n\} &= P_\theta\left\{|T_{n,\theta}^*| > \left[c\Delta n_2(\Delta)\right]\left[\frac{n}{n_2(\Delta)}\right]\right\} \\ &\leqslant c^{-r} c_r \epsilon^{-r/2}\left[n_2(\Delta)\right]^{r/2} n^{-r/2} \\ &= O(n^{-r/2}) \cdot \left[O\left(\left[n_2(\Delta)\right]^{r/2}\right)\right], \qquad r > 2, \end{aligned} \tag{9.6.15}$$

we obtain from (9.6.4), (9.6.5), (9.6.14), and (9.6.15) that for $n \geqslant n_2(\Delta) - 1$,

$$\begin{aligned} \tilde{P}_\lambda\{N(\Delta) > n\} &\leqslant P_\theta\left\{b\sigma^2(1 + 2\eta)/\Delta\sqrt{n} - \Delta\sqrt{n}\,(\lambda - \tfrac{1}{2}) < n^{-1/2}T_{n,\theta}^*\right. \\ &\qquad \left. < a\sigma^2(1 + 2\eta)/\Delta\sqrt{n} - \Delta\sqrt{n}\,(\lambda - \tfrac{1}{2})\right\} \qquad (\theta = \theta_0 + \Delta\lambda) \\ &= O\left((n^{-r/2}) \cdot \left[n_2(\Delta)\right]^{r/2}\right), \end{aligned} \tag{9.6.16}$$

where $\Delta^2[n_2(\Delta)]^{r/2} \leqslant \epsilon^{-r/2}\Delta^{2-r/2} \to 0$ as $\Delta \to 0$ (where we choose $r < 4$). Hence by (9.6.3), (9.6.13), and (9.6.16), it follows that

$$\Delta^2 \sum_{n > n_2(\Delta)} n\tilde{P}_\lambda\{N(\Delta) = n\} \to 0 \qquad \text{as} \quad \Delta \to 0. \tag{9.6.17}$$

Suppose now that in (9.4.10) we replace Z_n by $\tilde{Z}_{n,i}(\lambda) = \Delta[T^*_{n,\theta_0+\lambda\Delta} + n(\lambda - \frac{1}{2})\Delta]/\sigma^2(1 + (-1)^i\eta)$, $n \geqslant n_0(\Delta)$ and denote the corresponding stopping variable by $\tilde{N}_i(\Delta)$, for $i = 1,2$. Then, as above, it follows that as $\Delta \to 0$

$$\Delta^2\left[\sum_{n \leqslant n_1(\Delta)} + \sum_{n > n_2(\Delta)} n\tilde{P}_\lambda\{\tilde{N}_i(\Delta) = n\}\right] \to 0, \qquad \text{for} \quad i = 1,2. \tag{9.6.18}$$

On the other hand, as in (9.6.13), for each i $(= 0, 1, 2)$,

$$\begin{aligned}\sum_{n_1(\Delta) < n \leqslant n_2(\Delta)} n\tilde{P}_\lambda\{\tilde{N}_i(\Delta) = n\} &= (n_1(\Delta) + 1)\tilde{P}_\lambda\{\tilde{N}_i(\Delta) > n_1(\Delta)\} \\ &\quad + \sum_{n_1(\Delta) < n \leqslant n_2(\Delta)} \tilde{P}_\lambda\{\tilde{N}_i(\Delta) > n\} \\ &\quad - n_2(\Delta)\tilde{P}_\lambda\{\tilde{N}_i(\Delta) > n_2(\Delta)\},\end{aligned} \tag{9.6.19}$$

where we let $N(\Delta) = \tilde{N}_0(\Delta)$. Hence by using (9.6.3), (9.6.4), (9.6.5), (9.6.11), and (9.6.16), we obtain that as $\Delta \to 0$,

$$\begin{aligned}\Delta^2 \sum_{n_1(\Delta) < n \leqslant n_2(\Delta)} n\tilde{P}_\lambda\{\tilde{N}_1(\Delta) = n\} &- O\left(\left[n_1(\Delta)\right]^{-\delta}\right) \\ \leqslant \Delta^2 \sum_{n_1(\Delta) < n \leqslant n_2(\Delta)} & n\tilde{P}_\lambda\{N(\Delta) = n\} \\ \leqslant \Delta^2 \sum_{n_1(\Delta) < n \leqslant n_2(\Delta)} & n\tilde{P}_\lambda\{\tilde{N}_2(\Delta) = n\} + O\left(\left[n_1(\Delta)\right]^{-\delta}\right),\end{aligned} \tag{9.6.20}$$

where

$$\left[n_1(\Delta)\right]^{-\delta} \sim \epsilon^{-\delta}\Delta^{2\delta} \to 0 \qquad \text{as} \quad \Delta \to 0. \tag{9.6.21}$$

Hence it suffices to show that

$$\begin{aligned}\lim_{\eta \to 0}\left[\lim_{\Delta \to 0} \Delta^2 \sum_{n_1(\Delta) < n \leqslant n_2(\Delta)} n\tilde{P}_\lambda\{\tilde{N}_i(\Delta) = n\}\right] \\ = \zeta(\lambda, \sigma^2), \qquad \forall \lambda(\neq \tfrac{1}{2}) \in J.\end{aligned} \tag{9.6.22}$$

Under (9.5.1), (9.5.2), (9.5.4), (9.6.3), and (9.6.4), Theorem 9.5.1 also applies to $\{\tilde{Z}_{n,i}(\lambda),\, n \geqslant n_0(\Delta)\}$, and for $\eta(>0)$ arbitrarily close to 0, the OC function of the corresponding sequential procedure [based on $\{\tilde{N}_i(\Delta)\}$] will be arbitrarily close to $L(\lambda)$, $\lambda \in J$ (as $\Delta \to 0$), for $i = 1,2$. Further, (9.6.8) ensures the asymptotic (as $\Delta \to 0$) negligibility of the excess (of $\tilde{Z}_{\tilde{N}_i(\Delta),i}(\lambda)$) over the boundaries a or b, for $n_1(\Delta) \leqslant \tilde{N}_i(\Delta) \leqslant n_2(\Delta)$, $i = 1,2$. Hence for $\eta(>0)$ arbitrarily small, for each i $(= 1,2)$,

$$E_\lambda Z_{\tilde{N}_i(\Delta),i}(\lambda) \to aL(\lambda) + b\{1 - L(\lambda)\} \qquad \text{as} \quad \Delta \to 0. \tag{9.6.23}$$

On the other hand, since $\{T^*_{n,\theta}\}$ is a martingale, we obtain that $\{\tilde{Z}_{n,i}(\lambda) - n\Delta^2(\lambda - \frac{1}{2})/\sigma^2(1 + (-1)^i\eta),\, n \geqslant n_0\}$ is also a martingale. Also, by using

(9.6.15) and (9.6.16), it follows that

$$\liminf_{n\to\infty} \int_{[N>n]} |\tilde{Z}_{N,i}(\lambda)|\,d\tilde{P}_\lambda = 0, \qquad i=1,2. \tag{9.6.24}$$

Thus using the Wald lemma for martingales [see (2.2.12) and (2.2.13)], we obtain from above that as $\Delta\to 0$, $\eta\to 0$

$$\left\{\Delta^2(\lambda-\tfrac{1}{2})/\sigma^2\left(1+(-1)^i\eta\right)\right\}\tilde{E}_\lambda \tilde{N}_i(\lambda) \to bL(\lambda)+a\{1-L(\lambda)\} \tag{9.6.25}$$

for $i=1,2$, and by (9.6.22) and (9.6.25), the proof of (9.6.10) for $\lambda\neq\frac{1}{2}$ is complete.

For $\lambda=\frac{1}{2}$, in (9.6.16), both the upper and lower bounds for $n^{-1/2}T^*_{n,\theta}$ converge to 0, so that the rhs of (9.6.16) may not be $0(n^{-r/2})$, and hence the current proof may not work out properly. However, we may choose a sequence $\{\lambda_r\}$, say $\lambda_r=\frac{1}{2}+(-1)^r/r$, $r\geqslant 1$, follow the foregoing proof for each λ_r, and then arrive at (9.6.10) by using the L'Hospital rule. Alternatively, we may also use the Wald second lemma for submartingales and derive (9.6.10) for $\lambda=\frac{1}{2}$.

For every $\psi(n)$, not necessarily equal to n, we make certain additional assumptions. First, we assume that for every $r>1$,

$$\sum_{n\geqslant n_0} \frac{\{\psi(n+1)-\psi(n)\}}{[\psi(n)]^r} < \infty \qquad \text{where} \quad n_0=\inf\{n:\psi(n)>0\}. \tag{9.6.26}$$

Second, in (9.6.6) and (9.6.7), we replace n by $\psi(n)$ and in (9.6.3) and (9.6.4), we replace $n^{-1-\delta}$ by $[n\vee\psi(n)]^{-1-\delta}$ and $n(T_n-\theta)$ by $\psi(n)(T_n-\theta)$.Then, by the same line of proof, we find that

$$\lim_{\Delta\to 0}\left[\Delta^2\tilde{E}_\lambda\{\psi(N(\Delta))\}\right]=\zeta(\lambda,\sigma^2), \qquad \forall\lambda\in J. \tag{9.6.27}$$

But (9.6.27) fails to provide the limit for $\Delta^2\tilde{E}_\lambda N(\Delta)$ when $\psi(n)\neq n$.

We proceed to type D tests now. We replace (9.5.18) by the following: for some $\delta>0$ and every $\eta>0$, there exist a positive constant $K(<\infty)$ and an integer $n^*=n^*(\delta,\eta,K)$ such that

$$P_\theta\left\{\left|\frac{D_n}{\gamma(F)}-1\right|>\eta\right\}\leqslant Kn^{-1-\delta}, \qquad \forall n\geqslant n^*, \tag{9.6.28}$$

where, again, we consider the case of $\psi(n)\equiv n$. Second, we assume that for every $\theta=\theta_0+\lambda\Delta$, $\lambda\in J$, when θ obtains, there exists a martingale sequence $\{U^*_{n,\lambda}, n\geqslant n_0\}$, such that

$$P_\lambda\left\{|n\{\bar{U}_n-h_n(\theta)\}-U^*_{n,\lambda}|>e_n\right\}\leqslant Kn^{-1-\delta}, \qquad \forall n\geqslant n^*, \tag{9.6.29}$$

where K and δ are defined as in (9.6.28), $\{e_n\}$ satisfies (9.6.5) and $h_n(\theta)$ is defined as in the text preceding (9.4.16); $U^*_{n,\lambda}$ has expectation 0, and we assume that (9.6.7) and (9.6.8) hold for $U^*_{n,\lambda}$ replacing $T^*_{n,\theta}$. Finally, we assume that as $\Delta \to 0$, for $\theta = \theta_0 + \lambda\Delta$,

$$\max\{|h_n(\theta) - \Delta(\lambda - \tfrac{1}{2})\gamma(F)| : n_1(\Delta) \leqslant n \leqslant n_2(\Delta)\} = o(\Delta), \quad (9.6.30)$$

where the $n_i(\Delta)$ are defined in (9.6.11). Then we have the following.

THEOREM 9.6.2. *For type D sequential tests, under* (9.4.16), (9.5.1), (9.5.2), (9.5.16), *and* (9.6.28) *through* (9.6.30), $\psi(n) \equiv n$,

$$\lim_{\Delta \to 0} \{\Delta^2 \tilde{E}_\lambda N(\Delta)\} = \zeta(\lambda, \nu^2), \qquad \forall \lambda \in J, \quad (9.6.31)$$

where ν^2 *is defined by* (9.5.17).

The proof runs almost parallel to that of Theorem 9.6.1 and hence is left as an exercise. A result parallel to (9.6.27) also holds here for general $\psi(n)$.

In the rest of this section, we consider the various specific tests treated in Section 9.5 and study their asymptotic ASN functions. For the sequential tests for the population mean, treated in (a) and (b) of Section 9.5, σ_θ^2 is the population variance and (9.6.3) holds when F has a finite moment of order $r > 2$, $\{n(T_n - \theta)\}$ is itself a martingale so that (9.6.4) and (9.6.5) hold with $e_n \equiv 0$ and $V(n(T_n - \theta)) = \sigma_\theta^2$ while (9.6.7) follows from the existence of the rth moment, for some $r > 2$; the existence of the variance ensures (9.6.8). Hence (9.6.10) holds. The case of the test for the population quantile in (c) follows on the same line with $\sigma^2 = p_0(1 - p_0)$ or $\nu^2 = \sigma^2/f^2(\xi_{p_0})$ according as we apply the type C or type D test. For sequential tests for functionals of d.f.'s, treated in (d) and (e) of Section 9.5, (9.6.3) through (9.6.8) follow from the results studied in detail in Chapters 3, 7, and 8. We leave some of these computations as exercises.

Let us next consider the sequential rank tests for location, for which we need to verify Theorem 9.6.2. In this case, verification of (9.6.30) poses the least problem; since $\Delta^2 n_2(\Delta) \sim \epsilon^{-1}$, $\Delta^2 n_1(\Delta) \sim \epsilon$, both $n_1(\Delta)$ and $n_2(\Delta) \to \infty$ as $\Delta \to 0$, and hence we may use Theorems 5.4.2 and 5.7.1 and some routine analysis to prove (9.6.30). By virtue of the remark at the end of Section 5.4, (9.6.29) follows from Theorem 5.4.2. Verification of (9.6.28) is more involved. First, by Theorem 5.4.2 and the remark at the end of Section 5.4, it can be shown that by definitions in (9.5.22) and (9.5.23), for every $\eta > 0$, there exist a $\delta > 0$, positive numbers K_1, K_2, and an integer n^* such that for every $n \geqslant n^*$,

$$P\{\theta - n^{-1/2}\tau_{\alpha/2}A/\gamma(F) - K_1 n^{-1/2+\eta} < \hat{\theta}_{L,n} < \tilde{\theta}_{U,n}$$
$$< \theta + n^{-1/2}\tau_{\alpha/2}A/\gamma(F) + K_1 n^{-1/2+\eta}\} \geqslant 1 - K_2 n^{-1-\delta}. \quad (9.6.32)$$

Second, the proof of Theorem 5.5.4 can be amended to cover the case of $|b| \geqslant cn^{\eta}$, $c > 0$, $n(> 0)$ arbitrarily small, and the a.s. convergence result can be replaced by a probability $\geqslant 1 - O(n^{-1-\delta})$; Problems 9.6.4 and 9.6.5 are set to verify these details. As such, (9.6.28) follows in this case. Hence (9.6.30) holds with $\nu^2 = A^2/\gamma^2(F)$.

Let us next consider the sequential rank tests for regression treated in (g) of Section 9.5. In this case, $\psi(n) = C_n^2$, defined by (4.2). Thus if we assume that $n^{-1}C_n^2 \to C_0^2$ (known) as $n \to \infty$, when $\psi(n)$ is asymptotically linear in n; this case usually arises in the two-sample location/scale problems and also in a variety of simple regression problems. Under the above condition, one may proceed as in the case of sequential rank tests for location and obtain that (9.6.30) holds with $\nu^2 = A^2/[\nu^2(F)C_0^2]$. On the other hand, if $n^{-1}C_n^2$ does not converge to a limit, one needs additional regularity conditions on the $\{c_i\}$ as well as on the score functions, and we may refer to Ghosh and Sen (1977) for details.

For sequential rank tests for independence, treated in (h) of Section 9.5, we have type C tests, and verification of (9.6.4) through (9.6.8) can be made by using (6.5.6) and replacing the a.s. convergence result in (6.5.7) by a statement involving a probability (for a given n) greater than $1 - O(n^{-1-\delta})$, for some $\delta > 0$. The proof follows along the same line as in the case of Theorem 6.5.1, and hence we omit the details; refer to Bönner, Müller-Funk, and Witting (1979) for related results.

Finally, for sequential M-tests for location or regression, we note that whenever $E|\psi(X_i - \Delta c_i)|^r < \infty$, for some $r > 4$, the Markov inequality yields a power rate of convergence for s_N^2 to σ_0^2, so that by (8.4.20), we have a similar statement involving $s_N^{02} - \sigma_0^2$. Also, (8.4.22) and Theorem 8.4.1 provide the tools for verifying (9.6.4) through (9.6.8). Hence Theorem 9.6.1 holds for these tests as well.

9.7. ASYMPTOTIC EFFICIENCY OF SROT AND RELATED TESTS

By virtue of Theorems 9.5.1 and 9.5.2, the asymptotic OC function of a general class of type C and type D sequential tests is independent of the particular sequence $\{Z_n\}$ and this common form permits us to compare two or more competing tests for a common hypothesis testing problem. A very natural criterion for this comparison is the ASN, and Theorems 9.6.1 and 9.6.2 provide us with the desired results.

Suppose that for testing a null hypothesis H_0 versus an alternative H_1 [see (9.4.9) or (9.4.11)], we have two competing sequential tests based on two sequences of test statistics, say $\{Z_n^{(1)}\}$ and $\{Z_n^{(2)}\}$ and we denote the corresponding stopping variables by $\{N^{(1)}(\Delta)\}$ and $\{N^{(2)}(\Delta)\}$, respectively.

We assume that for both the tests, the asymptotic OC is given by $L(\lambda)$, $\lambda \in J$, defined in (9.5.5). We further assume that

$$\lim_{\Delta \to 0} \Delta^2 \tilde{E}_\lambda N^{(i)}(\Delta) = \zeta_i, \qquad i = 1, 2 \quad \text{both exist.} \tag{9.7.1}$$

Then the asymptotic relative efficiency (ARE) of the first procedure relative to the second one is defined as

$$e_{12} = \lim_{\Delta \to 0} \left\{ \frac{\tilde{E}_\lambda N^{(2)}(\Delta)}{\tilde{E}_\lambda N^{(1)}(\Delta)} \right\} = \frac{\zeta_2}{\zeta_1}. \tag{9.7.2}$$

From (9.6.9), (9.6.10) and (9.7.2), it follows that for both the tests being of type C, the ARE is equal to

$$e_{12} = \frac{\sigma_2^2}{\sigma_1^2}, \tag{9.7.3}$$

where σ_i^2 is the variance function appearing in $\{Z_n^{(i)}\}$, $i = 1, 2$. Similarly, for both being type D tests, we have

$$e_{12} = \frac{\nu_2^2}{\nu_1^2}, \tag{9.7.4}$$

and for the first one being type D and the second on being type C test we have

$$e_{12} = \frac{\sigma_2^2}{\nu_1^2}. \tag{9.7.5}$$

The last three expressions are useful for studying e_{12} in various specific problems. For example, suppose that θ is the location parameter of a (symmetric) d.f. and we frame H_0 and H_1 as in (9.4.9) with $\theta_0 = 0$. If θ is the only parameter involved in the d.f. F and the form of F is completely specified, then the SPRT is applicable to this problem. We denoting the corresponding stopping variable by $N^*(\Delta)$ and obtain by parallel arguments that

$$\lim_{\Delta \to 0} \Delta^2 \tilde{E}_\lambda N^*(\Delta) = \zeta\left(\lambda, \left[I(f)\right]^{-1}\right), \qquad \forall \lambda \in J. \tag{9.7.6}$$

where $I(f)$ is the Fisher information of F. Consider next the test in (a), (b) of Section 9.5, based on the sequence of sample means; here σ^2 = population variance of F. Then the ARE of this test with respect to the SPRT is given by (9.7.3) as

$$\frac{1}{\sigma^2 I(f)} \leqslant 1, \qquad \forall F, \tag{9.7.7}$$

(by the Rao-Cramér inequality), where the equality sign holds for certain exponential d.f.'s including the normal d.f. as a particular case. Similarly, the ARE of the sequential median test with respect to the mean test is

$$\sigma^2 \Big/ \frac{1}{4f^2(0)} = 4\sigma^2 f^2(0), \tag{9.7.8}$$

agrees with the Pitman ARE of the sign test with respect to Student's t-test. The ARE of the sequential rank test with respect to the mean test is given by

$$\sigma^2 \Big/ \left[\frac{A^2}{\gamma^2(F)} \right] = \frac{\sigma^2 \gamma^2(F)}{A^2}, \tag{9.7.9}$$

which also agrees with the Pitman ARE in the nonsequential case. As such, the optimal score function in this respect is $\phi^*(u) = -f'(F^{-1}(u)) / f(F^{-1}(u))$, $0 < u < 1$, in which case (9.7.9) is the reciprocal of (9.7.7) and is $\geqslant 1$. Also, for normal scores (9.7.9) is always $\geqslant 1$, with the equality sign holding only when F is also normal. Similar results hold for the regression problem, and we pose some of these as problems at the end of the chapter. We conclude this chapter with some remarks on the Pitman ARE in the sequential case where we need not verify (9.7.1) and the result follows under less restrictive conditions; this study is due to Sen and Ghosh (1980).

Let $\{T_n\}$ and $\{T_n^*\}$ be two sequences of real-valued statistics which are employed as in (9.4.10) or (9.4.13) for testing H_0 versus H_1 in (9.4.9). Assume that for both these tests, under the regularity conditions of Section 9.5, Theorem 9.5.1 or 9.5.2 holds and in this context, (9.5.4) or (9.5.16) is assumed to hold. Then, for both the procedures, the asymptotic OC function $L(\Delta)$ is the same and is defined by (9.5.5). We denote the corresponding stopping variables by $N(\Delta)$ and $N^*(\Delta)$, respectively. Assume that for $\theta = \theta_0 + \lambda\Delta$, as $\Delta \to 0$,

$$\Delta^2 \psi(N(\Delta)) \xrightarrow{\mathfrak{D}} \tau_\lambda \quad \text{and} \quad \Delta^2 \psi(N^*(\Delta)) \xrightarrow{\mathfrak{D}} \tau_\lambda^*, \tag{9.7.10}$$

where $\psi(n)$ is defined as in the discussion proceding (9.5.4) and τ_λ and τ_λ^* are nonnegative r.v.'s with d.f.'s

$$P_\lambda(t) = P\{\tau_\lambda \leqslant t\} \quad \text{and} \quad P_\lambda^*(t) = P\{\tau_\lambda^* \leqslant t\}, \qquad t \in [0, \infty). \tag{9.7.11}$$

Further, we assume that for any sequence $\{c_n\}$ of positive numbers with $\lim_{n\to\infty} c_n = c$ $(0 < c < \infty)$ ensures that

$$\lim_{n\to\infty} \left\{ \frac{\psi(nc_n)}{\psi(n)} \right\} = s(c) \text{ exists and is} \uparrow \text{in } c \in (0, \infty). \tag{9.7.12}$$

Note that if, in particular, $\psi(n) \equiv n$ (as is usually the case), then $s(c) \equiv c$.

Finally, suppose that there exists a positive number h, such that

$$P_\lambda^*(t) = P_\lambda(ht), \qquad \forall t \in [0,\infty) \quad \text{and every} \quad \lambda \in J, \tag{9.7.13}$$

where J is defined in (9.5.1). Then we have the following.

THEOREM 9.7.1. *Suppose that* $\{T_n\}$ *(and* $\{T_n^*\}$*) satisfy* (9.5.4) *with scale factors* σ *(and* σ^**) or* (9.5.16) *with scale factors* ν *(and* ν^**). Then, under* (9.5.1), (9.5.2), *and* (9.7.12), (9.7.13) *holds, and hence the asymptotic relative efficiency of* $\{T_n^*\}$ *with respect to* $\{T_n\}$ *is given by*

$$e(T^*,T) = s^{-1}\left(\left(\frac{\delta^*}{\delta}\right)^2\right), \tag{9.7.14}$$

where δ *(and* δ^**) are given by* σ *(and* σ^**) or* ν *(and* ν^**) according as the sequential procedures based on* $\{T_n\}$ *(and* $\{T_n^*\}$*) are of type C or type D, respectively.*

Remarks. Note that by Theorems 9.6.1 and 9.6.2 and (9.7.2) and (9.7.14), whenever $\psi(n) \equiv n$ and $\Delta^2 \tilde{E}_\lambda N(\Delta)$ (and $\Delta^2 \tilde{E}_\lambda N^*(\Delta)$) converge to limits, (9.7.14) agrees with (9.7.2). However, for (9.7.14) to hold, we do not need to assume that $\psi(n) \equiv n$ or impose the extra conditions under which (9.7.1) holds. In fact, (9.7.14) may hold even when $\psi(n)$ is not linear in n and when $\lim_{\Delta\to 0} \tilde{E}_\lambda N(\Delta)/\tilde{E}_\lambda N^*(\Delta)$ may not converge to $s^{-1}((\delta^*/\delta)^2)$. Thus, as in the nonsequential case, the ARE in (9.7.14) is solely based on the asymptotic distributions of the stopping times and not on their first moments; this eliminates the need of computing $\Delta^2 \tilde{E}_\lambda N(\Delta)$ and thereby avoids the extra regularity conditions needed for this.

Proof. For $\delta_1 = \delta_2 = \delta$, denote the expressions in (2.7.17) by $P_1(T \mid \gamma_1, \gamma_2, \delta)$. Also, replacing $X(t)$ by $-X(t)$ and interchanging γ_1 and γ_2 in (2.7.17), we denote the resulting probability by $P_2(T \mid \gamma_1, \gamma_2, \delta)$. Then, by (9.5.1), (9.5.2), (9.5.4), or (9.5.16), for all $t \in [0,\infty)$,

$$P_\lambda(t) = P_1\left(t \mid a\delta^{-1}, b\delta^{-1}, -(\lambda - \tfrac{1}{2})\delta\right) + P_2\left(t \mid a\delta^{-1}, b\delta^{-1}, -(\lambda - \tfrac{1}{2})\delta\right); \tag{9.7.15}$$

$$P_\lambda^*(t) = P_1\left(t \mid \frac{a}{\delta^*}, \frac{b}{\delta^*}, -(\lambda - \tfrac{1}{2})\delta^*\right) + P_2\left(t \mid \frac{a}{\delta^*}, \frac{b}{\delta^*}, -(\lambda - \tfrac{1}{2})\delta^*\right). \tag{9.7.16}$$

Hence to prove (9.7.13) it suffices to show that for each $j(=1,2)$,

$$P_j\left(t \mid a/\delta^*, b/\delta^*, -(\lambda - \tfrac{1}{2})\delta^*\right) = P_j\left((\delta^*/\delta)^2 t \mid a/\delta, b/\delta, -(\lambda - \tfrac{1}{2})\delta\right), \quad \forall t \geqslant 0. \tag{9.7.17}$$

Now, (9.7.17) follows directly from (2.7.17) by substituting for γ_1, γ_2, and δ the entries in (9.7.15) or (9.7.16); the details are left as an exercise. Thus (9.7.13) holds with $h = (\delta^*/\delta)^2$ and hence (9.7.14) follows from (9.7.10) and (9.7.13), where for (9.7.10), both (9.7.15) and (9.7.16) apply. Q.E.D.

A special case where (9.7.14) is easier to verify than (9.7.2) is the sequential testing for regression where C_n^2 may not behave linearly in n.

EXERCISES

9.2.1. Use (2.7.20) to derive the approximation in (9.2.23) (Strassen, 1967; Robbins and Siegmund, 1970).

9.2.2. Verify (9.2.34).

9.2.3. Verify (9.2.44).

9.2.4. For the model (9.2.45), show that if $\max_{1 \leqslant i \leqslant n}|c_{ni}^*| = O(n^{-1/2})$ and $\phi \in L^r$ for some $r > 2$ (in addition to being absolutely continuous), then $E[N \mid \beta \neq 0] < \infty$.

9.2.5. Use the fact that $\sup_x|F_n(x) - F_0(x)| \leqslant \max_{1 \leqslant j \leqslant m}|F_n(x_j) - F_0(x_j)| + \frac{1}{2}\epsilon$ (where $F(x_j) = \frac{1}{2}j\epsilon$ for $j = 1, \ldots, m \simeq 2/\epsilon$) and (9.2.28) to verify (9.2.49).

9.2.6. Show that for the general multivariate case, as $N \to \infty$, $[-\log P\{\sup_x|F_n(x) - F(x)| \geqslant (cn\log n)^{1/2}, \text{ for some } n \geqslant N\}]/(2c\log N) \to 1$ (Sen, 1973e).

9.2.7. Use the result in Problem 9.2.6, to show that for the multivariate two-sample problem, for the test based on the generalized Kolmogorov-Smirnov statistics, $E[N \mid H_0 \text{ not true}] < \infty$.

9.3.1. Prove Theorem 9.3.1.

9.3.2. Consider the model $F_i(x) = F(x - \beta_0 - \boldsymbol{\beta}'\mathbf{c}_i)$, $i = 1, \ldots, N$ where $\boldsymbol{\beta}$ is an unknown q-vector and the c_i are the given q-vectors. For testing $H_0: \boldsymbol{\beta} = \mathbf{0}$ versus $H_1: \boldsymbol{\beta} \neq \mathbf{0}$, consider the statistics $\mathbf{L}_k = \sum_{i=1}^k(\mathbf{c}_i - \bar{\mathbf{c}}_k) \cdot a_k(R_{ki})$, $k = 1, \ldots, N$ $[\mathbf{L}_0 = 0]$ where $\bar{\mathbf{c}}_k = k^{-1}\sum_{i=1}^k \mathbf{c}_i$ and the $\mathbf{L}_k$ are otherwise defined as in (4.2.1). Also, let $T_{Nk} = A_k^{-2}(\mathbf{L}_k'\mathbf{C}_N^{-1}\mathbf{L}_k)$, $k = 1, \ldots, N$ $[T_{N0} = 0]$, where $\mathbf{C}_N = \sum_{i=1}^N(\mathbf{c}_i - \bar{\mathbf{c}}_N)(\mathbf{c}_i - \bar{\mathbf{c}}_N)'$. Assume that $N^{-1}\mathbf{C}_N \to C_0$ (p.d.) as $N \to \infty$ and define $Z_N = \{Z_N(t), t \in [0, 1]\}$ by letting $Z_N(k/N) = T_{Nk}$ for $k = 0, 1, \ldots, N$ and completing the definition of $Z_N(t)$ by linear interpolation. Also, let $\mathbf{W} = \{\mathbf{W}(t), t \in [0, 1]\}$ be a q-variate Brownian motion (i.e., $\mathbf{W} = (W_1, \ldots, W_q)'$ where the W_j are independent copies of a Brownian motion on [0, 1]). Then, show that under H_0, $Z_N \xrightarrow{\mathscr{D}} \mathbf{W}'\mathbf{W}$ and incorporate this result in the formulation of some RST procedures based on the T_{Nk} (Sen (1978a).

9.3.3. Obtain results parallel to (9.3.12), (9.3.13), (9.3.15), and (9.3.16) for the two-sided tests in (9.3.4).

9.3.4. For the RST procedure in Problem 9.3.2, consider a sequence $\{K_N\}$ of alternative hypotheses, where under K_N, $\boldsymbol{\beta} = C_N^{-1/2}\boldsymbol{\lambda}$ for some (fixed) $\boldsymbol{\lambda} \in R^q$. Show that whenever the d.f. F has an absolutely continuous density function f with a finite Fisher information $I(f)$, the contiguity of the sequence of probability measures under $\{K_N\}$ to those under H_0 holds. Hence derive the asymptotic distribution of Z_N under $\{K_N\}$ and extend (9.3.12) and (9.3.15) to this vector case.

9.3.5. In problems 9.3.2 and 9.3.4, for the T_{Nk}, replace the $\mathbf{C}_N^{-1}$ by $\mathbf{C}_k^{-1}$, for $k = 1, \ldots, N$. In that case, show that for an initial sample size n_0 satisfying $N^{-1}n_0 \to t_0 > 0$, the weak convergence holds, where the limiting process is $\{t^{-1}\mathbf{W}(t)'\mathbf{W}(t),\ t_0 \leqslant t \leqslant 1\}$. Hence, or otherwise, extend the results in (9.3.13) and (9.3.16) to the vector case.

9.3.6. For the one-sample location problem, consider the RST procedures based on (i) signed rank statistics, (ii) sign statistics and, in each case, for a common sequence of Pitman alternatives, find the expression for the asymptotic power function and expected stopping time.

9.3.7. For the terminal sample size test, based on T_N alone, use (9.3.6) along with the asymptotic normality of T_N/σ_N (under H_0) to show that

$$-\left(\frac{2}{\lambda^2}\right)\log P\left\{\frac{T_N}{\sigma_N} > \lambda \,|\, H_0\right\} \to 1 \qquad \text{as} \quad \lambda \to \infty.$$

Further, for some (fixed) alternative hypothesis H_1, assume that there exists a real constant h (dependent on H_1), such that

$$\frac{N^{-1/2}T_N}{\sigma_N} \to h \qquad \text{a.s., as} \quad N \to \infty, \qquad \text{when } H_1 \text{ holds.}$$

Then, the asymptotic Bahadur slope for $\{T_N\}$ is equal to h^2. Consider next the one-sided test in (9.3.4) and as in Theorem 9.3.1, assume that under H_0, $\max_{k \leqslant N} T_k/\sigma_N$ converges in distribution to $\sup\{W(t)\colon t \in [0, 1]\}$. Since, for every $\lambda > 0$, $P\{\sup\{W(t)\colon 0 \leqslant t \leqslant 1\} > \lambda\} = 2P\{W(1) > \lambda\}$, show that, under H_0,

$$-(2/\lambda^2)\log P\left\{\max_{k \leqslant N} \frac{T_k}{\sigma_N} > \lambda \,|\, H_0\right\} \to 1 \qquad \text{as} \quad \lambda \to \infty.$$

Thus if we assume that under the same alternative H_1 there exists a real h^* (dependent on H_1) such that

$$N^{-1/2}\left\{\max_{k\leqslant N}\frac{T_k}{\sigma_N}\right\}\to h^* \qquad \text{a.s., as } N\to\infty, \text{ when } H_1 \text{ obtains,}$$

then the asymptotic Bahadur slope for the one-sided RST procedure is equal to h^{*2}. Hence the approximate Bahadur efficiency of the RST procedure with respect to the terminal sample size fixed-sample procedure is equal to $(h^*/h)^2$ ($\geqslant 1$). A similar result holds when in the above a.s. convergence results, $N^{-1/2}$ is replaced by some $q(N)$ and $q(N)$ satisfies certain smoothness conditions (Sen, 1978a).

9.3.8. Extend the results of Problem 9.3.7 to the case of two-sided tests.

9.3.9. For the particular case of the one-sample location problem, two-sample location or scale problem, or the regression problem, show that as the alternative hypothesis H_1 is chosen nearer to the null one, the Bahadur ARE approaches 1. Thus the limiting Bahadur ARE is equal to 1.

9.3.10. Extend the results of Problem 9.3.7 to the multiple regression problem in Exercise 9.3.2 or 9.3.4 [In this context, one needs to find an asymptotic expression for $-2\log P\{\sup_{0\leqslant t\leqslant 1}B(t)>\lambda\}$ (as $\lambda\to\infty$) where $\{B(t),\, t\geqslant 0\}$ is a Bessel process. Using a submartingale approach, this has been verified by Majumdar and Sen (1978a).]

9.4.1. Prove the inequalities in (9.4.5) and show that if the excess (of $\log\lambda_N$) over the boundaries are negligible, then these inequalities are approximate equalities (Wald, 1947).

9.4.2. Let $\{X_i,\ i\geqslant 1\}$ be i.i.d.r.v.'s with (i) a binomial $(1,p)$ and (ii) normal $(\mu,1)$ d.f. Then deduce the expression for $\log\lambda_n$ when (i) $H_0: p=p_0$ versus $H_1: p=p_1=\Delta+p_0$ and (ii) $H_0: \mu=0$ versus $H_1: \mu=\Delta>0$. Verify (9.4.7).

9.5.1. Provide a formal proof of Theorem 9.5.2.

9.5.2. Verify (9.5.27).

9.6.1. Provide a formal proof of Theorem 9.6.2.

9.6.2. Use (3.2.27) and (3.2.29) to verify (9.6.4) and (9.6.5) for the case of U-statistics and von Mises functionals, where $T^*_{n,\theta}=nV_{n,1}$ is defined by (3.2.26). Hence verify (9.6.6) and (9.6.8). Also verify (9.6.3) with the aid of (3.7.21).

9.6.3. Use (7.5.1) and (7.5.2) and a modified version of (7.5.3) [replacing the a.s. result by a statement with probability $\geqslant 1-O(n^{-s})$, for some

$s > 1$] to verify (9.6.4) and (9.6.5) for LCFOS. Also, by using the method of proof of Theorem 7.6.3, verify that (9.6.3) holds for LCFOS.

9.6.4. Verify (9.6.32).

9.6.5. Provide a proof of (5.5.4) where, for the range of b, replace c by cn^{η} for some $\eta > 0$ and also replace the a.s. statement by a probability $\geqslant 1 - O(n^{-s})$, for some $s > 1$. For this purpose, in (5.4.14), choose $\delta > \frac{1}{3}$ and make use of (2.6.29) instead of (2.6.31) [used in the proof of (5.5.4)] (Sen, 1980b).

9.7.1. Derive the expression for (9.7.9) when for the SROT, the following scores are used (i) *Wilcoxon scores*: $a_n(i) = i/(n + 1)$, $1 \leqslant i \leqslant n$, (ii) *normal scores*: $a_n(i) = \chi_1^{-1}(i/(n + 1))$, $i = 1, \ldots, n$, where $\chi_1(x)$ is the d.f. of the chi distribution with 1 degree of freedom, and (iii) $a_n(i) = \Pi^{-1}(i/(n + 1))$, $i = 1, \ldots, n$ where Π is the double exponential d.f.

9.7.2. Extend the results of Problem 9.7.1 to the case of sequential tests for regression (Ghosh and Sen, 1977).

9.7.3. For the regression test in Problem 9.7.2, to verify (9.7.1) one needs comparatively stronger conditions on the score functions. In this context, Ghosh and Sen (1977) have assumed that for the score function ϕ,

$$\left|\left(\frac{d}{du}\right)\phi(u)\right| \leqslant K\left[u(1 - u)\right]^{-1}, \qquad \text{for every } u \in (0, 1),$$

where K is a positive constant ($< \infty$). Show that the above condition holds for the Wilcoxon scores and normal scores as well as the exponential scores.

9.7.4. Use Theorem 9.7.1 to show that for the computation of the Pitman ARE in accordance with the definition in (9.7.14), we need not impose the condition in the preceding problem and the result continues to hold under the usual conditions pertaining to the validity of Theorem 9.5.2.

9.7.5. Provide the intermediate steps leading to the verification of (9.7.17).

CHAPTER 10

Asymptotic Theory of Nonparametric Sequential Estimation

10.1. INTRODUCTION

This chapter is devoted to the study of nonparametric sequential point as well as interval estimation procedures. There are certain estimation problems that cannot be solved by any procedure based on a prefixed sample size and require some sequential sampling scheme. Two basic problems of estimation arising in this context are the following. First, the problem of constructing a *confidence interval* for a parameter with the properties that the *width* of the interval is bounded from above by a given positive number and the *coverage probability* is equal to a preassigned number γ $(0 < \gamma < 1)$. The second problem is to find a *point estimator* of a parameter such that for some suitably defined *loss function*, the *risk* (expected loss) of the estimator is a minimum. In either case, the presence of nuisance parameters (when the form of the underlying d.f. is assumed to be known) or the lack of knowledge of the underlying d.f. leads to the fact that the desired sample size cannot be specified in advance. For the mean of a normal distribution (variance unknown), Dantzig (1940) proved the nonexistence of fixed-sample size procedures. Stein (1945) considered a two-stage procedure (for the confidence interval for the mean of a normal distribution with unknown variance) that satisfies both the requirements mentioned earlier. Though valid, Stein's procedure is not generally fully efficient and it depends crucially on the assumed normality of the underlying d.f. Chow and Robbins (1965) initiated a sequential confidence interval procedure for the mean of a distribution which remains valid for a broad class of underlying d.f.'s and is efficient in an asymptotic sense (to be explained later). We are interested in the various nonparametric analogues of this procedure for a wider class of estimation problems. While for the normal mean the Stein procedure can be adapted to have a bounded risk point estimator, the criticism leveled against the Stein interval estimation procedure applies to this procedure as well. Robbins (1959) considered a somewhat different approach to the sequential point estimation of the normal mean and this was later extended by Starr (1966), Starr and Woodroofe (1969), Woodroofe (1977), and others. Again, we are principally interested in some nonparametric analogues of this procedure for a wider class of point estimation problems. The various invariance principles considered in Part 1 enable us to formulate these procedures in a broader perspective and to provide asymptotically efficient solutions.

It is convenient to deal with the confidence interval problem first. As in Chapter 9, we consider two different types of bounded-length sequential confidence intervals, *type A*, based on natural estimators of the parameters under consideration, and *type B*, based on some (genuinely or asymptotically) distribution-free statistics yielding the desired estimators. These are considered in Sections 10.2 and 10.3, respectively. Besides these fixed-width confidence intervals, in Section 10.4 we also consider some confidence sequences for parameters based on the Darling-Robbins type procedures (as studied in Section 9.2). Section 10.5 deals with the problem of asymptotically risk-efficient sequential point estimation procedures based on U-statistics. For the particular case of location problem, parallel procedures based on R-estimators, M-estimators, and L-estimators are then studied in Section 10.6. Some general remarks (and open problems) are appended in the concluding section.

10.2. TYPE A SEQUENTIAL CONFIDENCE REGIONS

Let $\{X_i, i \geqslant 1\}$ be a sequence of independent r.v.'s defined on a common probability space $(\Omega, \mathscr{A}, P)$ and let $\mathbf{X}^{(n)} = (X_1, \ldots, X_n)$, for $n \geqslant 1$. Let θ be a real-valued parameter (a functional of the underlying d.f.) and Θ be the parameter space [an interval in $R = (-\infty, \infty)$]. Let $\hat{\theta}_{L,n}$ and $\hat{\theta}_{U,n}$ [both based on $\mathbf{X}^{(n)}$] be r.v.'s such that $\hat{\theta}_{L,n} \leqslant \hat{\theta}_{U,n}$, and on letting $I_n = \{a : \hat{\theta}_{L,n} \leqslant a \leqslant \hat{\theta}_{U,n}\}$ we have

$$P\{\theta \in I_n\} \geqslant 1 - \alpha, \qquad \text{for some specified} \quad \alpha \in (0,1). \qquad (10.2.1)$$

Note that I_n is a random interval and (10.2.1) specifies the *coverage probability* (that the interval I_n contains the true value of θ). In this setup, $1 - \alpha$ is termed the *confidence coefficient*, I_n *the confidence interval*, $\hat{\theta}_{L,n}$ and $\hat{\theta}_{U,n}$ *the lower and upper confidence limits* and the *width* of this confidence interval is equal to $\hat{\theta}_{U,n} - \hat{\theta}_{L,n}$. In many problems of practical interest, one wants to provide such a confidence interval for a parameter of interest satisfying the additional condition that for some preassigned $d(>0)$,

$$0 < \hat{\theta}_{U,n} - \hat{\theta}_{L,n} \leqslant 2d. \tag{10.2.2}$$

For example, if the X_i are i.i.d.r.v.'s with a normal d.f. with mean θ and variance σ^2, we may want to provide a confidence interval for θ with coverage probability $1-\alpha$ and prescribed width $2d$. Let $\bar{X}_n = n^{-1}\sum_{i=1}^{n} X_i$, $n \geqslant 1$ and let τ_ϵ be the upper $100\epsilon\%$ point of the standard normal d.f. If σ is known, we may set $n =$ smallest integer $\geqslant (\tau_{\alpha/2}\sigma/d)^2$ and with this choice of n, let $I_n = [\bar{X}_n - d, \bar{X}_n + d]$. Then $P\{\theta \in I_n\} = P\{\bar{X}_n - d \leqslant \theta \leqslant \bar{X}_n + d\} = P\{n^{1/2}|\bar{X}_n - \theta|/\sigma \leqslant n^{1/2}d/\sigma\} \geqslant 1-\alpha$ as $n^{1/2}d/\sigma \geqslant \tau_{\alpha/2}$. Thus both (10.2.1) and (10.2.2) hold. On the other hand, if σ is not specified, for any given n, $P\{\theta \in I_n\}$ depends on σ, and hence (10.2.1) may not hold for every σ. A similar situation arises if we use the Student t-statistic for setting the desired confidence interval for θ; there (10.2.1) holds but (10.2.2) fails to do so. In general, the joint distribution of $(\hat{\theta}_{L,n}, \hat{\theta}_{U,n})$ depends on the underlying d.f. of the X_i (which is usually not completely known) and it is not possible to prescribe a fixed-sample-size procedure for which both (10.2.1) and (10.2.2) hold. For the particular case of the normal d.f. with unknown σ, Dantzig (1940) proved the nonexistence of fixed-sample-size confidence intervals for the mean for which both (10.2.1) and (10.2.2) hold, while Stein (1945) considered a two-stage (sequential) procedure that meets both these requirements. The scope of the Stein procedure is somewhat limited to problems relating to (multi-) normal d.f.'s and generally this is not fully efficient. Chow and Robbins (1965) initiated a line of attack that is remarkably flexible, broad in scope, and efficient. Though they considered the case of the mean of a distribution, their procedure extends to more general parameters as well. We generally follow their line of attack framed in a more general setup.

It is desired to determine sequentially a *stopping variable* N [$= N(d)$, a positive integer value r.v.] and the corresponding $(\hat{\theta}_{L,N}, \hat{\theta}_{U,N})$ such that both (10.2.1) and (10.2.2) hold. In this section we consider sequential procedures, termed the *type A procedures*, based on an appropriate sequences of point estimators. Let $\{T_n = T(\mathbf{X}^{(n)})\}$ be a sequence of point estimators of θ and set

$$\hat{\theta}_{L,N} = T_N - d \quad \text{and} \quad \hat{\theta}_{U,N} = T_N + d, \tag{10.2.3}$$

so that (10.2.2) holds, and hence we need to determine $N(= N(d))$ and $\{T_n\}$ in such a way that (10.2.1) holds.

Toward this, as in Chapter 9, we assume that T_n has an asymptotic mean square $\sigma_\theta^2/\psi(n)$ where $\psi(n)$ is nonnegative, nondecreasing, and $\lim_{n\to\infty}\psi(n) = \infty$, and σ_θ^2 is unknown. Also let $\{s_n^2 = s^2(\mathbf{X}^{(n)})\}$ be a sequence of consistent estimators of σ_θ^2, and s_n^2 is properly defined for all $n \geqslant n_0$ where $n_0 \geqslant 1$. Then we define our stopping variable by

$$N = N(d) = \min\left\{k : k \geqslant n_0 \quad \text{and} \quad s_k^2 \leqslant \frac{\psi(k)d^2}{\tau_{\alpha/2}^2}\right\}. \tag{10.2.4}$$

Thus the sequential procedure consists in continuing sampling until N, defined by (10.2.4), is obtained and then choosing the confidence interval

$$I_N = [T_N - d, T_N + d]. \tag{10.2.5}$$

We like to show that (10.2.5) satisfies (10.2.1) in the *asymptotic sense* where we let $d \to 0$. In this context, we make the following assumptions:

$$s_n^2 \to \sigma_\theta^2 \qquad \text{a.s.,} \quad \text{as} \quad n \to \infty, \tag{10.2.6}$$

$$\lim_{n\to\infty} a_n = a \quad \Rightarrow \quad \lim_{n\to\infty} \frac{\psi(na_n)}{\psi(n)} = s(a) \tag{10.2.7}$$

where $s(a)$ is $\uparrow$ in a with $s(1) = 1$. We next define

$$n(d) = \min\{k : k \geqslant n_0 \quad \text{and} \quad \psi(k) \geqslant d^{-2}\tau_{\alpha/2}^2\sigma_\theta^2\}, \qquad d > 0. \tag{10.2.8}$$

Consider then a stochastic process $W_d = \{W_d(t),\ t \in [0, T]\}$ where $T(> 1)$ is a finite positive number and

$$W_d(t) = \psi(k_d(t))[T_{k_d(t)} - \theta]/\sigma_\theta\psi^{1/2}(n(d)), \qquad t \in [0, T]; \tag{10.2.9}$$

$$k_d(t) = \max\{k : \psi(k) \leqslant t\psi(n(d))\}, \qquad t \in [0, T], \tag{10.2.10}$$

and let $W = \{W(t),\ t \in [0, T]\}$ be a standard Wiener process on $[0, T]$. Then we assume that as $d \downarrow 0$,

$$W_d \overset{\mathfrak{D}}{\to} W, \qquad \text{in the } J_1\text{-topology on} \quad D[0, T]. \tag{10.2.11}$$

Actually, we do not exactly need (10.2.11). It will be enough to assume that

$$W_d(1) \overset{\mathfrak{D}}{\to} W(1); \quad \sup_{t : |t-1| < \delta} |W_d(t) - W_d(1)| \overset{P}{\to} 0 \qquad \text{as} \quad \delta \downarrow 0, \tag{10.2.12}$$

and (10.2.12) follows from (10.2.11).

Note that by (10.2.4), whenever $N > n_0$,

$$s_{N-1}^2 > \frac{\psi(N-1)d^2}{\tau_{\alpha/2}^2} \geqslant \left[\frac{\psi(N-1)}{\psi(N)}\right]s_N^2, \tag{10.2.13}$$

and hence by (10.2.6), (10.2.7), and (10.2.13), we conclude that

$$N = N(d) \quad \text{is finite a.s. for every } d > 0 \text{ and } N(d) \text{ is } \searrow \text{ in } d. \tag{10.2.14}$$

Moreover, by (10.2.6), (10.2.7), (10.2.8), and (10.2.13),

$$\frac{\psi(N(d))}{\psi(n(d))} \to 1 \qquad \text{a.s., as} \quad d \to 0, \tag{10.2.15}$$

$$\frac{N(d)}{n(d)} \to 1 \qquad \text{a.s., as} \quad d \to 0. \tag{10.2.16}$$

Also $W_d(1) \xrightarrow{\mathfrak{D}} W(1)$ ensures that as $d \to 0$,

$$P\left\{|T_{n(d)} - \theta| \leqslant \tau_{\alpha/2}\sigma_\theta / \psi^{1/2}(n(d))\right\} \to 1 - \alpha,$$

so that by (10.2.4), (10.2.12), and (10.2.15), we have

$$P\left\{|T_{N(d)} - \theta| \leqslant d\right\} \to 1 - \alpha \qquad \text{as} \quad d \to 0. \tag{10.2.17}$$

Hence by (10.2.3), (10.2.5), and (10.2.17), we conclude that

$$\lim_{d \downarrow 0} P\left\{\theta \in I_{N(d)}\right\} = 1 - \alpha \qquad (\textit{asymptotic consistency}). \tag{10.2.18}$$

Note that by (10.2.4)

$$\begin{aligned} \psi(N-1) &\leqslant \psi(n_0 - 1) I_{[N = n_0]} + \tau_{\alpha/2}^2 d^{-2} s_{N-1}^2 I_{[N > n_0]} \\ &\leqslant \psi(n_0 - 1) I_{[N = n_0]} + d^{-2}\tau_{\alpha/2}^2 \Big(\sup_{n \geqslant n_0} s_n^2\Big), \end{aligned} \tag{10.2.19}$$

where I_A stands for the indicator function of the set A. Also, by (10.2.7), $\psi(n)/\psi(n-1) \to 1$ as $n \to \infty$. Hence, for every (fixed) $d(>0)$,

$$E\Big(\sup_{n \geqslant n_0} s_n^2\Big) < \infty \Rightarrow E\psi(N(d)) < \infty. \tag{10.2.20}$$

In many cases, $\{s_n^2, n \geqslant n_0\}$ forms a reverse martingale sequence, so that by the Doob inequality (2.2.8)

$$E\Big(\sup_{n \geqslant n_0} s_n^2\Big) \leqslant (e/(e-1))\left\{1 + E\left(s_{n_0}^2 \log s_{n_0}^2\right)\right\}, \tag{10.2.21}$$

and hence for $E\psi(N(d)) < \infty$, it suffices to assume that $E(s_n^2 \log s_n^2) < \infty$ for some $n \geqslant 1$. More generally, if

$$s_n^2 = \sum_{g=1}^{m} \alpha_{n,h} s_{n,g}^2 \tag{10.2.22}$$

where $m(\geqslant 1)$ is a positive integer, $\{s_{n,h}^2;\ n \geqslant n_0\}$ is a reverse martingale for each h and $\sup_n \max_{1 \leqslant h \leqslant m} |\alpha_{n,h}| < \infty$, then $\max_{1 \leqslant h \leqslant m} E(s_{n,h}^2 \log s_{n,h}^2) < \infty$ $\Rightarrow E\psi(N(d)) < \infty$. Later, we shall see that for U-statistics and von Mises functionals this characterization works out very well. Further, if we assume that

$$\liminf_n n^{-1}\psi(n) \geqslant \psi_0 > 0, \tag{10.2.23}$$

then $E\psi(N(d)) < \infty \Rightarrow EN(d) < \infty$, and hence, under (10.2.23), $E(\sup_{n \geqslant n_0} s_n^2) < \infty$ also ensures that $EN(d) < \infty$, $\forall d > 0$. We may note also that

$$EN(d) = \sum_{n \geqslant n_0} nP\{N(d) = n\} = n_0 + \sum_{n \geqslant n_0} P\{N(d) > n\}, \tag{10.2.24}$$

where by (10.2.4), for $n \geqslant n_0$,

$$\begin{aligned} P\{N(d) > n\} &\leqslant P\{s_n^2 > \psi(n)d^2/\tau_{\alpha/2}^2\} \\ &\leqslant (\tau_{\alpha/2}^2/d^2)^r E\big[(s_n^2/\psi(n))^r\big] \qquad (r > 0). \end{aligned} \tag{10.2.25}$$

Hence, if for some $r > 0$,

$$\sum_{n \geqslant n_0} E\big(\big[s_n^2/\psi(n)\big]^r\big) < \infty, \tag{10.2.26}$$

then $EN(d) < \infty$, $\forall d > 0$. There are situations (see, Exercise 10.2.4) where (10.2.26) may hold under less stringent regularity conditions than (10.2.20) or (10.2.21).

In some situations, we may have an inequality of the form

$$\begin{gathered} s_n^2 \leqslant (n-m)^{-1}Z_n^*;\ Z_n^* = \sum_1^n Z_i, \\ 0 \leqslant m \leqslant n_0 - 1 \qquad \text{and} \qquad \forall n \geqslant n_0(\geqslant 2), \end{gathered} \tag{10.2.27}$$

where the Z_i are nonnegative i.i.d.r.v. with $EZ_1 = \sigma_\theta^2(> \infty)$. Then we have the following theorem where $n(d)$ is defined by (10.2.8).

THEOREM 10.2.1. *For $\psi(n) \nearrow$ in n, under* (10.2.27), *for every $d > 0$,*

$$EN(d) \leqslant m + \psi^{-1}\left(\psi(n_0) + \left(\frac{n_0}{(n_0 - m)}\right)\psi(n(d))\right) (< \infty). \tag{10.2.28}$$

Further, under (10.2.6), (10.2.7), *and* (10.2.27),

$$\lim_{d \downarrow 0} \frac{EN(d)}{n(d)} = 1 \qquad (\textit{asymptotic efficiency}) \tag{10.2.29}$$

Proof. Note that by (10.2.19) and (10.2.27), for $N = N(d)$,

$$\begin{aligned} (N-m)\psi(N-m) &\leqslant (N-m)\psi(N-1) \\ &\leqslant (N-m)\psi(n_0 - 1) + d^{-2}\tau_{\alpha/2}^2 Z_{N-1}^* \\ &\leqslant (N-m)\psi(n_0 - 1) + d^{-2}\tau_{\alpha/2}^2 Z_N^*. \end{aligned} \tag{10.2.30}$$

Let us first assume that $EN(d)\psi(N(d)) < \infty (\Rightarrow EN(d) < \infty$ and $E\psi(N(d)) < \infty)$. Then by definition of Z_N^* and (2.2.13) (where we take $X_n = Z_n^* - nEZ_1$, $n \geqslant 1$),

$$EZ_{N(d)}^* = \left[EN(d)\right]EZ_1 = \sigma_\theta^2 EN(d), \tag{10.2.31}$$

while $t\psi(t)$ being a convex function of t, by the Jensen inequality

$$E\{[N(d)-m]\psi(N(d)-m)\} \geqslant \{EN(d)-m\}\psi(EN(d)-m), \tag{10.2.32}$$

so that by (10.2.8), (10.2.30), (10.2.31), and (10.2.32), we have

$$\begin{aligned}&\{EN(d)-m\}\psi(EN(d)-m)\\ &\quad\leqslant \{EN(d)-m\}\psi(n_0-1)+\sigma_\theta^2 d^{-2}\tau_{\alpha/2}^2 EN(d)\\ &\quad\leqslant \{EN(d)-m\}\psi(n_0-1)+\{EN(d)\}\psi(n(d)).\end{aligned} \tag{10.2.33}$$

Thus

$$\{EN(d)-m\}\{\psi(EN(d)-m)-\psi(n_0-1)\} \leqslant \{EN(d)\}\psi(n(d)) \tag{10.2.34}$$

and (10.2.28) follows from (10.2.34) by noting that

$$\frac{EN(d)}{E\{N(d)-m\}} \leqslant \frac{n_0}{n_0-m}. \tag{10.2.35}$$

Next, if we do not assume that $EN\psi(N) < \infty$, we define (for every $k \geqslant n_0$), $N_k(d) = N(d) \wedge k$. Then, for every $k < \infty$, $EN_k(d)\psi(N_k(d)) < k\psi(k) < \infty$ and hence, proceeding as before, we find that

$$EN_k(d) \leqslant m + \psi^{-1}\left(\psi(n_0) + \left(\frac{n_0}{n_0-m}\right)\psi(n(d))\right). \tag{10.2.36}$$

Also, $N_k(d) \uparrow$ to $N(d)$ as $k\uparrow\infty$. Thus (10.2.28) follows from (10.2.36) and the monotone convergence theorem.

Since, by (10.2.4), $\psi(N) \geqslant \tau_{\alpha/2}^2 d^{-2} s_N^2$ and by (10.2.16), $N(d)/n(d) \to 1$ a.s., as $d \to 0$, by Fatou's lemma

$$\liminf_{d\to 0} \frac{EN(d)}{n(d)} \geqslant 1. \tag{10.2.37}$$

Let now $\{n_0(d)\}$ be a sequence of positive integers such that

$$n_0(d)\to\infty \quad \text{but} \quad \frac{\psi(n_0(d))}{\psi(n(d))}\to 0 \quad \text{as} \quad d\to 0, \tag{10.2.38}$$

and let us define

$$N^*(d) = N(d) \vee n_0(d), \qquad d > 0. \tag{10.2.39}$$

Then $N^*(d) \geqslant N(d)$, $\forall d > 0$. On the other hand, using (10.2.28) for $N^*(d)$, we obtain that for every d (> 0),

$$EN^*(d) \leqslant m + \psi^{-1}\left(\psi(n_0(d)) + \frac{n_0(d)}{n_0(d) - m}\psi(n(d))\right). \quad (10.2.40)$$

Since $m(\geqslant 0)$ is fixed, by (10.2.7), (10.2.8), (10.2.38), and (10.2.40),

$$\limsup_{d\to 0} \frac{EN^*(d)}{n(d)} \leqslant 1. \quad (10.2.41)$$

Hence $\limsup_{d\to 0} EN(d)/n(d) \leqslant \limsup_{d\to 0} EN^*(d)/n(d) \leqslant 1$. Q.E.D.

The bound in (10.2.28) is not necessarily the sharpest one. For example, for $\psi(n) = n$, $n_0(\geqslant 2)$, $m = 1$, a sharper bound is available (see Exercise 10.2.5).

In the absence of (10.2.27), the asymptotic efficiency can be proved under alternative conditions. Note that by (10.2.19),

$$\psi(N(d)) = \psi(N(d) - 1)\left[\frac{\psi(N(d))}{\psi(N(d) - 1)}\right]$$

$$\leqslant \left[\frac{\psi(N(d))}{\psi(N(d) - 1)}\right]\left\{\psi(n_0 - 1) + d^{-2}\tau_{\alpha/2}^2 \sup_n s_n^2\right\}. \quad (10.2.42)$$

Let $n^*(\geqslant n_0)$ be such that $\psi(n)/\psi(n-1) \leqslant 2$, $\forall n \geqslant n^*$ and let $N^*(d) = N(d) \vee n^*$. It follows that

$$\psi(N^*(d)) \leqslant \left[2\psi(n_0 - 1) + 2\psi(n(d))\left(\sup_n s_n^2/\sigma_\theta^2\right)\right]. \quad (10.2.43)$$

By (10.2.16), (10.2.43), and the dominated covergence theorem, we conclude that

$$E\left(\sup_n s_n^2\right) < \infty \Rightarrow \frac{E\psi(N^*(d))}{\psi(n(d))} \to 1 \quad \text{as} \quad d \to 0. \quad (10.2.44)$$

On the other hand, $N(d) \leqslant N^*(d)$, $\forall d > 0$, so that $E\psi(N(d)) \leqslant E\psi(N^*(d))$, $\forall d > 0$. Hence $\limsup_{d\to 0} E\psi(N(d))/\psi(n(d)) \leqslant 1$. Further, by (10.2.4), (10.2.15) and the Fatou's lemma, $\liminf_{d\to 0} E\psi(N(d))/\psi(n(d)) \geqslant 1$, so that

$$E\left(\sup_n s_n^2\right) < \infty \Rightarrow \lim_{d\downarrow 0} \frac{E\psi(N(d))}{\psi(n(d))} = 1. \quad (10.2.45)$$

If $\psi^{-1}(t)$ is a concave function of t, then by (10.2.16) and (10.2.43),

$$E\left\{\psi^{-1}\left(\sup_n s_n^2\right)\right\} < \infty \Rightarrow \limsup_{d\downarrow 0} \frac{EN^*(d)}{n(d)} \leqslant 1, \quad (10.2.46)$$

so that by (10.2.37) and (10.2.46) [and the fact that $EN(d) \leqslant EN^*(d)$, $\forall d > 0$], we obtain

$$E\left\{\psi^{-1}\left(\sup_n s_n^2\right)\right\} < \infty \Rightarrow \limsup_{d\downarrow 0} \frac{EN(d)}{n(d)} = 1. \tag{10.2.47}$$

Let us also denote by

$$n_1(d) = n(d)(1-\epsilon) \quad \text{and} \quad n_2(d) = n(d)(1+\epsilon), \qquad \epsilon > 0. \tag{10.2.48}$$

Then, we have

$$\frac{EN(d)}{n(d)} = \frac{1}{n(d)}\left[\sum_{n \leqslant n_1(d)} + \sum_{n_1(d) < n \leqslant n_2(d)} + \sum_{n > n_2(d)} nP\{N(d) = n\}\right] \tag{10.2.49}$$

where

$$\frac{\left(\sum_{n \leqslant n_1(d)} nP\{N(d) = n\}\right)}{n(d)} \leqslant (1-\epsilon)P\{N(d) < n_1(d)\} \to 0, \quad \text{by (10.2.15);} \tag{10.2.50}$$

$$\left|\sum_{n_1(d) < n \leqslant n_2(d)} \frac{nP\{N(d) = n\}}{n(d)} - 1\right| \leqslant \epsilon \tag{10.2.51}$$

and

$$\sum_{n > n_2(d)} \frac{nP\{N(d) = n\}}{n(d)} = \left[\frac{\{n_2(d) + 1\}}{n(d)}\right] P\{N(d) > n_2(d)\} + \sum_{n > n_2(d)} \frac{P\{N(d) > n\}}{n(d)}. \tag{10.2.52}$$

Hence, to show that (10.2.29) holds, it suffices to show that

$$\lim_{d \to 0}\left[\sum_{n \geqslant n_2(d)} \frac{P\{N(d) > n\}}{n(d)}\right] = 0, \qquad \forall \epsilon > 0. \tag{10.2.53}$$

Now, as in (10.2.25), by (10.2.7) and (10.2.48), for every $n \geqslant n_2(d)$,

$$P\{N(d) > n\} \leqslant P\left\{s_n^2 > \frac{\psi(n)d^2}{\tau_{\alpha/2}^2}\right\}$$

$$\leqslant P\left\{\frac{s_n^2}{\sigma_\theta^2} > \frac{\psi(n)}{\psi(n(d))}\right\} \left(\text{as } \frac{\psi(n)}{\psi(n(d))} \geqslant s(1+\epsilon) = 1+\eta\right)$$

$$= P\left\{\frac{s_n^2}{\sigma_\theta^2} - 1 > \eta\right\}, \qquad \eta > 0. \tag{10.2.54}$$

Hence for (10.2.53) it suffices to assume that for every $\eta > 0$,

$$P\left\{\left|\frac{s_n^2}{\sigma_\theta^2} - 1\right| > \eta\right\} < c_\eta n^{-r}, \qquad \forall n \geqslant n_2(d), \tag{10.2.55}$$

for some $r > 1$ where $c_\eta(<\infty)$ depends on η. Usually, (10.2.55) demands the existence of $E[n^{1/2}|s_n^2 - \sigma_\theta^2|]^{2r}$, for some $r > 1$, and is more restrictive than $E(\sup_n s_n^2) < \infty$.

Note that by (10.2.4) and (10.2.8), for $N(d) > n_0$ and $n(d) > n_0$,

$$\frac{\psi(N(d)-1)d^2}{\tau_{\alpha/2}^2} < s_{N(d)}^2 \leqslant \frac{\psi(N(d))d^2}{\tau_{\alpha/2}^2}, \tag{10.2.56}$$

$$\frac{\psi(n(d)-1)d^2}{\tau_{\alpha/2}^2} < \sigma_\theta^2 \leqslant \frac{\psi(n(d))d^2}{\tau_{\alpha/2}^2}, \tag{10.2.57}$$

where, by (10.2.15), $\psi(N(d))/\psi(n(d)) \to 1$ a.s. as $d\downarrow 0$ and $P\{N(d) > n_0\} \to 1$ as $d\downarrow 0$. Let us now consider a sequence $\{g(n)\}$ of positive numbers such that $g(n)$ is $\nearrow$ in n, $\lim_{n\to\infty} g(n) = +\infty$ and the following holds: (i)

$$\frac{g(n(d))\{\psi(n(d)) - \psi(n(d)-1)\}}{\psi(n(d))} \to 0 \quad \text{as} \quad d\downarrow 0, \tag{10.2.58}$$

(ii) there exists a positive (and finite) constant γ^*, such that as $d\downarrow 0$,

$$\frac{g(n(d))\{s_{n(d)}^2/\sigma_\theta^2 - 1\}}{\gamma^*} \xrightarrow{\mathfrak{D}} \mathfrak{N}(0,1), \tag{10.2.59}$$

and (iii) for every $\epsilon > 0$ and $\eta > 0$, there exist an $n_0(= n_0(\epsilon, \eta))$ and a

$\delta(>0)$, such that for every $n \geqslant n_0$,

$$P\left\{\max_{m:|\psi(m)/\psi(n)-1|<\delta} \frac{g(n)|s_m^2 - s_n^2|}{\sigma_\theta^2} > \epsilon\right\} < \eta. \tag{10.2.60}$$

From (10.2.59), (10.2.60), and (10.2.8), we conclude that as $d\downarrow 0$,

$$\frac{g(n(d))\{s_{N(d)}^2/\sigma_\theta^2 - 1\}}{\gamma^*} \xrightarrow{\mathfrak{D}} \mathfrak{N}(0,1), \tag{10.2.61}$$

while, by (10.2.56), (10.2.57), and (10.2.58), we conclude that as $d\downarrow 0$,

$$\frac{g(n(d))(s_{N(d)}^2/\sigma_\theta^2 - 1)}{\gamma^*} = \frac{g(n(d))(\psi(N(d))/\psi(n(d)) - 1)}{\gamma^*} + o(1) \quad \text{a.s.} \tag{10.2.62}$$

From (10.2.61) and (10.2.62), we arrive at the following.

THEOREM 10.2.2. *Under* (10.2.15), (10.2.58), (10.2.59), *and* (10.2.60), *as* $d\downarrow 0$,

$$\frac{g(n(d))\{\psi(N(d))/\psi(n(d)) - 1\}}{\gamma^*} \xrightarrow{\mathfrak{D}} \mathfrak{N}(0,1). \tag{10.2.63}$$

It is also possible to replace $g(n(d))$ by $g(N(d))$ in (10.2.63). In general, verification of (10.2.59) and (10.2.60) demands extra regularity conditions. We next consider various specific problems of special interest and illustrate the sequential procedures and their properties for them.

10.2.1. General Estimable Parameter, I

Let $\{X_i,\ i \geqslant 1\}$ be i.i.d.r.v. with a d.f. F and consider an estimable parameter $\theta = \theta(F) = Eg(X_1, \ldots, X_m)$ where g is a (symmetric) kernel of degree $m(\geqslant 1)$. As in (3.2.3) and (3.2.4), we consider the U-statistics U_n and the von Mises functional $\theta(F_n)$ corresponding to θ and desire to provide a bounded-length sequential confidence interval for θ. Here we define $m^2 s_n^2$ as in (3.7.5), so that the corresponding $\sigma_\theta^2 = m^2\zeta_1(F)$ and $\psi(n) = n$. Thus

$$N(d) = \min\left\{k : k \geqslant m \quad \text{and} \quad m^2 s_k^2 \leqslant \frac{kd^2}{\tau_{\alpha/2}^2}\right\}, \tag{10.2.64}$$

$$n(d) \sim d^{-2} m^2 \zeta_1(F)\tau_{\alpha/2}^2. \tag{10.2.65}$$

Here (10.2.7) holds and (10.2.6) follows from (3.7.7). Further, (10.2.11) follows from Theorem 3.3.1. Hence both (10.2.14) and (10.2.16) hold, as does (10.2.18). As in Section 3.7, the estimator s_n^2 in (3.7.5) can be expressed as a linear combination of several U-statistics (which are all reverse martingales), and hence (10.2.22) holds. Thus if we assume that for the kernel g, $E(g^2 \log g^2) < \infty$, then $\max_{1 \leqslant h \leqslant m} E(s_{n,h}^2 \log s_{n,h}^2) < \infty$ and hence, via (10.2.20) and (10.2.21), $EN(d) < \infty$, $\forall d > 0$, and by (10.2.45), $EN(d)/n(d) \to 1$ as $d \to 0$.

Consider next the asymptotic normality of the stopping time $N(d)$. Again, note that s_n^2 is a linear combination of U-statistics where the first coefficient is equal to $1 + O(n^{-1})$, while the rest are all $O(n^{-1})$. Hence if we assume that $Eg^4 < \infty$, then, by an appeal to Theorem 3.3.1 (where $Eg^4 < \infty$ ensures the existence of the second moments of the kernels appearing in the components of s_n^2), we conclude that (10.2.59) and (10.2.60) hold. Further, here, $\psi(n) = n$ and $g(n) = n^{1/2}$, and hence (10.2.58) holds. Thus from Theorem 10.2.2 we conclude that as $d \downarrow 0$,

$$\frac{\sqrt{n(d)}\,\{N(d)/n(d) - 1\}}{\gamma^*} \xrightarrow{\mathfrak{D}} \mathfrak{N}(0, 1), \tag{10.2.66}$$

where γ^{*2} is the asymptotic variance of $\sqrt{n}\,(s_n^2/\zeta_1(F) - 1)$.

The case of the population mean $\mu = \theta(F) = \int x\,dF(x)$ is a particular case of $g(x) = x$. Here, $T_n = \bar{X}_n = n^{-1}\sum_{i=1}^n X_i$ and $s_n^2 = (n-1)^{-1} \cdot \sum_{i=1}^n (X_i - \bar{X}_n)^2$. As $s_n^{-1} \leqslant (n-1)^{-1} \cdot \sum_{i=1}^n (X_i - \mu)^2$, (10.2.27) holds with $Z_i = (X_i - \mu)^2$ and $EZ_i = \sigma^2 = E(X_i - \mu)^2$. Hence Theorem 10.2.1 provides (10.2.29) under the sole assumption that $\sigma^2 < \infty$. Also, (10.2.66) holds when EX^4 exists and then $\gamma^{*2} = E(X - \mu)^4/\sigma^4 - 1$.

10.2.2. General Estimable Parameter, II

In Chapter 7, we studied LCFOS (linear combinations of functions of order statistics), both types I and II, which are estimators of suitable functionals of the underlying d.f. If we denote such a functional by $\theta(F)$, we may provide bounded length sequential confidence interval for the same. We may define s_n^2 as in (7.6.34) (where $\hat{\sigma}_n^2$ has been used to denote this estimator). Then $\psi(n) = n$ and Theorem 7.6.3 ensures (10.2.6). Further, (10.2.11) follows from Theorems 7.4.2 through 7.4.7. Hence (10.2.14), (10.2.16), and (10.2.18) hold for the related sequential procedure in (10.2.4). Here it may be difficult to verify (10.2.45) or to apply directly Theorem 10.2.1, but (10.2.55) can be verified under general conditions (see Exercise 10.2.7). Further, Theorem 7.6.5 may be incorporated in the verification of (10.2.59) and (10.2.60). Finally, (10.2.58) holds as $g(n) = n^{1/2}$ and $\psi(n) = n$. In some practical applications (e.g., the estimation of location, scale, or the regression parameter), the LCFOS provide estimators which are more robust than the alternative ones based on the least squares principle or the maximum likelihood principle and therefore the sequential procedures based on these LCFOS will be robust competitors to their parametric counterparts.

In Chapter 8, in (8.6.4), we considered certain functionals of the underlying d.f. and the corresponding point estimators are given by (8.6.3). Here also we may be interested in providing a bounded length sequential confidence interval for the parameter in (8.6.4) (see Sen, 1973d). We define $T_n = Z_n$ by (8.6.3) and (8.7.21) and $s_n^2 = \hat{\sigma}_n^2$ by (8.7.22). Theorems 8.7.1 through 8.7.3 ensure then (10.2.11), while $\psi(n)$ being equal to n, (10.2.7) holds. Further, (8.7.24) ensures (10.2.6), while (8.7.25) ensures (10.2.55). Hence (10.2.14), (10.2.16), (10.2.18), and (10.2.29) all hold in this case. Finally, the asymptotic normality in (10.2.59) and (10.2.60) can be verified in this case by invoking the asymptotic normality of sample quantiles and the continuity of ψ (see Exercise 10.2.10). Hence (10.2.66) also holds. The theory applies to the M-estimators as well.

10.2.3. Type A Sequential Confidence Regions

We now proceed to extend the theory to the case of more than one parameter and intend to construct general confidence sets or regions having similar properties.

Suppose that $\boldsymbol{\theta}$ is a $p(\geqslant 1)$-vector and, as before, let $\mathbf{X}^{(n)} = (X_1, \ldots, X_n)$ for $n \geqslant 1$. Then we consider a sequence $\{\mathbf{T}_n = \mathbf{T}(\mathbf{X}^{(n)})\}$ of estimators (p-vectors) of $\boldsymbol{\theta}$ and we need to construct a region I_n (a closed subspace of the p-dimensional Euclidean space R^p) such that (10.2.1) holds and analogous to (10.2.2), we have

$$\textit{diameter of } I_n \leqslant 2d \qquad \textit{for some prefixed} \quad d(>0), \tag{10.2.67}$$

where the *diameter* may be defined in a meaningful way. For example, if we choose I_n to be an ellipsoid in R^p (with center $\boldsymbol{\theta}$) the diameter of I_n is the maximum diameter of this ellipsoid. Alternatively, if I_n is taken as a rectangle in R^p, the diameter is the width of the largest side of this rectangle. There are other possibilities too. We consider here specifically the case of ellipsoidal confidence regions; the other cases follow on parallel lines.

We assume that $\mathbf{T}_n$ has an asymptotic dispersion matrix $[\psi(n)]^{-1}\boldsymbol{\Sigma}_{\boldsymbol{\theta}}$ where $\psi(n)\uparrow\infty$ as $n\to\infty$ and $\boldsymbol{\Sigma}_{\boldsymbol{\theta}}$ is positive semidefinite (with finite elements). Also, we assume that there exists a sequence $\{\mathbf{S}_n = \mathbf{S}(\mathbf{X}^{(n)}), n \geqslant n_0\}$ of consistent estimators of $\boldsymbol{\Sigma}_{\boldsymbol{\theta}}$, where $n_0(\geqslant 1)$ is a suitable positive integer, and as $n \to \infty$,

$$[\psi(n)]^{1/2}(\mathbf{T}_n - \boldsymbol{\theta}) \xrightarrow{\mathfrak{D}} \mathfrak{N}(\mathbf{0}, \boldsymbol{\Sigma}_{\boldsymbol{\theta}}). \tag{10.2.68}$$

If $\boldsymbol{\Sigma}_{\boldsymbol{\theta}}$ were known, denoting by $ch_1(\boldsymbol{\Sigma}_{\boldsymbol{\theta}})$ the largest characteristic root of $\boldsymbol{\Sigma}_{\boldsymbol{\theta}}$, choosing $n = n(d)$, where

$$n(d) = \text{smallest } k(\geqslant n_0): \quad ch_1(\boldsymbol{\Sigma}_{\boldsymbol{\theta}}) \leqslant d^2\psi(k)/\chi^2_{r,\alpha} \tag{10.2.69}$$

[$\chi^2_{r,\alpha}$ being the upper $100\alpha\%$ point of the chi square d.f. with r degrees of freedom and $r(\geqslant 1)$ being the assumed rank of $\boldsymbol{\Sigma}_{\boldsymbol{\theta}}$] and using the inequality that

$$\sup_{\mathbf{l} \neq \mathbf{0}} \frac{\{\mathbf{l}'(\mathbf{T}_n - \boldsymbol{\theta})\}^2}{(\mathbf{l}'\mathbf{l})} \leqslant \left[\sup_{\mathbf{l} \neq \mathbf{0}} \frac{(\mathbf{l}'\boldsymbol{\Sigma}_{\boldsymbol{\theta}}\mathbf{l})}{(\mathbf{l}'\mathbf{l})}\right](\mathbf{T}_n - \boldsymbol{\theta})'\boldsymbol{\Sigma}_{\boldsymbol{\theta}}^-(\mathbf{T}_n - \boldsymbol{\theta})$$

$$= \frac{ch_1(\boldsymbol{\Sigma}_{\boldsymbol{\theta}})[\psi(n)(\mathbf{T}_n - \boldsymbol{\theta})'\boldsymbol{\Sigma}_{\boldsymbol{\theta}}^-(\mathbf{T}_n - \boldsymbol{\theta})]}{\psi(n)}, \tag{10.2.70}$$

we find that $P\{\boldsymbol{\theta} \in I^*_{n(d)}\} \geqslant 1 - \alpha$, where $I^*_n = \{\mathbf{a} \in R^p : (\mathbf{T}_n - \mathbf{a})'\boldsymbol{\Sigma}_{\boldsymbol{\theta}}^- \cdot (\mathbf{T}_n - \mathbf{a}) \leqslant \chi^2_{r,\alpha}/\psi(n)\}$ has asymptotically (as $d \downarrow 0$) the coverage probability $1 - \alpha$ and by (10.2.69), the maximum diameter of $I^*_{n(d)}$ is equal to $2d$. As such, if $\boldsymbol{\Sigma}_{\boldsymbol{\theta}}$ is not known, as in (10.2.4), we replace, in (10.2.69) and (10.2.70), $\boldsymbol{\Sigma}_{\boldsymbol{\theta}}$ by $\mathbf{S}_n$ and define the stopping variable

$$N = N(d) = \min\left\{k : k \geqslant n_0 \quad \text{and} \quad ch_1(\mathbf{S}_k) \leqslant \frac{d^2\psi(k)}{\chi^2_{r,\alpha}}\right\}, \tag{10.2.71}$$

and the desired confidence region is $I_{N(d)}$, where

$$I_n = \{\boldsymbol{\theta} : |\mathbf{l}'(\mathbf{T}_n - \boldsymbol{\theta})| \leqslant (\mathbf{l}'\mathbf{l})^{1/2}d, \forall \mathbf{l} \neq \mathbf{0}\}, \qquad \text{for} \quad n \geqslant n_0. \tag{10.2.72}$$

Note that for $I^*_{n(d)}$ to have the coverage probability, we have made use of the fact that for large n, $\psi(n)(\mathbf{T}_n - \boldsymbol{\theta})'\boldsymbol{\Sigma}_{\boldsymbol{\theta}}^-(\mathbf{T}_n - \boldsymbol{\theta})$ [or $\psi(n)(\mathbf{T}_n - \boldsymbol{\theta})' \cdot \mathbf{S}_n^-(\mathbf{T}_n - \boldsymbol{\theta})$] has [by virtue of (10.2.57)] the chi square d.f. with r degrees of freedom.

Parallel to (10.2.6), we need to assume here that

$$\mathbf{S}_n \to \boldsymbol{\Sigma}_{\boldsymbol{\theta}} \quad \text{a.s.,} \quad \text{as} \quad n \to \infty, \tag{10.2.73}$$

while (10.2.7) remains unchanged. Further, in addition to (10.2.68), we need to assume that the *tightness part* of (10.2.12) for each of the p marginal processes constructed from the p marginal sequences $\{T_{n,j}\}$, $j = 1, \ldots, p$ (where $\mathbf{T}_n = (T_{n,1}, \ldots, T_{n,p})'$ and $\boldsymbol{\theta} = (\theta_1, \ldots, \theta_p)'$) holds. Then (10.2.7) and (10.2.73) ensure (10.2.14) and, defining $n(d)$ by (10.2.69), we have as in (10.2.13) through (10.2.16), that $N(d)/n(d) \to 1$ a.s., as $d \downarrow 0$. Hence by

(10.2.68) we have for $d \downarrow 0$,

$$(N(d))(\mathbf{T}_{N(d)} - \boldsymbol{\theta})'\mathbf{S}_{N(d)}^{-}(\mathbf{T}_{N(d)} - \boldsymbol{\theta}) \xrightarrow{\mathfrak{D}} \chi_r^2, \tag{10.2.74}$$

so that using (10.2.70) [with $\boldsymbol{\Sigma}_{\boldsymbol{\theta}}$ replaced by $\mathbf{S}_{N(d)}$], (10.2.71), and (10.2.72), we conclude that $P\{\boldsymbol{\theta} \in I_{N(d)} \,|\, \boldsymbol{\theta}\} \to 1 - \alpha$ as $d \downarrow 0$. We may virtually repeat the arguments in (10.2.29) through (10.2.55), provided we replace everywhere s_n^2 by $ch_1(\mathbf{S}_n)$. In view of the inequality

$$0 < ch_1(\mathbf{S}_n) \leqslant \text{trace } \mathbf{S}_n = \text{sum of the diagonal elements of } \mathbf{S}_n, \tag{10.2.75}$$

the existence of the $r(>0)$th moment of the individual elements of $\mathbf{S}_n$ ensures the same for $ch_1(\mathbf{S}_n)$, and hence for (10.2.26) or (10.2.55) to hold [for s_n^2 and σ_θ^2 being replaced by $ch_1(\mathbf{S}_n)$ and $ch_1(\boldsymbol{\Sigma}_{\boldsymbol{\theta}})$, respectively], we do not need more stringent regulatory conditions. It may also be noted that unlike (10.2.20) and (10.2.21), $\{ch_1(\mathbf{S}_n)\}$ may not form a reverse martingale sequence, though a reverse submartingale characterization is often possible. Nevertheless, whenever it is a reverse submartingale [or as in (10.2.22), it can be expressed as a finite mixture of reverse submartingales], (10.2.21) continues to hold for s_n^2 being replaced by $ch_1(\mathbf{S}_n)$. Now, parallel to (10.2.56) and (10.2.57), we have bounds for $ch_1(\mathbf{S}_{N(d)})$ and $ch_1(\boldsymbol{\Sigma}_{\boldsymbol{\theta}})$, where $\tau_{\alpha/2}^2$ has to be replaced by $\chi_{r,\alpha}^2$. Thus parallel to (10.2.59) and (10.2.60) we need to assume that as $d \downarrow 0$,

$$g(n)\left\{ \frac{ch_1(\mathbf{S}_{N(d)})}{ch_1(\boldsymbol{\Sigma}_{\boldsymbol{\theta}})} - 1 \right\} \xrightarrow{\mathfrak{D}} Z, \tag{10.2.76}$$

where Z has a nondegenerate d.f. G and then proceeding as in (10.2.61) and (10.2.62), we conclude that under (10.2.76),

$$g(n)\left\{ \frac{\psi(N(d))}{\psi(n(d))} - 1 \right\} \xrightarrow{\mathfrak{D}} Z, \quad \text{as} \quad d \downarrow 0. \tag{10.2.77}$$

Verification of (10.2.76) is, in general, possible (by using the asymptotic multinormality of $g(n)[\mathbf{S}_n - \boldsymbol{\Sigma}_{\boldsymbol{\theta}}]$), but Z may not have a normal d.f.

The treatment of the rectangular case for general $\boldsymbol{\Sigma}_{\boldsymbol{\theta}}$ is quite involved. Instead of $\chi_{r,\alpha}^2$, in (10.2.69), the percentile point of the maximum of r correlated normal variables (with the covariance matrix $\boldsymbol{\Sigma}_{\boldsymbol{\theta}}$) has to be used. In general, the computation of these percentile points is very cumbersome. However, in many practical applications, $\boldsymbol{\Sigma}_{\boldsymbol{\theta}}$ has some special forms, and, in such cases, the procedure simplifies considerably. We pose some of these as problems.

We conclude this section with the remark that the weak convergence result in Problem 7.∗.4 may be incorporated in providing a bounded percentage-width confidence interval for the population size N; see Darling and Robbins (1967d).

10.3. TYPE B SEQUENTIAL CONFIDENCE REGIONS

The basic problem is the same as in (10.2.1) and (10.2.2). In this case, however, we use some distribution-free statistics to provide robust, distribution-free confidence intervals. Thus, unlike (10.2.3), $\hat{\theta}_{L,n}, \hat{\theta}_{U,n}$ are not based here on the point estimator T_n.

We conceive of a sequence $\{U_n = U(\mathbf{X}^{(n)})\}$ of real-valued statistics satisfying the following conditions:

(a) There is a suitable null hypothesis (H_0) relating to a specific value of θ, such that, under H_0, U_n is a distribution-free statistic for every $n \geqslant 1$.

(b) For every $n(\geqslant 1)$ and θ, there exists a transformation $\mathbf{X}^{(n)} \to \mathbf{X}_\theta^{(n)}$ such that when θ holds, $\mathbf{X}_\theta^{(n)}$ has the same d.f. as of $\mathbf{X}^{(n)}$ under H_0.

(c) For every given $\mathbf{X}^{(n)}$, $U_{n,\theta} = U(\mathbf{X}_\theta^{(n)})$ is monotone in θ.

By virtue of (a), there exist, for every $n(\geqslant 1)$, two constants, say, $U_{n,\alpha}^{(1)}, U_{n,\alpha}^{(2)}$, depending on α, such that

$$P\left\{U_{n,\alpha}^{(1)} \leqslant U_n \leqslant U_{n,\alpha}^{(2)} \mid H_0\right\} \geqslant 1 - \alpha \tag{10.3.1}$$

(where for large n, the $\geqslant$ may be replaced by $=$ under fairly general regularity conditions). Let us then define

$$\hat{\theta}_{L,n} = \inf\left\{\theta : U_{n,\theta} \in \left[U_{n,\alpha}^{(1)}, U_{n,\alpha}^{(2)}\right]\right\},$$
$$\hat{\theta}_{U,n} = \sup\left\{\theta : U_{n,\theta} \in \left[U_{n,\alpha}^{(1)}, U_{n,\alpha}^{(2)}\right]\right\}; \tag{10.3.2}$$

$$I_n = \{\theta : \hat{\theta}_{L,n} \leqslant \theta \leqslant \hat{\theta}_{U,n}\}. \tag{10.3.3}$$

Then, by virtue of (b), (c), and (10.3.1),

$$P\{\theta \in I_n\} \geqslant 1 - \alpha, \qquad \forall n \geqslant 1. \tag{10.3.4}$$

The sequential procedure consists in defining the *stopping variable* $N(d)$ as

$$N(d) = \min\{n : \hat{\theta}_{U,n} - \hat{\theta}_{L,n} \leqslant 2d\} \tag{10.3.5}$$

and then letting the confidence interval as

$$I_{N(d)} = \left\{\theta : \hat{\theta}_{L,N(d)} \leqslant \theta \leqslant \hat{\theta}_{U,N(d)}\right\}. \tag{10.3.6}$$

Let us assume that (10.2.7) holds and $[\psi(n)]^{1/2}U_n$ (under H_0) has asymptotically a normal distribution with mean 0 and variance A^2. Further, we assume that for some $\nu(0 < \nu < \infty)$,

$$\left[\psi(n)\right]^{1/2}(\hat{\theta}_{L,n} - \theta) + \nu\tau_{\alpha/2} \xrightarrow{\mathfrak{D}} \mathfrak{N}(0, 1), \tag{10.3.7}$$

$$\left[\psi(n)\right]^{1/2}(\hat{\theta}_{U,n} - \hat{\theta}_{L,n}) - 2\nu\tau_{\alpha/2} \xrightarrow{P} 0 \qquad \text{as} \quad n \to \infty; \tag{10.3.8}$$

for every $\epsilon > 0$ and $\eta > 0$, there exist a $\delta(>0)$ and an integer n^0 such that

$$P\left\{\max_{m:|m-n|<\delta n}\left[\psi(n)\right]^{1/2}|\hat{\theta}_{L,n}-\hat{\theta}_{L,m}|>\epsilon\right\}<\eta, \qquad \forall n \geqslant n^0, \tag{10.3.9}$$

$$P\left\{\max_{m:|m-n|<\delta n}\left[\psi(n)\right]^{1/2}|\hat{\theta}_{U,n}-\hat{\theta}_{U,m}|>\epsilon\right\}<\eta, \qquad \forall n \geqslant n^0. \tag{10.3.10}$$

Finally, let us define

$$n(d)=\min\left\{n:\psi(n)\geqslant d^{-2}\tau_{\alpha/2}^2\nu^2\right\}, \qquad d>0. \tag{10.3.11}$$

Note that by (10.3.5),

$$\begin{aligned}\psi(N(d)-1)&\left[\hat{\theta}_{U,N(d)-1}-\hat{\theta}_{L,N(d)-1}\right]\\ &>\left[\psi(N(d)-1)/\psi(N(d))\right]2d\psi(N(d))\\ &\geqslant\left[\psi(N(d)-1)/\psi(N(d))\right]\psi(N(d))\left[\hat{\theta}_{U,N(d)}-\hat{\theta}_{L,N(d)}\right]\end{aligned} \tag{10.3.12}$$

and hence, by (10.2.7), (10.3.8), (10.3.11), and (10.3.12),

$$\psi(N(d))/\psi(n(d))\xrightarrow{P}1\left(\Rightarrow N(d)/n(d)\xrightarrow{P}1\right) \qquad \text{as} \quad d\downarrow 0. \tag{10.3.13}$$

Also, by (10.3.7), (10.3.9), and (10.3.13),

$$\left[\psi(N(d))\right]^{1/2}(\hat{\theta}_{L,N(d)}-\theta)+\tau_{\alpha/2}\nu\xrightarrow{\mathfrak{D}}\mathfrak{N}(0,1) \qquad \text{as} \quad d\downarrow 0 \tag{10.3.14}$$

and, similarly, by (10.3.8) through (10.3.10) and (10.3.13),

$$\left[\psi(N(d))\right]^{1/2}(\hat{\theta}_{U,N(d)}-\hat{\theta}_{L,N(d)})\xrightarrow{P}2\nu\tau_{\alpha/2} \qquad \textit{as} \quad d\downarrow 0. \tag{10.3.15}$$

From (10.3.6), (10.3.14), and (10.3.15), we conclude that

$$\lim_{d\downarrow 0}P\left\{\theta\in I_{N(d)}\right\}=1-\alpha \qquad (\textit{asymptotic consistency}). \tag{10.3.16}$$

We may also remark that if we replace in (10.3.8) "$\xrightarrow{P}0$" by "a.s.$\to$0," then in (10.3.13) we also have strong convergence. Further, as in (10.2.24), we have for every $d>0$,

$$EN(d)=1+\sum_{n\geqslant 1}P\left\{N(d)>n\right\}, \tag{10.3.17}$$

where by (10.3.5),

$$\begin{aligned}P\left\{N(d)>n\right\}&\leqslant P\left\{\hat{\theta}_{U,n}-\hat{\theta}_{L,n}>2d\right\}\\ &\leqslant(2d)^{-r}E(\hat{\theta}_{U,n}-\hat{\theta}_{L,n})^r \qquad (r>0).\end{aligned} \tag{10.3.18}$$

Thus, for $EN(d)<\infty$, it suffices to assume that

$$\sum_{n\geqslant 1}E(\hat{\theta}_{U,n}-\hat{\theta}_{L,n})^r<\infty \qquad \text{for some} \quad r>0. \tag{10.3.19}$$

Further, we may also rewrite $P\{\hat{\theta}_{U,n}-\hat{\theta}_{L,n}>2d\}$ as

$$P\left\{\left[\psi(n)\right]^{1/2}(\hat{\theta}_{U,n}-\hat{\theta}_{L,n})-2\nu\tau_{\alpha/2}>2d\left[\psi(n)\right]^{1/2}-2\nu\tau_{\alpha/2}\right\}, \tag{10.3.20}$$

where by (10.3.12), $d[\psi(n)]^{1/2} - \nu\tau_{\alpha/2}$ is positive for all $n \geqslant n(d)$ and it goes to $+\infty$ as $n \to \infty$. Thus an alternative sufficient condition for $EN(d) < \infty$ ($\forall d > 0$) is that for some $r > 1$ and every $\eta > 0$, there exist a finite c_η and an n_η such that

$$P\{|[\psi(n)](\hat{\theta}_{U,n} - \hat{\theta}_{L,n}) - 2\nu\tau_{\alpha/2}| > \eta\} < c_\eta n^{-r}, \qquad \forall n \geqslant n_\eta. \quad (10.3.21)$$

In fact, we may virtually repeat the steps in (10.2.48) through (10.2.56) and show that under (10.3.21),

$$\lim_{d \downarrow 0} \frac{EN(d)}{n(d)} = 1 \qquad (\textit{asymptotic efficiency}). \quad (10.3.22)$$

Let us next consider the asymptotic normality of the stopping time $N(d)$. Note that by (10.3.5), letting $L_n = \hat{\theta}_{U,n} - \hat{\theta}_{L,n}$, $n \geqslant 1$,

$$L_{N(d)}\psi(N(d)) \leqslant 2d\psi(N(d)) < L_{N(d)-1}\psi(N(d)), \quad (10.3.23)$$

where by (10.3.8), $\sqrt{\psi(n)}\, L_n \xrightarrow{P} 2\lambda$ and $\lambda = \nu\tau_{\alpha/2}$. Thus for every real y and any sequence $\{g(n)\}$ of positive numbers,

$$P_\theta\left\{ g(n(d))\left\{ d\sqrt{\psi(N(d))} - \lambda \right\}/\lambda \leqslant y \right\}$$
$$\leqslant P_\theta\left\{ g(n(d))\left\{ \sqrt{\psi(N(d))}\, L_{N(d)} - 2\lambda \right\}/\lambda \leqslant 2y \right\} \quad (10.3.24)$$

and

$$P_\theta\left\{ g(n(d))\left\{ d\sqrt{\psi(N(d)-1)} - \lambda \right\}/\lambda \geqslant y \right\}$$
$$\leqslant P_\theta\left\{ g(n(d))\left\{ \sqrt{\psi(N(d)-1)}\, L_{N(d)-1} - 2\lambda \right\}/\lambda \geqslant 2y \right\}. \quad (10.3.25)$$

Hence if we assume that there exist a nondecreasing sequence $\{g(n)\}$ of positive numbers (with $\lim_{n\to\infty} g(n) = \infty$) and a positive constant ν^* ($< \infty$) such that as $n \to \infty$,

$$P_\theta\left\{ g(n)\left(\sqrt{\psi(n)}\, L_n - 2\lambda \right)/2\lambda \leqslant y \right\} \to \Phi(y/\nu^*), \qquad \forall y \in E \quad (10.3.26)$$

(where Φ is the standard normal d.f.) and for every $\epsilon > 0$ and $\eta > 0$, there exist a positive number $n_0 = n_0(\epsilon, \eta)$ and a $\delta (> 0)$ such that for every $n \geqslant n_0$,

$$P_\theta\left\{ \max_{m\,:\,|\psi(m)/\psi(n) - 1| < \delta} g(n)|\sqrt{\psi(m)}\, L_m - \sqrt{\psi(n)}\, L_n| > \epsilon \right\} < \eta, \quad (10.3.27)$$

then from (10.3.24) through (10.3.27), we find that as $d \downarrow 0$,

$$g(n(d))\left\{ \lambda^{-1} d\sqrt{\psi(N(d))} - 1 \right\} \xrightarrow{\mathcal{D}} \mathcal{N}(0, \nu^{*2}). \quad (10.3.28)$$

We may note at this stage that by (10.3.11) $\lambda^{-1}d \sim \sqrt{\psi(n(d))}$, as $d\downarrow 0$. Thus if we assume that there exists a positive $\beta(<\infty)$, such that as $n\to\infty$ and for every $m: |m/n-1|\to 0$,

$$\sqrt{\psi(m)/\psi(n)} - 1 = \beta(m/n-1)\{1+o(1)\}, \qquad (10.3.29)$$

then, from (10.3.28) and (10.3.29), we arrive at the following. Under (10.3.13), (10.3.26), (10.3.27), and (10.3.29), as $d\downarrow 0$,

$$g(n(d))\{N(d)/n(d)-1\} \xrightarrow{\mathfrak{D}} \mathfrak{N}(0, \nu^{*2}/\beta^2). \qquad (10.3.30)$$

Note that in the usual case of $\psi(n)=n$, (10.3.29) holds with $\beta=\frac{1}{2}$. Also, (10.3.26) and (10.3.27) would follow from an invariance principle for the sequence $\{g(n)\sqrt{\psi(k)}\,L_k;\ k\leqslant n\}$.

We now proceed to consider some specific problems of special interest and, with the aid of the principal results in Chapters 4, 5, 7, and 8, verify the foregoing conditions in these contexts.

10.3.1. Population Quantile

Let $\{X_i,\ i\geqslant 1\}$ be i.i.d.r.v. and a d.f. F, defined on the real line $(-\infty,\infty)$. For some p: $0<p<1$, we assume that $F(x)=p$ has a unique solution ξ_p and in some neighborhood of ξ_p, F is absolutely continuous with a continuous and positive p.d.f. f. We desire to provide a bounded length (sequential) confidence interval for $\xi_p=\theta$.

Let $U_n = n^{-1}\sum_{i=1}^n c(\theta - X_i)$ [where $c(u)=1$ or 0 according as u is $\geqslant$ or <0] and $\psi(n)\equiv n$. Then, under $H_0:\theta=0$, nU_n has a binomial (n,p) distribution (whatever be the underlying F), so condition (a) holds. Also, $\mathbf{X}_\theta^{(n)} = \mathbf{X}^{(n)} - \theta\mathbf{1}_n$ and $U_{n,\theta}$ is $\searrow$ in θ. Hence (b) and (c) hold. Further, (10.3.1) holds with $r_n = nU_{n,\alpha}^{(1)}$ and $s_n = nU_{n,\alpha}^{(2)}$ with $P\{B(n,p)<r_n\}\leqslant\frac{1}{2}\alpha$, $P\{B(n,p)>s_n\}\leqslant\frac{1}{2}\alpha$, where $B(n,p)$ has the binomial d.f. with parameters (n,p). Thus if $X_{n,1}\leqslant\cdots\leqslant X_{n,n}$ are the order statistics corresponding to $X_1,\ldots,X_n$, then

$$\hat\theta_{L,n} = X_{n,r_n} \quad\text{and}\quad \hat\theta_{U,n} = X_{n,s_n}; \qquad (10.3.31)$$

$$\frac{r_n - np}{\sqrt{np(1-p)}} \to -\tau_{\alpha/2}, \quad \frac{s_n-np}{\sqrt{np(1-p)}} \to \tau_{\alpha/2}. \qquad (10.3.32)$$

Also, if F_n is the empirical d.f. based on $\mathbf{X}^{(n)}$, then $F_n(\hat\theta_{L,n}) = n^{-1}r_n$ and $F_n(\hat\theta_{U,n}) = n^{-1}s_n$. Let $\tilde X_n = X_{n,[np]+1}$ be the sample p-quantile. Then, by an appeal to Theorem 7.3.1, we have $n^{1/2}(\tilde X_n - \theta)f(\theta) + n^{1/2}(p - F_n(\theta))\to 0$ a.s., as $n\to\infty$, $n^{1/2}(\hat\theta_{L,n} - \tilde X_n)f(\theta) + \tau_{\alpha/2}\sqrt{p(1-p)} \to 0$ a.s., as $n\to\infty$ and

$n^{1/2}(\hat{\theta}_{U,n} - \tilde{X}_n)f(\theta) - \tau_{\alpha/2}\sqrt{p(1-p)} \to 0$ a.s., as $n \to \infty$. Thus (10.3.7) and (10.3.8) hold with $\nu^2 = p(1-p)/f^2(\theta)$. Further, $\{F_n(\theta) - F(\theta);\ n \geqslant 1\}$ is a reverse martingale, and hence by using the Kolmogorov inequality, we find

$$P\left\{\max_{m\,:\,n \leqslant m \leqslant n+[\delta n]} n^{1/2}|F_m(\theta) - F_n(\theta)| > \epsilon\right\}$$

$$\leqslant \epsilon^{-2}E\left\{n\left[F_{n+[n\delta]}(\theta) - F_n(\theta)\right]^2\right\} \leqslant \frac{\delta p(1-p)}{\epsilon^2} \to 0 \quad \text{as} \quad \delta \downarrow 0, \tag{10.3.33}$$

and a similar result holds for $n - [n\delta] \leqslant m \leqslant n$. Using (10.3.33) along with the three a.s. convergence results stated above, it follows that both (10.3.9) and (10.3.10) hold here with $\psi(n) \equiv n$. Hence (10.3.16) holds. Further, using (7.3.9) instead of Theorem 7.3.1, it follows that (10.3.21) holds, and hence (10.3.22) also holds with $\nu^2 = p(1-p)/f^2(\theta)$. Finally, (10.3.30) holds with $g(n) = n^{1/4}$, $\beta = \frac{1}{2}$, and $\nu^{*2} = (4p(1-p)\tau^2_{\alpha/2})^{-1/2}$. Actually, the sample quantile is a particular case of M-estimators, treated in Chapter 8 (here the score function has only one jump), and hence verification of (10.3.26) and (10.3.27) can be made more easily by using general results on such M-estimators (to be considered later). Therefore, we pose this as an exercise.

10.3.2. Location Parameter of a Symmetric d.f.

Let $\{X_i,\ i \geqslant 1\}$ be i.i.d.r.v. with an absolutely continuous d.f. F, symmetric about its median θ. We intend to provide a bounded length confidence interval for θ.

We incorporate the signed rank statistics of Chapter 5 to derive the estimators $(\hat{\theta}_{L,n}, \hat{\theta}_{U,n})$ as well as $N(d)$. Define $U_n = n^{-1}S_n$ where S_n is defined by (5.2.1) and also define $S_n(b)$, $-\infty < b < \infty$ as in (5.5.1). Then, under $H_0: \theta = 0$, S_n is distribution-free with mean 0 and variance nA_n^2, where A_n^2 is defined by (5.2.5). Further, under H_0,

$$n^{-1/2}S_n \sim \mathfrak{N}(0, A^2); \qquad A^2 \text{ defined by (5.3.1).} \tag{10.3.34}$$

Also $\mathbf{X}_\theta^{(n)} = \mathbf{X}^{(n)} - \theta\mathbf{1}_n$ and $S_n(b) = nU_{n,b}$ is $\searrow$ in b: $-\infty < b < \infty$. Thus, conditions (a), (b), and (c) and (10.3.1) all hold with

$$n^{1/2}U_{n,\alpha}^{(j)} \to (-1)^j A\tau_{\alpha/2} \qquad \text{as} \quad n \to \infty, \qquad \text{for } j = 1, 2; \tag{10.3.35}$$

$$\begin{aligned} \hat{\theta}_{L,n} &= \sup\left\{b : S_n(b) > nU_{n,\alpha}^{(2)}\right\}, \\ \hat{\theta}_{U,n} &= \inf\left\{b : S_n(b) < nU_{n,\alpha}^{(1)}\right\}. \end{aligned} \tag{10.3.36}$$

Note that, by definition, for any real c, $P\{\hat{\theta}_{L,n} < \theta - n^{-1/2}c\} = P\{\hat{\theta}_{L,n}$

$< -n^{-1/2}c \mid H_0\} = P\{S_n(-n^{-1/2}c) \leqslant nU_{n,\alpha}^{(2)} \mid H_0\}$ and a similar case holds for $\{\hat{\theta}_{U,n}\}$. Hence using Theorem 5.3.4, it follows that for every $\epsilon > 0$, there exist a positive $c_\epsilon(< \infty)$ and an n_ϵ such that for $n \geqslant n_\epsilon$,

$$P\{\theta - n^{-1/2}c_\epsilon \leqslant \hat{\theta}_{L,n} < \hat{\theta}_{U,n} \leqslant \theta + n^{-1/2}c_\epsilon\} \geqslant 1 - \epsilon. \quad (10.3.37)$$

As such, by (10.3.36) and (10.3.37) and Theorem 5.5.2, we conclude that (10.3.7) holds. Similarly, by (10.3.37) and Theorems 5.5.2 and 5.5.3, it follows that (10.3.8) through (10.3.10) all hold. As a result, (10.3.16) holds.

In the above development it is sufficient to assume that the score function $\phi(u)$ in nondecreasing and square integrable inside I and the d.f. F has an absolutely continuous p.d.f. f with a finite Fisher information $I(f)$. For (10.3.22), demanding more than weak convergence, we need to verify (10.3.21), and this, in turn, needs more stringent regularity conditions. Assume that F is symmetric and has bounded (a.e.) first and second derivatives with a finite $I(f)$. Also, recall that $\phi(u) = \phi^*(\frac{1}{2}(1+u))$, $0 < u < 1$ where $\phi^*(u)$ is $\nearrow$ and skew-symmetric in $u \in I$. We assume that

$$|(\partial^r/\partial u^r)\phi(u)| \leqslant K(1-u)^{-\delta-r}, \quad r = 0, 1, 2; \quad 0 < u < 1, \quad (10.3.38)$$

where $K(< \infty)$ and $\delta(< \frac{1}{6})$ are positive constants. Let

$$g(n) = n^{-1/2}(\log n). \quad (10.3.39)$$

Then, by repeating the proof of Theorem 5.5.4 (see Theorem A.4.2), we arrive at the following. Under $H_0: \theta = 0$ and the conditions stated above, for every $s(< (1-2\delta)/2\delta)$ and $\epsilon > 0$ there exist a positive constant $k_s(\epsilon)$ and a sample size $n_s(\epsilon)$ such that for every $n \geqslant n_s(\epsilon)$,

$$P\left\{\sup_{b : |b| \leqslant g(n)} |S_n(b) - S_n(0) + nbB^*(F)| > \epsilon\sqrt{n}\right\} \leqslant k_s(\epsilon)n^{-s}, \quad (10.3.40)$$

where

$$B^*(F) = \int_{-\infty}^{\infty} \phi^*(F(x))\{-f'(x)/f(x)\}\, dF(x). \quad (10.3.41)$$

We also note that under $H_0: \theta = 0$,

$$\begin{aligned} E[\exp\{tS_n(0)\}] &= E\{E[\exp\{tS_n(0)\} \mid \mathbf{R}_n^+]\} \\ &= E\left\{\prod_{i=1}^{n} \tfrac{1}{2}[\exp(ta_n(R_{ni}^+)) + \exp(-ta_n(R_{ni}^+))]\right\} \\ &\leqslant E\left[\prod_{i=1}^{n} \exp\{\tfrac{1}{2}t^2a_n^2(R_{ni}^+)\}\right] \\ &= \exp\left\{\tfrac{1}{2}t^2\sum_{i=1}^{n} a_n^2(i)\right\} = \exp\left\{\frac{n}{2}t^2A_n^2\right\}. \end{aligned} \quad (10.3.42)$$

Therefore, for every $c > 0$,

$$P\{|S_n(0)| > cn^{1/2}(\log n)^{1/2}\}$$
$$\leqslant 2\Big[\inf_{t>0} \exp\big(-ctn^{1/2}(\log n)^{1/2}\big)E\{\exp(tS_n(0))\}\Big]$$
$$= 2\exp\left\{-\frac{\frac{1}{2}c^2(\log n)}{A_n^2}\right\}. \tag{10.3.43}$$

The rhs of (10.3.43) can be made $O(n^{-s})$ (for any $s > 0$) by proper choice of c. Hence by using (10.3.40) and (10.3.43) along with the definitions of $\hat{\theta}_{L,n}, \hat{\theta}_{U,n}$, it follows that for every $s > 0$, there exist positive numbers $(c_s^{(1)}, c_s^{(2)})$ and a sample size n_s such that for $n \geqslant n_s$,

$$P\{n^{1/2}(\hat{\theta}_{L,n} - \theta) + \tau_{\alpha/2}A/B^*(F) < -c_s^{(1)}(\log n)\} \leqslant c_s^{(2)}n^{-s}, \tag{10.3.44}$$

$$P\{n^{1/2}(\hat{\theta}_{U,n} - \theta) - \tau_{\alpha/2}A/B^*(F) > c_s^{(1)}(\log n)\} \leqslant c_s^{(2)}n^{-s}. \tag{10.3.45}$$

Thus, letting $s > 1$ and using (10.3.40), (10.3.44), and (10.3.45), we find that (10.3.21) holds with $\nu = A/B^*(F)$. Hence (10.3.22) also holds.

To verify (10.3.30), we need to strengthen (10.3.40) into an invariance principle for $\{S_k(N^{-1/2}t) - S_k(0) + N^{-1/2}tkB^*(F)\} = W_N^*(k/N, t)$, say, for $k \leqslant N$ and $|t| \leqslant K(< \infty)$. The weak convergence of W_N^* to some two-dimensional time-parameter Gaussian function enables us to verify (10.3.26) and (10.3.27); here also, $g(n)$ will be $n^{1/2}$. This may be termed the second-order linearity theorem for signed rank statistics. Antille (1972) has studied such a result for a class of bounded score functions. The main difficulty lies in some sort of expansion of $S_k(N^{-1/2}t) - S_k(0)$ retaining terms of the first and second orders. We pose some of these problems as exercises. For some related results on the Wilcoxon two-sample process, we may refer to Jurečková (1973).

As in Section 8.2, we may employ M-estimators for obtaining bounded-length sequential confidence intervals for θ. In (8.2.6), we take the c_i all equal to 1 [so that we have the location model in (8.2.1)] and the resulting solution in (8.2.7) (the M-estimator of θ) is denoted by $\tilde{\theta}_N$ for $N \geqslant 1$. Moreover, in (8.3.13), replacing $\hat{\Delta}_N$ by $\tilde{\theta}_N$ and all the c_i by 1, we denote the resulting estimator by s_N^{02}. Further, we define $\tilde{\theta}_{NU}$, $\tilde{\theta}_{NL}$, and L_N as in (8.3.19) and (8.3.20) with the same modification for the c_i and C_N. Then the sequential confidence interval for θ is $I_{N_d} = (\tilde{\theta}_{N_dL}, \tilde{\theta}_{N_dU})$, where the stopping variable N_d is defined by

$$N_d = \min\{N \geqslant n_0 : L_N \leqslant 2d\}, \tag{10.3.46}$$

and n_0 is the initial sample size ($\geqslant 2$).

In this case, (10.3.7) through (10.3.11) follow from Theorems 8.3.1 and 8.3.2 and (8.3.21). Further, here $\psi(n) \equiv n$, so that (10.3.13) and (10.3.16) follow from the above results. By (8.4.20) and the Markov inequality [on $(s_n^2 - \sigma^2)$], we obtain that under the condition that $E_0[\psi(X_i)]^6 < \infty$,

$$P\{|s_n^{02} - \sigma^2| > \epsilon\} \leqslant c(\epsilon)n^{-3/2}, \qquad \forall n \geqslant n_0(\epsilon), \tag{10.3.47}$$

where $c(\epsilon)$ is finite for every $\epsilon > 0$. Then, by (10.3.47), (8.4.1), (8.4.2), (8.4.19) and proceeding as in (8.4.6) through (8.4.10), we conclude that (8.4.10) holds for both $n^{1/2}(\tilde{\theta}_{nL} - \theta)$ and $n^{1/2}(\tilde{\theta}_{nU} - \theta)$. Hence by (8.4.15) we conclude that in this case (10.3.21) holds and this, in turn, ensures (10.3.22).

To verify (10.3.30), in this case, because $\psi(n) \equiv n$ and (10.3.13) holds, we need to verify only (10.3.26) and (10.3.27), where $g^2(n)$ may be equal to n or $n^{1/2}$, depending on the nature of the score function ψ. For this purpose, we need to consider an invariance principle for the two-dimensional time-parameter process $\{S_k(N^{-1/2}t) - S_k(0) + N^{-1/2}tk\gamma : k \leqslant N, |t| \leqslant K\}$, where the normalizing factor ($N^{-1/2}$ or $N^{-3/4}$) depends on the nature of the score function ψ. In fact, Problems 8.3.4 and 8.3.6 relate to this. Thus if the score function is absolutely continuous on any bounded interval in R and the other regularity conditions on ψ hold (the conditions on the c_{ni} and d_{ni} hold here as $c_{ni} = d_{ni} = n^{-1/2}$, $i = 1, \ldots, n$; $n \geqslant 1$) then (10.3.30) holds with $g(n) = n^{1/2}$. On the other hand, if ψ admits of a step function satisfying the regularity conditions of the Problem 8.3.6, then (10.3.30) holds with $g(n) = n^{1/4}$.

At this stage it is appropriate to include some discussion of the comparison of rival procedures for a common problem. For example, for the problem of finding a bounded-length sequential confidence interval for the location of a symmetric d.f., we have considered procedures based on the signed statistics, signed rank statistics, and M-estimators. The smallness of the ASN [i.e., $EN(d)$] seems to be a natural criterion for this purpose [as otherwise the procedures all lead to the same (asymptotic) coverage probability and the same width of the intervals]. First, suppose that we confine ourselves to the location parameter of a d.f. symmetric about its median. If the d.f. admits of the mean (μ) and variance σ^2, then we may employ a type A procedure based on the sequence of sample means $\{\overline{X}_n\}$ and variances $\{s_n^2\}$. If $N_m(d)$ is the corresponding stopping variable, we have then

$$\lim_{d\downarrow 0} d^2 EN_m(d) = \sigma^2\tau_{\alpha/2}^2. \tag{10.3.48}$$

We may also consider a type B procedure based on the sample quantiles. If $N_q(d)$ denote the corresponding stopping variable, from Section 10.3.1, we conclude that (for $p = \frac{1}{2}$),

$$\lim_{d\downarrow 0} d^2 EN_q(d) = \tau_{\alpha/2}^2/4f^2(\theta). \tag{10.3.49}$$

We may also employ a type B procedure based on rank order estimators and on denoting by $N_r(d)$, the corresponding stopping variable, we obtain, from Section 10.3.2,

$$\lim_{d\downarrow 0} d^2 EN_r(d) = \tau^2_{\alpha/2} A^2 (B^*(F))^{-2}. \tag{10.3.50}$$

From this we conclude that the asymptotic relative efficiency (ARE) of the median procedure with respect to the mean is

$$e_{q,m} = \lim_{d\downarrow 0} \frac{EN_m(d)}{EN_q(d)} = 4\sigma^2 f^2(\theta), \tag{10.3.51}$$

and for the rank procedure with respect to the mean is

$$e_{r,m} = \lim_{d\downarrow 0} \frac{EN_m(d)}{EN_r(d)} = \sigma^2 \left(\frac{B^*(F)}{A} \right)^2. \tag{10.3.52}$$

These expressions agree with the conventional Pitman ARE results, discussed in detail in Section 9.7. A similar case holds with the M-estimators. Hence the Pitman ARE results, conceived originally for the nonsequential case, remain valid for the sequential confidence interval problem too. Referring back to Section 9.7, we conclude therefore that the ARE of the rank procedure based on the normal scores signed rank statistics with respect to the Chow-Robbins procedure is bounded from below by 1, where this lower bound is achieved only when the underlying d.f. is normal. This clearly indicates the supremacy of the nonparametric over the Chow-Robbins procedure. Also, asymptotically optimal rank procedures (for specified type of d.f.'s) can be obtained under very general conditions.

10.3.3. Regression Parameter

Consider the simple regression model: $X_i = \beta_0 + \beta c_i + e_i$, $i \geqslant 1$, where β_0, β are unknown parameters, $\{c_i, i \geqslant 1\}$ is a sequence of known (regression) constants and the e_i are i.i.d.r.v. with a continuous d.f. F, defined on $(-\infty, \infty)$. We next provide a bounded length (sequential) confidence interval for the regression parameter β.

In this case, we take $U_n = C_n^{-2} L_n$ where L_n is a suitable linear rank statistic, defined by (4.2.1), and C_n^2 is defined by (4.2.9). Then, under H_0: $\beta = 0$, L_n is distribution-free with 0 mean, variance $C_n^2 A_n^2$, where A_n^2 is defined by (4.2.6) and

$$L_n / A_n C_n \xrightarrow{\mathfrak{D}} \mathfrak{N}(0, 1). \tag{10.3.53}$$

Let us define $U_n(b) = C_n^{-2} L_n(b)$ where $L_n(b)$ is defined by (4.5.1). Then $\mathbf{X}_\theta^{(n)} = \mathbf{X}^{(n)} - \theta \mathbf{c}_n$ where $\mathbf{c}_n = (c_1, \ldots, c_n)$ and $U_n(b)$ is $\searrow$ in $b: -\infty < b$

$< \infty$. Thus all the conditions (a), (b), and (c) and (10.3.1) hold with $\psi(n) = C_n^2$ and

$$C_n U_{n,\alpha}^{(j)} \to (-1)^j A \tau_{\alpha/2} \quad \text{as} \quad n \to \infty, \qquad \text{for} \quad j = 1, 2, \tag{10.3.54}$$

where A is defined by (4.3.12). Also, letting $\beta = \theta$, we have here

$$\begin{aligned} \hat{\theta}_{L,n} &= \sup\{ b : L_n(b) > C_n^2 U_{n,\alpha}^{(2)} \}, \\ \hat{\theta}_{U,n} &= \inf\{ b : L_n(b) < C_n^2 U_{n,\alpha}^{(1)} \}. \end{aligned} \tag{10.3.55}$$

By arguments similar to those leading to (10.3.37), we obtain that for every $\epsilon > 0$, there exist a $c_\epsilon (< \infty)$ and a sample size n_ϵ, such that for $n \geqslant n_\epsilon$,

$$P\{\theta - C_n^{-1} c_\epsilon \leqslant \hat{\theta}_{L,n} < \hat{\theta}_{U,n} \leqslant \theta + C_n^{-1} c_\epsilon\} \geqslant 1 - \epsilon. \tag{10.3.56}$$

As such, by (10.3.54) and (10.3.56) and Theorem 4.5.2, we conclude that (10.3.7) holds and, in addition, Theorem 4.5.3 leads to (10.3.8) through (10.3.10). Hence (10.3.16) holds under general regularity conditions.

For (10.3.22) to hold here also, we need more stringent regularity conditions. We assume that the d.f. F has bounded (a.e.) first and second derivatives. Also, parallel to (10.3.38), we assume here that

$$|(d/du)\phi(u)| \leqslant K[u(1-u)]^{-3/2+\delta}, \quad \forall u \in (0,1), \tag{10.3.57}$$

where K and δ $(>\frac{1}{4})$ are positive constants. We also assume that

$$\lim_{x \to \pm\infty} \phi(F(x)) f(x) \tag{10.3.58}$$

are finite. Finally, in addition to (10.2.7) and (10.2.23), we assume that

$$\overline{\lim} \left\{ \max_{1 \leqslant i \leqslant n} \frac{n(c_i - \bar{c}_n)^2}{C_n^2} \right\} < \infty. \tag{10.3.59}$$

Then we have the following result whose proof is given in the Appendix (see Theorem A.4.1).

Under $\beta = 0$ and the regularity conditions mentioned above, for every $s (< 2\delta/(1-2\delta))$ and $\epsilon (> 0)$, there exist a sample size $N_{s\epsilon}$ and a positive constant $K_{s\epsilon} (< \infty)$, such that for every $N \geqslant N_{s\epsilon}$ and any (given) $C (< \infty)$,

$$P\left\{ \sup_{b : |b| < C \log N} |\tilde{L}_N(b) - \tilde{L}_N(0) + bB(F)| > \epsilon \right\} \leqslant K_{s\epsilon} N^{-s}, \tag{10.3.60}$$

where $B(F) = \int_{-\infty}^{\infty} (d/dx)\phi(F(x))\, dF(x)$ and $\tilde{L}_N(b)$ is defined as in Section 4.5.

Let $c_{ni}^* = (c_i - \bar{c}_n)/C_n$, $1 \leqslant i \leqslant n$, so that $\sum_{i=1}^n c_{ni}^* = 0$, $\sum_{i=1}^n c_{ni}^{*2} = 1$ and by (10.3.48), $\max_{1 \leqslant i \leqslant n} |c_{ni}^*| = O(n^{-1/2})$. Also, let $L_n^* = C_n^{-1} L_n$

$= \sum_{i=1}^{n} c_{ni}^{*}[a_n(R_{ni}) - \bar{a}_n]$, $n \geqslant 1$. Then, by the Hoeffding (1963) inequality,

$$E\{\exp(tL_n^*) \mid H_0\} \leqslant \prod_{i=1}^{n} E\{\exp(tc_{ni}^*[a_n(R_{ni}) - \bar{a}_n]) \mid H_0\}, \qquad \forall t > 0, \tag{10.3.61}$$

where

$$\begin{aligned} E\{\exp(tc_{ni}^*[a_n(R_{ni}) - \bar{a}_n]) \mid H_0\} &= n^{-1} \sum_{j=1}^{n} \exp\{tc_{ni}^*[a_n(j) - \bar{a}_n]\} \\ &= 1 + tc_{ni}^* n^{-1} \sum_{j=1}^{n} [a_n(j) - \bar{a}_n] \\ &\quad + \tfrac{1}{2} t^2 c_{ni}^{*2} n^{-1} \sum_{j=1}^{n} [a_n(j) - \bar{a}_n]^2 + O(|tc_{ni}^*|^3) \\ &= 1 + \tfrac{1}{2} t^2 c_{ni}^{*2} \cdot \frac{n-1}{n} A_n^2 + O(n^{-1/2}|t| c_{ni}^{*2} t^2), \qquad \forall 1 \leqslant i \leqslant n. \end{aligned} \tag{10.3.62}$$

Hence from (10.3.61) and (10.3.62), we obtain that for $|t| = o(n^{1/6})$,

$$\begin{aligned} E\{\exp(tL_n^*) \mid H_0\} &\leqslant \left[\frac{1}{n} \sum_{i=1}^{n} E\{\exp(tc_{ni}^*[a_n(R_{ni}) - \bar{a}_n]) \mid H_0\} \right]^n \\ &= \left[1 + \tfrac{1}{2} t^2 \frac{n-1}{n^2} A_n^2 + O(n^{-1/2}|t^3|) \frac{1}{n} \right]^n \\ &= \exp\{\tfrac{1}{2} t^2 A_n^2\}\{1 + o(1)\}. \end{aligned} \tag{10.3.63}$$

Thus using (10.3.63) and proceeding as in (10.3.43), we obtain that for every $c > 0$,

$$P_0\{|L_n^*| > c \log n\} \leqslant 2 \exp\{-\tfrac{1}{2} c^2 (\log n)^2 / A_n^2\}. \tag{10.3.64}$$

As such, by using (10.3.60) and (10.3.64) along with the definitions of $(\hat{\theta}_{L,n}, \hat{\theta}_{U,n})$, it follows that for every $s > 0$, there exist positive numbers $(c_s^{(1)}, c_s^{(2)})$ and a sample size n_s, such that for every $n \geqslant n_s$,

$$P\{C_n(\hat{\theta}_{L,n} - \theta) + \tau_{\alpha/2} A / B(F) < -c_s^{(1)} \log n\} \leqslant c_s^{(2)} n^{-s}, \tag{10.3.65}$$

$$P\{C_n(\hat{\theta}_{U,n} - \theta) - \tau_{\alpha/2} A / B(F) > c_s^{(1)} \log n\} \leqslant c_s^{(2)} n^{-s}. \tag{10.3.66}$$

Thus, letting $s > 1$ and using (10.3.60), (10.3.65), and (10.3.66), we conclude that (10.3.21) holds here with $\nu = A / B(F)$. As a result, (10.3.22) also holds.

To verify (10.3.30), we need to study here an invariance principle for $\{L_k(C_N^{-1}t) - L_k(0) + C_N^{-1} t C_k^2 B(F),\ |t| \leqslant K^*(< \infty) \text{ and } k \leqslant N\}$. For the case of Wilcoxon scores, such a result is due to Jurečková (1973) and her treatment holds for bounded scores as well. For unbounded scores, the problem has not yet been attacked in its full generality; Hušková (1981) has provided some solutions. We pose some of these as problems.

As in the case of the location model, here also, one may use the M-estimators [point as well as the confidence limits in (8.3.19) and (8.3.20)] to set bounded length sequential confidence intervals for the regression parameter Δ in the model (8.2.1). The procedure is very similar to the one in (10.3.46) and with the direct replacement of the location estimators by the regression ones, the results proved for the location case remain true for the regression case as well. Hence the details are omitted.

It is possible to extend the ARE results in (10.3.48) through (10.3.52) to the regression case too. We employ the least squares estimator instead of the sequence to sample means. The only difference here will be that EN has to be replaced by $E\psi(n)$ and, as a result, by (10.2.7), in the final step, the ARE will be $s^{-1}(e)$, where e is the ARE in the location parameter case and $s(\ )$ is defined by (10.2.7).

10.3.4. Type B Sequential Confidence Regions

We may proceed as in Section 10.2.3. Here, instead of (10.3.1), we need to have a simultaneous confidence region for a (vector) $\mathbf{U}_n$ and that each element of $\mathbf{U}_n$ satisfies the condition (a), (b), and (c), stated in the beginning of this section. The simultaneous confidence region for $\mathbf{U}_n$ along with the coordinatewise monotonicity property of $U_{n,\boldsymbol{\theta}}$ leads to a simultaneous confidence region for $\boldsymbol{\theta}$. The sequential procedure consists of choosing such a sequence of simultaneous confidence regions and defining a stopping variable as the minimum sample size for which the diameter of the confidence regions is $\leqslant 2d$, for some $d > 0$. Here also, the "uniform continuity in probability" [see (10.3.9) through (10.3.10)] of the coordinatewise estimators ensure the same for the vector and hence (10.3.16) can be established under parallel regularity conditions. For (10.3.22), we need to verify that (10.3.21) holds for the vector of coordinatewise estimators. We may refer to Ghosh and Sen (1973) for certain specific problems of special interest.

10.4. CONFIDENCE SEQUENCES FOR PARAMETERS

We may utilize the theory developed in Section 9.2 for constructing confidence sequences for some meaningful parameters. Here also it will be convenient to deal with exact distribution-free and asymptotically distribution-free procedures under separate headings.

10.4.1. Exact Distribution-Free Procedures

We conceive of a sequence $\{X_i,\ i \geqslant 1\}$ of independent r.v.'s, defined on a common probability space $(\Omega, \mathcal{Q}, P)$, and we let $\mathbf{X}^{(n)} = (X_1, \ldots, X_n)$, $\forall n \geqslant 1$. Let θ be a parameter of interest and we conceive of a sequence $\{J_n\}$ of intervals such that as $n \to \infty$, the length of $J_n \to 0$ and

$$P\{\theta \in J_n \quad \text{for every} \quad n \geqslant m\} \geqslant 1 - \alpha_m, \qquad (10.4.1)$$

where α_m can be made close to 0 by taking m large. For this purpose, as in Section 10.3, we conceive of a sequence $\{U_n\}$ of statistics, satisfying the conditions (a), (b), and (c), and instead of (10.3.1), we conceive of a sequence $\{J_n^*\}$ of intervals, such that

$$P\{U_n \in J_n^*, \quad \text{for every} \quad n \geqslant m \mid H_0\} \geqslant 1 - \alpha_m, \qquad (10.4.2)$$

where the length of $J_n^* \to 0$ as $n \to \infty$. Let

$$\hat{\theta}_{L,n} = \inf\{\theta : U_{n,\theta} \in J_n^*\}, \hat{\theta}_{U,n} = \sup\{\theta : U_{n,\theta} \in J_n^*\}; \qquad (10.4.3)$$

$$J_n = \{\theta : \hat{\theta}_{L,n} \leqslant \theta \leqslant \hat{\theta}_{U,n}\}, \qquad n \geqslant n_0, \qquad (10.4.4)$$

where n_0 is the minimum sample size for which U_n is defined properly. Then, from (10.4.2), (10.4.3), and (10.4.4), it follows that (10.4.1) holds with $\{J_n\}$, defined by (10.4.4). Since under H_0, $\{U_n\}$ is a distribution-free sequence, (10.4.2) remains valid for a broad class of d.f.'s, and hence we have an exact distribution-free procedure. The construction of the sequence $\{J_n^*\}$ is greatly facilitated by the almost sure invariance principles for $\{U_n\}$. We consider first the estimation of the median (location parameter) of a symmetric distribution. For this purpose we use the signed rank statistics $\{S_n\}$, defined by (5.2.1). We define $S_n(b)$ as in (5.5.1) and note that for monotonic score functions, $S_n(b)$ is also a monotonic function of b. Thus if we define J_n^* as in (9.2.30) and let $U_n = S_n$, then (10.4.2) holds. Further, because of the monotonicity of $U_{n,\theta}$ in θ, we have no problem in defining $\hat{\theta}_{L,n}$ and $\hat{\theta}_{U,n}$ in (10.4.3); in this context, see (10.3.28) where we need to replace $U_{n,\alpha}^{(j)}$ by $(-1)^j A_n (2n \log\log n)^{1/2}$ for $j = 1, 2$. The choice of J_n^* in (9.2.30) fails to provide us with an (asymptotic) expression for α_m in (10.4.2). We may obtain such an asymptotic expression if we choose a comparatively wider sequence of $\{J_n^*\}$, as we did in (9.2.22). Thus if we let $J_n^* = \{x : |x| \leqslant A_n (cn \log n)^{1/2}\}$, for some $c > 0$, then α_m is asymptotically equal to the rhs of (9.2.23) for $m = n_0$. Corresponding to any specified $\alpha : 0 < \alpha < 1$, we can choose c and m (m at least moderately large) so that $\alpha_m \leqslant \alpha$. A similar case holds for the estimation of a population quantile based on the sample order statistics, as has been treated in Section 10.3.1, where we need to change r_n and s_n in (10.3.24).

Let us consider the estimation of the regression parameter, treated in Section 10.3.3. With the same model discussed there, we may provide a confidence sequence to the regression coefficient β. We define L_n as in (4.2.1) and $L_n(b)$ as in (4.5.1), so that $L_n(b)$ is nonincreasing in b for monotonic score functions. With these, we can follow on the same line as in the one sample location problem (treated earlier) with the only change that $(2n \log\log n)$ or $(cn \log n)$ has to be replaced by $(2C_n^2 \log\log C_n^2)$ or $(cC_n^2 \log C_n^2)$, respectively.

10.4.2. Asymptotically Distribution-free Procedures

The goal is to set up a sequence J_n satisfying (10.4.1). But here we may not have a sequence $\{U_n\}$ of distribution-free statistics. We conceive of a sequence $\{\hat{\theta}_n\}$ of point estimators of θ and as in Section 10.2, we assume that the asymptotic mean square of $\hat{\theta}_n$ is $\sigma_\theta^2/\psi(n)$, $n \geqslant n_0$ where $\psi(n)$ is $\nearrow$ in n, $\lim_{n\to\infty}\psi(n) = \infty$ and there exists a sequence $\{s_n^2\}$ of strongly consistent estimators of σ_θ^2. In such a case, we let

$$J_n = \{x : |x - \hat{\theta}_n| \leqslant c_n s_n/\psi^{1/2}(n)\}, \tag{10.4.6}$$

where $\{c_n\}$ is a suitable sequence of positive numbers for which (10.4.1) holds true. For example, if $\{\hat{\theta}_n\}$ satisfies the conditions for the embedding of some Wiener process, we can take $c_n > (2 \log\log n)^{1/2}$. For U-statistics, von Mises functionals, indeed such a choice is possible by virtue of the a.s. invariance principles, studied in Chapter 3. Similarly, for linear combinations of order statistics and the extrema of sample functions, studied in Chapters 7 and 8, almost sure invariance principles hold, and we may take the same solution. Finally, for the M-estimators of location or regression, we employ a combination of the two types of procedure described before. In such a case, we define $S_N(t)$, $\hat{\Delta}_N$ and s_N^{02} as in (8.2.6), (8.2.7), and (8.3.13), respectively. Note that for the model (8.2.1) when Δ obtains, $S_N(\Delta)$ has independent summands with mean 0 and the invariance principles of Chapter 2 apply under the usual regularity conditions on the constants $\{c_i\}$ and the score function ψ (as assumed in Section 8.3). Thus by (8.3.15), (8.3.18), and the invariance principles for the $S_N(\Delta)$, on defining

$$J_n^* = \{x : |x| \leqslant d_n s_n^0\}, \tag{10.4.7}$$

where $\{d_n\}$ has the same role as $\{c_n\}$ in (10.4.6), we obtain that for large n_0,

$$P\left\{\frac{S_n(\Delta)}{C_n} \in J_n^*, \qquad \forall n \geqslant n_0 \,|\, \Delta\right\} \simeq 1 - \alpha, \tag{10.4.8}$$

where C_n^2 is defined by (8.2.16). We then proceed as in (8.3.19) but use the critical values $\pm d_n C_n s_n^0$ (instead of $\pm \tau_{\alpha/2} C_N s_N^0$) and denote the two statistics by $\tilde{\Delta}_{nU}$ and $\tilde{\Delta}_{nL}$, respectively. Then from (10.4.8) we conclude that asymptotically (as $n_0 \to \infty$),

$$P\{\tilde{\Delta}_{nL} \leqslant \Delta \leqslant \tilde{\Delta}_{nU}, \qquad \forall n \geqslant n_0 \,|\, \Delta\} \simeq 1 - \alpha. \tag{10.4.9}$$

For the choice of $\{d_n\}$, we may proceed as in the case of rank procedures for the location or regression problem, treated earlier.

10.5. SEQUENTIAL POINT ESTIMATION BASED ON U-STATISTICS

To motivate the procedure, we start with the normal theory model. Let $\{X_i, i \geqslant 1\}$ be a sequence of i.i.d.r.v.'s with a normal d.f. for which both the mean μ and variance σ^2 are unknown. Our problem is to estimate μ. For every $n(\geqslant 1)$, let $\overline{X}_n = n^{-1}\sum_{i=1}^{n} X_i$. Now $\overline{X}_n$ is an unbiased estimator of μ and $\text{Var}(\overline{X}_n) = n^{-1}\sigma^2$. Let $c(>0)$ be the cost per unit observation and consider the *loss* due to estimation of μ by $\overline{X}_n$:

$$L_n(a,c) = a\left(\overline{X}_n - \mu\right)^2 + cn, \qquad \text{where } a(>0) \text{ is a constant.} \tag{10.5.1}$$

The expected loss or *risk* is therefore

$$R_n(a,c) = EL_n(a,c) = an^{-1}\sigma^2 + cn. \tag{10.5.2}$$

In this setup, one would naturally like to choose n in such a way that for given (a,c), $R_n(a,c)$ is minimized for the specific choice of n; this is termed the *minimum risk estimation* (MRE). If σ is specified, then the MRE is achieved by a sample of size $n^* = n^*(a,c) = (a/c)^{1/2}\sigma$ with the corresponding risk $R_{n^*} = 2(ac)^{1/2}\sigma = 2cn^*$, where it should be noted that in this definition n^* is not necessarily an integer, and hence the desired solution should be either the integer just greater than or just less than n^*, when n^* is itself not so. Note that n^* depends on the unknown σ, and hence no fixed sample size procedure leads to the MRE simultaneously for all σ. The same conclusion holds if in (10.5.1) we take the loss as $a|\overline{X}_n - \mu|^t + cn^s$, where t and s are some positive numbers or when the underlying d.f. is not necessarily normal.

Note that for any two-stage (like that of the Stein procedure), multistage, or sequential procedure based on a *stopping time* N and the point estimator $\overline{X}_N$, whenever the event $[N = n]$ and $\overline{X}_n$ are stochastically independent (this is typically the case with normal d.f.'s where the stopping variable is based on the sample variances which are independent of the sample means), we have

$$\frac{\left\{aE\left(\overline{X}_N - \mu\right)^2 + cEN\right\}}{(2cn^*)} = \frac{\left\{a\sigma^2E(N^{-1}) + cEN\right\}}{(2cn^*)}$$
$$= \tfrac{1}{2}E\left(\frac{n^*}{N} + \frac{N}{n^*}\right), \tag{10.5.3}$$

where, $\frac{1}{2}(x + x^{-1}) \geqslant 1 \ \forall x \geqslant 0$, and the equality sign holds only when

$x = 1$. Thus unless $N = n^*$ with probability 1, (10.5.3) is strictly greater than 1, so that $\bar{X}_N$ is not an MRE of μ. This criticism is particularly applicable to the Stein procedure where for some initial sample of size $n_0(\geqslant 2)$, N is defined as

$$N = \max\left\{n_0, \left[\left(\frac{a}{c}\right)^{1/2} s_{n_0}\right] + 1\right\}, \tag{10.5.4}$$

$s_{n_0}^2$ is the sample variance of this initial sample and $(n_0 - 1)^{1/2}(N/n^*)$ has a nondegenerate distribution on the positive part of the real line (see Problem 10.5.1). For $c\downarrow 0$, the distribution of $(n_0 - 1)^{1/2}N/n^*$ converges to the chi distribution with $n_0 - 1$ degrees of freedom, and hence for a fixed n_0, even when c is small, N/n^* does not converge to 1 with probability 1. Hence the Stein two-stage procedure is not risk-efficient (even asymptotically when $c\downarrow 0$) when n_0 is fixed. Note that if in the Stein procedure the initial sample size n_0 is made to depend on c [i.e., $n_0 = n_0(c)$] in such a way that

$$n_0(c) \to \infty \qquad \text{but} \quad c^{1/2} n_0(c) \to 0 \qquad \text{as} \quad c\downarrow 0 \tag{10.5.5}$$

[e.g., $n_0(c) = [a^{1/2}c^{-\gamma}]$, for some $\gamma \in (0, \frac{1}{2})$], then (10.5.3) converges to 1 as $c\downarrow 0$, so that the Stein procedure becomes *asymptotically risk-efficient* (as $c\downarrow 0$). This feature of the Stein procedure has been observed by Mukhopadhyay (1980) and in this context, one needs to use the fact that if Z_c has the chi distribution with m_c degrees of freedom where $m_c \to \infty$ as $c\downarrow 0$, then both $m_c^{-1}EZ_c$ and $m_c EZ_c^{-1}$ converge to 1 as $c\downarrow 0$ (see Problem 10.5.2). However, for nonnormal d.f.'s (where $\bar{X}_n$ and $[N = n]$ are not necessarily independent), the simplified formula on the rhs of (10.5.3) may not hold and a more elaborate analysis may be needed to establish the asymptotic risk-efficiency of the Stein procedure under (10.5.5).

For the normal mean problem, a sequential procedure based on the updated versions of the sample variances has been considered by Robbins (1959) and extended further by Starr (1966) and Starr and Woodroofe (1969), among others. The latter authors have also considered the case of the mean of a gamma distribution. For d.f.'s of unspecified forms, Ghosh and Mukhopadhyay (1979) have proposed a sequential procedure for the mean and established its *asymptotic risk efficiency* under appropriate regularity conditions. Chow and Yu (1981) have obtained the same result under weaker conditions, while Sen and Ghosh (1981) have studied the asymptotic risk efficiency of a general class of sequential point estimation procedures based on U-statistics and valid under regularity conditions not more stringent than those needed for the sample means. In this section we consider their procedure. In the next section we confine ourselves to the specific problem of location parameter of a symmetric d.f. and consider some robust nonparametric procedures based on L-, M-, and R-estimators of location.

For an *estimable parameter* $\theta(F)$, defined as in (3.2.1), for a sample of size $n[\geqslant m$, the *degree* of $\theta(F)]$, the *U-statistic* U_n is defined as in (3.2.4). Then U_n is an unbiased estimator of $\theta(F)$ with variance σ_n^2 defined by (3.2.15). Note that by the reverse martingale property in Theorem 3.2.1, $\sigma_n^2 - \sigma_{n+1}^2 = V(U_n - U_{n+1}) \geqslant 0$, so that σ_n^2 is $\downarrow$ in $n(\geqslant m)$.

As in (10.5.1), we consider the *loss* incurred in estimating $\theta(F)$ by U_n:

$$L_n(a,c) = a(U_n - \theta(F))^2 + cn; \qquad a > 0, \qquad c > 0, \tag{10.5.6}$$

and our object is to minimize the corresponding *risk*

$$R_c(n; a, F) = EL_n(a,c) = a\sigma_n^2 + cn \tag{10.5.7}$$

by a proper choice of n. Note that $\Delta R_c(n; a, F) = R_c(n+1; a, F) - R_c(n; a, F) = c - a(\sigma_n^2 - \sigma_{n+1}^2)$, $n \geqslant m$, while $\Delta^2 R_c(n; a, F) = R_c(n+2; a, F) - 2R_c(n+1; a, F) + R_c(n; a, F) = a(\sigma_{n+2}^2 - 2\sigma_{n+1}^2 + \sigma_n^2)$. Thus whenever, $\Delta^2 R_c(n; a, F) \geqslant 0$, $R_c(n; a, F)$ is a convex function of n. Based on (3.2.14), the proof of this convexity follows by standard steps and is left as an exercise (see Problem 10.5.4). Hence there exists an $n_c^*[= n^*(a, c; F)]$ such that

$$R_c(n_c^*; a, F) = a\sigma_{n_c^*}^2 + cn_c^* = \min_{n \geqslant m} R_c(n; a, F), \tag{10.5.8}$$

where n_c^* (being a positive integer) need not be unique; we may settle with the smaller of the two consecutive integers if it is not unique. From (3.2.15), (10.5.7), and (10.5.8), it follows that n_c^* depends on a, c, m as well as the unknown parameters ζ_d, $1 \leqslant d \leqslant m$ (which are functionals of the underlying d.f. F). Hence in the absence of knowledge of these ζ_d, $1 \leqslant d \leqslant m$, no fixed sample size leads to MRE simultaneously for all ζ_d, $1 \leqslant d \leqslant m$, so that a sequential procedure may be desirable to achieve this goal.

If we assume that $\theta(F)$ is stationary of order 0, then, by (3.2.16), $\sigma_n^2 = m^2 n^{-1}\zeta_1 + \xi_{(n)}$ where $\xi_{(n)} \downarrow 0$ as $n \to \infty$, and hence for small c, on letting

$$n_c^0 \sim \left(\frac{am^2\zeta_1}{c}\right)^{1/2} \qquad \text{and} \qquad R_c^0(F) = 2cn_c^0, \tag{10.5.9}$$

we obtain that n_c^*/n_c^0 and $R_c(n_c^*; a, F)/R_c^0(F)$ both converge to 1 as $c \downarrow 0$. Hence for small c we can interchange the role of n_c^* and n_c^0. For the sequential point estimation procedure, we define the jackknife estimator V_n^* as in (3.7.22) and proceed as follows.

Let $n_0(\geqslant m+1)$ be an initial sample size, $\gamma(>0)$ be a suitable constant, to be specified later, and define *stopping number* N_c by

$$N_c = \min\left\{n \geqslant n_0 : n \geqslant \left(\frac{a}{c}\right)^{1/2}\left((V_n^*)^{1/2} + n^{-\gamma}\right)\right\}. \tag{10.5.10}$$

U_{N_c} is then the point estimator of $\theta(F)$ and the corresponding risk is

$$R_c^*(a) = aE\{U_{N_c} - \theta(F)\}^2 + cEN_c. \tag{10.5.11}$$

As a measure of the *relative efficiency* of U_{N_c} with respect to $U_{n_c^*}$, we then consider the following:

$$e(c,a) = \frac{R_c^0(F)}{R_c^*(a)}. \tag{10.5.12}$$

Then U_{N_c} is *risk-efficient* if $e(c,a) = 1$. However, this is not generally the case, but as $c \downarrow 0$, under fairly general conditions, (10.5.12) converges to 1. We term the sequential procedure as *asymptotically risk-efficient* if

$$\lim_{c \downarrow 0} e(c,a) = 1, \tag{10.5.13}$$

for all F belonging to a class of d.f.'s. It may be remarked that in the case of the interval estimation problem, the asymptotic consistency and efficiency of the procedures in Sections 10.2 and 10.3 have been established when the width of the interval is made to converge to 0. In this section, in the dual problem of point estimation, the same feature holds under the setup that $c \downarrow 0$, that is, the cost of unit observation is negligible. We have the following theorems.

THEOREM 10.5.1. *If $\theta(F)$ is stationary of order 0, $E|g|^{2+\delta} < \infty$ for some $\delta > 0$ and in (10.5.10), $\gamma \in (0, \delta^2/2(2+\delta))$, then (10.5.13) holds. Further, without any restriction on $\gamma(>0)$,*

$$\lim_{c \downarrow 0}\left(\frac{EN_c}{n_c^*}\right) = 1 \quad \textit{and} \quad \frac{U_{N_c} - \theta(F)}{\sigma_{n_c^*}} \xrightarrow{\mathcal{D}} \mathfrak{N}(0,1) \quad \textit{as} \quad c \downarrow 0. \tag{10.5.14}$$

THEOREM 10.5.2. *If $E|g|^{4+\delta} < \infty$ for some $\delta > 0$ and in (10.5.10), $\gamma > \frac{1}{2}$, then, as $c \downarrow 0$,*

$$\frac{2m^2\zeta_1(N_c - n_c^0)}{(\nu^2 n_c^0)^{1/2}} \xrightarrow{\mathcal{D}} \mathfrak{N}(0,1), \tag{10.5.15}$$

where

$$\nu^2 = \lim_{n\to\infty}\{n\,\mathrm{Var}(V_n^*)\}. \tag{10.5.16}$$

For the proofs of the theorems, we need the following lemma, which is proved first.

Lemma 10.5.3. If $E|g|^{2r} < \infty$, for some $r \geqslant 1$ and $\theta(F)$ is stationary of order 0, then, for every $\epsilon : 0 < \epsilon < 1$, as $c \downarrow 0$,

$$P\{N_c \leqslant n_c^*(1-\epsilon)\} = O(c^{r^*/\{2(1+\gamma)\}}). \tag{10.5.17}$$

where $r^* = r/2$ or $r - 1$ according as r is $\geqslant 2$ or $1 \leqslant r < 2$.

Proof. Note that by (3.7.5), (3.7.9), and (3.7.24),

$$V_n^* - m^2\zeta_1 = m^2\big(U_n^{*(0)}\big) + \sum_{d=0}^{m} e_{nd} U_n^{(d)}, \tag{10.5.18}$$

where $U_n^{*(0)}$ (has expectation 0) and all the $U_n^{(d)}$ are U-statistics with finite absolute moment of order $r(\geqslant 1)$, and the e_{nd} are all $O_e(n^{-1})$. Also, by (3.7.21),

$$E|V_n^* - m^2\zeta_1|^r = O(n^{-r^*}), \qquad \forall n \geqslant m. \tag{10.5.19}$$

Further, by (10.5.10), $N_c \geqslant b^{1/(1+\gamma)}$, with probability 1, where $b^2 = a/c$. Let $n_{1c} = [b^{1/(1+\gamma)}]$ and $n_{2c} = n_c^*(1-\epsilon)$ and choose c so small that $n_{1c} \leqslant n_{2c}$. Then, by (10.5.10), for small c,

$$\begin{aligned}
P\{N_c \leqslant n_{2c}\} &\leqslant P\{V_n^* < b^{-2}n^2, \quad \text{for some} \quad n : n_{1c} \leqslant n \leqslant n_{2c}\} \\
&\leqslant P\{V_n^* < b^{-2}n_{2c}^2, \quad \text{for some} \quad n : n_{1c} \leqslant n \leqslant n_{2c}\} \\
&\leqslant P\Big\{V_n^* - m^2\zeta_1 \leqslant m^2\zeta_1\{(1-\epsilon)^2 - 1\}, \\
&\qquad \text{for some} \quad n : n_{1c} \leqslant n \leqslant n_{2c}\Big\} \\
&\leqslant P\Big\{|V_n^* - m^2\zeta_1|/m^2\zeta_1 \geqslant \epsilon(2-\epsilon), \\
&\qquad \text{for some} \quad n : n_{1c} \leqslant n \leqslant n_{2c}\Big\} \\
&\leqslant P\Big\{\max_{n_{1c} \leqslant n \leqslant n_{2c}} m^2|U_n^{*(0)}|/\zeta_1 \geqslant \epsilon\Big\} \\
&\quad + \sum_{d=0}^{m} P\Big\{\max_{n_{1c} \leqslant n \leqslant n_{2c}} |e_{nd}U_n^{(d)}|/m^2\zeta_1 \geqslant \tfrac{1}{2}\epsilon\Big\}.
\end{aligned} \tag{10.5.20}$$

By the reverse submartingale property of the $|U_n^{(d)}|^r$ and $|U_n^{*(0)}|^r$, (10.5.19), the Hájek-Rènyi inequality in (2.2.5) and the fact that $|ne_{nd}|$ are all bounded, we conclude that the rhs of (10.5.20) is $O(n_{1c}^{-r^*}) = O(b^{-r^*/(1+\gamma)}) = O(c^{r^*/2(1+\gamma)})$. Q.E.D.

Let us now proceed to the proof of the theorems. Note that by (10.5.10),

$$bV_{N_c}^{*1/2} \leqslant N_c \leqslant n_0 + b\big(V_{N_c-1}^{*1/2} + (N_c - 1)^{-\gamma}\big). \tag{10.5.21}$$

Hence dividing all sides of (10.5.21) by n_c^0 and using (3.7.24), (3.7.7), and

(10.5.9), we obtain

$$\frac{N_c}{n_c^0} \to 1 \quad \text{a.s.,} \quad \text{as} \quad c \downarrow 0. \tag{10.5.22}$$

By virtue of the comment after (10.5.9), we may also replace n_c^0 by n_c^* in (10.5.22). Further, by (10.5.18), $\sup_{n \geqslant n^*} V_n^* \leqslant m^2 \sup_{n \geqslant n^*} |U_n^{*(0)}| + K\sum_{d=0}^{m} \sup_{n \geqslant n^*} |U_n^{(d)}|$, where K is a generic constant and n^* is any positive integer $\geqslant m$. Let now $p = 2 + \delta$, so that $E|g|^p < \infty$ and let $2 < q = p\alpha$, for some $\alpha < 1$. Then, by (2.2.6), the Kolmogorov inequality for reverse submartingales, we have for every $\lambda > 0$,

$$P\left\{ \sup_{n \geqslant n^*} |U_n^{(d)}|^{q/2} > \lambda \right\} \leqslant \lambda^{-p/q} E|U_{n^*}^{(d)}|^{p/2} \leqslant K^* \lambda^{-p/q} = K^* \lambda^{-1-\eta}, \tag{10.5.23}$$

where $\eta = (1 - \alpha)/\alpha > 0$ and K^* is a finite positive constant ($d = 0, \ldots, m$) and a similar inequality holds for the case of the $U_n^{*(0)}$. Thus

$$E\left[\sup_{n \geqslant n^*} V_n^{*q/2} \right] \leqslant K^{**} \left(E\left[\sup_{n \geqslant n^*} |U_n^{*(0)}|^{q/2} \right] + \sum_{d=0}^{m} E\left[\sup_{n \geqslant n^*} |U_n^{(d)}|^{q/2} \right] \right)$$
$$\leqslant K^0 \left[1 + \int_1^\infty (\lambda^{-1-\eta})\, d\lambda \right] < \infty, \qquad \forall n^* \geqslant 2m, \tag{10.5.24}$$

where K^{**} and K^0 are finite positive constants and we have made use of the fact that for a nonnegative r.v. Z, $EZ \leqslant 1 + \int_1^\infty P(Z \geqslant z)\, dz$. Since $q/2 > 1$, by (10.5.21), (10.5.22), and (10.5.24) and the dominated ergodic theorem, it follows that $EN_c/n_c^* \to 1$ as $c \downarrow 0$. The second part of (10.5.14) is a direct consequence of (10.5.22) and Theorem 3.2.1. To show that (10.5.13) holds, we note that, by virtue of (10.5.9), (10.5.11), and (10.5.14) (first part), it is sufficient to show that

$$\lim_{c \downarrow 0} \left\{ \frac{aE[U_{N_c} - \theta(F)]^2}{(cn_c^*)} \right\} = 1. \tag{10.5.25}$$

Further, proceeding as in (10.5.9) and using (3.2.16), we obtain that as $c \downarrow 0$, $\{aE[U_{n_c^*} - \theta(F)]^2 / (cn_c^*)\} \to 1$, and hence it suffices to show that

$$\lim_{c \downarrow 0} \left\{ n_c^* E\left[(U_{N_c} - \theta(F))^2 - (U_{n_c^*} - \theta(F))^2 \right] \right\} = 0. \tag{10.5.26}$$

By using (3.2.48), Lemma 10.5.3, the Hölder inequality and the maximal inequality for nonnegative (sub)martingales, we have, on defining n_{1c} as in before,

$$E\left[(U_{N_c} - \theta(F)]^2 I(N_c \leqslant n_c^*(1 - \epsilon)) \right]$$
$$\leqslant \{p/(p-1)\}^2 (E|U_{n_{1c}} - \theta(F)|^p)^{2/p} (P\{N_c \leqslant n_c^*(1 - \epsilon)\})^{(p-2)/p}$$
$$= O(c^h), \qquad \text{where} \quad h > \tfrac{1}{2}. \tag{10.5.27}$$

Similarly, $E[(U_{n_c^*} - \theta(F))^2 I(N_c \leqslant n_c^*(1-\epsilon))] = O(c^{h'})$, where $h' > \frac{1}{2}$. Further, a very similar treatment holds for the upper tail $[N_c \geqslant n_c^*(1+\epsilon)]$. Hence to prove (10.5.26), it suffices to show that for ϵ arbitrarily close to 0,

$$n_c^* E\Big[\big\{(U_{N_c} - \theta(F))^2 - (U_{n_c^*} - \theta(F))^2\big\} I(|N_c/n_c^* - 1| < \epsilon)\Big] \to 0 \quad \text{as} \quad c\downarrow 0. \tag{10.5.28}$$

Toward this, we note that the lhs of (10.5.28) is bounded from above by

$$2n_c^* \left\{ E\Big[\max_{n\,:\,|n/n_c^* - 1| < \epsilon} (U_n - U_{n_c^*})^2\Big] E\Big[\max_{n\,:\,|n/n_c^* - 1| < \epsilon} (U_n - \theta(F))^2\Big]\right\}^{1/2}, \tag{10.5.29}$$

where, by using the reverse martingale property of U-statistics and (2.2.8), along with (3.2.16), we conclude that (10.5.29) can be made arbitrarily small by choosing ϵ adequately small. This completes the proof of Theorem 10.5.1.

To prove Theorem 10.5.2, we note that by virtue of (10.5.16), (10.5.18), and (10.5.22) and Theorem 3.2.1, as $c\downarrow 0$,

$$\frac{N_c^{1/2}\left(V_{N_c}^* - m^2\zeta_1\right)}{\nu} \xrightarrow{\mathcal{D}} \mathfrak{N}(0,1), \tag{10.5.30}$$

where $V_{N_c}^*$ may also be replaced by $V_{N_c-1}^*$. Therefore, by the Mann-Wald theorem,

$$\frac{2N_c^{1/2}\left(\sqrt{V_{N_c}^*} - m\zeta_1^{1/2}\right)}{\left(\nu/m\zeta_1^{1/2}\right)} \xrightarrow{\mathcal{D}} \mathfrak{N}(0,1), \qquad \text{as } c\downarrow 0, \tag{10.5.31}$$

where again $V_{N_c}^*$ may be replaced by $V_{N_c-1}^*$. Since $\gamma > \frac{1}{2}$, by (10.5.22), $(N_c - 1)^{-\gamma} n_c^{*1/2} \to 0$ a.s., as $c\downarrow 0$. Hence (10.5.15) follows from (10.5.9), (10.5.21), (10.5.22), and (10.5.31). Q.E.D.

The results of this section can be extended to the vector case as well as to functions of several estimable parameters. We pose some of these extensions as problems at the end of this chapter.

10.6. SEQUENTIAL POINT ESTIMATION OF LOCATION BASED ON SOME R-, L-, and M-ESTIMATORS

In this section, we confine ourselves to the estimation of location of a symmetric d.f. and consider asymptotically risk-efficient sequential procedures based on rank statistics, linear combinations of order statistics and M-estimators. Let $\{X_i,\ i \geqslant 1\}$ be a sequence of i.i.d.r.v.'s with d.f. $F_\theta(x) = F(x - \theta)$, $x \in R$, $\theta \in R$, where F is symmetric about 0 and is unknown and θ is the unknown location parameter to be estimated. Based on $X_1, \dots, X_n$, let $T_n = T(X_1, \dots, X_n)$ be a suitable estimator of θ and we assume that for some $n_0(\geqslant 1)$,

$$\sigma_n^2 = nE(T_n - \theta)^2 \qquad \text{exists for all} \quad n \geqslant n_0; \tag{10.6.1}$$

$$\sigma_n^2 \to \sigma^2 \qquad \text{as } n \to \infty, \quad \text{where} \quad 0 < \sigma < \infty. \tag{10.6.2}$$

As in (10.5.6) and (10.5.7) [replacing U_n and $\theta(F)$ by T_n and θ, respectively], consider the loss function $L_n(a,c)$ and the risk function $R_c(n; a, F)$. We would like to minimize the risk by a proper choice of n. This optimal choice of n depends, in general, on the unknown F (through σ_n^2 or σ^2), for fixed c or even asymptotically as $c \downarrow 0$. As in (10.5.9), in the asymptotic (as $c \downarrow 0$) case, the optimal choice n_c^0 (or n_c^*) and the corresponding risk $R_c^0(F)$ are given by

$$n_c^0 \sim \sigma\left(\frac{a}{c}\right)^{1/2} \qquad \text{and} \quad R_c^0(F) \sim 2\sigma(ac)^{1/2}. \tag{10.6.3}$$

Finally, we assume that there exists a sequence $\{\hat{\sigma}_n\}$ of consistent estimators of σ and, as in (10.5.10), consider a *stopping number* N_c defined by

$$N_c = \min\left\{n \geqslant n_0 : n \geqslant (a/c)^{1/2}(\hat{\sigma}_n + n^{-h})\right\}, \qquad c > 0, \tag{10.6.4}$$

where n_0 is an initial sample size and $h(>0)$ is an arbitrary constant. Then the sequential point estimator of θ is T_{N_c} and its risk are given by

$$R_c^*(a) = aE(T_{N_c} - \theta)^2 + cEN_c, \qquad c > 0. \tag{10.6.5}$$

Our general objective is to provide results parallel to those in Theorems 10.5.1 and 10.5.2 when for $\{T_n\}$ we choose appropriate rank estimators, M-estimators, and L-estimators, defined in Chapters 5, 8, and 7, respectively.

First consider the rank (R-) estimators $\hat{\theta}_{n(R)}$ defined by (5.6.3) and (5.6.4). We want to verify (10.6.1) and (10.6.2) where we have for $\sigma = \sigma_{(R)}$

$$\sigma_{(R)}^2 = \frac{A^2}{\gamma_{(R)}^2(F)}, \tag{10.6.6}$$

and A^2 and $\gamma_{(R)}(F)$ are defined by (5.3.1) and (9.5.20), respectively. [Note that in (9.5.20) the subscript (R) was suppressed.] Finally, as in (9.5.25), let $D_{n(R)} = D_n$ be the estimator of $\gamma_{(R)}(F)$, so that we have

$$\hat{\sigma}_{n(R)} = A / D_{n(R)}, \qquad n \geqslant 2; \tag{10.6.7}$$

the corresponding stopping number in (10.6.4) is denoted by $N_c(R)$.

Next we consider the M-estimators and define $\hat{\theta}_{n(M)}$ as in (8.2.7) and (8.2.8) [where in (8.2.6), we take all the c_i equal to 1]. Define $\gamma = \gamma_{(M)}$ as in (8.2.18) and σ_0 as in (8.2.13). Then for (10.6.2) we have

$$\sigma_{(M)} = \sigma_0 / \gamma_{(M)}. \tag{10.6.8}$$

Further, we define $D_{n(M)} = \hat{\gamma}_n$ as in (8.3.22) (with the c_i all equal to 1) and s_n^0 as in (8.3.13). Then we take

$$\hat{\sigma}_{n(M)} = s_n^0 / D_{n(M)} \tag{10.6.9}$$

and denote the corresponding stopping number in (10.6.4) by $N_c(M)$.

Finally, we consider a LCFOS, defined by (7.2.1) and (7.4.1) with $g(x) \equiv x$. This being an estimator of θ, for symmetric F, we take $c_{ni} = c_{nn-i+1}$ for $i = 1, \ldots, n$ and $\sum_{i=1}^{n} c_i = 1$. Then, for (10.6.2), $\sigma^2 = \sigma_{(L)}^2$ is defined by (7.4.3) and its estimator $\hat{\sigma}_{n(L)}^2 = \hat{\sigma}_n^2$ by (7.6.34). Finally, the stopping number in (10.6.4) is denoted by $N_c(L)$; the point estimator is denoted by $\hat{\theta}_{n(L)}$.

Before we proceed to establish the asymptotic risk efficiency of these procedures [in the sense of (10.5.11)], we need to verify that (10.6.1) and (10.6.2) hold. Also, we need to establish some *uniform integrability* properties of these estimators which will be needed in the sequel. For all these results, suitable regularity conditions must be specified. These results are mainly adapted from Sen (1980e) and Jurečková and Sen (1981c). In each case, we assume that the d.f. F has an absolutely continuous p.d.f. f [where $f(x) = f(-x)$, $\forall x \in R$] that is strongly unimodal with a finite Fisher information $I(f)$ and for some $a > 0$,

$$E|X_1|^a = \int_{-\infty}^{\infty} |x|^a \, dF(x) < \infty. \tag{10.6.10}$$

For the R-estimator, we assume that the score function satisfies (10.3.38). Then we have the following.

THEOREM 10.6.1. *Under* (10.6.10) *and for* $A^2 > 0$, *for every* $k(>0)$, *there exists a positive integer* n_{0k}, *such that* $E|\hat{\theta}_{n(R)}|^k < \infty$, $\forall n \geqslant n_{0k}$. *Further, if* (10.3.38) *holds for some* $\delta < (4 + 2\tau)^{-1}$, $\tau > 0$, *then for every* $k < 2(1 + \tau)$,

$$\lim_{n\to\infty} E\left\{ n^{k/2} |\hat{\theta}_{n(R)} - \theta|^k \right\} = \sigma_{(R)}^k E|Z|^k, \tag{10.6.11}$$

where $\sigma_{(R)}$ *is defined by* (10.6.6) *and* Z *has the standard normal d.f.*

Proof. Note that (see Problem 10.6.1) under (10.6.10), for any positive k,

$$E|X_{n,r}|^k < \infty, \qquad \forall (n,r) : k/a \leqslant r \leqslant n - k/a + 1. \tag{10.6.12}$$

Further, the scores $a_n(i)$, in the definition of $S_n(b)$, are nonnegative and nondecreasing and $A > 0$. Thus there exists a sequence $\{k_n\}$ of positive integers such that $n/2 < k_n < n$, $\sum_{\{i<k_n\}} a_n(i) \leqslant \sum_{\{i>k_n\}} a_n(i) < \sum_{\{i\leqslant k_n\}} a_n(i)$ and $n^{-1}k_n \to \alpha$, for some $\alpha \in (\frac{1}{2}, 1)$. Hence by (5.6.3) and (5.6.4), $S_n(X_{n,k_n}) < 0$, so that $\hat{\theta}_{n(R)} < X_{n,k_n}$, with probability 1. Similarly, $\hat{\theta}_{n(R)} > X_{n,n-k_n+1}$, with probability 1. These inequalities along with (10.6.12) ensure that $E|\hat{\theta}_{n(R)}|^k < \infty$, for all $n \geqslant n_{0k}$. To prove (10.6.12), note that for every $k > 0$, writing $Y_n = n^{1/2}(\hat{\theta}_{n(R)} - \theta)$,

$$E|Y_n|^k = E\{|Y_n|^k I(|Y_n| \leqslant \log n)\} + E\{|Y_n|^k I(|Y_n| > \log n)\}$$
$$= J_{n1} + J_{n2}, \qquad \text{say.} \tag{10.6.13}$$

For $q > k$, by the Hölder inequality

$$|J_{n2}| \leqslant n^{k/2}\{E|\hat{\theta}_{n(R)} - \theta|^q\}^{k/q}\{P(|Y_n| > \log n)\}^{1-k/q} \tag{10.6.14}$$

where we have already shown that $E|\theta_{n(R)} - \theta|^q$ exists for every $n \geqslant n_{0q}$. Proceeding as in (10.3.40) through (10.3.45), we have

$$n^{k/2}\{P(|Y_n| > \log n)\}^{1-k/q} = O(n^{k/2-(1-k/q)(1+\tau)}). \tag{10.6.15}$$

Since $k < 2(1 + \tau)$, by choosing q adequately large, the rhs of (10.6.15) can be made to converge to 0 as $n \to \infty$. Hence $J_{n2} \to 0$ as $n \to \infty$. Also, for $\epsilon > 0$,

$$J_{n1} = E\{|Y_n|^k I(|Y_n| < \log n) I(\omega_n \leqslant \epsilon)\} + E\{|Y_n|^k I(|Y_n| < \log n) I(\omega_n > \epsilon)\}$$
$$= J_{n11} + J_{n22}, \qquad \text{say,} \tag{10.6.16}$$

where

$$\omega_n = \sup_{b\,:\,|b| \leqslant n^{-1/2}(\log n)} n^{-1/2}\{|S_n(b) - S_n(0) + nbB^*(F)|\} \tag{10.6.17}$$

and, by (10.3.40), for every $\epsilon > 0$, as n increases,

$$P_0\{\omega_n > \epsilon\} = O(n^{-(1+\tau)}). \tag{10.6.18}$$

Therefore, by (10.6.16) and (10.6.18), $J_{n12} \leqslant (\log n)^k\{O(n^{-(1+\tau)})\} \to 0$ as $n \to \infty$. On the other hand, for $|Y_n| \leqslant \log n$ and $\omega_n < \epsilon$, by (10.3.40), $B^*(F)Y_n = n^{-1/2}S_n(0) + R_n, |R_n| < 2\epsilon$ and for $S_n(0)$, (10.3.42) and (10.3.43) yield that as $n \to \infty$,

$$E\{|n^{-1/2}S_n(0)|^k I(|Y_n| \leqslant \log n) I(\omega_n < \epsilon)\} \to A^k E|Z|^k. \tag{10.6.19}$$

Hence the proof of (10.6.11) is completed by noting that $\sigma_{(R)} = A/B^*(F)$.

Q.E.D.

For the M-estimator, we assume that (8.2.10) through (8.2.14) hold and in addition $\psi(x) = \psi(C)\operatorname{sgn} x$, for $|x| > C$, where $C(> 0)$ is some finite number. Thus $\psi(x)$ is bounded and admits of a finite number of jumps. This boundedness condition can be eliminated if in (10.6.10) it is assumed that a is large; however, from consideration of robustness an assumption will be somewhat restrictive, and hence it is preferred to work with bounded scores for which (10.6.10) suffices. In this case, we define $\gamma = \gamma_{(M)}$ as in (8.2.19).

THEOREM 10.6.2. *Under* (10.6.10) *and the assumptions made above, for every* $k(>0)$ *there exists a positive number* n_{0k}, *such that* $E|\hat{\theta}_{n(M)}|^k$ *exists, for every* $n \geqslant n_{0k}$. *Further, for every* $k(>0)$,

$$\lim_{n\to\infty} E\left\{n^{k/2}|\hat{\theta}_{n(M)} - \theta|^k\right\} = \sigma_{(M)}^k E|Z|^k, \tag{10.6.20}$$

where $\sigma_{(M)}$ *and* Z *are defined as in* (10.6.8) *and* (10.6.11).

Proof. For the given $C(>0)$, we define $c_1 > C$ and note that

$$\begin{aligned} E_\theta\left\{n^{k/2}|\hat{\theta}_{n(M)} - \theta|^k\right\} &= \int_0^\infty kt^{k-1}P_0\{\sqrt{n}|\hat{\theta}_{n(M)}| > t\} \\ &= \left\{\int_0^{c_1\sqrt{n}} + \int_{c_1\sqrt{n}}^\infty\right\} kt^{k-1}P_0\left\{\sqrt{n}\,|\hat{\theta}_{n(M)}| > t\right\}dt \\ &= J_{n1} + J_{n2}, \qquad \text{say.} \end{aligned} \tag{10.6.21}$$

Note that for every $t > 0$,

$$P_0\left\{\sqrt{n}\,|\hat{\theta}_{n(M)}| > t\right\} = 2P_0\left\{\sqrt{n}\,\hat{\theta}_{n(M)} > t\right\}, \tag{10.6.22}$$

$$\begin{aligned} P_0\left\{\sqrt{n}\,\hat{\theta}_{n(M)} > t\right\} &\leqslant P_0\left\{S_n\left(\frac{t}{\sqrt{n}}\right) \geqslant 0\right\} \\ &= P_0\left\{n^{-1/2}S_n\left(\frac{t}{\sqrt{n}}\right) + \mu_n(t) \geqslant -\mu_n(t)\right\}, \end{aligned} \tag{10.6.23}$$

where $S_n(\cdot)$ is defined by (8.2.6) and

$$\begin{aligned} -\mu_n(t) &= -E_0 n^{-1/2}S_n\left(\frac{t}{\sqrt{n}}\right) = -E_0\psi(X_1 - n^{-1/2}t)n^{1/2} \\ &= E_0\left[\psi(X_1) - \psi(X_1 - n^{-1/2}t)\right]n^{1/2}. \end{aligned} \tag{10.6.24}$$

Since ψ is skew-symmetric, f is symmetric, unimodal, and positive on $[0, C + c_1]$ and $\psi(C) > 0$, it follows that for every $t: 0 < t \leqslant c_1\sqrt{n}$ (see Problem 10.6.2),

$$-\mu_n(t) \geqslant 2f(C + c_1)\psi(C)\cdot t. \tag{10.6.25}$$

On the other hand, the $S_n(\cdot)$ involve independent and bounded valued r.v., so that by the Bernstein inequality and (10.6.25), (10.6.23) is bounded from above by

$$\exp\{-c_2t^2\}, \qquad \text{where} \quad c_2 > 0. \tag{10.6.26}$$

Hence by (10.6.21), (10.6.22), (10.6.23), and (10.6.24),

$$J_{n1} \leqslant 2k\int_0^\infty \exp\{-c_2t^2\}t^{k-1}dt < \infty, \qquad \text{uniformly in} \quad n \geqslant 1. \tag{10.6.27}$$

To handle J_{n2}, we note that because of the boundedness of ψ, for every $t > 0$,

$$P_0\{n^{1/2}\hat{\theta}_{n(M)} > t\} \leqslant P_0\{X_{n,m+1} \geqslant -C + n^{-1/2}t\}, \qquad (10.6.28)$$

where, for simplicity, we take $n = 2m$; a very similar case holds for $n = 2m + 1$. Since for $t \geqslant c_1\sqrt{n}$, $c_1 > C$, $-C + t/\sqrt{n} > 0$, the rhs of (10.6.28) is

$$\begin{aligned} &P_0\{F_n(-C + t/\sqrt{n}) - F(-C + t/\sqrt{n}) \leqslant m/n - F(-C + t/\sqrt{n})\} \\ &\quad \leqslant \{\rho(-C + t/\sqrt{n})\}^n, \end{aligned} \qquad (10.6.29)$$

by the Hoeffding (1963) inequality, where

$$\rho\left(-C + \frac{t}{\sqrt{n}}\right) = 4F\left(-C + \frac{t}{\sqrt{n}}\right)\left[1 - F\left(-C + \frac{t}{n}\right)\right] \in (0, 1),$$

$$\forall t \geqslant c_1\sqrt{n}. \qquad (10.6.30)$$

Thus if we define a as in (10.6.10) and let $n_{0k} = [k/a] + 1$, then for every $n \geqslant n_{0k}$,

$$\begin{aligned} J_{n2} &\leqslant 2k\int_{c_1\sqrt{n}}^{\infty} t^{k-1}\left[\rho(-C + t/\sqrt{n})\right]^n dt \\ &= 2kn^{k/2}\int_{c_1}^{\infty} u^{k-1}\left[\rho(-C + u)\right]^n du \\ &\leqslant Kn^{k/2}\left\{\sup_{v \geqslant c_1 - C} v^{k-a}\left[\rho(v)\right]^{n-1}\right\} \\ &\quad \times \left\{\int_{c_1 - C}^{\infty} v^{a-1}\{1 - F(v)\}\, dv\right\}, \end{aligned} \qquad (10.6.31)$$

where $K(< \infty)$ is independent of n, for $v \geqslant c_1 - C > 0$, $(v + C)^{k-1} \leqslant a_k(v^{k-1} + C^{k-1}) \leqslant a_k^* v^{k-1}$ and a_k, a_k^* are finite positive constants depending on k, C, and c_1 (but not on n), and $\rho(v) \leqslant 4[1 - F(v)]$. Now (10.6.10) ensures that the integral on the rhs of (10.6.31) is finite, while, by using the fact that under (10.6.10), $\sup\{|x|^a F(x)[1 - F(x)] : x \in E\} < E_0|X_1|^a = \nu_a < \infty$, we obtain that for $n \geqslant n_{0k}$,

$$\sup_{v \geqslant c_1 - C} v^{k-a}\left[\rho(v)\right]^{n-1} n^{k/2} = \sup_{v \geqslant c_1 - C} v^{k-a}\left[1 - F(v)\right]^{n-1}\left[F(v)\right]^{n-1} 4^n n^{k/2} \qquad (10.6.32)$$

is uniformly (in n) bounded and it converges to 0 as $n \to \infty$. Thus J_{n2} exists for every $n \geqslant n_{0k}$ and it converges to 0 as $n \to \infty$ (see Problem 10.6.4 in this context).

Having this uniform integrability established, we complete the proof of (10.6.20) by an appeal to (8.3.11) (with the c_i all equal to 1) and the moment convergence of $n^{-1/2}S_n(0)$; since $\psi(X_i)$ is bounded with probability 1, the latter holds. (See Problem 10.6.5 in this context.) Q.E.D.

For the L-estimator $\hat{\theta}_{n(L)}$, we assume that $c_{ni} = c_{nn-i+1} \geqslant 0$, $i = 1, \ldots, n$; $\sum_{i=1}^{n} c_{ni} = 1$ and there exist an α_0 $(0 < \alpha_0 < \frac{1}{2})$ and a sequence $\{k_n\}, k_n > 0$, such that

$$c_{ni} = c_{nn-i+1} = 0 \quad \text{for} \quad i \leqslant k_n \quad \text{where} \quad n^{-1}k_n \to \alpha_0 \quad \text{as} \quad n \to \infty. \tag{10.6.33}$$

Also, we let $J_n(t) = nc_{ni}$ for $(i-1)/n < t \leqslant i/n$, $i = 1, \ldots, n$, and assume that $\lim_{n\to\infty} J_n(t) = J(t)$ exists for every $t \in [0, 1]$ where $J(t)$ has bounded variation on $[0, 1]$ and $J(t) = J(1-t) \geqslant 0$, $0 \leqslant t \leqslant 1$ with $\int_0^1 J(t)\,dt = 1$. Finally, (10.6.10) is assumed to hold here and the finite Fisher information may be replaced by $\sup\{|f'(x)| : F^{-1}(\alpha_0) \leqslant x \leqslant F^{-1}(1-\alpha_0)\} < \infty$.

THEOREM 10.6.3. *The results of Theorem* 10.6.2 *also hold for the L-estimator, where* $\sigma_{(M)}$ *has to be replaced by* $\sigma_{(L)}$.

Proof. Note that for every $k > 0$.

$$\begin{aligned} E_\theta\left\{n^{k/2}|\hat{\theta}_{n(L)} - \theta|^k\right\} &= E_0\left\{n^{k/2}|\hat{\theta}_{n(L)}|^k\right\} \\ &= E_0\left(\sum_{i=k_n}^{n-k_n+1} c_{ni}\sqrt{n}\left[X_{n,i} - F^{-1}\left(\frac{i}{n+1}\right)\right]\right)^k \\ &\leqslant \sum_{i=k_n}^{n-k_n+1} c_{ni} E_0\left(\sqrt{n}\,|X_{n,i} - F^{-1}(i/(n+1))|\right)^k. \end{aligned} \tag{10.6.34}$$

The uniform integrability result then directly follows by using the uniform integrability of the order statistics (see Problem 10.6.6) and (10.6.34). The desired result [parallel to (10.6.20)] follows then by using (8.5.13) and (8.5.14) along with the moment convergence of $n^{-1/2}\sum_{i=1}^{n} J^*(F(X_i))$, where J^* is bounded. Q.E.D.

To study the properties of the nonparametric sequential point estimators of location, besides the moment convergence results of these point estimators (studied in the preceding theorems), we also need some rates of convergence of the variance estimators $\sigma_{n(R)}$, $\sigma_{n(M)}$, and $\sigma_{n(L)}$. First, note that by (10.3.38) through (10.3.45), for $\sigma_{n(R)}$ defined by (10.6.7) and for n

adequately large, for every $\epsilon > 0$, there exists a positive constant $c(\epsilon)(< \infty)$ such that

$$P\{|\sigma_{n(R)}/\sigma_{(R)} - 1| > \epsilon\} \leqslant c(\epsilon)n^{-s}, \tag{10.6.35}$$

where $s = 1 + \tau$, $\tau(> 0)$ is defined as in Theorem 10.6.1. Similarly, by (8.4.20), (8.4.22), the Markov inequality for s_n^2 (using its $2k$th order central moment for some positive integer k) and the fact that by virtue of the boundedness of the score function ψ, moments of all finite order exists, we have for every $s > 1$ and n adequately large, $\forall\epsilon > 0$,

$$P\{|\sigma_{n(M)}/\sigma_{(M)} - 1| > \epsilon\} \leqslant c_\epsilon n^{-s}, \qquad \text{where } c_\epsilon < \infty. \tag{10.6.36}$$

Finally, for the estimator $\sigma_{n(L)}$, by virtue of the boundedness of J and J^*, we may virtually repeat the proof of Theorem 7.6.4 [but use (2.6.29) whenever needed] and that for every $\epsilon, s > 0$ there exist a positive number $n_{s\epsilon}$ and a constant $c_{s\epsilon}(< \infty)$ such that for every $n \geqslant n_{s\epsilon}$,

$$P\left\{\left|\frac{\sigma_{n(L)}}{\sigma_{(L)}} - 1\right| > \epsilon\right\} \leqslant c_{s\epsilon}n^{-s}; \tag{10.6.37}$$

we leave the details of the proof as exercise (see Problem 10.6.7).

THEOREM 10.6.4. *For the stopping number N_c in* (10.6.4) *and the procedures based on the nonparametric estimators of location, under the assumed regularity conditions, as $c\downarrow 0$,*

$$\frac{N_c}{n_c^0} \xrightarrow{P} 1, \qquad EN_c/n_c^0 \to 1, \tag{10.6.38}$$

$$\frac{\sqrt{n_c^0}(\hat{\theta}_{N_c} - \theta)}{\sigma} \xrightarrow{\mathfrak{D}} \mathfrak{N}(0, 1), \tag{10.6.39}$$

where

$$n_c^0 \sim (a/c)^{1/2}\sigma \tag{10.6.40}$$

and σ is given by $\sigma_{(R)}$, $\sigma_{(M)}$, and $\sigma_{(L)}$ in the case of R-, M- and L-estimators.

Proof. For some arbitrary $\epsilon(> 0)$, let $n_{1c} = [b^{1/(1+h)}]$, $n_{2c} = [n_c^0(1 - \epsilon)]$ and $n_{3c} = [n_c^0(1 + \epsilon)]$, where h is defined as in (10.6.4) and $b^2 = a/c$. Then by (10.6.4), $N_c \geqslant n_{1c}$, with probability 1. Also, as $c\downarrow 0$, $n_{1c} < n_{2c}$ and by

(10.6.4), (10.6.35), (10.6.36) and (10.6.37),

$$\begin{aligned}
P\{N_c \leqslant n_{2c}\} &= P\{\hat{\sigma}_n < b^{-1}n \quad \text{for some} \quad n : n_{1c} \leqslant n \leqslant n_{2c}\} \\
&\leqslant P\left\{\left|\frac{\hat{\sigma}_n}{\sigma} - 1\right| \geqslant \epsilon, \quad \text{for some} \quad n : n_{1c} \leqslant n \leqslant n_{2c}\right\} \\
&\leqslant \sum_{n=n_{1c}}^{n_{2c}} P\{|\hat{\sigma}_n/\sigma - 1| \geqslant \epsilon\} \\
&= O\big((n_{1c})^{s-1}\big) = O\big(c^{(s-1)/\{2(1+h)\}}\big) \\
&\to 0 \quad \text{as } c\downarrow 0 \quad (\text{as } s > 1). \qquad (10.6.41)
\end{aligned}$$

In a similar manner, for $n \geqslant n_{3c}$, by (10.6.35)–(10.6.37),

$$\begin{aligned}
P\{N_c > n\} &= P\left\{k < (a/c)^{1/2}(\hat{\sigma}_k + k^{-h}), \quad \forall k \in [n_0, n]\right\} \\
&\leqslant P\left\{n < (a/c)^{1/2}(\hat{\sigma}_n + n^{-h})\right\} = P\left\{\hat{\sigma}_n > b^{-1}n - n^{-h}\right\} \\
&= P\left\{\hat{\sigma}_n - \sigma > b^{-1}(n - n_c^0) - n^{-h}\right\} \\
&\leqslant P\{|\hat{\sigma}_n - \sigma| > \eta\} \qquad (\text{where } \eta > 0) \\
&= O(n^{-s}), \qquad \text{where } s > 1. \qquad (10.6.42)
\end{aligned}$$

The last formula leads us to

$$\sum_{n \geqslant n_{3c}} P\{N_c > n\} = O\big((n_{3c})^{-s+1}\big) \to 0 \qquad \text{as } c\downarrow 0. \qquad (10.6.43)$$

The first part of (10.6.38) follows from (10.6.41) and (10.6.42), while the second part follows from the fact that $|EN_c/n_c^0 - 1| \leqslant (1-\epsilon)P(N_c \leqslant n_{2c}) + \epsilon + \sum_{n \geqslant n_{3c}} P(N_c > n)$ and that (10.6.43) ensures the rest. Finally, (10.6.39) follows from (10.6.38) and the invariance principles for the R-, M-, and L-estimators studied in Section 8.5; the verification of the Anscombe condition is easily made by using the tightness part of these invariance principles. Q.E.D.

With $R_c^0(F)$ and $R_c^*(a)$ defined by (10.6.3) and (10.6.5), respectively, we define the *relative risk-efficiency* as in (10.5.12) and intend to study the *asymptotic risk-efficiency* as in (10.5.13).

THEOREM 10.6.5. *For the sequential procedure based on the R-estimators, if* (10.3.38) *holds for some* $\delta < (4 + 2\tau)^{-1}$ *where* $\tau > 1 + 2h$ *with* $h(>0)$ *de-*

fined in (10.6.4), *then the asymptotic risk efficiency holds. For the procedures based on the M- and L-estimators,* (10.5.13) *holds under no extra condition.**

Proof. We only provide the proof for the rank procedure; very similar proofs hold for the two other cases and these are left as exercises (see Problems 10.6.8 and 10.6.9). By virtue of (10.6.2), (10.6.5), and (10.6.38), it suffices to show that

$$\lim_{c\downarrow 0}\left\{a\left[\hat{\theta}_{N_c}-\theta\right]^2/(cn_c^0)\right\}=1. \tag{10.6.44}$$

Now, by Theorem 10.6.1 and (10.6.35), for every $k\in(2,2+2\tau)$,

$$\begin{aligned}E\left\{(\hat{\theta}_{N_c}-\theta)^2 I(N_c\leqslant n_{2c})\right\} &= \sum_{n=n_{1c}}^{n_{2c}} E\left\{(\hat{\theta}_n-\theta)^2 I(N_c=n)\right\}\\ &\leqslant \sum_{n=n_{1c}}^{n_{2c}}\left(E|\hat{\theta}_n-\theta|^k\right)^{2/k}\left(P\{N_c=n\}\right)^{1-2/k}\\ &\leqslant\left(\sum_{n=n_{1c}}^{n_{2c}} E|\hat{\theta}_n-\theta|^k\right)^{2/k}\left(P\{N_c\leqslant n_{2c}\}\right)^{1-2/k}\\ &=\left(O\left(n_{1c}^{-(k-2)/2}\right)\right)^{2/k}\cdot\left(O\left(c^{\tau/2(1+h)}\right)\right)^{1-2/k}\\ &=O\left(c^{(k-2)(1+\tau)/\{2k(1+h)\}}\right).\end{aligned} \tag{10.6.45}$$

Note that $1+\tau>2(1+\mathrm{h})$, while for $\tau=1+\xi$, $\mathrm{k}=2(1+\tau)-\eta$, $0<\eta<2\xi$, $(k-2)/k=\frac{1}{2}+(2\xi-\eta)/(8+4\xi-2\eta)>\frac{1}{2}$, so that

$$\lim_{c\downarrow 0}\left[(cn_c^0)^{-1}E\left\{(\hat{\theta}_{N_c}-\theta)^2 I(N_c\leqslant n_{2c})\right\}\right]=0. \tag{10.6.46}$$

Similarly, using (10.6.42) along with Theorem 10.6.1, we obtain

$$\lim_{c\downarrow 0}\left[(cn_c^0)^{-1}E\left\{(\hat{\theta}_{N_c}-\theta)^2 I(N_c\geqslant n_{3c})\right\}\right]=0. \tag{10.6.47}$$

Further, by using Theorem 10.6.1 along with (10.6.41), (10.6.42), and the Hölder inequality, we have

$$\lim_{c\downarrow 0}\left[(cn_c^0)^{-1}E\left\{(\hat{\theta}_{n_c^0}-\theta)^2 I(N_c\leqslant n_{2c}\ \text{ or }\ N_c\geqslant n_{3c})\right\}\right]=0, \tag{10.6.48}$$

$$\lim_{c\downarrow 0}\left[(cn_c^0)^{-1}E\left\{(\hat{\theta}_{n_c^0}-\theta)^2 I(n_{2c}<N_c<n_{3c})\right\}\right]=1. \tag{10.6.49}$$

Actually, (10.6.49) follows from (10.6.3) and (10.6.48). Hence to prove

*Note that for the rank procedure, the choice of h in (10.6.4) and δ in the score function in (10.3.38) are related, while h is unrestricted for the other two procedures for which the scores are assumed to be bounded.

(10.6.44), it suffices to show that

$$\lim_{c\downarrow 0}\left[\left(cn_c^0\right)^{-1}E\left\{\left[(\hat{\theta}_{N_c}-\hat{\theta}_{n_c^0})^2 I(n_{2c}<N_c<n_{3c})\right]\right\}\right]=0. \quad (10.6.50)$$

As in the proof of Theorem 10.6.1, we write $Y_n = n^{1/2}(\hat{\theta}_n - \theta)$ and $B^*(F) \cdot Y_n = n^{-1/2}S_n(0) + R_n$, so that on writing $E_0R_n^2 = E_0(R_n^2I(|Y_n| \leqslant \log n)) + E_0[R_n^2I(|Y_n| > \log n)] \leqslant E_0[R_n^2I(|Y_n| \leqslant \log n)] + 2(B^*(F))^2E_0[Y_n^2I(|Y_n| > \log n)] + 2E_0[n^{-1}S_n^2(0)I(|Y_n| > \log n)]$, where for simplicity (and without any loss of generality), we have taken $\theta = 0$ and E_0 denotes the expectation under $\theta = 0$, and proceeding then as in (10.6.14) through (10.6.19) we obtain by some standard steps that

$$E_0R_n^2 = O(n^{-1-h}), \quad (10.6.51)$$

where h is defined in (10.6.4). Hence

$$E_0\left\{\max_{n_{2c}\leqslant n\leqslant n_{3c}} R_n^2\right\} \leqslant \sum_{n=n_{2c}}^{n_{3c}} E_0R_n^2 = O\left(n_{2c}^{-h}\right) = O\left(c^{h/2}\right)\to 0 \quad \text{as} \quad c\downarrow 0. \quad (10.6.52)$$

Thus it suffices to show that by choosing $\epsilon(>0)$ sufficiently small and allowing $c\downarrow 0$,

$$\left(n_c^0\right)^{-1}E_0\left[\left(S_{N_c}(0)-S_{n_c^0}(0)\right)^2 I(n_{2c}\leqslant N_c\leqslant n_{3c})\right] \quad \text{can be made to} \to 0. \quad (10.6.53)$$

Since the $S_n(0)$ involves martingale differences with mean 0 and finite moments up to the order $k(=2+\lambda$, for some $\lambda > 0)$, by the Hölder inequality and the fact that $(n_{3c} - n_{2c})/n_c^0 \sim 2\epsilon$, the proof of (10.6.53) follows by some standard steps (see Problem 10.6.10). Q.E.D.

Theorems 10.6.4 and 10.6.5 enable us to compare the performance of the competing procedures based on (i) sample means and variances, (ii) rank estimators, (iii) M-estimators and (iv) L-estimators, either in terms of the corresponding n_c^0 or in terms of the squared risk. In terms of either of the definitions, by virtue of (10.6.3), the relative efficiency of the rank procedure with respect to the parametric procedure, based on the sample means and variance, is given by (10.3.52). As such, if we use the normal scores statistics for the rank procedure, then this ARE is bounded from below by 1 where the lower bound is attained only when F is itself a normal d.f. This clearly explains the relative advantages of the rank procedure over the classical Robbins-Starr parametric one. In particular, for d.f. with heavy tails, the Robbins-Starr procedure performs poorly and the use of non-parametric procedures is highly recommended. Similar conclusions hold for the M-procedure and the L-procedure. Since these ARE results are in agreement with those in the sequential testing and confidence intervals problems, details are not given here.

We conclude this section with the remark that by (10.6.4),

$$b\hat{\sigma}_{N_c} \leqslant N_c \leqslant n_0 + b\left(\hat{\sigma}_{N_c - 1} + (N_c - 1)^{-h}\right); \qquad b^2 = a/c, \quad (10.6.54)$$

where, by (10.6.38), n_0^2/n_c^0 and $b(N_c - 1)^{-h}/\sqrt{n_c^0}$ both converge to 0 (a.s.) as $c \downarrow 0$; for the latter we need $h > \frac{1}{2}$. Hence, whenever in (10.6.4), we let $h > \frac{1}{2}$ and for some $\beta : 0 < \beta < \infty$, we assume that

$$\frac{\sqrt{n_c^0}\,(\hat{\sigma}_{N_c} - \sigma)}{\beta} \xrightarrow{\mathfrak{D}} \mathfrak{N}(0, 1), \qquad \text{as} \quad c \downarrow 0, \qquad (10.6.55)$$

then, proceeding as in the proof of Theorem 10.5.2, we obtain that as $c \downarrow 0$,

$$(N_c - n_c^0)/\sqrt{n_c^0} \xrightarrow{\mathfrak{D}} \mathfrak{N}\left(0, \frac{\beta^2}{\sigma^2}\right). \qquad (10.6.56)$$

Also, for (10.6.55) to hold, it suffices to verify an invariance principle for the $\hat{\sigma}_n$ or more simply to show that as $n \to \infty$,

$$\frac{n^{1/2}(\hat{\sigma}_n - \sigma)}{\beta} \xrightarrow{\mathfrak{D}} \mathfrak{N}(0, 1) \qquad (10.6.57)$$

and for every $\epsilon > 0$,

$$\lim_{\delta \downarrow 0} P\left\{ \sup_{m : |m/n - 1| < \delta} n^{1/2} |\hat{\sigma}_m - \hat{\sigma}_n| > \epsilon \right\} = 0. \qquad (10.6.58)$$

For L-estimators, Theorem 7.6.5, and Problem 7.6.3 deal with (10.6.55) and (10.6.57) and (10.6.58). For M-estimators, Problem 8.3.4 relates to (10.6.55) when the score function ψ is assumed to be bounded and absolutely continuous inside the real line. Thus for both the L- and M-estimators, the asymptotic normality of the (standardized form of the) stopping number N_c holds under the assumed regularity conditions when the score function is assumed to be absolutely continuous. For score functions having finitely many jump discontinuities, Problem 8.3.7 depicts that (10.6.55) holds with $\sqrt{n_c^0}$ being replaced by $(n_c^0)^{1/4}$. Hence in this case, in (10.6.56), in the denominator on the lhs, $(n_c^0)^{1/2}$ has to be replaced by $(n_c^0)^{3/4}$. Finally, for the rank estimators, the situation is very similar to that in Section 10.3 [see the remarks following (10.3.45)], where we have discussed the issue and posed some exercises. These results also pertain to the point estimation problem.

10.7. SOME GENERAL COMMENTS

In the preceding section, we considered the problem of asymptotically risk-efficient estimation of location parameter based on nonparametric statistics. It is possible to extend the theory to the problem of estimating the regression parameter in a simple regression model, such as in (4.2.12). For a regression model, under the normality assumption on the d.f. F, the optimal estimator is the least squares estimator and as this is linear in the observations, the theory developed in Section 10.5 can be readily adapted to suit this purpose. For these least squares estimators, even when the d.f. F is not normal but possesses finite moments up to a certain order, similar results hold (see Problems 10.7.1 and 10.7.2). Further, the invariance principles developed in Chapter 8 can be incorporated along with appropriate moment-convergence results on M-estimators of regression parameters to study the asymptotic risk efficiency of point estimation procedures based on such M-estimators. To this end, we pose Problems 10.7.3 and 10.7.4. For rank estimators considered in Sections 4.5 and 10.3, besides the invariance principles studied earlier and Theorem A.4.1, one needs to study some moment convergence results parallel to those in Theorem 10.6.1. The basic inequality in the proof of Theorem 10.6.1 $[X_{n,k_n} \geqslant \hat{\theta}_{n(R)} \geqslant X_{n,n-k_n+1}]$ for the location case needs some reframing in the regression case. In the absence of such an inequality, the proof of the uniform integrability of the rank estimator of regression becomes complicated. This is posed as an open problem (see Problem 10.7.5). Finally, for the L-estimators, this requires a complete reworking with the invariance principles and moment convergence results with nonidentically distributed r.v.'s and is beyond the scope of the treatment of Chapter 7.

In Section 10.3, we interpreted the asymptotic efficiency in terms of (10.3.22) (and a similar case holds for the type A procedures in Section 10.2). In this context, it may be of some interest to compare $EN(d)$ and $n(d)$ by introducing the *regret function*

$$\Pi(d) = EN(d) - n(d), \qquad d > 0 \tag{10.7.1}$$

and studying its nature as $d \downarrow 0$. It may be noted that (10.3.22) may hold, yet (10.7.1) may tend to ∞ as $d \downarrow 0$. If $\Pi(d)$ is asymptotically (as $d \downarrow 0$) finite, then the sequential procedure on an average entails only a finite excess over the fixed sample size $n(d)$. Actually, if $\Pi(d) = o(n(d))$, then (10.3.22) holds. Thus a study of (10.7.1) is more informative than (10.3.22). However, for general nonlinear statistics, this entails very complicated analysis. For the case of the sequential point estimation problem, instead of the definition of the risk efficiency in (10.5.12) and (10.5.13), one may consider the following regret function:

$$\Pi^*(c) = R_c^*(a) - R_c^0(F)$$
$$= c\left[EN_c - n_c^*\right] + \left(\frac{a}{n_c^*}\right)\left[n_c^* E\left(\hat{\theta}_{N_c} - \theta\right)^2 - \sigma_{n_c^*}^2\right], \quad (10.7.2)$$

where we use the notations in (10.6.1) and (10.6.2). Usually, by virtue of (10.6.3), as $c \downarrow 0$,

$$\sigma_{n_c^*}^2 = \sigma^2 + O(c^{1/2}), \qquad n_c^* = \left(\frac{a}{c}\right)^{1/2}\sigma\left[1 + o(1)\right], \qquad (10.7.3)$$

so that $R_c^0(F)$ is also of the order $c^{1/2}$. Thus if we need to study the nature of $\Pi^*(c)$, we aim it to be $O(c)$, as $c \downarrow 0$. Sufficient conditions for this are

$$E(N_c - n_c^*) = O(1) \quad \text{and} \quad n_c^* E\left(\hat{\theta}_{N_c} - \theta\right)^2 = \sigma_{n_c^*}^2 + O(c^{1/2}), \quad \text{as } c \downarrow 0. \qquad (10.7.4)$$

The first condition in (10.7.4) is comparable to (10.7.1), while the second one is extra and is more difficult to verify. Ideally, it would be desirable to have an asymptotic expansion for $\Pi^*(c)$. Lai and Siegmund (1977, 1979) and Woodroofe (1977) have considered some nonlinear renewal theory which may be used in some simpler problems (like the Chow-Robbins confidence interval for the mean or the Robbins-Starr-Woodroofe sequential point estimation of the mean of a normal d.f. or scale parameter of a gamma d.f.) to provide such an asymptotic expansion. However, the series of regularity conditions needed to be verified for the applicability of these theorems are not so conveniently verifiable for the bulk of nonparametric problems treated in this chapter. The main difficulty stems from the remainder terms in the asymptotic expressions for the nonparametric estimators [viz., (8.5.1), (8.5.4), and (8.5.14)] on which such refined analysis imposes so many restrictions that the scope of these theorems would remain confined to only a small subclass of such nonparametric statistics. We therefore pose this study for nonparametrics as an open one.

In the beginning of Section 10.5, we have discussed the Stein two-stage procedure and stated that if the initial sample size is made to depend on c in such a way that as $c \downarrow 0$, it goes to ∞, though at a slower rate compared to n_c^0, then (10.5.13) may be achieved under certain conditions; for normal d.f. this has been proved by Mukhopadhyay (1980) and the result continues to hold for nonnormal F under appropriate moment conditions. A very similar conclusion holds for the bounded-length confidence interval problem for the mean of a distribution where the Stein two-stage procedure provides a valid solution for normal F. However, in terms of the regret function in (10.7.1) or (10.7.2), this two-stage procedure does not achieve the general goal, even if the initial sample size is made to depend on c, satisfying (10.5.5). For normal F, this result has been proved by Ghosh and Mukhopadhyay (1981) and a similar conclusion holds for other F as well. Thus from the point of *first-order efficiency*, the two-stage procedure with initial sample size dependent on c (or d in the confidence interval problem) may work out well, but is unlikely to do so with respect to the *second-order efficiency*. The sequential procedures considered in this chapter have been shown to be first-order efficient and, under additional regularity conditions, they perform well with respect to the regret function criteria in (10.7.1) and (10.7.2). Again, verification of the details is left as an open problem.

For both the bounded length confidence interval problem and the sequential point estimation problem, the theory developed here is of asymptotic nature (where d and c are made to converge to 0). A natural question arises: How good are these asymptotic results for finite values of d or c? Some numerical studies have been made for some simple problems and they advocate that the asymptotic results work out well for moderately small values of d and c. More work in this direction is desired. Further, some corrections (for moderate values of d or c) to the asymptotic expressions to increase the accuracy of these approximations will constitute an important contribution to this area. However, in the nonparametric case this will depend on the underlying d.f. F (unknown) and hence their applicability may be more limited.

EXERCISES

10.2.1. For normal F with unknown mean μ and variance σ^2, let $\bar{X}_n = n^{-1}\sum_{i=1}^{n} X_i$ and $s_n^2 = (n-1)^{-1}\sum_{i=1}^{n}(X_i - \bar{X}_n)^2$, $n \geqslant 2$ and for some initial sample of size $n_0(\geqslant 2)$, define the stopping number N as $\max\{n_0, [t_{n_0-1}^2 s_{n_0}^2/d^2] + 1\}$, where t_m is the upper $100\alpha/2\%$ point of the Student t-distribution with m degrees of freedom. Consider then the confidence interval $[\bar{X}_N - d, \bar{X}_N + d]$ for μ. Show that the coverage probability is $\geqslant 1-\alpha$, for all $\mu, \sigma(>0)$ and further, as $d \downarrow 0$, this converges to $1-\alpha$ (Stein, 1945).

10.2.2 (10.2.1 continued). For the same problem, show that $t_{n_0-1}^2\sigma^2/d^2 \leqslant EN \leqslant n_0 + t_{n_0-1}^2\sigma^2/d^2$ and hence $EN/n(d)$ [where $n(d) = [\tau_{\alpha/2}^2\sigma^2/d^2] + 1$] does not converge to 1 as $d \downarrow 0$ when n_0 is kept fixed.

10.2.3 (10.2.2 continued). Let now $n_0 = n_0(d)$ be such that $n_0(d) \to \infty$ but $d^2 n_0(d) \to 0$ as $d \downarrow 0$. Then, show that for the Stein procedure and normal F, $EN/n(d) \to 1$ and $d \downarrow 0$. Comment on the case where F is not necessarily normal (Mukhopadhyay, 1980).

10.2.4. Let $X_i = \beta_0 + \beta i + e_i$, $i \geqslant 1$, where the e_i are i.i.d.r.v. with a d.f. F

having mean 0 and a finite variance σ^2. Show that for this problem, $\psi(n) = n(n^2 - 1)/12 \sim n^3/12$, and hence (10.2.26) holds for $r = 1$.

10.2.5 (10.2.4 continued). Show that $s_n^2 = (n-2)^{-1}\sum_{i=1}^{n}(X_i - \bar{X}_n - \hat{\beta}_n(i - n + 1/2))^2$ (where $\hat{\beta}_n$ is the least squares estimator of β based on $X_1, \ldots, X_n$), $n \geqslant 3$ constitute a sequence of estimators of σ^2 for which both (10.2.6) and (10.2.27) (for $m = 2$) hold. Hence (10.2.29) holds.

10.2.6. For U-statistics and von Mises functionals, verify (10.2.55) under appropriate moment conditions on the basic kernel $g(X_1, \ldots, X_m)$.

10.2.7. For LCFOS, define s_n^2 as in (7.6.34). Follow the lines of proof of Theorem 7.6.3 [but use (2.6.29) whenever needed] to show that (10.2.55) holds when in (7.6.18), $\{F(x)[1 - F(x)]\}^{1/2}$ is replaced by $\{F(x)[1 - F(x)]\}^{1/4}$.

10.2.8. Consider an L-estimator of location of a symmetric d.f. where the weights are all nonnegative and they add up to 1. Characterize the *trimmed mean* (i.e., $(n-2m)^{-1}\sum_{i=m+1}^{n-m} X_{n,i}$) and the *Winsorized mean* [i.e., $n^{-1}(m+1)(X_{n,m+1} + X_{n,n-m}) + n^{-1}(X_{n,m+2} + \cdots + X_{n,n-m-1})$] as special cases and verify (10.2.55) for the s_n^2. Note that here the score function is bounded.

10.2.9. Consider the asymptotically best estimator (based on LCFOS) of location and scale parameters of a distribution with a specified form of the p.d.f. (see Problem 7.4.1) and in this case verify (10.2.55).

10.2.10. Express $\hat{\sigma}_n^2$ in (8.7.22) as $\{\psi_{01}^2(X_{n,r}, F_n(X_{n,r}))F_n(X_{n,r})[1 - F_n(X_{n,r})] - \psi_{01}^2(X_{n,r}, F(X_{n,r}))F(X_{n,r})[1 - F(X_{n,r})]\} + \{\psi_{01}^2(X_{n,r}, F(X_{n,r}))F(X_{n,r})[1 - F(X_{n,r})] - \psi_{01}^2(x_0, F(x_0))F(x_0)[1 - F(x_0)]\} + \psi_{01}^2(x_0, F(x_0))F(x_0)[1 - F(x_0)]$ and for the first two terms, proceed as in the proof of Theorem 8.7.1, to show that (10.2.59) and (10.2.60) hold in this case under some differentiability conditions on ψ_{01}. [Use the tightness condition (implicit) in Theorem 2.6.1 in this respect (to replace $n^{1/2}[F_n(X_{n,r}) - F(X_{n,r})]$ by $n^{1/2}[F_n(x_0) - F(x_0)]$, in probability, whenever needed).]

10.2.11. Let $\{X_{ij}, i \geqslant 1\}$ be i.i.d.r.v. with mean μ_j and variance σ^2, for $j = 1, \ldots, k(\geqslant 2)$, all these sequences are independent. We are interested in the set of all contrasts $\boldsymbol{\phi} = \{\phi(\mathbf{l}) = \sum_{j=1}^{k} l_j\mu_j$ where $\sum_{j=1}^{k} l_j = 0\}$. With the aid of the theory in Section 10.2.3, construct both ellipsoidal and rectangular confidence regions for $\boldsymbol{\phi}$.

10.2.12. Suppose that $\{\mathbf{X}_i = (X_{i1}, \ldots, X_{ik})', i \geqslant 1\}$ be a sequence of i.i.d.r.v.'s with mean vector $\boldsymbol{\mu}$ and dispersion matrix $\boldsymbol{\Sigma} = (1-\rho)\mathbf{I} + \rho\mathbf{1}\mathbf{1}'$ where $-(k-1)^{-1} \leqslant \rho < 1$. For the set of contrasts $\boldsymbol{\Phi} = \{\mathbf{l}'\boldsymbol{\mu} : \mathbf{l}'\mathbf{1} = 0$ and $\mathbf{l}'\mathbf{l} = 1\}$, obtain both ellipsoidal and rectangular confidence regions.

10.3.1. For the signed rank statistic $S_N(b)$, defined by (5.5.1) with the particular score function $a_N(i) = i/(N+1)$, $i = 1, \ldots, n$ (i.e., Wilcoxon scores), construct a process

$$W_N = \Big\{ W_N(t) = S_N\big(t/\sqrt{N}\big) - S_N(0) + \sqrt{N}\, t \int_{-\infty}^{\infty} f^2(x)\,dx - ES_N\big(t/\sqrt{N}\big), \qquad -k \leqslant t \leqslant k \Big\},$$

where $k(>0)$ is any finite number. Let $c^2 = 4[\int_{-\infty}^{\infty} f^3(x)\,dx - (\int_{-\infty}^{\infty} f^2(x)\,dx)^2]$ and assume that $0 < c < \infty$. Also, let $W = \{W(t) = tZ, -k \leqslant t \leqslant k\}$, where Z has the normal distribution with mean 0 and variance c^2. Show that W_N weakly converges to W. Construct a two-dimensional time parameter process from W_k, $k \leqslant N$ and extend the invariance principle to this case (Antille, 1972).

10.3.2. Consider the process $\{[L_k(C_N^{-1}t) - L_k(0) + C_N^{-1}C_k^2 tB(F)]/C_N = W_N^*(k/N, t)$, for $|t| \leqslant K^*(<\infty)$ and $k \leqslant N\}$, where the $L_k(b)$ are defined as in (4.5.1). Suppose now that (i) $\max\{(c_i - \bar{c}_N)^4/\sum_{i=1}^N (c_i - \bar{c}_N)^4 : 1 \leqslant i \leqslant N\} \to 0$ as $N \to \infty$, (ii) the score function $\phi(u) = u$, $0 \leqslant u \leqslant 1$ (i.e., Wilcoxon scores are used), and (iii) the regularity conditions of Theorem 4.5.3 hold. Then, on denoting by $A_N^{*2} = \sum_{i=1}^N (c_i - \bar{c}_N)^4/(\sum_{i=1}^N (c_i - \bar{c}_N)^2)^2$, show that $(A_N^*)^{-1} W_N^*(\cdot, \cdot)$ converges weakly to a two-dimensional time parameter Gaussian function (Jurečková, 1973). [Actually, Jurečková considered the case of $\{(A_N^*)^{-1} W_N^*(1, t),\ t \in [-K^*, K^*]\}$, but her treatment holds for the above process as well.]

10.3.3. Show that if the score function ϕ is bounded and has bounded and continuous first- and second-order derivatives, then the results of the preceding two problems hold for the corresponding processes.

10.3.4. Show that (10.3.50) is minimized when the score function ϕ^* is given by $\phi^*(u) = -f'(F^{-1}(u))/f(F^{-1}(u))$, $0 < u < 1$. Comment on the cases where f is normal, double exponential, logistic, and uniform p.d.f.

10.3.5. For the regression problem in Problems 10.2.1 and 10.2.2, use the rank procedure in Section 10.3 and verify that the needed regularity conditions all hold.

10.3.6. For the problem 10.2.11, use the rank order procedure in Section 10.3.4 and comment on its asymptotic efficiency relative to the parametric procedure in Problem 10.2.11 (Ghosh and Sen, 1973).

10.3.7. For the problem 10.2.12, consider rank order procedures and study the allied efficiency results.

10.5.1. Express the stopping variable N in (10.5.4) as $N = n_0 \cdot I([(a/c)^{1/2} s_{n_0}] + 1 \leqslant n_0) + ([a/c)^{1/2} s_{n_0}] + 1) I([(a/c)^{1/2} s_{n_0}] + 1 > n_0)$ and use the fact that $(n_0 - 1)^{1/2}s_{n_0}/\sigma$ has the chi d.f. with $(n_0 - 1)$ degrees of freedom to show that for every (fixed) n_0 and σ, N has a nondegenerate distribution. Hence $(n_0 - 1)^{1/2}N/n_c^*$ has a nondegenerate distribution, which, for $c\downarrow 0$, converge to the chi d.f. with $n_0 - 1$ degrees of freedom.

10.5.2. Let Z_m have the chi distribution with m degrees of freedom. Show that as $m \to \infty$, $m^{-1}EZ_m$ and mEZ_m^{-1} both converge to 1.

10.5.3. Use the representation for N_c in Problem 10.5.1 to show that under (10.5.5), $EN_c/n_c^* \downarrow 1$ as $c\downarrow 0$, even when F is not normal but possesses a finite second moment. However, for nonnormal F, the simplification in (10.5.3) may not hold. What can you say about the asymptotic risk-efficiency of the Stein two-stage procedure when (10.5.5) holds?

10.5.4. Use (3.2.14) to show that

$$\zeta_d = \sum_{i=0}^{d} \binom{d}{i} \delta_{d-i},$$

where the δ_i are all $\geqslant 0$. Hence use (3.2.15) and the above to show that $\Delta^2 R_c(n; a, F)$, defined after (10.5.7), is given by

$$\sum_{i=1}^{m} \binom{m}{i}\binom{n}{m}^{-1}\binom{n-i}{m-i}\delta_i \left[\frac{(n-i+1)(n-i+2)}{(n+1)(n+2)} - 2\frac{(n-i+1)}{(n+1)} + 1 \right] \geqslant 0$$

(Sen and Ghosh, 1981).

10.5.5. Define $\xi_{(n)}$ as in the discussion preceding (10.5.9) and show that $\xi_{(n)} \downarrow 0$ as $n\uparrow\infty$.

10.5.6. Show that for every $\epsilon > 0$, $n_c^* E[(U_{N_c} - \theta(F))^2 I(N_c \geqslant n_c^*(1+\epsilon))] \to 0$ as $c\downarrow 0$.

Hints. Use the Hölder inequality along with (10.5.19).

10.5.7. Let $\mathbf{X}_i$, $i \geqslant 1$ be i.i.d.r.v.'s with mean (vector) $\boldsymbol{\mu}$ and a positive definite dispersion matrix $\boldsymbol{\Sigma}$. Let $\overline{\mathbf{X}}_n = n^{-1}\sum_{i=1}^{n}\mathbf{X}_i$, $n \geqslant 1$, $\mathbf{S}_n = (n-1)^{-1}\sum_{i=1}^{n}(\mathbf{X}_i - \overline{\mathbf{X}}_n)(\mathbf{X}_i - \overline{\mathbf{X}}_n)'$, $n \geqslant 2$. Consider the loss function (in estimating $\boldsymbol{\mu}$ by $\overline{\mathbf{X}}_n$) $L_n = a(\overline{\mathbf{X}}_n - \boldsymbol{\mu})'\mathbf{A}(\overline{\mathbf{X}}_n - \boldsymbol{\mu}) + cn$, so that the risk function is $R_n(a,c) = n^{-1}\mathrm{Tr}(\mathbf{A}\boldsymbol{\Sigma}) + cn$, where $\mathbf{A}$ is some positive definite matrix. By analogy with (10.5.10), define the stopping number $N_c = \min\{n \geqslant 2 : n \geqslant (a/c)^{1/2}(\mathrm{Tr}(\mathbf{A}\mathbf{S}_n) + n^{-\gamma})^{1/2}\}$, where note that $n_c^* \sim \{(a/c)\mathrm{Tr}(\mathbf{A}\boldsymbol{\Sigma})\}^{1/2}$ as $c\downarrow 0$. Characterize $\mathrm{Tr}(\mathbf{A}\mathbf{S}_n)$ as a U-statistic and apply the results in Section 10.5 to establish the asymptotic risk efficiency of this sequential procedure when F is not necessarily normal (Ghosh, Sinha and Mukhopadhyay, 1976; Sen and Ghosh, 1981).

10.5.8. In the preceding problem, the stopping variable involves a linear function of several U-statistics. Consider now a general function $g(\mathbf{U}_n)$ of a vector of U-statistics and under suitable regularity conditions on g and $\mathbf{U}_n$ extend the results in Problem 10.5.7 to such functions.

10.5.9. Instead of the loss function L_n in (10.5.6), consider the following $L_n(a,c) = a|U_n - \theta(F)|^s + cn^t$, where s and t are positive numbers. Obtain the asymptotic expression for n_c^* in this case and, under appropriate moment conditions on the basic kernel, extend the results in Theorems 10.5.1 and 10.5.2 for the current situation.

10.6.1. Let $c_a(x) = |x|^a F(x)[1 - F(x)]$, $x \in E$. Show that under (10.6.10), $c_a(x)$ is bounded for every $x \in E$ and $c_a(x) \to 0$ as $x \to +\infty$ or $-\infty$. Note that

$$E|X_{n,r}|^k = \int_{-\infty}^{\infty} |x|^k \binom{n-1}{r-1} n\left[F(x)\right]^{r-1}\left[1 - F(x)\right]^{n-r} dF(x)$$
$$= n\binom{n-1}{r-1}\int_{-\infty}^{\infty} |x|^a \left\{|x|^{k-a}\left[F(x)\right]^{r-1}\left[1 - F(x)\right]^{n-r}\right\} dF(x).$$

Hence let $s = \min(r-1, n-r)$ and verify that the integral on the rhs is bounded from above by

$$n\binom{n-1}{r-1}\left\{\sup_x c_a(x)\right\}^{k/a-1}\int_{-\infty}^{\infty} |x|^a\, dF(x) < \infty,$$

for every $s \geqslant k/a - 1$ (Sen, 1959).

10.6.2. Write the rhs of (10.6.24) as $\int_0^C n^{1/2}[F(x + t/\sqrt{n}) - F(x) + F(-x + t/\sqrt{n}) - F(-x)]d\psi(x) = \int_0^C n^{1/2}[F(x + t/\sqrt{n}) - F(-x - t/\sqrt{n})]d\psi(x)$ (using the symmetry of F and the skew-symmetry of ψ), and use the unimodality of f to verify (10.6.25) (Jurečková and Sen, 1981c).

10.6.3. Show that the rhs of (10.6.23) is bounded from above by $\inf_{\theta > 0}\{\exp(\theta\mu_n(t))E[\exp\{\theta n^{-1/2}S_n(t/\sqrt{n}) + \mu_n(t)\}]\}$. Use (10.6.25) to verify (10.6.22) (Jurečková and Sen, 1981c).

10.6.4. Choose $c_1 > C$ and use the results in Problem 10.6.1 to verify (10.6.32).

10.6.5. Let $X_1, \ldots, X_n$ be independent r.v.'s with means equal to 0 and finite variances $\sigma_1^2, \ldots, \sigma_n^2$, respectively. Let $Y_n = (X_1 + \cdots + X_n)/s_n$ where $s_n^2 = \sum_{i=1}^n \sigma_i^2$. If $E|X_i|^k < \infty$, for every $i(=1, \ldots, n)$ and some positive integer $k(>3)$, then $|EY_n^r - \int_{-\infty}^{\infty} x^r\, d\Phi(x)| \to 0$ as $n \to \infty$, for every $r \leqslant k$ where $\Phi(x)$ is the standard normal d.f. (von Bahr, 1965).

10.6.6. Use the tools in the Problem 10.6.1 to show that under (10.6.10), for every $\{k_n\}$, satisfying (10.6.33) and every k (fixed),

$$E_0\left(\sqrt{n}\left|X_{n,i} - F^{-1}\left(\frac{i}{n+1}\right)\right|\right)^k < C < \infty,$$

uniformly in $i : k_n \leqslant i \leqslant n - k_n + 1$ and $n \geqslant n_0$, where C may depend on the d.f. F and α_0 in (10.6.33).

10.6.7. Provide a proof of (10.6.37).

10.6.8. Use the results in Theorems 10.6.2 and 10.6.4 and proceed as in the proof of Theorem 10.6.5 to establish (10.5.13) for the M-procedure.

10.6.9 Use similarly the results in Theorems 10.6.3 and 10.6.4 to establish (10.5.13) for the L-procedure.

10.6.10. Show that the lhs of (10.6.53) is bounded from above by $(n_c^0)^{-1}(\sum_{n=n_{2c}}^{n_{3c}} E_0[S_n(0) - S_{n_c^0}(0)]^4)^{1/2}(P\{n_{2c} \leqslant N_c \leqslant n_{3c}\})^{1/2} \leqslant (n_c^0)^{-1}(\sum_{n=n_{2c}}^{n_{3c}} O(|n - n_c^0|))^{1/2} = O((n_{3c} - n_{2c})/n_c^0) = O(2\epsilon)$, thus verifying (10.6. 53) (Sen, 1980e).

10.7.1. Consider the regression model: $X_i = \beta_0 + \beta c_i + e_i$, $i \geqslant 1$, where the c_i are specified constants (not all equal) and the e_i are i.i.d.r.v.'s with 0 mean and a finite and positive variance σ^2. The least square estimator of β is $\hat{\beta}_n = C_n^{-2}(\sum_{i=1}^n (c_i - \bar{c}_n)X_i)$, $n \geqslant 2$ and the estimator of σ^2 is $s_n^2 = (n-2)^{-1}\sum_{i=1}^n (X_i - \bar{X}_n - \hat{\beta}_n(c_i - \bar{c}_n))^2$, $n \geqslant 3$, where $\bar{c}_n = n^{-1}\sum_{i=1}^n c_i$, $\bar{X}_n = n^{-1}\sum_{i=1}^n X_i$, $n \geqslant 1$ and $C_n^2 = \sum_{i=1}^n (c_i - \bar{c}_n)^2$, $n \geqslant 2$. Show that parallel to (10.6.1) and (10.6.2), here, for $\sigma_n^2 = C_n^2 E(\hat{\beta}_n - \beta)^2 = \sigma^2$, $n \geqslant 2$. Also, $\hat{\sigma}_n = s_n$, $n \geqslant 3$ and, further, if $C_n^2 \sim dn^h$ (asymptotically) for some $d > 0$ and $h > 0$, then, parallel to (10.6.3), we have $n_c^0 \sim (ah\sigma^2/dc)^{1/(h+1)}$ and $R_c^0(F) \sim \{(h+1)/h\}c(\sigma^2 ah/dc)^{1/(h+1)}$. Thus, here, parallel to (10.6.4), consider the stopping number $N_c = \min\{n \geqslant 3 : C_n^2 \geqslant \{d^{1/(h+1)} \cdot (ahc^{-1})^{h/(h+1)}\}(s_n^{2h/(h+1)} + n^{-h'})\}$, for some $h' > 0$. With n_c^0 specified as above, obtain results parallel to (10.6.38) and (10.6.39).

10.7.2 (10.7.1 continued). Show that under appropriate moment conditions on F and the c_i, for some $r > 1$ and every $\epsilon > 0$, there exist a positive number $c_{r\epsilon}(< \infty)$ and an integer $n_{r\epsilon}$, such that for every $n \geqslant n_{r\epsilon}$, $P\{|s_n - \sigma| > \epsilon\} \leqslant c_{r\epsilon}n^{-r}$. Hence, or otherwise, establish the asymptotic risk-efficiency of the sequential procedure.

10.7.3. Consider the same regression model as in Problem 10.7.1, but let $\beta_0 = 0$. For this model, consider the M-estimators, described in Section 8.2, where we define C_n^2 and the variance $\sigma^2(= \nu^2)$ as in (8.2.16) and (8.2.24), respectively. Also, define s_n^0 and $\hat{\gamma}_n$ as in (8.3.13) and (8.3.22), and let $\hat{\sigma}_{n(M)}$ be defined as in (10.6.9). As in the problem 10.7.1, define the stopping variable N_c and verify (10.6.38) and (10.6.39) for the same.

10.7.4 (10.7.3 continued). Use Theorem 8.4.1, (8.4.20), (8.4.22) and the moment-convergence results on the $S_n(t)$ and verify (10.5.13) for the M-procedure for regression.

10.7.5. For the two-sample location problem, viewed as a particular case of the simple regression problem [see (4.2.11)], show that (10.6.11) holds for the rank estimator of β, under the same regularity conditions.

Hints. Use an inequality very similar to the one-sample case and then use the moment convergence properties of the individual sample order statistics.

CHAPTER 11

Nonparametric Time-Sequential Procedures

11.1. INTRODUCTION

A natural characteristic of clinical trials and life testing experimentations is that the observations are gathered sequentially in time, that is, the smallest observation comes first, the second smallest second, and so on, until the largest one emerges last. Because of practical limitations (on the cost and duration of experimentation), it is not always possible to continue experimentation until all the individuals have responded and then to draw statistical inference; usually, the experiment is curtailed either after a specified length of time or after a prespecified proportion (or number) of individuals have responded. In statistical terminology, such procedures are termed respectively the *truncated* or *censored procedures*. In the truncated scheme, the duration of experimentation is prefixed, but the number of respondents is a random variable. On the other hand, in the censored case, the number of respondents is prefixed, but the duration of the experimentation is a random variable. In either case, the experiment remains somewhat incomplete in the sense that for a part of the experimental units, the only information known is that their responses do not occur in the tenure of the experiment, that is, these are *censored observations*. Besides, the responses being the order statistics from a sample of prespecified size are not independently or identically distributed, and hence a somewhat different approach is needed in dealing with their distribution theory. The martingale approach stressed throughout this monograph remains accessible in this context too and will be adopted. In this chapter, we study truncated and censored nonparametric life testing procedures, and, in this context, we show that under a natural setup, a *progressively censoring scheme* (PCS) appears to be more practicable and leads to suitable time-sequential tests and estimates for the life testing problems. The main body of the theory deals with such progressively censored nonparametric procedures.

Along with the preliminary notions, the basic differences between a censored and a complete experiment (from the point of view of statistical analysis) are discussed in Section 11.2. Both parametric and nonparametric procedures are considered in this context. The PCS is introduced in Section 11.3. The next section is devoted to the study of the distribution theory of the related test statistics. Asymptotic power properties of the PCS procedures are considered in Section 11.5. Some additional problems and related tests are presented in Section 11.6 and the concluding section deals with some quasi-nonparametric procedures.

11.2. PRELIMINARY NOTIONS

We motivate the statistical analysis of a censored (or truncated) experiment by a parametric model first. Let $\{X_i, i \geqslant 1\}$ be a sequence of i.i.d.r.v.'s with a p.d.f. $f_\theta(x), x \in R^+$ and $\theta \in \Theta \subset R$. We desire to test for $H_0: \theta = \theta_0$ (specified) against one- or two-sided alternatives. For $n(\geqslant 1)$ items under a life test, the observable r.v.'s are $X_{n,1}, \ldots, X_{n,n}$, the order statistics corresponding to $X_1, \ldots, X_n$, along with other tagging variables. In a censored plan, for some fixed $r(\leqslant n)$, the experiment is terminated at $X_{n,r}$ and the test for H_0 is then based on $(X_{n,1}, \ldots, X_{n,r})$. The (joint) p.d.f. of $\mathbf{Z}_n^{(r)} = (X_{n,1}, \ldots, X_{n,r})$ is

$$n^{[r]} \prod_{i=1}^{r} f_\theta(X_{n,i})\left[1 - F_\theta(X_{n,r})\right]^{n-r} = p_\theta(\mathbf{Z}_n^{(r)}), \qquad \text{say}, \qquad (11.2.1)$$

where (11.2.1) is defined over the domain $\{X_{n,1} \leqslant \cdots \leqslant X_{n,n}\}$ and F_θ is the d.f. corresponding to the p.d.f. f_θ. Now the $X_{n,i}$ are neither independently nor identically distributed r.v.'s, and hence (11.2.1) differs from the usual models where the joint p.d.f. is the product of a number of p.d.f.'s of independent r.v.'s, and the classical theory of statistical inference based on independent r.v.'s is not directly applicable here. Nevertheless, $p_\theta(\mathbf{Z}_n^{(r)})$ can be expressed as the product of conditional densities of the $X_{n,k}$, given $\mathbf{Z}_n^{(k-1)}$, for $k = 1, \ldots, r$, namely,

$$p_\theta(\mathbf{Z}_n^{(r)}) = \prod_{k=1}^{r} \left\{ \frac{p_\theta(\mathbf{Z}_n^{(k)})}{p_\theta(\mathbf{Z}_n^{(k-1)})} \right\} = \prod_{k=1}^{r} p_\theta^*(X_{n,k} \mid \mathbf{Z}_n^{(k-1)}), \qquad (11.2.2)$$

where $p_\theta(\mathbf{Z}_n^{(0)}) = 1$, $p_\theta^*(X_{n,k} \mid \mathbf{Z}_n^{(k-1)})$ is the conditional p.d.f. of $X_{n,k}$ given $\mathbf{Z}_n^{(k-1)}$, for $k = 1, \ldots, n$, and then some martingale theory can be developed as follows. For testing a simple null hypothesis $H_\theta : \theta = \theta_0$ against a simple alternative $H_1 : \theta = \theta_1$, the likelihood ratio test statistic is

$$\mathcal{L}_{n,r} = \frac{p_{\theta_1}(\mathbf{Z}_n^{(r)})}{p_{\theta_0}(\mathbf{Z}_n^{(r)})} = \mathcal{L}_{n,r-1}\left\{\frac{p_{\theta_1}^*(X_{n,r}|\mathbf{Z}_n^{(r-1)})}{p_{\theta_0}^*(X_{n,r}|\mathbf{Z}_n^{(r-1)})}\right\}, \quad \text{by (11.2.2).} \quad (11.2.3)$$

Thus, if $\mathcal{B}_{nk} = \mathcal{B}(\mathbf{Z}_n^{(k)})$ be the sigma field generated by $\mathbf{Z}_n^{(k)}, 0 \leqslant k \leqslant n$, then, by (11.2.2) and (11.2.3), for every $k : 1 \leqslant k \leqslant n$,

$$\begin{aligned} E_{\theta_0}(\mathcal{L}_{n,k} \mid \mathcal{B}_{nk-1}) &= \mathcal{L}_{n,k-1}\int_{X_{n,k-1}}^{\infty} p_{\theta_1}(x \mid \mathbf{Z}_n^{(k-1)})\,dx \\ &= \mathcal{L}_{n,k-1} \quad \text{a.e.,} \quad \text{where} \quad \mathcal{L}_{n,0} = 1. \end{aligned} \quad (11.2.4)$$

Thus, under $H_0 : \theta = \theta_0$, for every n, $\{\mathcal{L}_{n,k}, \mathcal{B}_{nk}; 0 \leqslant k \leqslant n\}$ is a martingale, and hence the central limit theorems for martingales developed in Chapter 2 (Section 2.4) may be used to study the (asymptotic) distribution theory of $\mathcal{L}_{n,r}$. This approach needs the verification of the underlying regularity conditions and thereby demands extra conditions on the p.d.f. f_θ. Also note that by (11.2.1) and (11.2.3),

$$\log \mathcal{L}_{n,r} = \sum_{i=1}^{r} \log\left\{\frac{f_{\theta_1}(X_{n,i})}{f_{\theta_0}(X_{n,i})}\right\} + (n-r)\log\left\{\frac{[1 - F_{\theta_1}(X_{n,r})]}{[1 - F_{\theta_0}(X_{n,r})]}\right\}, \quad (11.2.5)$$

which, for given θ_0, θ_1 and a specified form of f (and F), is a linear combination of functions of the order statistics $X_{n,1}, \ldots, X_{n,r}$. For some simple forms of f_θ (e.g., $f_\theta(x) = \theta^{-1}\exp(-x/\theta), x \in R^+$), (11.2.5) can be expressed as a simple linear function of these order statistics and its exact distribution theory can be developed (see Problem 11.2.1 in this context). However, in general, the form of (11.2.5) may not be simple enough for the derivation of the exact distribution theory of $\log \mathcal{L}_{n,r}$ (or $\mathcal{L}_{n,r}$), and we may have to rely on the asymptotic theory developed in Chapter 7. Verification of the needed regularity conditions to incorporate the asymptotic theory of LCFOS in this context requires, in turn, some extra regularity conditions on f_θ. For testing $H_0 : \theta = \theta_0$ against a composite alternative (e.g., $H_1 : \theta > \theta_0$ or $H_2 : \theta < \theta_0$ or $H^* : \theta \neq \theta_0$), θ_1 is not specified and one may work with the locally most powerful (LMP) test statistic $T_{n,r}$ where

$$T_{n,r} = \left(\frac{\partial}{\partial\theta}\right)\log p_\theta(\mathbf{Z}_n^{(r)})\bigg|_{\theta=\theta_0} = \sum_{i=1}^{r} \dot{f}_{\theta_0}(X_{n,i}) + (n-r)\dot{G}_{\theta_0}(X_{n,r}) \quad (11.2.6)$$

where, for a nonnegative $a_\theta(\cdot), \dot{a}_\theta(\cdot) = (\partial/\partial\theta)\log a_\theta(\cdot)$ and $G_\theta(x) = 1 - F_\theta(x)$. By (11.2.2) and (11.2.6), we have for every $r : 1 \leqslant r \leqslant n$,

$$T_{n,r} = T_{n,r-1} + \dot{p}_{\theta_0}^*(X_{n,r} \mid \mathbf{Z}_n^{(r-1)}), \qquad (T_{n,0} = 0), \quad (11.2.7)$$

so that if one assumes that differentiation (with respect to θ) is permissible under the integral sign, one obtains that for every r $(= 1, \ldots, r)$,

$$E_\theta\left\{\dot{p}^*_{\theta_0}(X_{n,r}\,|\,\mathbf{Z}_n^{(r-1)})\,|\,\mathcal{B}_{nr-1}\right\} = \left[\left(\frac{\partial}{\partial\theta}\right)\int_{X_{n,r-1}}^{\infty} p^*_\theta\left(x\,|\,\mathbf{Z}_n^{(r-1)}\right)dx\right]_{\theta=\theta_0}$$

$$= \left(\frac{\partial}{\partial\theta}\right)(1) = 0. \qquad (11.2.8)$$

Thus, under $H_0:\theta=\theta_0$, for every $n\{T_{n,k}, \mathcal{B}_{nk}; 0 \leqslant k \leqslant n\}$ is a martingale (see Problem 11.2.4 in this context). Again, martingale theorems developed in Section 2.4 may be used to study the (asymptotic) distribution theory of $\{T_{n,k}\}$ when suitable regularity conditions are assumed (see Sen, 1976f and Gardiner and Sen, 1978, in this context). In general, it is comparatively simpler to verify these regularity conditions than those for the $\{\mathfrak{L}_{n,k}\}$. The scope of this parametric inference will be somewhat limited to the basic assumption that the actual p.d.f. is of the specified form f_θ and generally lacks the robustness against departures from this basic assumption. In most life testing problems, one may not have a very precise idea about the true form of the underlying p.d.f., and hence the use of parametric procedures may lead to lack of validity. Moreover, in most clinical trials and life testing problems, one really faces a comparative study where in a parametric model usually there is a vector of unknown parameters some of which may appear as nuisance parameters in the context of the particular hypothesis testing one has in mind. In the presence of such nuisance parameters, the parametric procedure may become more complicated. For these reasons, we like to concentrate on suitable nonparametric procedures that remain valid for a broad class of d.f.'s. Later on, we will discuss some quasi-nonparametric procedures too.

Keeping in mind the multisample location, scale, or the regression problem, we conceive of n independent r.v.'s $X_1, \ldots, X_n$ with continuous d.f.'s $F_1, \ldots, F_n$, respectively, all defined on the real line R. For testing the null hypothesis

$$H_0: F_1 = \ldots = F_n = F \qquad \text{(unknown)} \qquad (11.2.9)$$

against suitable (location, scale, or regression alternatives), we considered in Chapter 4 (as well as in Chapters 9 and 10) a linear rank statistic L_n [see (4.2.1) and (4.2.2)] of the form $\sum_{i=1}^n (c_i - \bar{c}_n)a_n(R_{ni})$ where $c_1, \ldots, c_n$ are given constants, $\bar{c}_n = n^{-1}\sum_{i=1}^n c_i$, R_{ni} is the rank of X_i among $X_1, \ldots, X_n$ and the scores $a_n(i)$ are defined suitably; the choice of the c_i and the $a_n(i)$ depends on the alternatives we have in mind. Under H_0 in (11.2.9), L_n is a distribution-free statistic and from the results of Chapter 4 we have

$$E(L_n\,|\,H_0) = 0 \qquad \text{and} \quad E\left(L_n^2\,|\,H_0\right) = C_n^2 A_n^2, \qquad (11.2.10)$$

where

$$C_n^2 = \sum_{i=1}^{n} (c_i - \bar{c}_n)^2, \qquad A_n^2 = (n-1)^{-1} \sum_{i=1}^{n} \left[a_n(i) - \bar{a}_n \right]^2,$$
$$\bar{a}_n = n^{-1} \sum_{i=1}^{n} a_n(i). \tag{11.2.11}$$

Further (see Section 4.3), under very general conditions, as $n \to \infty$,

$$\frac{L_n}{A_n C_n} \xrightarrow{\mathfrak{D}} \mathfrak{N}(0, 1) \qquad \text{when } H_0 \text{ holds.} \tag{11.2.12}$$

Tests based on L_n demand complete knowledge of the R_{ni}, that is, of the relative magnitudes of all the n observations. In a single point censoring scheme, experimentation continues until r of the n individuals respond. That is, if $X_{n,1} \leqslant \ldots \leqslant X_{n,n}$ be the order statistics corresponding to $X_1, \ldots, X_n$, then the experiment is terminated when $X_{n,r}$ is observed. Thus we have the knowledge of $X_{n,1}, \ldots, X_{n,r}$ (as well as the indicator variables relating to these order statistics), while it is only known that the remaining $n - r$ censored observations have responses greater than $X_{n,r}$. By virtue of the assumed continuity of the F_i, ties among the observations may be neglected, in probability. Then we may rewrite L_n as

$$L_n = \sum_{i=1}^{n} (c_{S_{ni}} - \bar{c}_n) a_n(i), \tag{11.2.13}$$

where $(S_{n1}, \ldots, S_{nn})$ is the vector of *antiranks*, that is, with probability 1,

$$R_{nS_{ni}} = S_{nR_{ni}} = i \qquad \text{and} \qquad X_{n,i} = X_{S_{ni}} \qquad \text{for} \quad i = 1, \ldots, n. \tag{11.2.14}$$

Now, in the case under study, we know $\mathbf{S}_n^{(r)} = (S_{n1}, \ldots, S_{nr})$ and the unordered collection $(S_{nr+1}, \ldots, S_{nn})$ [as the S_{ni} form a permutation of $(1, \ldots, n)$]. One way of adjusting L_n is to consider the average score for the $n - r$ censored observations, that is,

$$a_n^*(r) = (n-r)^{-1} \sum_{i=r+1}^{n} a_n(j) \qquad (\text{defined for every } 0 \leqslant r \leqslant n-1) \tag{11.2.15}$$

and attach this average score to each censored observation. This leads us to

$$L_{n,r} = \sum_{i=1}^{r} (c_{S_{ni}} - \bar{c}_n) a_n(i) + a_n^*(r) \sum_{i=r+1}^{n} (c_{S_{ni}} - \bar{c}_n)$$
$$= \sum_{i=1}^{r} (c_{S_{ni}} - \bar{c}_n) \left[a_n(i) - a_n^*(r) \right], \qquad 1 \leqslant r \leqslant n-1, \tag{11.2.16}$$

while for $r = n - 1$, censoring has no effect, and hence

$$L_{n,n-1} = L_{n,n} = L_n. \tag{11.2.17}$$

By letting, conventionally, $a_n^*(n) = 0$, we are able to define $L_{n,r}$ by (11.2.16), for every $1 \leqslant r \leqslant n$. Also, we let

$$L_{n,0} = 0, \qquad \forall n \geqslant 1. \tag{11.2.18}$$

Writing $b_n(i,r) = a_n(i)$, for $1 \leqslant i \leqslant r$ and $= a_n^*(r)$ for $r+1 \leqslant i \leqslant n, [b_0(i,0) = 0, \forall 1 \leqslant i \leqslant n]$, we have from (11.2.16) that $L_{n,r} = \sum_{i=1}^n (c_{S_{ni}} - \bar{c}_n) b_n(i,r) = \sum_{i=1}^n (c_i - \bar{c}_n) b_n(R_{ni}; r)$, so that $L_{n,r}$ is also a linear rank statistic, and hence

$$E(L_{n,r} \mid H_0) = 0, \qquad E(L_{n,r}^2 \mid H_0) = C_n^2 A_{n,r}^2 \tag{11.2.19}$$

where

$$\begin{aligned} A_{n,r}^2 &= A_n^2 - \frac{1}{n-1} \sum_{j=r+1}^{n} \left[a_n(j) - a_n^*(r) \right]^2, \qquad 0 \leqslant r \leqslant n-2, \\ &= A_n^2, \qquad r = n-1, n; \end{aligned} \tag{11.2.20}$$

note that $A_{n,0}^2 = 0$. Also, parallel to (11.2.12) under H_0 in (11.2.9),

$$\left(\frac{L_{n,r}}{A_{n,r} C_n} \right) \xrightarrow{\mathfrak{D}} \mathfrak{N}(0,1) \qquad \text{as} \quad n \to \infty. \tag{11.2.21}$$

For a single point truncation scheme, experimentation stops at a prefixed time point (T), and let

$$r(T) = (\text{number of } X_{n,i} \leqslant T, \qquad 1 \leqslant i \leqslant n). \tag{11.2.22}$$

Then, $r(T)$ is a nonnegative integer-valued r.v. $(0 \leqslant r(T) \leqslant n)$ and

$$P\{r(T) = r \mid H_0\} = \binom{n}{r} [F(T)]^r [1 - F(T)]^{n-r}, \qquad r = 0, 1, \ldots, n. \tag{11.2.23}$$

We define

$$L_{n,r(T)} = L_{n,r} \qquad \text{when} \quad r(T) = r : 0 \leqslant r \leqslant n. \tag{11.2.24}$$

Note that the d.f. in (11.2.23) depends on F; hence under H_0 in (11.2.9), $L_{n,r(T)}$ is not genuinely distribution-free. But under H_0, $(R_{n1}, \ldots, R_{nn})$ are distributed independently of $(X_{n,1}, \ldots, X_{n,n})$, and hence given $r(T) = r$, conditionally $L_{n,r(T)}$ is distribution-free with mean and variance given by (11.2.19). Thus $L_{n,r(T)}$ provides a conditionally distribution-free test for H_0. Note that by (11.2.23), under H_0,

$$\frac{r(T)}{n} \to F(T), \qquad \text{in probability} \quad \text{as} \quad n \to \infty. \tag{11.2.25}$$

Hence by Theorem 4.3.3 it can be shown that (11.2.21) also holds for $L_{n,r(T)}$.

In most practical problems, unless the survival distribution is at least roughly known, fixed point truncation or censoring may lead to considerable loss of efficiency relative to cost. This drawback is particularly felt in exploratory studies. A too early termination of the experiment may lead to an inadequately small set of observations and thus increases the risk of an incorrect decision. Unnecessary prolongation of the experiment may lead to loss of valuable time with practically no extra gain in efficiency. For this reason, we consider in the rest of this chapter a progressive censoring scheme (PCS) which allows us to monitor the experiment from the beginning until a statistically valid decision (with prescribed risks) is made. The theory of nonparametric testing under PCS is mainly adopted from Chatterjee and Sen (1973a), as further extended by Sen (1976a, b, f, 1979b, e), Majumdar and Sen (1977, 1978a, b), Halperin and Ware (1974), Gardiner and Sen (1978), Sinha and Sen (1979a, b), and Davis (1978), among others. Cox (1972, 1975) has considered some quasi-nonparametric models for survival analysis and extensions of these to PCS are due to Sen (1981a); we shall also refer to these later on.

11.3. NONPARAMETRIC TESTING UNDER PROGRESSIVE CENSORING

In a life-testing type situation, continuous monitoring of the process is needed to record the $X_{n,i}, S_{ni}; i \geqslant 1$. As such, it seems unrealistic to wait until $X_{n,r}$ (for a fixed r) is observed or the prefixed time T is reached, and then decide on the acceptance or rejection of H_0. Naturally, one would like to review the results as the experiment progresses, and thereby have the scope for terminating the experiment at an early stage depending on the accumulated evidence from the observations. Such flexibility would increase the efficiency of the procedure relative to cost and would be particularly relevant in long-term studies of an exploratory nature.

Operationally, the procedure consists of observing the sequence $\{L_{n,k}, k \geqslant 0\}$ as time passes on (computing $L_{n,k}$ at time $X_{n,k}, 1 \leqslant k \leqslant n$), and at each failure time $X_{n,k}$, to make a decision as to the termination of the experimentation or continuation up to the next failure time. Suppose that r is some fixed positive integer ($\leqslant n$); we may even let $r = n$. Let

$$D_{n,r}^{+} = \max_{0 \leqslant k \leqslant r} L_{n,k} \quad \text{and} \quad D_{n,r} = \max_{0 \leqslant k \leqslant r} |L_{n,k}|. \tag{11.3.1}$$

Note that by definition in (11.2.16) and (11.3.1), both $D_{n,r}^{+}$ and $D_{n,r}$ are functions of $(S_{n1}, \ldots, S_{nr})$, when the c_i and $a_n(i)$ are given. Also, under H_0 in (11.2.9), $(S_{n1}, \ldots, S_{nn})$ assumes all possible permutations of $(1, \ldots, n)$ with equal probability $(n!)^{-1}$. Hence under H_0, both $D_{n,r}^{+}$ and $D_{n,r}$ are genuinely distribution-free. Therefore, for every preassigned level of significance $\alpha (0 < \alpha < 1)$ and (r, n), there exist positive numbers $D_{n,r,\alpha}^{*}$ and $D_{n,r,\alpha}$

such that

$$\alpha_n^+ = P\{D_{n,r}^+ > D_{n,r,\alpha}^+ \mid H_0\} \leqslant \alpha < P\{D_{n,r}^+ \geqslant D_{n,r,\alpha}^+ \mid H_0\}, \quad (11.3.2)$$

$$\alpha_n = P\{D_{n,r} > D_{n,r,\alpha} \mid H_0\} \leqslant \alpha < P\{D_{n,r} \geqslant D_{n,r,\alpha} \mid H_0\}, \quad (11.3.3)$$

where α_n^+, α_n, $D_{n,r,\alpha}^+$, and $D_{n,r,\alpha}$ depend on $c_1, \ldots, c_n$ and $a_n(1), \ldots, a_n(n)$. The next section provides simple asymptotic expressions for these constants.

By virtue of (11.3.1) through (11.3.3), we are now in a position to formulate the test procedure as follows.

Continue experimentation so long as at the observed failure time $X_{n,k}$, the corresponding $L_{n,k}$ (or $|L_{n,k}|$) is $\leqslant D_{n,r,\alpha}^+$ (or $\leqslant D_{n,r,\alpha}$). If, for the first time, for some $k = N(\leqslant r)$, $L_{n,N}$ (or $|L_{n,N}|$) is $> D_{n,r,\alpha}^+$ (or $D_{n,r,\alpha}$), stop experimentation when $X_{n,N}$ is observed and reject H_0 in (11.2.1). If no such $N(\leqslant r)$ exists, stop experimentation of $X_{n,r}$ along with the acceptance of H_0. In this formulation, r may even be taken to be equal to n.

In this setup, $X_{n,N}$ is the *stopping time* and N is the *stopping number* for the time-sequential procedure. Some extensions of this simple PCS will be considered in a later section. Note that unlike the case of the sequential tests in Chapter 9, here, $N \leqslant n$ and $X_{n,N} \leqslant X_{n,n}$, with probability 1.

11.4. ASYMPTOTIC DISTRIBUTIONS OF $D_{n,r}^+$ AND $D_{n,r}$

For a given $r(1 \leqslant r \leqslant n)$, we define

$$k_n^{(r)}(t) = \max\{k : A_{n,k}^2 \leqslant tA_{n,r}^2\}, \qquad 0 \leqslant t \leqslant 1, \quad (11.4.1)$$

so that $\{k_n^{(r)}(t), t \in I\}$ is a nondecreasing, right-continuous, and nonnegative integer-valued sequence. Let

$$W_n^{(r)}(t) = \frac{L_{n,k_n^{(r)}(t)}}{A_{n,r}C_n}, \qquad 0 \leqslant t \leqslant 1, \quad (11.4.2)$$

and let $W_n^{(r)} = \{W_n^{(r)}(t), t \in I\}$. Then for every $n(\geqslant r \geqslant 1)$, $W_n^{(r)}$ belongs to the $D[0,1]$ space endowed with the J_1-topology. Note that by (11.3.1) and (11.4.2),

$$\frac{D_{n,r}^+}{A_{n,r}C_n} = \sup_{t\in I} W_n^{(r)}(t) \quad \text{and} \quad \frac{D_{n,r}}{A_{n,r}C_n} = \sup_{t\in I}\left|W_n^{(r)}(t)\right|. \quad (11.4.3)$$

Therefore, we proceed to study first the asymptotic behavior of the process $W_n^{(r)}$, where we let $n \to \infty$ and $r \to \infty$ with

$$\lim_{n\to\infty} \frac{r}{n} = p : 0 < p \leqslant 1. \quad (11.4.4)$$

In this context, we assume that

$$\lim_{n\to\infty}\left\{\max_{1\leqslant i\leqslant n}\frac{(c_i-\bar{c}_n)^2}{C_n^2}\right\}=0, \tag{11.4.5}$$

and that the scores $a_n(1), \ldots, a_n(n)$ are defined by

$$a_n(i)=\phi\left(\frac{i}{n+1}\right) \quad \text{or} \quad E\phi(U_{ni}), \qquad i=1,\ldots,n, \tag{11.4.6}$$

where $U_{n1} < \ldots < U_{nn}$ are defined as in (4.2.2) and

$$\phi(u)=\phi_1(u)-\phi_2(u), \qquad \phi_j(u)\nearrow \text{ in } u(\epsilon I), \qquad j=1,2, \tag{11.4.7}$$

$$\int_0^1 \phi_j^2(u)\,du<\infty, \qquad j=1,2. \tag{11.4.8}$$

For convenience of notations, we let

$$\phi_n(u)=a_n(i), \qquad \text{for } \frac{i-1}{n}<u\leqslant\frac{i}{n}, \qquad i=1,\ldots,n. \tag{11.4.9}$$

Then, under (11.4.6) through (11.4.8),

$$\int_0^1\left[\phi_n(u)-\phi(u)\right]^2du\to 0 \qquad \text{as} \quad n\to\infty. \tag{11.4.10}$$

Let $\mathcal{B}_n^{(k)}$ be the sigma field generated by $\mathbf{S}_n^{(k)}=(S_{n1},\ldots,S_{nk})$ for $k=1,\ldots,n$; $\mathcal{B}_n^{(0)}$ is the trivial sigma field. Then $\mathcal{B}_n^{(k)}$ is nondecreasing in $k(\leqslant n)$.

Lemma 11.4.1. Under H_0 in (11.2.9), for every $n(\geqslant 1)$, $\{L_{n,k},\mathcal{B}_n^{(k)};\ 0\leqslant k\leqslant n\}$ is a martingale.

Proof. By (11.2.13), we have for every $k(\leqslant n)$,

$$E(L_n\,|\,\mathcal{B}_n^{(k)},H_0)=\sum_{i=1}^n a_n(i)E\left\{(c_{S_{ni}}-\bar{c}_n)\,|\,\mathcal{B}_n^{(k)},H_0\right\}. \tag{11.4.11}$$

Now, given $\mathcal{B}_n^{(k)}, S_{n1},\ldots,S_{nk}$ are held fixed, so that

$$E\left\{(c_{S_{ni}}-\bar{c}_n)\,|\,\mathcal{B}_n^{(k)},H_0\right\}=(c_{S_{ni}}-\bar{c}_n), \qquad \text{for} \quad i=1,\ldots,k. \tag{11.4.12}$$

Also, given $\mathcal{B}_n^{(k)}$, $(S_{nk+1},\ldots,S_{nn})$ assumes all the $(n-k)!$ permutations of the complement of $\mathbf{S}_n^{(k)}$ [in $\{1,\ldots,n\}$] with the equal (conditional) probability $[(n-k)!]^{-1}$. Hence for $i: k<i\leqslant n$,

$$E\left\{(c_{S_{ni}}-\bar{c}_n)\,|\,\mathcal{B}_n^{(k)},H_0\right\}=\frac{1}{n-k}\sum_{j=k+1}^n(c_{S_{nj}}-\bar{c}_n). \tag{11.4.13}$$

From (11.4.11) through (11.4.13), we obtain by a few routine steps that

$$E(L_n \mid \mathfrak{B}_n^{(k)}, H_0) = \sum_{i=1}^{k} a_n(i)[c_{S_{ni}} - \bar{c}_n] + \left[\frac{1}{n-k}\sum_{i=k+1}^{n}(c_{S_{ni}} - \bar{c}_n)\right]\left[\sum_{i=k+1}^{n} a_n(i)\right]$$
$$= L_{n,k}, \quad \text{by } (11.2.16). \tag{11.4.14}$$

This shows that $L_{n,0}(=0), L_{n,1}, \ldots, L_{n,n-1}, L_{n,n}(=L_n)$ form a martingale sequence (closed on the right by L_n) when H_0 holds. Q.E.D.

THEOREM 11.4.2. *Under* (11.2.9) *and* (11.4.4) *through* (11.4.8), *as* $n \to \infty$,

$$W_n^{(r)} \xrightarrow{\mathfrak{D}} W, \quad \text{in the } J_1\text{-topology on } D[0,1], \tag{11.4.15}$$

where $W = \{W(t), t \in I\}$ *is a standard Wiener process on* I.

Proof. We need to show that the f.d.d.'s of $\{W_n^{(r)}\}$ converge to the corresponding ones of W and that $W_n^{(r)}$ is tight. For the first part, let us define

$$\tilde{\phi}_p(u) = \begin{cases} \phi(u), & 0 < u \leqslant p, \\ \phi_p^* = (1-p)^{-1}\int_p^1 \phi(t)\,dt, & p < u < 1, \end{cases} \tag{11.4.16}$$

and $\tilde{a}_n(\cdot)$ as in (11.4.6) with ϕ replaced by $\tilde{\phi}_p$, $\tilde{L}_{n,p}$ by (11.2.2) with $a_n(\cdot)$ replaced by $\tilde{a}_n(\cdot)$, and for $p \in I$, we define

$$\nu(p) = \int_0^1 \tilde{\phi}_p^2(u)\,du - \left(\int_0^1 \tilde{\phi}_p(u)\,du\right)^2$$
$$= \int_0^1 \phi^2(u)\,du + (1-p)(\phi_p^*)^2 - \left(\int_0^1 \phi(u)\,du\right)^2. \tag{11.4.17}$$

Note that $\nu(0) = 0, \nu(1) = A^2$ and $\nu(p)$ is $\nearrow$ in $p \in I$. Further note that, by definition, for every $k: 2 \leqslant k \leqslant n$,

$$A_{n,k}^2 - A_{n,k-1}^2 = \frac{n-k}{(n-1)(n-k+1)}\left[a_n(k) - a_n^*(k)\right]^2, \tag{11.4.18}$$

where (11.4.6) through (11.4.8) imply that as $n \to \infty$

$$\max_{1 \leqslant i \leqslant n} \frac{a_n^2(i)}{n} \to 0 \quad \text{and} \quad \max_{1 \leqslant i \leqslant n} \frac{[a_n^*(i)]^2}{n} \to 0. \tag{11.4.19}$$

As a result,

$$\max_{1 \leq k \leq n} \left[A_{n,k}^2 - A_{n,k-1}^2 \right] \to 0 \quad \text{as} \quad n \to \infty, \tag{11.4.20}$$

and along the lines of Section 4.3, it can be shown that under H_0,

$$\left[\frac{r}{n} \to p \right] \Rightarrow \left[A_{n,r}^2 \to \nu(p) \right], \qquad \forall p \in I; \tag{11.4.21}$$

$$\left[\frac{r}{n} \to p \right] \Rightarrow C_n^{-1} \left| L_{n,r} - \tilde{L}_{n,p} \right| \xrightarrow{p} 0, \qquad C_n^{-1} \left| \tilde{L}_{n,p} - \tilde{L}_{n,p}^* \right| \xrightarrow{p} 0, \tag{11.4.22}$$

(see Problems 11.4.1 and 11.4.2), where

$$\tilde{L}_{n,p}^* = \sum_{i=1}^{n} (c_i - \bar{c}_n) \tilde{\phi}_p(F(X_i)). \tag{11.4.23}$$

Further note that $E(\tilde{L}_{n,p}^* \mid H_0) = 0, \forall p \in I$ and $\forall p, p' \in I$.

$$\begin{aligned} E(\tilde{L}_{n,p}^* \tilde{L}_{n,p'}^* \mid H_0) &= C_n^2 \left[\int_0^1 \phi_p(u) \phi_{p'}(u)\, du - \left(\int_0^1 \phi(u)\, du \right)^2 \right] \\ &= C_n^2 \left[\int_0^1 \tilde{\phi}_{p \wedge p'}^2(u)\, du - \left(\int_0^1 \phi(u)\, du \right)^2 \right] = C_n^2 \nu(p \wedge p'). \end{aligned} \tag{11.4.24}$$

Also, a direct application of the multivariate central limit theorem yields that for every $m(\geq 1)$ and $(0 \leq) s_1 < \dots < s_m (\leq 1)$, $C_n^{-1}[\tilde{L}_{n,s_1}^*, \dots, \tilde{L}_{n,s_m}^*]$ has asymptotically a multivariate normal distribution with null mean vector and dispersion matrix

$$((\nu(s_j \wedge s_l)))_{j,l=1,\dots,m}. \tag{11.4.25}$$

Consequently, on letting

$$s_j = \sup\{ q : \nu(q) \leq t_j \nu(p) \}, \qquad j = 1, \dots, m, \tag{11.4.26}$$

we obtain from (11.4.21) and the above that

$$C_n^{-1} A_{n,r}^{-1} \left[\tilde{L}_{n,s_1}^*, \dots, \tilde{L}_{n,s_m}^* \right] \xrightarrow{\mathfrak{D}} \left[W(t_1), \dots, W(t_m) \right]. \tag{11.4.27}$$

On the other hand, by (11.4.1), (11.4.2), (11.4.20), (11.4.21), (11.4.22) and (11.4.26), it follows that under H_0,

$$\max_{1 \leq j \leq m} \left| W_n^{(r)}(t_j) - C_n^{-1} A_{n,r}^{-1} \tilde{L}_{n,s_j}^* \right| \xrightarrow{p} 0, \tag{11.4.28}$$

and hence the proof of the convergence of the f.d.d.'s of $\{W_n^{(r)}\}$ to those of W follows from (11.4.27) and (11.4.28). By virtue of Lemma 11.4.1, (11.4.20), and the convergence of the f.d.d.'s of $\{W_n^{(r)}\}$ to those of W, the proof of the tightness of $W_n^{(r)}$ follows directly from Corollary 2.4.4.1.

Q.E.D.

By virtue of (11.3.1), (11.4.3), Theorem 11.4.2 and (2.7.1) and (2.7.3), we obtain the following.

THEOREM 11.4.3. *Under the hypothesis of Theorem* 11.4.2, *for every* $\lambda > 0$,

$$P\left\{\frac{D_{n,r}^{+}}{A_{n,r}C_n} \geqslant \lambda\right\} \to \left(\frac{2}{\pi}\right)^{1/2}\int_{\lambda}^{\infty}\exp\left(-\frac{1}{2}t^2\right)dt = 2\left[1-\Phi(\lambda)\right], \quad (11.4.29)$$

$$P\left\{\frac{D_{n,r}}{A_{n,r}C_n} \geqslant \lambda\right\} \to 1 - \sum_{k=-\infty}^{\infty}(-1)^k\left[\Phi((2k+1)\lambda) - \Phi((2k-1)\lambda)\right], \quad (11.4.30)$$

where $\Phi(x)$ *is the standard normal distribution function.*

Thus, if D_{α}^{+} and D_{α} be, respectively, the upper $100\alpha\%$ points of the distributions of $\sup_{0\leqslant t\leqslant 1}W(t)$ and $\sup_{0\leqslant t\leqslant 1}|W(t)|$ (for which the rhs of (11.4.29) and (11.4.30) are both equal to α for $\lambda = D_{\alpha}^{+}$ and D_{α}), we have from (11.3.2) and (11.3.3) and (11.4.29) and (11.4.30) that as $n\to\infty$,

$$\alpha_n^{+} \to \alpha, \qquad \alpha_n \to \alpha, \qquad \frac{D_{n,r,\alpha}^{+}}{A_{n,r}C_n} \to D_{\alpha}^{+} \quad \text{and} \quad \frac{D_{n,r,\alpha}}{A_{n,r}C_n} \to D_{\alpha}. \quad (11.4.31)$$

We now proceed to study the behavior of $W_n^{(r)}$ when H_0 may not hold. As in (4.3.36) through (4.3.56), we shall consider here contiguous (to H_0) alternatives for which simple asymptotic solutions exist and provide interesting tools for the study of the asymptotic properties of $D_{n,r}^{+}$ and $D_{n,r}$.

Consider a triangular array $\{X_{ni}, 1 \leqslant i \leqslant n; n \geqslant 1\}$ of row-wise independent r.v.'s where for some nonzero (finite) θ,

$$F_{n,i}(x) = P\{X_{ni} \leqslant x\} = F(x - \theta d_{ni}), \qquad 1 \leqslant i \leqslant n, \quad (11.4.32)$$

$\{d_{ni}, 1 \leqslant i \leqslant n; n \geqslant 1\}$ is a triangular array of constants satisfying

$$\sup_n \sum_{i=1}^{n} d_{ni}^2 < \infty \qquad \text{and} \qquad \max_{1\leqslant i\leqslant n} d_{ni}^2 \to 0 \quad \text{as} \quad n\to\infty, \quad (11.4.33)$$

and the d.f. F has an absolutely continuous p.d.f. $f(x)$ with a finite Fisher information $I(f) = \int_{-\infty}^{\infty}[f'/f]^2\,dF$. We denote by $P_{n,k}$ the joint d.f. of $(X_{n1}, \ldots, X_{nk})$ when the d_{ni} in (11.4.32) are all equal to 0 and by $Q_{n,k}$ when (11.4.32) and (11.4.33) hold for $k \leqslant n$. Then, as in Section 4.3,

$$\left[\, Q_{n,k} \quad \text{is contiguous to} \quad P_{n,k}, \qquad \forall k \leqslant n\right]. \quad (11.4.34)$$

Let us now define

$$\psi(u) = \frac{-f'(F^{-1}(u))}{f(F^{-1}(u))}, \qquad 0 < u < 1, \qquad A_{\psi}^2 = I(f), \quad (11.4.35)$$

$$\rho_1(t;p) = \frac{\left(\int_0^1 \tilde{\phi}_t(u)\psi(u)\,du\right)}{\left(\nu(p)A_{\psi}^2\right)^{1/2}}, \qquad \forall 0 \leqslant t \leqslant p \leqslant 1, \quad (11.4.36)$$

where $\tilde{\phi}_t$ is defined by (11.4.16) and $\nu(p)$ by (11.4.17). Finally, we assume that

$$\lim_{n\to\infty} C_n^{-1} \sum_{i=1}^{n} (c_i - \bar{c}_n) d_{ni} = \rho_2^* \tag{11.4.37}$$

exists and on defining $s[= s(t)]$ as in (11.4.26), we let

$$\mu = \{ \mu(t) = \theta\rho_2^* A_\psi \rho_1(s(t); p), \qquad t \in I \}. \tag{11.4.38}$$

THEOREM 11.4.4. *Under the assumptions made above, when* $\{ Q_n \}$ *holds,*

$$W_n^{(r)} - \mu \xrightarrow{\mathfrak{D}} W, \qquad \textit{in the } J_1\textit{-topology on} \quad D[0,1]. \tag{11.4.39}$$

Proof. By the contiguity in (11.4.34) and the tightness of $\{ W_n^{(r)} \}$ under H_0 (as has been proved in Theorem 11.4.3), the tightness of $\{ W_n^{(r)} \}$ under $\{ Q_n \}$ follows directly by using Theorem 4.3.4. Hence it suffices to establish only the convergence of the f.d.d.'s. Note that by the contiguity of $\{ Q_{n,n} \}$ to $\{ P_{n,n} \}$, (11.4.28) continues to hold under $\{ Q_{n,n} \}$ as well. Thus for the convergence of f.d.d.'s, it suffices to work with $C_n^{-1} A_{n,r}^{-1} [\tilde{L}_{n,s_1}^*, \ldots, \tilde{L}_{n,s_m}^*]$. Under $\{ Q_{n,n} \}, C_n^{-1} A_{n,r}^{-1} \tilde{L}_{n,s_j}^*$ has asymptotically mean $\theta\rho_2^* A_\psi \cdot \rho_1(s_j, p)$ and their asymptotic dispersion matrix agrees with (11.4.25); the (joint) asymptotic normality of $(C_n^{-1} A_{n,r}^{-1} [\tilde{L}_{n,s_1}^*, \ldots, \tilde{L}_{n,s_m}^*])$ follows directly by using (4.3.28). Hence the convergence of the f.d.d.'s is established. Q.E.D.

In Theorem 11.4.4, we considered contiguous alternatives relating to the location/regression model. For the sacle problem, the details can be worked out similarly; however, the drift function μ will be different (see Problem 11.4.3).

11.5. ASYMPTOTIC POWER OF THE TESTS BASED ON $D_{n,r}^+$ AND $D_{n,r}$

Having studied the asymptotic null distributions of $D_{n,r}^+$ and $D_{n,r}$, we are now in a position to use Theorem 11.4.4 for the study of the asymptotic power of these tests. Here too we consider contiguous location/regression alternatives only.

For regression/location problems, one usually takes $\phi(u)$ to be monotonic and we assume that $\phi(u)$ is $\nearrow$ in $u(\epsilon I)$. Also, we assume that $\psi(u)$ is $\nearrow$ in $u(\epsilon I)$; this is true whenever the p.d.f. f is strongly unimodal. Finally, keeping in mind the case of the one-sided test based on $D_{n,r}^+$, we confine ourselves here to the model (11.4.32) where $\theta\rho_2^* > 0$; for the two-sided test this is not necessary.

Let $\gamma_n^+(\theta)$ and $\gamma_n(\theta)$ be the power of the tests based on $D_{n,r}^+$ and $D_{n,r}$, respectively, when (11.4.32) is the true model. Then, by (11.4.3) and Theorem 11.4.4, it follows that under the hypothesis of Theorem 11.4.4,

$$\lim_{n\to\infty} \gamma_n^+(\theta) = \gamma^+(\theta) \qquad \text{and} \qquad \lim_{n\to\infty} \gamma_n(\theta) = \gamma(\theta) \tag{11.5.1}$$

both exist and

$$\gamma^+(\theta) = P\{W(t) + \theta\rho_2^* A_\psi \rho_1(s(t), p) > D_\alpha^+ \quad \text{for some} \quad 0 \leqslant t \leqslant 1\}, \tag{11.5.2}$$

$$\gamma(\theta) = P\{|W(t) + \theta\rho_2^* A_\psi \rho_1(s(t), p)| \geqslant D_\alpha \quad \text{for some} \quad 0 \leqslant t \leqslant 1\}. \tag{11.5.3}$$

To simplify the expressions for $\gamma^+(\theta)$ and $\gamma(\theta)$, we define $\psi(u)$ as in (11.4.35) and let

$$\tilde{\psi}_t(u) = \begin{cases} \psi(u), & 0 < u \leqslant t \\ \psi_t^* = (1-t)^{-1}\int_t^1 \psi(w)\,dw, & t < u < 1, \end{cases} \qquad \forall t \in I; \tag{11.5.4}$$

$$\nu^*(p) = \int_0^1 \tilde{\psi}_p^2(u)\,du, \qquad \forall p \in I \qquad (\Rightarrow \nu^*(1) = A_\psi^2 = I(f)), \tag{11.5.5}$$

$$\rho(\tilde{\phi}_t, \tilde{\psi}_t) = \frac{\int_0^1 \tilde{\phi}_t(u)\tilde{\psi}_t(u)\,du}{[\nu(t)\nu^*(t)]^{1/2}}, \qquad t \in I. \tag{11.5.6}$$

Note that ϕ and ψ are both nondecreasing, and hence $\rho(\tilde{\phi}_t, \tilde{\psi}_t) \geqslant 0 \forall t \in I$ and $\rho(\tilde{\phi}_t, \tilde{\psi}_t) = 1$ if $\tilde{\phi}_t \equiv \tilde{\psi}_t$ (up to a multiplicative constant).

From (11.4.36) and the above we have

$$\begin{aligned} A_\psi \rho_1(s(t), p) &= \frac{\left(\int_0^1 \tilde{\phi}_{s(t)}(u)\psi(u)\,du\right)}{\nu^{1/2}(p)} = \frac{\nu^{1/2}(s(t))}{\nu^{1/2}(p)} \cdot \frac{\left(\int_0^1 \tilde{\phi}_{s(t)}(u)\psi(u)\,du\right)}{\nu^{1/2}(s(t))} \\ &= \left[\frac{\nu(s(t))}{\nu(p)}\right]^{1/2} [\nu^*(s(t))]^{1/2} \frac{\left(\int_0^1 \tilde{\phi}_{s(t)}(u)\tilde{\psi}_{s(t)}(u)\,du\right)}{[\nu(s(t))\nu^*(s(t))]^{1/2}} \\ &= t^{1/2}[\nu^*(p)]^{1/2}\left\{\frac{\nu^*(s(t))}{\nu^*(p)}\right\}^{1/2} \rho(\tilde{\phi}_{s(t)}, \tilde{\psi}_{s(t)}). \end{aligned} \tag{11.5.7}$$

Recall that $\nu(s(t))/\nu(p) = t$. If we define

$$s^*(t) = \max\left\{q : \frac{\nu^*(q)}{\nu^*(p)} \leqslant t\right\}, \qquad t \in I, \tag{11.5.8}$$

then by the monotonicity of $\nu^*(u)$ (in $u \in I$),

$$\left[s(t) \leqslant s^*(t)\right] \Rightarrow \left[\frac{\nu^*(s(t))}{\nu^*(p)} \leqslant t\right]. \tag{11.5.9}$$

Let $\mathcal{G}$ be the class of all score functions $\{\phi(u), u \in I\}$ such that if $s_\phi = \{s_\phi(t), t \in I$, for the score function $\phi\}$, then

$$s_\phi \prec s_\psi \quad \text{if} \quad s_\phi(t) \leqslant s_\psi(t) = s^*(t), \qquad \forall t \in I. \tag{11.5.10}$$

From the above, we conclude that

$$[\phi \in \mathcal{G}] \Rightarrow \left[A_\psi \rho_1(s_\phi(t), p) \leqslant t\left[\nu^*(p)\right]^{1/2}\right], \tag{11.5.11}$$

where the quality sign holds when $\phi = \psi$. For the score function $\phi \equiv \psi$, (11.5.2) reduces to

$$P\left\{W(t) > D_\alpha^+ - \phi\rho_2^*\left[\nu^*(p)\right]^{1/2} t \quad \text{for some} \quad t \in [0,1]\right\} \tag{11.5.12}$$

and we are in a position to apply (2.7.16) and (2.7.17). For any other $\phi(\epsilon\mathcal{G})$, the asymptotic power is bounded from above by (11.5.12). Of course, for $\phi \notin \mathcal{G}$, (11.5.11) may not hold and we do not have the asymptotic optimality of the score function ψ. Similar manipulations can be made for (11.5.3). However, the expressions are not readily usable to characterize the asymptotically optimal score functions as here these will be two parallel curves and boundary crossing probabilities are not that precisely known.

For contiguous alternatives, the optimality of ψ (within the class $\mathcal{G}$) has been studied earlier. We are, however, not able to get a single measure of efficiency of the progressive censoring test based on $D_{n,r}^+$ (or $D_{n,r}$) with respect to the single-point censoring test based on $L_{n,r}$; the relative performance depends on the level of significance as well as on θ. For this reason, we shall consider now the (approximate) Bahadur efficiency measure to study some related results.

Assume that $\{c_i\}$ satisfy

$$n^{-1}C_n^2 \to C_0^2 : 0 < C_0 < \infty; \max_{1 \leqslant i \leqslant n} \frac{|c_i - \bar{c}_n|}{\sqrt{n}} \to 0. \tag{11.5.13}$$

Also,

$$c_{ni}^* = \frac{c_i - \bar{c}_n}{C_n}, 1 \leqslant i \leqslant n \left(\Rightarrow \sum_1^n c_{ni}^* = 0, \sum_1^n (c_{ni}^*)^2 = 1\right). \tag{11.5.14}$$

Let then $\{X_i, i \geqslant 1\}$ are independent r.v.'s with continuous d.f.'s $\{F_i, i \geqslant 1\}$ where

$$F_i(x) = F(x - \beta_0 - \theta d_i), \qquad i \geqslant 1, \tag{11.5.15}$$

and the $\{d_i\}$ satisfy the conditions:

$$(c_i - \bar{c}_n)(d_i - \bar{d}_n) \geqslant 0, \qquad 1 \leqslant i \leqslant n, \tag{11.5.16}$$

$$\max_{1 \leqslant i \leqslant n} \frac{|d_i - \bar{d}|}{\sqrt{n}} \to 0, \limsup_n |\bar{d}_n| < \infty \tag{11.5.17}$$

$$\lim_{n\to\infty} n^{-1} \sum_{i=1}^{n} (c_i - \bar{c}_n)(d_i - \bar{d}_n) = \rho_2^* C_0, 0 < \rho_2^* < \infty. \tag{11.5.18}$$

Further, the d.f. F has an absolutely continuous p.d.f. f with a finite Fisher information $I(f)$. Let

$$\overline{H}_\theta^{(n)}(x) = n^{-1} \sum_{i=1}^{n} F_i(x), \qquad G_\theta^{(n)}(x) = n^{-1/2} \sum_{i=1}^{n} c_{ni}^* F_i(x) \tag{11.5.19}$$

and assume that

$$\lim_{n\to\infty} \overline{H}_\theta^{(n)}(x) = H_\theta(x) \qquad \text{and} \quad G_\theta^{(n)}(x) = G_\theta(x) \tag{11.5.20}$$

exist, $\forall x \in (-\infty, \infty)$.

If we let

$$H_n(x) = n^{-1} \sum_{i=1}^{n} I(X_i \leqslant x) \qquad \text{and} \quad G_n(x) = n^{-1/2} \sum_{i=1}^{n} c_{ni}^* I(X_i \leqslant x). \tag{11.5.21}$$

and $a_n(i) = \phi_n(u)$, $(i-1)/n < u \leqslant i/n$, $1 \leqslant i \leqslant n$, $\phi_n^*(k/n) = a_n^*(k)$, $1 \leqslant k \leqslant n-1$, then we have for each r $(1 \leqslant r \leqslant n)$,

$$n^{-1/2} C_n^{-1} L_{n,r} = \int_{-\infty}^{X_{n,r}} \left\{ \phi_n(H_n(x)) - \phi_n^*\left(\frac{r}{n}\right) \right\} dG_n(x). \tag{11.5.22}$$

Let us define

$$\tau_\theta(z) = \int_{-\infty}^{z} \{ \phi(H_\theta(x)) - \phi^*(H_\theta(z)) \} dG_\theta(x), \qquad -\infty < z < \infty. \tag{11.5.23}$$

$$\tau_{\theta,p}^{+} = \sup\{ \tau_\theta(z) : -\infty < z < H^{-1}(p) \}, \qquad 0 < p \leqslant 1. \tag{11.5.24}$$

Note that by using the Schwarz inequality, we have for every $q < k$,

$$n^{-1} C_n^{-2} (L_{n,k} - L_{n,q})^2 \leqslant \frac{1}{n} \sum_{q+1}^{k} \left[a_n(i) - a_n^*(q) \right]^2 + \frac{n-k}{n} \left[a_n^*(k) - a_n^*(q) \right]^2, \tag{11.5.25}$$

and hence for every $\epsilon > 0$, there exists a $\delta(>0)$, such that

$$\max_{q:|k-q|\leqslant \delta n} n^{-1/2}C_n^{-1}|L_{n,q} - L_{n,k}| < \epsilon \qquad \text{for every} \quad k \in [1,n]. \tag{11.5.26}$$

As such, using Theorem 4.6.1 (for finitely many $k : k/n \to j\delta, j = 1, \ldots, m_\delta$ where $\delta(m_\delta - 1) < 1 \leqslant \delta m_\delta$) and (11.5.26), it follows that

$$\max_{1 \leqslant k \leqslant r} \{n^{-1/2}C_n^{-1}L_{n,k}\} \to \tau_{\theta,p}^{+} \qquad \text{a.s.,} \quad \text{as} \quad n \to \infty \tag{11.5.27}$$

where $r/n \to p : 0 < p \leqslant 1$, and a similar result holds for $\{|L_{n,k}|\}$.

From (11.4.21), (11.4.29), and (11.5.27), we conclude that the *approximate Bahadur slope* for $D_{n,r}^{+}/A_{n,r}C_n$ is

$$(\tau_{\theta,p}^{+})^2/\nu(p), \tag{11.5.28}$$

where both τ and ν depend on the score function ϕ. Also note that $C_n^{-1}A_{n,r}^{-1}L_{n,r} \xrightarrow{\mathfrak{D}} \mathfrak{N}(0,1)$ under H_0 (i.e., when $d_i = 0, \forall i$) and $n^{-1/2}C_n^{-1}L_{n,r} \to \tau_\theta(H_\theta^{-1}(p))$ a.s., as $n \to \infty$. Hence by similar arguments it follows that the approximate Bahadur slope for the single-point censoring test based on $L_{n,r}$ is

$$\left[\tau_\theta\left(H_\theta^{-1}(p)\right)\right]^2/\nu(p). \tag{11.5.29}$$

We may remark that for (11.5.29), we assume that $\tau_\theta(H_\theta^{-1}(p))$ is positive (as otherwise, the power of the test will cease to go to 1 as $n \to \infty$). Then, by (11.5.23) and (11.5.24), we have

$$\tau_\theta\left(H_\theta^{-1}(p)\right) \leqslant \tau_{\theta,p}^{+}, \tag{11.5.30}$$

so that from (11.5.28) through (11.5.30), we conclude that the Bahadur (asymptotic relative) efficiency of the progressive censoring test with respect to the single-point censoring test (using the same score function ϕ) is

$$\tau_\theta(P,S) = \left[\tau_{\theta,p}^{+}/\tau_\theta\left(H_\theta^{-1}(p)\right)\right]^2 \geqslant 1, \qquad \forall\theta. \tag{11.5.31}$$

Thus, from this point of view, progressive censoring is more effective than the single-point censoring schemes. Besides, the expected stopping time is smaller in the former case, resulting in savings of cost and time of experimentation.

Note that ϕ and ψ are both assumed to be nondecreasing and (see Problem 11.5) it can be shown that as $\theta \to 0$,

$$\tau_\theta\left(H_\theta^{-1}(t)\right) = \theta\rho_2^*\left(\int_0^1 \tilde{\phi}_t(u)\psi(u)\,du\right) + o(\theta), \tag{11.5.32}$$

where ρ_2^* is defined by (11.5.18). Hence, recalling that $\int_0^1 \tilde{\phi}_t(u)\psi(u)\,du$ is $\nearrow$ in

$t \in (0,1)$, we obtain from (11.5.23), (11.5.24) and (11.5.32) that

$$\lim_{\theta\to 0} \frac{\tau_\theta\left(H_\theta^{-1}(p)\right)}{\tau_{\theta,p}^{+}} = 1, \qquad (11.5.33)$$

so that in the light of the limiting Bahadur efficiency, the progressive censoring and the single-point censoring are equally efficient.

To compare two progressive censoring tests based on two different score functions, say $\phi_{(1)}$ and $\phi_{(2)}$, we define $\tilde{\phi}_{(j)p}$ as in (11.4.16). Then by (11.5.29), (11.5.32), and (11.5.33), the limiting Bahadur efficiency of the test based on $\phi_{(1)}$ with respect to the other one based on $\phi_{(2)}$ is

$$e(\phi_{(1)}, \phi_{(2)}) = \frac{\rho^2(\tilde{\phi}_{(1)p}, \tilde{\psi}_p)}{\rho^2(\tilde{\phi}_{(2)p}, \tilde{\psi}_p)}, \qquad (11.5.34)$$

where $\tilde{\psi}_p$ is defined by (11.5.4) and $\rho(\tilde{\phi}_{(j),p}, \tilde{\psi}_p)$, $j = 1, 2$ as in (11.5.6). Clearly, from (11.5.34), the optimal score function is $\phi \equiv \psi$, so that

$$e(\psi, \phi) = \frac{1}{\rho^2(\tilde{\phi}_p, \tilde{\psi}_p)} \geqslant 1, \qquad \forall \phi. \qquad (11.5.35)$$

Also, (11.5.35) does not require the condition that $\phi \in \mathcal{G}$.

To compare two progressive censoring tests based on the same score function (ϕ) but on two terminating censoring numbers r_1 and r_2 where $r_j/n \to p_j$, $j = 1, 2$, $0 < p_2 \leqslant p_1 \leqslant 1$, the limiting Bahadur efficiency is

$$e_\phi(p_1, p_2) = \left[\frac{\nu(p_2)}{\nu(p_1)}\right]\left[\frac{\left(\int_0^1 \tilde{\phi}_{p_1}(u)\psi(u)\,du\right)}{\left(\int_0^1 \tilde{\phi}_{p_2}(u)\psi(u)\,du\right)}\right]^2 \qquad (11.5.36)$$

where $\nu(p_1) \geqslant \nu(p_2)$ and $\left(\int_0^1 \tilde{\phi}_{p_1}(u)\psi(u)\,du\right) \geqslant \left(\int_0^1 \tilde{\phi}_{p_2}(u)\psi(u)\,du\right)$. If $\phi \equiv \psi$, then (11.5.36) reduces to

$$e_\psi(p_1, p_2) = \frac{\nu(p_1)}{\nu(p_2)} \geqslant 1, \qquad \forall 0 < p_2 < p_1 \leqslant 1, \qquad (11.5.37)$$

so that larger is the value of p, the better should be the test. On the other hand, if $\phi \neq \psi$, the first factor on the rhs of (11.5.36) is $\leqslant 1$ while the second factor is $\geqslant 1$ and we do not know about the product.

Finally, we like to study the relative performance of the tests from the point of view of *stopping times*. We recall that at each failure $X_{n,k}$, we compute the value of $L_{n,k}$ and if for $k = N$, the smallest positive integer ($\leqslant r$), $L_{n,N}$ is $\geqslant D_{n,r,\alpha}^{+}$, we stop the experiment along with the rejection of H_0. Thus the *stopping number* is $N(\leqslant r)$ and the *stopping time* is $X_{n,N}$.

Note that, by definition,

$$P\{N > q\} = P\left\{\frac{N}{n} > \frac{q}{n}\right\}$$

$$= \begin{cases} P\{L_{n,k} \leqslant D_{n,r,\alpha}^{+}, \forall k \leqslant q\}, & 0 \leqslant q \leqslant r-1 \\ 0, & q \leqslant r \end{cases}; \quad (11.5.38)$$

$$E\left(\frac{N}{n}\right) = \sum_{q=0}^{r-1} n^{-1} P\left\{\frac{N}{n} > \frac{q}{n}\right\} \quad (11.5.39)$$

As in (11.4.28), we note that

$$\frac{q}{n} \to u_0 \Rightarrow \frac{\nu(u_0)}{\nu(p)} = t_0 \quad \text{where} \quad 0 \leqslant u_0 \leqslant p \Rightarrow 0 \leqslant t_0 \leqslant 1. \quad (11.5.40)$$

Thus if we consider the case of contiguous alternatives in (11.4.32) and (11.4.33) and introduce the notations and assumptions as in (11.4.35) through (11.4.38), we have by (11.5.40) and Theorem 11.4.4,

$$\lim_{n\to\infty} P\left\{\frac{N}{n} > u_0\right\} = P\{W(t) + \theta\rho_2^* A_\psi \rho_1(s_\psi(t), p) \leqslant D_\alpha^+,$$

$$\forall 0 \leqslant t \leqslant t_0\} \quad (11.5.41)$$

Note that as in (11.5.7) through (11.5.11),

$$[\phi \in \mathcal{G}] \Rightarrow \left[\rho_1(s_\phi(t), p) A_\psi \leqslant t[\nu^*(p)]^{1/2}\right], \quad (11.5.42)$$

and hence

$$P\{W(t) + \theta\rho_2^* A_\psi \rho_1(s_\phi(t), p) \leqslant D_\alpha^+, \forall 0 \leqslant t \leqslant t_0\}$$

$$\geqslant P\left\{W(t) + \theta\rho_2^*[\nu^*(p)]^{1/2} t, \forall 0 \leqslant t \leqslant t_0\right\}, \qquad \forall \phi \in \mathcal{G}, \quad (11.5.43)$$

where the equality sign holds when $\phi = \psi$. Thus from (11.5.38) through (11.5.43), we conclude that for contiguous alternatives, in accordance with the criterion of minimum $E(N/n)$, within the class $\mathcal{G}, \psi$ is an optimal score function.

Let us now consider the case of the stopping time $X_{n,N}$. Note that by definition,

$$E(X_{n,N}) = \sum_{k=1}^{r} E\{X_{n,k} I_{[N=k]}\}, \qquad \text{where} \quad r = [np] + 1, \qquad 0 < p \leqslant 1. \quad (11.5.44)$$

Under H_0, the vector of ranks and the order statistics are stochastically independent. Hence $I_{[N=k]}$, being an $\mathcal{R}_k$-measurable function, is independent of $X_{n,k}$, so that if $\xi_{n,k} = EX_{n,k}$ exists for $k = 1, \ldots, r$, then, for

$0 < p < 1$, we write

$$E_0(X_{n,N}) = \sum_{k=1}^{r} P\{N = k\}\xi_{n,k} = \sum_{k=1}^{r} (\xi_{n,k} - \xi_{n,k-1})P\{N \geqslant k\}, \quad (11.5.45)$$

where $\xi_{n,0} = 0$. Noting that asymptotically $n(\xi_{n,k} - \xi_{n,k-1})f(\xi_{n,k})$ converges to 1, so that by using (11.5.38) and (11.5.45), we obtain that as $n \to \infty$,

$$E_0(X_{n,N}) \to \int_0^{\xi_p} \{f(x)\}^{-1} P\left\{ W(t) \leqslant D_\alpha^+, \quad \forall t \leqslant \frac{\nu(F(x))}{\nu(p)} \right\} dF(x)$$

$$= \int_0^{\xi_p} P\left\{ W(t) \leqslant D_\alpha^+, \qquad \forall t \leqslant \frac{\nu(F(x))}{\nu(p)} \right\} dx; \ \xi_p = F^{-1}(p). \quad (11.5.46)$$

The situation is somewhat different when H_0 does not hold; in that case, $X_{n,k}$ and $[N = k]$ are not independent. However, one may write for some arbitrary $\epsilon > 0$,

$$E(X_{n,N}) = \sum_{k=1}^{r} E\{X_{n,k} I_{[N=k]} I(|X_{n,k} - \xi_{n,k}| \leqslant \epsilon)$$

$$+ \sum_{k=1}^{r} E\{X_{n,k} I_{[N=k]} I(|X_{n,k} - \xi_{n,k}| > \epsilon)\}. \quad (11.5.47)$$

For $\epsilon(>0)$ sufficiently small, the first term on the rhs of (11.5.47) can be asymptotically taken as $\sum_{k=1}^{r} \xi_{n,k} P\{N = k\} = \sum_{k=1}^{r} (\xi_{n,k} - \xi_{n,k-1}) P\{N \geqslant k\}$, which by (11.5.41) converges to

$$\int_0^{\xi_p} P\left\{ W(t) + \theta\rho_2^* A_\psi \rho_1(s_\psi(t), p) \leqslant D_\alpha^+, \quad \forall t \leqslant \frac{\nu(F(x))}{\nu(p)} \right\} dx. \quad (11.5.48)$$

The second term on the rhs of (11.5.47) is bounded from above by

$$\sum_{k=1}^{r} E\{|X_{n,k}| I(|X_{n,k} - \xi_{n,k}| > \epsilon) \quad (11.5.49)$$

Thus if for every $\epsilon > 0$

$$\max_{k \leqslant r} nE\{|X_{n,k}| I(|X_{n,k} - \xi_{n,k}| > \epsilon)\} \to 0, \qquad \text{as} \quad n \to \infty, \quad (11.5.50)$$

then, by (11.5.47) and (11.5.49), $EX_{n,N}$ converges to the expression in (11.5.48). To this end, we pose the proof of (11.5.50) (under appropriate regularity conditions) as an exercise (see Problem 11.5.2). Thus for $r = [np] + 1$, $0 < p < 1$, (11.5.48) provides an asymptotic expression for $EX_{n,N}$ and in view of (11.5.43), again the asymptotic optimality results with respect to the criterion EN are carried over to the case when one chooses $EX_{n,N}$ as the criterion.

For $p = 1$, that is, $r = n$, $\xi_p = +\infty$, so that (11.5.46) does not converge and the same thing is true of (11.5.48). Actually, for any d.f. with an infinite upper extremity, $EX_{n,n} \to \infty$, as $n \to \infty$, and hence for $p = 1$, $EX_{n,N}$ may not converge to any limit when $n \to \infty$, no matter whether we stick to H_0 or some contiguous alternatives. Since in this case a complete experimentation demands the knowledge of $X_{n,n}$, the expected time duration for the same is $\xi_{n,n}$. As such, one may consider the asymptotic limit of $(EX_{n,N})/\xi_{n,n}$ and compare the same for different procedures in a PCS. In this case, we let $r^* = n - k_n$, where $k_n \to \infty$ but $n^{-1}k_n \to 0$ as $n \to \infty$, and we express $(EX_{n,N})/\xi_{n,n}$ as

$$\frac{(EX_{n,N})}{\xi_{n,n}} = (\xi_{n,n})^{-1}\left\{ \sum_{k=1}^{r^*} E\{X_{n,k}I_{[N=k]}\} + E(X_{n,r^*+1}I_{[N r^*]}) + \sum_{k=r^*+1}^{n} E((X_{n,k} - X_{n,r^*+1})I_{[N=k]}) \right\}. \quad (11.5.51)$$

While for the first two terms on the rhs of (11.5.51), the treatment for the case of $p < 1$ holds, by using the fact that $X_{n,k}$ is $\nearrow$ in k and the Cauchy-Schwarz inequality, the last term can be bounded (in absolute value) by

$$\left\{(\xi_{n,n})^{-1}\left(E(X_{n,n} - X_{n,r^*+1})^2\right)^{1/2}\right\}(P\{N > r^*\})^{1/2}. \quad (11.5.52)$$

Thus whenever, by choosing $(1/n)k_n$ sufficiently small, $E(X_{n,n} - X_{n,r^*+1})^2/\xi_{n,n}^2$ can be made arbitrarily small, (11.5.52) can be neglected, and hence by adapting the proofs for the case of $p < 1$ for the other two terms in (11.5.51) and referring to (11.5.43), we again conclude that the asymptotic optimality results hold even for $p = 1$. We conclude this section with some exercises (see Problem 11.5.3 and 11.5.4) relating to the asymptotic negligibility of the first factor in (11.5.52).

11.6. SOME ADDITIONAL PROBLEMS AND RELATED TESTS

Throughout earlier sections, we confined ourselves to tests based on rank order statistics. Hájek (1963) considered an elegant extension of the classical Kologorov-Smirnov (KS)-type tests for regression alternatives; this has been systematically studied in Hájek and Šidák (1967, Chapters V and VI). Sinha and Sen (1979a, b) used some weighted versions of these tests and stressed their uses in PCS. We may formulate these procedures as follows. Though these tests are not rank order tests, they are genuinely distribution-free under the null hypothesis of identity of the underlying d.f. As in the earlier section, we conceive of n independent r.v. $X_1, \ldots, X_n$ with continuous d.f. $F_1, \ldots, F_n$, all defined on the real line and consider the regression model

$$F_i(x) = F(x - \beta_0 - \beta_1 c_i) \qquad i = 1, \ldots, n, \qquad x \in R = (-\infty, \infty), \tag{11.6.1}$$

where the $\beta_j, j = 0, 1$ are unknown parameters and the c_i are known constants, not all equal. We intend to test (in a PCS) the null hypothesis that the d.f. $F_1, \ldots, F_n$ are all identical, that is, $\beta_1 = 0$. We define $\bar{c}_n$ and C_n^2 as in (11.2.11) and consider the empirical processes:

$$S_n(x) = n^{-1} \sum_{i=1}^{n} c(x - X_i), \qquad x \in R, \tag{11.6.2}$$

$$H_n(x) = C_n^{-1} \sum_{i=1}^{n} (c_i - \bar{c}_n) c(x - X_i), \qquad x \in R, \tag{11.6.3}$$

where $c(t)$ is equal to 1 or 0 according as t is $\geqslant$ or < 0. Hájek (1963) considered a test for the identity of the F_i against the regression alternative based on

$$K_n^+ = \sup_x H_n(x) \qquad \text{and} \qquad K_n = \sup_x |H_n(x)| \tag{11.6.4}$$

(for the one- and two-sided alternatives respectively. Sinha and Sen (1979a) considered the weighted test statistics

$$K_{nw}^+ = \sup_x \frac{H_n(x)}{w(S_n(x))} \qquad \text{and} \qquad K_{nw} = \sup_x \frac{|H_n(x)|}{w(s_n(x))}, \tag{11.6.5}$$

where $w = \{w(t), 0 < t < 1\}$ is some suitable weight function. Further, in either case, instead of the supremum being taken over the entire real line, one can take it over any semi-infinite interval with right censoring. Hájek was able to establish the weak convergence of the process H_n to a Brownian bridge and thereby to incorporate (2.6.21) and (2.6.22) in the computation of the asymptotic critical values of K_n^+ and K_n; for small samples, these are to be evaluated by enumerating all possible permutations of the ranks of the X_i among themselves. The recent tabulations of the percentile points of the truncated or censored KS-statistics by Koziol and Byar (1975) and Schey (1977) provide the asymptotic critical values of K_n^+ and K_n in the truncated case. Problems 11.6.1 and 11.6.2 relate to some of these results.

We may mention the following martingale characterizations due to Sinha and Sen (1979a). Let $\mathcal{F}_{nx}$ be the sigma field generated by $\{c(y - X_i), 1 \leqslant i \leqslant n, y \leqslant x\}$, $x \in R$, so that $\mathcal{F}_{nx}$ is nondecreasing in x.

Lemma 11.6.1. Under $H_0: F_1 = \cdots = F_n = F$, for every $n (\geqslant 1)$, both $\{S_n(x)/[1 - F(x)], x \in R\}$ and $\{H_n(x)/[1 - F(x)], x \in R\}$ are martingales.

Proof. Note that for every $y \leqslant x$, under H_0,

$$E\left\{ \frac{H_n(x)}{1 - F(x)} \middle| \mathcal{F}_{ny} \right\} = C_n^{-1} \sum_{i=1}^{n} (c_i - \bar{c}_n) \frac{E\{c(x - X_i) | \mathcal{F}_{ny}\}}{[1 - F(x)]}. \quad (11.6.6)$$

Also, for every $y \leqslant x$ and $i = 1, \ldots, n$,

$$\begin{aligned} E\{c(x - X_i) | \mathcal{F}_{ny}\} &= P\{X_i \leqslant x | H_0, \mathcal{F}_{ny}\} \\ &= 1 \cdot I_{[X_i \leqslant y]} + I_{[X_i > y]} \frac{\{F(x) - F(y)\}}{[1 - F(y)]} \\ &= \frac{1 - F(x)}{1 - F(y)} I_{[X_i \leqslant y]}. \end{aligned} \quad (11.6.7)$$

so that by (11.6.6) and (11.6.7), we obtain that

$$\begin{aligned} E\left\{ \frac{H_n(x)}{[1 - F(x)]} \middle| \mathcal{F}_{ny} \right\} &= \frac{C_n^{-1} \sum_{i=1}^{n} (c_i - \bar{c}_n) c(y - X_i)}{[1 - F(y)]} \\ &= \frac{H_n(y)}{[1 - F(y)]} \quad \text{a.e.} \end{aligned} \quad (11.6.8)$$

A similar proof holds for $\{S_n(x)/[1 - F(x)], x \in R\}$. Q.E.D.

The convergence of finite dimensional distributions of $S_n(\cdot)$ and $H_n(\cdot)$ (to some appropriate multinormal distributions) follows directly by using the classical central limit theorem, as either of these involves independent summands, while by Corollary 2.4.4.1 and Lemma 11.6.1 the tightness of either of these processes follows from the convergence of their finite dimensional laws. Actually, Theorem 2.4.8 can be used to establish the weak convergence of $H_n(\cdot)/w(S_n(\cdot))$ when $(1 - t)^{-1}w(t)$ satisfies the conditions stated [on $q(t)$] before (2.4.25). Problems 11.6.3 and 11.6.4 relate to these.

Now, as in Section 11.3, we may formulate some PCS tests based on K_{nw}^+ and K_{nw}. We continue experimentation as long as at the kth failure $X_{n,k}$, $H_n(X_{n,k})/w(S_n(X_{n,k}))$ (or $|H_n(X_{n,k})|/w(S_n(X_{n,k}))$) is less than $D_{nw,r,\alpha}^+$ (or $D_{nw,r,\alpha}$), $k \geqslant 1$. If, for the first time, for some $k = N(\leqslant r)$, $H_n(X_{n,N})/w(S_n(X_{n,N}))$ [or $|H_n(X_{n,N})|/w(S_n(X_{n,N}))$] is $\geqslant D_{nw,r,\alpha}^+$ [or $D_{nw,r,\alpha}$], stop experimentation when $X_{n,N}$ is observed along with the rejection of H_0 and if no such $N(\leqslant r)$ exists, stop experimentation at the rth failure $X_{n,r}$ along with the acceptance of H_0. In this formulation, r is prefixed and may even be taken equal to n. Also, the constants $D_{nw,r,\alpha}^+$ and $D_{nw,r,\alpha}$ are to be so chosen that the sizes of these PCS tests are equal to α. The weak convergence results discussed earlier provide the large sample solutions. Lemma 11.6.1 or the proof of the weak convergence results based on it does not hold when the F_i are not identical. Nevertheless, we may conceive of some local (contiguous) alternatives and, as in (11.4.32) through (11.4.38), we may utilize the impact of contiguity to prove the weak convergence to some drifted Gaussian functions under such alternatives. The details are left as exercises (see Problems 11.6.5 and 11.6.6). This enables us to study the asymptotic power properties of these procedures for local (contiguous) alternatives. Usually, these KS-type tests are not very highly efficient (as compared to the PCS tests based on the linear rank statistics) when one has specifically location, scale, or regression alternatives in mind. The situation is very comparable to the KS tests versus the rank tests in the classical two sample location or scale problem, where the former is valid and consistent against a broader class of alternatives while the latter is more efficient for certain specific alternatives. One of the advantages of the KS-type tests is their adaptability to grouped data involving ordered categories, where in (11.6.4) and (11.6.5), one needs to replace the suprema over the entire real line by maxima over the set of points constituting the endpoints of the different class intervals. Since this is a subset of the whole real line, the critical values for the ungrouped case may still be used for the grouped case; however, this will result in a conservative property of the tests (i.e., their actual sizes will be less than the specified α and a similar loss occurs in the power of the tests). These losses become negligible when the number of the class intervals is not small and the individual class frequencies are not large. PCS tests based on linear rank statistics can also be adopted for such grouped data with ordered categories (see Majumdar and Sen 1977). In such a case, the possible termination points of experimentation are the endpoints of these ordered class intervals, and, in view of the ties among the within class observations, one needs to modify the definitions of the $L_{n,k}$ in (11.2.16) and their variances in (11.2.19) by incorporating *average rank scores* (i.e., taking the averages of the rank scores for the observations in the same class interval and allotting the same to each one). The tests are then no longer genuinely distribution-free, but, are still conditionally distribution-free. From the PCS point of view (11.3.1), even with the adjustment for the average rank scores, may not be adaptable, because the variance of $L_{n,n}$ needed in this construction will not be known in advance. This difficulty can be eliminated by using a studentized version. Problem 11.6.7 relates to this formulation. Again, the theory of these tests rests on the weak convergence results parallel to those in Section 11.4 and Problems 11.6.8 and 11.6.9 relate to the same.

So far, we have considered the case of simple regression models only. In the PCS, tests for the multiple regression models can also be based on linear rank statistics and weighted empirical d.f. Toward this end, we assume that $X_1, \ldots, X_n$ are independent r.v. with distribution functions $F_1, \ldots, F_n$, respectively, where

$$F_i(x) = F(x - \beta_0 - \boldsymbol{\beta}'\mathbf{c}_i), \qquad x \in R,\ i = 1, \ldots, n, \tag{11.6.9}$$

the d.f. F is not known, β_0 and $\boldsymbol{\beta}' = (\beta_1, \ldots, \beta_p)$ are known parameters, $p \geqslant 1$, and the $\mathbf{c}_i = (c_{i1}, \ldots, c_{ip})$ are known vectors of regression constants. The several-sample location model is a special case of (11.6.9). Our concern is to test (under a PCS) for the null hypothesis

$$H_0 : F_1 = \cdots = F_n = F \qquad \text{that is,} \quad \boldsymbol{\beta} = \mathbf{0} \qquad \text{under (11.6.9).} \tag{11.6.10}$$

For this purpose, in (11.2.13) through (11.2.16), we replace the c_i by $\mathbf{c}_i$ and denote the resulting vector by $\mathbf{L}_{n,r}$, $r = 1, \ldots, n$. Then, parallel to (11.2.9), we have for each $r(= 1, \ldots, n)$,

$$E(\mathbf{L}_{n,r} \mid H_0) = \mathbf{0} \qquad \text{and} \quad \mathrm{Var}(\mathbf{L}_{n,r} \mid H_0) = A_{n,r}^2 \mathbf{C}_n, \tag{11.6.11}$$

where $A_{n,r}^2$ is defined by (11.2.20) and

$$\mathbf{C}_n = \sum_{i=1}^{n} (\mathbf{c}_i - \bar{\mathbf{c}}_n)(\mathbf{c}_i - \bar{\mathbf{c}}_n)', \qquad \bar{\mathbf{c}}_n = n^{-1} \sum_{i=1}^{n} \mathbf{c}_i. \tag{11.6.12}$$

We assume that there exists a positive number n_0, such that for every $n \geqslant n_0$,

$$\mathbf{C}_n \quad \text{is positive definite} \quad \text{(p.d.)}, \tag{11.6.13}$$

$$\overline{\lim_{n \to \infty}} \left\{ \max_{1 \leqslant i \leqslant n} (\mathbf{c}_i - \bar{\mathbf{c}})' \mathbf{C}_n^{-1} (\mathbf{c}_i - \bar{\mathbf{c}}_n) = 0. \right. \tag{11.6.14}$$

Let us then define

$$T_{n,k} = \left(\mathbf{L}_{n,k}' \mathbf{C}_n^{-1} \mathbf{L}_{n,k}\right)^{1/2}, \qquad k = 1, \ldots, n \tag{11.6.15}$$

and, conventionally, we let $T_{n,0} = 0$. In passing, we may remark that the $T_{n,k}$ in (11.6.15) remain invariant under any reparameterization of the model (11.6.9), that is, if instead of $\boldsymbol{\beta}'\mathbf{c}_i$ one takes $\boldsymbol{\Delta}'\mathbf{d}_i$ where $\boldsymbol{\beta} = \mathbf{B}\boldsymbol{\Delta}$ and $\mathbf{d}_i = (\mathbf{B}')^{-1}\mathbf{c}_i, i \geqslant 1$ for any nonsingular $\mathbf{B}$, then, the $T_{n,k}$ remain invariant under any choice of $\mathbf{B}$. This is important because in many situations (e.g., the several-sample problems), the model (11.6.9) can be framed in more than one (nonunique) way. Majumdar and Sen (1978a) considered the following type of tests based on the $T_{n,k}$:

(i) ***Kolmogorov-Smirnov (KS)-Type Statistics.*** Here, we define

$$K_{n,r} = \max_{0 \leqslant k \leqslant r} \left\{ A_{n,r}^{-1} T_{n,k} \right\} = \max_{0 \leqslant k \leqslant r} \left\{ A_{n,r}^{-1} \left(\mathbf{L}_{n,k}' \mathbf{C}_n^{-1} \mathbf{L}_{n,k}\right)^{1/2} \right\}. \tag{11.6.16}$$

(ii) ***Cramer-von Mises (CvM)-Type Statistics.*** Here, we define

$$M_{n,r} = \sum_{k=0}^{r-1} \frac{\lambda_k^* T_{n,k}^2}{A_{n,r}^2} = A_{n,r}^{-2} \sum_{k=0}^{r-1} \lambda_k^* (\mathbf{L}'_{n,k} \mathbf{C}_n^{-1} \mathbf{L}_{n,k}), \qquad (11.6.17)$$

where for each k $(= 0, \ldots, r-1)$,

$$\lambda_k^* = A_{n,r}^{-2} \left[\frac{(n-k-1)}{(n-1)(n-k)} \right] \left[a_n(k+1) - a_n^*(k+1) \right]^2, \quad (11.6.18)$$

and, in either case, r is defined as in Sections 11.3 and 11.4 [see (11.4.4)]. Actually, parallel to (11.4.2), if we consider the vector-valued process

$$\mathbf{W}_n^{(r)}(t) = A_{n,r}^{-1} \mathbf{C}_n^{-1/2} \mathbf{L}_{n,k_n(t)}, \qquad t \in [0,1], \qquad (11.6.19)$$

where $k_n(t)$ is defined by (11.4.1), then on letting $Y_{n,r}^2(t) = [\mathbf{W}_n^{(r)}(t)]' \cdot [\mathbf{W}_n^{(r)}(t)]$, $t \in [0,1]$, we obtain from (11.6.15) through (11.6.19) that

$$K_{n,r} = \sup_{t \in I} Y_{n,r}(t) \qquad \text{and} \quad M_{n,r} = \int_0^1 Y_{n,r}^2(t)\, dt \qquad (11.6.20)$$

which possibly explains the appropriateness of the terms KS and CvM types.

The PCS testing procedure is then very similar to the one described in Section 11.3, where we need to replace the $L_{n,k}$ by the $T_{n,k}$ and thereby, instead of (11.3.1), we need to consider the test statistics in (11.6.16) and (11.6.17). In (11.3.2) and (11.3.3), we need to use the critical values of $K_{n,r}$ or $M_{n,r}$. The asymptotic theory developed in Section 11.4 extends to the vector case as well: Lemma 11.4.1 holds without any change in arguments, while the proof of Theorem 11.4.2 in the vector case follows on parallel lines; we pose this as an exercise (see Problem 11.6.10). The extension of Theorem 11.4.4 to the vector case is also left as an exercise (see Problem 11.6.11). The next two problems (Problems 11.6.12 and 11.6.13) relate to the asymptotic powers for these PCS tests for some contiguous alternatives.

Extensions of the tests in (11.6.4) and (11.6.5) to the multiple regression case are due to Sinha and Sen (1979b). Here, in (11.6.3), we define

$$\mathbf{H}_n(x) = \mathbf{C}_n^{-1/2} \sum_{i=1}^{n} (\mathbf{c}_i - \bar{\mathbf{c}}_n) c(x - X_i), \qquad x \in R, \qquad (11.6.21)$$

and in (11.6.5), we take

$$K_{nw}^* = \sup_x \frac{[\mathbf{H}_n(x)]'[\mathbf{H}_n(x)]}{w^2(S_n(x))}. \qquad (11.6.22)$$

The rest of the PCS testing procedure is as in the earlier part of this section, where we need to replace the critical values of K_{nw} (or K_{nw}^{+}) by those of K_{nw}^{*}. Lemma 11.6.1 goes directly to the vector case, and hence the weak convergence of $[\mathbf{H}_n(x)]'[\mathbf{H}_n(x)]$ follows on parallel lines; Problem 11.6.14 is to ask for the details of the verification.

PCS tests also arise in some other contexts. One such case is the problem of testing stochastic independence in the bivariate case when one of the variates is under PCS. Thus if (X_i, Y_i), $i = 1, \ldots, n$ be n i.i.d.r.v. with the d.f. $F(x, y)$, defined on the real plane, to test for $H_0: F(x, y) = F(x, \infty) \cdot F(\infty, y), \forall (x, y) \in R^2$, PCS would be relevant if the Y_i are observable at the start of the experiment while the primary variables $X_1, \ldots, X_n$ are subject to monitoring as in Sections 11.2 and 11.3. Here, if $Y_1, \ldots, Y_n$ are conditionally fixed, the problem may be interpreted as that of testing the identity of the conditional d.f.'s of $X_1, \ldots, X_n$, given $Y_1, \ldots, Y_n$. If we let

$$c_{ni}^{*} = \frac{Y_i - \overline{Y}_n}{\left[\sum_{i=1}^{n} \left(Y_i - \overline{Y}_n\right)^2\right]^{1/2}}, \qquad i = 1, \ldots, n; \qquad \overline{Y}_n = n^{-1} \sum_{i=1}^{n} Y_i, \tag{11.6.23}$$

then we may define the (mixed-) rank statistics as in (11.2.16). If we assume that $V(Y_1) < \infty$, then (i) $n^{-1}\sum_{i=1}^{n}(Y_i - \overline{Y}_n)^2 \to V(Y_1)$ a.s. and (ii) $n^{-1/2}\max_{1 \leqslant i \leqslant n}|Y_i - \overline{Y}_n| \to 0$ a.s., as $n \to \infty$, so that by (11.6.23),

$$\max_{1 \leqslant i \leqslant n} |c_{ni}^{*}| \to 0 \quad \text{a.s., as} \quad n \to \infty. \tag{11.6.24}$$

As such, results of Sections 11.3 through 11.5 hold for this case too.

Instead of using a mixed rank statistic, one could have used a pure rank statistic as follows. Let $R_{ni}(Q_{ni})$ be the rank of $X_i(Y_i)$ among $X_1, \ldots, X_n(Y_1, \ldots, Y_n)$. Using the scores $a_n(i)$ and $b_n(i)$ for the X and Y variables, we let

$$\begin{aligned} L_{n,r} &= \sum_{i=1}^{r} \left[a_n(i) - a_n^{*}(r) \right] b_n(S_{ni}^{*}), \qquad 1 \leqslant r \leqslant n-2 \\ &= L_n, \qquad \text{for} \quad r = n-1, n, \end{aligned} \tag{11.6.25}$$

where $a_n^{*}(r)$ is defined by (11.2.15) and $(S_{n1}^{*}, \ldots, S_{nn}^{*})$ are the rank of the Y_i corresponding to the ordered X_i, that is,

$$S_{ni}^{*} = Q_{nS_{ni}}, \qquad 1 \leqslant i \leqslant n, \tag{11.6.26}$$

$S_{n1}, \ldots, S_{nn}$ being the antiranks of $X_1, \ldots, X_n$. Under the hypothesis of independence, $(S_{n1}^{*}, \ldots, S_{nn}^{*})$ assumes all possible permutations of $(1, \ldots, n)$ with the common probability $(n!)^{-1}$ and hence the theory of Section 11.4 holds. Further, using the invariance principles of Chapter 6, the case of contiguous alternatives can be treated similarly.

In actual practice, batch arrivals often characterize the models and introduce more complications. For a batch arrival model, not all the n observations enter into the study at the same time. Rather, they enter at different points of time, possibly in batches. Thus at any point of monitoring, we get a composite picture of the number of observations entering into the scheme (at times earlier to it) as well as the failure times occurring in the given period. To fit this phenomenon, one needs to extend the definition of $L_{n,r}$ to a two-parameter process. Such a process has been developed in Sen (1976a) and incorporated in Majumdar and Sen (1978b). We pose some of these developments as exercises.

Analysis of covariance (ANOCOVA) models are not uncommon in life testing problems. Suitable progressively censored nonparametric tests for ANOCOVA based on a general class of rank order statistics are due to Sen (1979e, 1981d). The basic formulation rests on a multivariate generalization of the theory developed in Sections 11.2 through 11.5. In this setup, the procedure, like the usual multivariate rank procedures, can only be conditionally distribution-free. It is of course asymptotically distribution-free. In the following, we outline some rank ANOCOVA procedures under PCS for the multiple regression model.

Let $\mathbf{X}_i^* = (X_{0i}, X_{1i}, \ldots, X_{pi})' = (X_{0i}, \mathbf{X}_i')$, $i = 1, \ldots, n$ be independent r.v. with continuous d.f. F_i^*, $i = 1, \ldots, n$, where the X_{0i} are the primary variates with marginal d.f. F_{0i}, defined on R and the $\mathbf{X}_i$ are the concomitant variates with marginal (p-variate) d.f. F_i, defined on E^p, for some $p \geqslant 1$. Let $F_i^0(y \mid \mathbf{x})$ be the conditional d.f. of X_{0i} given $\mathbf{X}_i = \mathbf{x}$, $i = 1, \ldots, n$, and, as is usually the case in an ANOCOVA model, we assume that $F_1 = \ldots = F_n = F$ (unknown). Our basic problem is to test for the null hypothesis

$$H_0 : F_1^0 = \cdots = F_n^0 = F^0 \qquad \text{(unknown)} \tag{11.6.27}$$

against an alternative that they are not all the same. Keeping in mind the usual one-way ANOCOVA model, we conceive of the model

$$F_i^0(y \mid \mathbf{x}) = F^0(y - \boldsymbol{\beta}'(\mathbf{c}_i - \bar{\mathbf{c}}_n) \mid \mathbf{x}), i = 1, \ldots, n, \qquad x \in R, \tag{11.6.28}$$

where $\boldsymbol{\beta}$ stands for the vector of treatment effects and the $\mathbf{c}_i$ are defined as they were previously, so that under (11.6.28) the null hypothesis in (11.6.27) reduces to

$$H_0 : \boldsymbol{\beta} = \mathbf{0} \qquad \text{against} \quad H_1 : \boldsymbol{\beta} \neq \mathbf{0}. \tag{11.6.29}$$

We assume that the concomitant variates $\mathbf{X}_1, \ldots, \mathbf{X}_n$ are observable at the beginning of experimentation, but the primary variates are not. Let $Z_{n1}^0, \ldots, Z_{nn}^0$ be the ordered r.v.'s corresponding to $X_{01}, \ldots, X_{0n}$ and we define the *antiranks* $S_1, \ldots, S_n$ by

$$Z_{nj}^0 = X_{0S_j}, \qquad \text{for} \quad j = 1, \ldots, n. \tag{11.6.30}$$

Then, at the kth failure Z_{nk}^0, the observable r.v.'s are $\mathbf{Q}_i = (S_i, Z_{ni}^0, \mathbf{X}_{S_i})$, $i = 1, \ldots, k$, for $k = 1, \ldots, n$. It may be remarked that the complementary sets of covariates are known but their antiranks are not. Let R_{ji} be the rank of X_{ji} among $X_{j1}, \ldots, X_{jn}$, for $i = 1, \ldots, n$; $j = 0, \ldots, p$. These yield the *rank collection matrix* $\mathbf{R}_n = ((R_{ji}))$ (of order $(p+1) \times n$) and we permute the columns of $\mathbf{R}_n$ in such a way that the first row is in the natural order; this matrix is known as the *reduced rank-collection matrix* $\mathbf{R}_n^* = ((R_{ji}^*))$. Note that by (11.6.30), for each i $(= 1, \ldots, n)$,

$$R_{0S_i} = R_{0i}^* = i \qquad \text{and} \quad R_{jS_i} = R_{ji}^*, \ 1 \leqslant j \leqslant p. \tag{11.6.31}$$

For each $j(= 0, 1, \ldots, p)$, let $\{a_{n,j}(i), \ i = 1, \ldots, n\}$ be a set of scores, as can be defined in Sections 11.2 and 11.3. We consider then the matrix of rank statistics

$$\mathbf{L}_n = \sum_{i=1}^{n} (\mathbf{c}_i - \bar{\mathbf{c}}_n)\left[a_{n,0}(R_{0i}), \ldots, a_{n,p}(R_{pi})\right] \tag{11.6.32}$$

(or order $q \times (p+1)$). Let now $\mathcal{P}_n$ be the permutational probability measure generated by the $n!$ (conditionally under H_0) equally likely column permutations of $\mathbf{R}_n$ (given R_n^*) and let $\mathbf{S}_{n,k} = (S_1, \ldots, S_k)$, for $k = 1, \ldots, n$. Then we define

$$\begin{aligned} \mathbf{L}_{n,k} &= E_{\mathcal{P}_n}(\mathbf{L}_n \mid \mathbf{S}_{n,k}) \\ &= \sum_{i=1}^{k} (\mathbf{c}_{S_i} - \bar{\mathbf{c}}_n)\left[a_{n,0}(i) - a_{n,0}^*(k), a_{n,j}(R_{ji}^*) - a_{n,j}^*(k), \qquad 1 \leqslant j \leqslant p\right] \end{aligned} \tag{11.6.33}$$

where

$$a_{n,j}^*(k) = \begin{cases} (n-k)^{-1}\left\{n\bar{a}_{n,j} - \sum_{i=1}^{k} a_{n,j}(R_{ji}^*)\right\}, & 1 \leqslant k \leqslant n-1, \\ 0, & k = n \end{cases} \tag{11.6.34}$$

and

$$\bar{a}_{n,j} = n^{-1} \sum_{i=1}^{n} a_{n,j}(i), \qquad \text{for} \quad j = 0, 1, \ldots, p. \tag{11.6.35}$$

Without any loss of generality, we may set $\bar{a}_{n,j} = 0$, for $j = 0, 1, \ldots, p$. Let then

$$\mathbf{C}_n = \sum_{i=1}^{n} (\mathbf{c}_i - \bar{\mathbf{c}}_n)(\mathbf{c}_i - \bar{\mathbf{c}}_n)' \tag{11.6.36}$$

and assume that $\mathbf{C}_n$ is p.d. For every $k(1 \leqslant k \leqslant n)$, we define a $(p+1) \times (p+1)$ matrix $\mathbf{V}_{n,k} = ((v_{njj'}^{(k)}))$, by letting for each $j, j'(=0, 1, \ldots, p)$,

$$v_{njj'}^{(k)} = n^{-1}\left\{ \sum_{i=1}^{k} a_{n,j}(R_{ji}^*)a_{n,j'}(R_{j'i}^*) + (n-k)a_{n,j}^*(k)a_{n,j'}^*(k) \right\}. \quad (11.6.37)$$

Then, if we roll out $\mathbf{L}_{n,k}$ in (11.6.33) into a qp-vector, it follows that

$$E_{\mathcal{P}_n}(\mathbf{L}_{nk}) = 0, \qquad \text{and} \quad V_{\mathcal{P}_n}(\mathbf{L}_{n,k}) = \mathbf{C}_n \otimes \mathbf{V}_{n,k}, \qquad k = 1, \ldots, n. \quad (11.6.38)$$

To eliminate the effects of the concomitant variates, we fit a linear regression of the primary variate rank statistics on the concomitant part and work with the residuals. With this in mind, we set

$$\underset{q(p+1)\times 1}{\mathbf{L}'_{n,k}} = \left(\underset{q\times 1}{\mathbf{L}_{n,k}^{(0)}}, \underset{qp\times 1}{\mathbf{L}_{n,k}^{(*)}}\right)', \quad \mathbf{V}_{n,k} = \begin{pmatrix} v_{n00}^{(k)} & \mathbf{v}_{n0}^{(k)} \\ \mathbf{v}_{n0}^{(k)\prime} & \mathbf{V}_{n,k}^* \end{pmatrix}, \qquad k = 1, \ldots, n, \quad (11.6.39)$$

so that the residuals are defined by

$$\mathbf{L}_{n,k}^{0} = \mathbf{L}_{n,k}^{(0)} - \mathbf{v}_{n0}^{(k)}(\mathbf{V}_{n,k}^*)^{-}\mathbf{L}_{n,k}^{(*)}, \qquad \text{for} \quad k = 1, \ldots, n. \quad (11.6.40)$$

By (11.6.38), (11.6.39), and (11.6.40), we find

$$E_{\mathcal{P}_n}(\mathbf{L}_{n,k}^{0}) = \mathbf{0}, \qquad \text{and} \quad V_{\mathcal{P}_n}(\mathbf{L}_{n,k}^{0}) = \mathbf{C}_n \cdot v_{n,k}^{**}, \qquad k = 1, \ldots, n, \quad (11.6.41)$$

where

$$v_{n,k}^{**} = v_{n00}^{(k)} - \mathbf{v}_{n0}^{(k)}(\mathbf{V}_{n,k}^*)^{-}\mathbf{v}_{n0}^{(k)\prime}, \qquad \text{for} \quad k = 1, \ldots, n. \quad (11.6.42)$$

Finally, let

$$T_{nk}^2 = \left\{\mathbf{L}_{n,k}^{0\prime}\mathbf{C}_n^{-1}\mathbf{L}_{n,k}^{0}\right\}/v_{n,k}^{**}, \qquad \text{for} \quad k = 1, \ldots, n. \quad (11.6.43)$$

Then, T_{nk} is the covariate adjusted rank statistic (for testing H_0) based on $Q_1, \ldots, Q_k$, for $k = 1, \ldots, n$. We let $T_{nk} = 0$ whenever $v_{n,k}^{**} = 0$. Then, as in Sections 11.2 and 11.3, we may consider the following PCS testing procedure for the ANOCOVA problem.

Continue experimentation as long as $k \geqslant n_0$ and $T_{nk} \leqslant l_n^{(\alpha)}$; if for the first time, for some $k = N(\epsilon[n_0, r])$, $T_{nN} > l_n^{(\alpha)}$ then at the Nth failure Z_{nN}^0 we stop experimentation along with the rejection of H_0. If no such k exists, then the experimentation is stopped at the preplanned rth failure Z_{nr}^0 along with the acceptance of H_0. Here, n_0 is some initial starting number, small compared to $r(=[np]+1$, for some $p \in (0, 1])$ and $l_n^{(\alpha)}$ is the critical value corresponding to the desired level of significance $\alpha: 0 < \alpha < 1$. As in the text following (11.6.15), the T_{nk} remain invariant under any reparameterization, and hence we may set, without any loss of generality, $\mathbf{c}_i - \bar{\mathbf{c}}_n = \mathbf{c}_{ni}$, $i = 1, \ldots, n$ where

$$\sum_{i=1}^{n} \mathbf{c}_{ni} = \mathbf{0} \quad \text{and} \quad \sum_{i=1}^{n} \mathbf{c}_{ni}\mathbf{c}'_{ni} = \mathbf{I}_q. \tag{11.6.44}$$

We assume that

$$\max_{1 \leqslant i \leqslant n} (\mathbf{c}'_{ni}\mathbf{c}_{ni}) \to 0 \quad \text{as} \quad n \to \infty. \tag{11.6.45}$$

Further, we note that by definition in (11.6.33), under H_0, $\{\mathbf{L}_{n,k}, 0 \leqslant k \leqslant n\}$ is a martingale sequence, so that by (11.6.38), $\mathbf{V}_{n,k+1} - \mathbf{V}_{n,k}$ is positive semidefinite (p.s.d.) for every $k(= 0, \ldots, n-1)$. Hence it follows that

$$v_{n,k}^{**} \quad \text{is nondecreasing in} \quad k(0 \leqslant k \leqslant n). \tag{11.6.46}$$

We assume that the scores $a_{n,j}(i)$ are generated by the score function $\phi_j(u), 0 < u < 1$, for $j = 0, 1, \ldots, p$, which satisfy the regularity conditions (on ϕ) in Section 11.4. Let then $F_{[j]}$, $F_{[jl]}$, and $F_{[0jl]}$ be, respectively, the marginal d.f. of X_{ji}, the joint d.f. of (X_{ji}, X_{li}), and the trivariate (if $p \geqslant 2$) d.f. of (X_{0i}, X_{ji}, X_{li}), for $0 \leqslant j \leqslant p$, $0 \leqslant j \neq l \leqslant p$ and $1 \leqslant j \neq l \leqslant p$, respectively, all under H_0. Also, we assume that $F_{[0]}(x) = t$ has a unique solution ω_t for every $t \in (0, 1)$. Let

$$\bar{\phi}_{0t} = (1-t)^{-1}\int_t^1 \phi_0(u)\,du, \qquad 0 \leqslant t < 1, \qquad \bar{\phi}_{01} = 0, \tag{11.6.47}$$

$$\bar{\phi}_{jt} = (1-t)^{-1}\int_{\omega_t}^{\infty}\int_{-\infty}^{\infty} \phi_j(F_{[j]}(y))\,dF_{[0j]}(x, y), \;;$$

$$t \in [0, 1], \qquad 1 \leqslant j \leqslant p, \tag{11.6.48}$$

$$\nu_{00}(t) = \int_0^t \phi_0^2(u)\,du + (1-t)\bar{\phi}_{0t}^2, \qquad t \in [0, 1], \tag{11.6.49}$$

$$\nu_{0j}(t) = \nu_{j0}(t) = \int_{-\infty}^{\omega_t}\int_{-\infty}^{\infty} \phi_0(F_{[0]}(x))\phi_j(F_{[j]}(y))\,dF_{[0j]}(x, y)$$

$$+ (1-t)\bar{\phi}_{0t}\bar{\phi}_{jt}, t \in [0, 1], \qquad j = 1, \ldots, p; \tag{11.6.50}$$

$$\nu_{jl}(t) = \int_{-\infty}^{\omega_t}\int_{-\infty}^{\infty}\int_{-\infty}^{\infty} \phi_{ij}(F_{[j]}(y))\phi_l(F_{[l]}(z))\,dF_{[0jl]}(x, y, z)$$

$$+ (1-t)\bar{\phi}_{jt}\bar{\phi}_{lt}, \; 1 \leqslant j, l \leqslant p, \qquad t \in [0, 1]. \tag{11.6.51}$$

$$\boldsymbol{\nu}(t) = ((\nu_{jl}(t))) = \begin{pmatrix} \nu_{00}(t) & \boldsymbol{\nu}_{0*}(t) \\ \boldsymbol{\nu}_{*0}(t) & \boldsymbol{\nu}_{**}(t) \end{pmatrix}, \qquad t \in [0, 1] \tag{11.6.52}$$

$$\nu_{00}^{**}(t) = \nu_{00}(t) - \boldsymbol{\nu}_{0*}(t)(\boldsymbol{\nu}_{**}(t))^{-}\boldsymbol{\nu}_{*0}(t), \qquad t \in [0, 1]. \tag{11.6.53}$$

Then, by using the decomposition in Section 6.5 (dealing with rank statistics of the type $v_{njj'}$ and using the weak convergence results instead of the a.s. convergence), it follows that under the assumed regularity conditions, as $n \to \infty$,

$$\max_{k \leqslant n} \left\{ \left\| \mathbf{V}_{n,k} - \boldsymbol{\nu}\left(\frac{k}{n}\right) \right\| \right\} \xrightarrow{P} 0, \tag{11.6.54}$$

where $\|\cdot\|$ stands for the maximum element. Thus by (11.6.42), (11.6.53), and (11.6.54), we conclude that

$$\max_{k \leqslant n} \left\{ \left| v_{n,k}^{**} - \nu_{00}^{**}\left(\frac{k}{n}\right) \right| \right\} \xrightarrow{P} 0, \qquad \text{as} \quad n \to \infty. \tag{11.6.55}$$

Also, the d.f. F is assumed to be continuous and the score functions ϕ_j, $0 \leqslant j \leqslant p$ are all assumed to be absolutely continuous inside $(0, 1)$. Hence as $\delta \downarrow 0$,

$$\sup\{\|\boldsymbol{\nu}(t) - \boldsymbol{\nu}(s)\| : 0 \leqslant s \leqslant t \leqslant s + \delta \leqslant 1\} \to 0, \tag{11.6.56}$$

$$\sup\{|\nu_{00}^{**}(t) - \nu_{00}^{**}(s)| : 0 \leqslant s \leqslant t \leqslant s + \delta \leqslant 1\} \to 0, \tag{11.6.57}$$

so that, by (11.6.55) and (11.6.57), we have, as $\delta \downarrow 0$,

$$\max\{|v_{n,k}^{**} - v_{n,q}^{**}| : |q - k| \leqslant \delta n\} \xrightarrow{P} 0, \qquad \text{as} \quad n \to \infty. \tag{11.6.58}$$

If in (11.4.11), we replace the $a_n(i)$ by $[a_{n,0}(i), a_{n,1}(R_{1i}^*), \ldots, a_{n,p}(R_{pi}^*)]'$ $(1 \leqslant i \leqslant n)$, then we may virtually repeat the proof in (11.4.11) through (11.4.14) and obtain that for every k $(= 0, \ldots, n)$,

$$E\{\mathbf{L}_n \mid \mathcal{B}_n^{(k)}, R_n^*, H_0\} = \mathbf{L}_{n,k} \qquad \text{a.e.} \tag{11.6.59}$$

Given this martingale property and (11.6.54), under (11.6.45) and the assumed regularity conditions on the score functions, the asymptotic normality of $\mathbf{C}_n^{-1/2}\mathbf{L}_{n,k_r}, r = 1, \ldots, s$ (where $n^{-1}k_r \to t_r \in [0, 1]$, for $r = 1, \ldots, s(\geqslant 1)$) follows by using the martingale central limit theorem of Section 2.4, where we first consider the permutational distribution (given $\mathbf{R}_n^*$) and subsequently pass on to the unconditional case. The details are left for Problem 11.6.20. By virtue of this result and (11.6.40), (11.6.54), and (11.6.55), we conclude that under H_0 and the assumed regularity conditions, for every (fixed) $s(\geqslant 1)$ and $t_1, \ldots, t_s$ belonging to $[0, 1]$, for $k_r/n \to t_r, r = 1, \ldots, s, \{\mathbf{C}_n^{-1/2}\mathbf{L}_{n,k_r}^0, r = 1, \ldots, s\}$ is asymptotically (multi-) normally distributed with null mean vectors and dispersion matrix

$$\mathbf{I}_q \otimes ((\nu_{00}^{**}(t_r \wedge t_{r'}))). \tag{11.6.60}$$

Now, for every (r, n) satisfying (11.4.4) and $\epsilon(0 < \epsilon < 1)$, we consider a q-variate stochastic process ${}_\epsilon\mathbf{W}_{n,r} = \{\mathbf{W}_{n,r}(t), \epsilon \leqslant t \leqslant 1\}$ by letting

$$\mathbf{W}_{n,r}(t) = (v_{n,k}^{**})^{-1/2}\mathbf{C}_n^{-1/2}\mathbf{L}_{n,k}^0 \quad \text{for} \quad k/r \leqslant t < (k+1)/r, t \in [\epsilon, 1]. \tag{11.6.61}$$

Note that ${}_{\epsilon}\mathbf{W}_{n,r}$ belongs to the space $D^q[\epsilon, 1]$. Let ${}_{\epsilon}\mathbf{W} = \{\mathbf{W}(t),\ \epsilon \leqslant t \leqslant 1\}$ be a q-variate Gaussian function with no drift and covariance function

$$E\{[\mathbf{W}(s)][\mathbf{W}(t)]'\} = \mathbf{I}_q \cdot \{\nu_{00}^{**}(p(s \wedge t))/\nu_{00}^{**}(p(s \vee t))\}^{1/2},$$
$$s, t \in [\epsilon, 1]. \tag{11.6.62}$$

The convergence of the finite dimensional distributions of ${}_{\epsilon}\mathbf{W}_{n,r}$ to those of ${}_{\epsilon}\mathbf{W}$ (under H_0) follows from the discussion made before (11.6.60). Further, by using (11.6.55) through (11.6.58) and the basic martingale property in (11.6.59), the tightness of ${}_{\epsilon}\mathbf{W}_{n,r}$ can also be established; we leave the details as an exercise (see Problem 11.6.21). By (11.6.43) and (11.6.61), we obtain by letting $n_0 = \epsilon r$ that $T_{nk}^2 = [\mathbf{W}_{n,r}(t)]'[\mathbf{W}_{n,r}(t)]$ when $t = k/r$, $k = n_0, \ldots, r$. Thus we have

$$\max_{n_0 \leqslant k \leqslant r} T_{nk}^2 = \sup\{[\mathbf{W}_{n,r}(t)]'[\mathbf{W}_{n,r}(t)] : \epsilon \leqslant t \leqslant 1\}. \tag{11.6.63}$$

Therefore, under H_0,

$$\max_{n_0 \leqslant k \leqslant r} T_{nk} \xrightarrow{\mathfrak{D}} \left(\sup\{[\mathbf{W}(t)]'[\mathbf{W}(t)] : \epsilon \leqslant t \leqslant 1\}\right)^{1/2}, \tag{11.6.64}$$

where writing

$$\eta = \frac{\nu_{00}^{**}(p\epsilon)}{\nu_{00}^{**}(p)}, \tag{11.6.65}$$

and making use of (11.6.46) and (11.6.55), we may apply a transformation on the time parameter in W and thereby rewrite the rhs of (11.6.64) as, say,

$$B_\eta^* = \left(\sup\{t^{-1}B^2(t) : \eta \leqslant t \leqslant 1\}\right)^{1/2} \tag{11.6.66}$$

where $B = \{B(t),\ 0 \leqslant t \leqslant 1\}$ is a q-parameter Bessel process, defined by (2.3.50). Hence for the asymptotic critical value $l_n^{(\alpha)}$, we need to find the percentile points of the distribution of B_η^*. These percentile points have been tabulated in Majumdar (1977), Majumdar and Sen (1977), and De Long (1981). These values are sensitive to the choice of η, for small values of η, but are fairly stable when η is not small. Since, in the current context, the $\nu_{00}^{**}(t)$, $t \in [\epsilon, 1]$ may not be known, we estimate η by substituting the estimators $v_{n,k}^{**}$ in (11.6.65), and using the fact that $k \geqslant n_0$ and by (11.6.42), $v_{n,r}^{**} \leqslant v_{n00}^{(r)}$, we consider the lower bound

$$\eta_L = \frac{v_{n,n_0}^{**}}{v_{n00}^{(r)}}, \tag{11.6.67}$$

where $v_{n00}^{(r)}$ is known in advance and v_{n,n_0}^{**} is known at the n_0th failure $Z_{nn_0}^0$, and hence given r and n_0, η_L can be specified at the failure $Z_{nn_0}^0$ and the

PCS testing procedure can be carried out by using the critical values of the statistic $B^*_{\eta_L}$.

As in Theorem 11.4.4, the weak convergence of ${}_\epsilon\mathbf{W}_{n,r}$ (to some drifted Gaussian functions) can also be established under local (contiguous) alternatives and the same can be incorporated in the description of the asymptotic power function of the test based on the T_{nk}. We leave these as exercises (see Problems 11.6.22 and 11.6.23).

11.7. SOME QUASI-NONPARAMETRIC PCS TESTS

All the procedures considered in the previous sections share one feature: they are based solely on the vector of ranks or antiranks, disregarding any information contained in the associated vector of order statistics or any plausible pattern of the effects of the covariates. It is possible to incorporate additional information relating to the vector of order statistics and also some models relating to the covariates and to base some PCS tests on these models. These are presented briefly in this section.

First, we consider some PCS testing procedures based on some quantile processes incorporating some information contained in the set of order statistics. To demonstrate this, let us conceive of a model such that $X_1, \ldots, X_n$ are independent r.v.'s with absolutely continuous d.f.'s $F_1, \ldots, F_n$, all defined on the real line R and assume that F_i admits of a p.d.f. f_i where

$$f_i(x) = f(x; \Delta(c_i - \bar{c}_n)), \qquad x \in R, \qquad i = 1, \ldots, n, \qquad (11.7.1)$$

where $c_1, \ldots, c_n$ are given constants (not all equal), $\bar{c}_n = n^{-1}\sum_{i=1}^n c_i$ and Δ is an unknown parameter. We intend to test $H_0 : \Delta = 0$ versus $H_1 : \Delta \neq$ (or $>$ or $<$) 0. We define C_n^2 and the c^*_{ni} as in (11.2.11) and (11.5.14), respectively. Then, the likelihood function for $\mathbf{S}_n^{(k)} = (S_{n1}, \ldots, S_{nk})$ and $\mathbf{X}_n^{(k)} = (X_{n,1}, \ldots, X_{n,k})$ is

$$\prod_{i=1}^{k} f(X_{n,i}; \Delta(c_{S_{ni}} - \bar{c}_n)) \prod_{i=k+1}^{n} \left[1 - F(X_{n,k}; \Delta(c_{S_{nk}} - \bar{c}_n))\right] \qquad (11.7.2)$$

Let us denote this by

$$\begin{aligned} g(x) &= -\left(\frac{\partial}{\partial\theta}\right)\log f(x;\theta)\big|_{\theta=0} \qquad \text{and} \\ \bar{G}(x) &= \left[1 - F(x;0)\right]^{-1}\int_x^\infty g(z)\,dF(z;0). \end{aligned} \qquad (11.7.3)$$

Assuming the usual Cramér-regularity conditions on the p.d.f. $f(\cdot;\theta)$, differentiating the logarithm of (11.7.2) with respect to θ and evaluating the same at $\theta = 0$, we obtain the statistic

$$T_{nk} = C_n^{-1}\left\{-\frac{\partial}{\partial\theta}\log(11.7.2)\,|_{\theta=0}\right\}$$

$$= \sum_{i=1}^{k} c^*_{nS_{ni}}\left[g(X_{n,i}) - \bar{G}(X_{n,k})\right], \qquad \text{for} \quad k = 1, \ldots, n \quad (11.7.4)$$

and, conventionally, we let $T_{n0} = 0$. Note that under the usual Cramér-regularity conditions, $\int_R g(x)\,dF(x;0) = 0$, so that by (11.7.3),

$$\bar{G}(x) = -\left[1 - F(x;0)\right]^{-1}\int_{-\infty}^{x} g(z)\,dF(z;0), \qquad \text{for} \quad x \in R. \quad (11.7.5)$$

We estimate the $\bar{G}(x)$, $x \in R$ by substituting the sample d.f. $F_n^*(x) = n^{-1}\sum_{i=1}^{n}u(x - X_i)$ for $F(x)$ [where $u(t)$ is 1 or 0 according as t is $\geqslant$ or < 0], so that

$$\bar{G}_n(X_{n,i}) = -\frac{1}{(n-i)}\sum_{j=1}^{i} g(X_{n,j}) \qquad \text{for} \quad i = 1, \ldots, n-1 \quad (11.7.6)$$

and, conventionally, we let $\bar{G}_n(x) = g(X_{n,n})$ for $x \geqslant X_{n,n}$ and 0, for $x < X_{n,1}$. Then, substituting (11.7.6) in (11.7.4), we obtain a related partial sequence

$$T^*_{nk} = \sum_{i=1}^{k} c^*_{nS_{ni}}\left[g(X_{n,1}) - \bar{G}_n(X_{n,k})\right]$$

$$= \sum_{i=1}^{k} g(X_{n,i})\left[c^*_{nS_{ni}} + (n-k)^{-1}\sum_{j=1}^{k} c^*_{nS_{nj}}\right], \qquad k = 1, \ldots, n-1$$

$$(11.7.7)$$

and $T^*_{n0} = 0$, $T^*_{nn} = T^*_n = T^*_{nn-1}$.

Note that the statistics in (11.7.7) are linear combinations of functions of order statistics with stochastic coefficients depending on the antiranks and the censoring stage (k). This makes it intuitively appealing to make use of the results in Chapter 7. However, such an adaptation may require stringent regularity conditions, which we avoid in the following alternative approach (due to Sen, 1979b).

Under $H_0: \Delta = 0$, the X_i are i.i.d.r.v.'s and hence $\mathbf{S}_n^{(n)}$ and $\mathbf{X}_n^{(n)}$ are stochastically independent. Further, $\mathbf{S}_n^{(n)}$ has a discrete uniform distribution over the set of $n!$ permutations of $(1, \ldots, n)$. With respect to this uniform probability measure, it is easy to show that $\{T^*_{nk}, k = 0, 1, \ldots, n\}$ forms a martingale sequence. As such, by an appeal to Theorem 2.4.7, we have an invariance principle relating to the partial sequence in (11.7.7). Thus all the results in Section 11.4 relating to the linear rank statistics can be extended to these progressively censored linear combinations of functions of order statistics, and, as in (11.3.1), PCS tests can be based on $\max_{1 \leqslant i \leqslant n} T^*_{ni}$ or $\max_{1 \leqslant i \leqslant n} |T^*_{ni}|$; the asymptotic distribution theory is the same as in (11.2.29) through (11.4.30). The main difference between this approach and the procedures considered in Section 11.4 lies in the fact that the rank based procedures remain invariant under any monotone transformation on the X_i, the procedure based on the T^*_{nk} does not. Further, unlike the rank procedures, it is not strictly distribution-free [under H_0]. The form of $g(\cdot)$ in (11.7.7) depends on the $f(\cdot\,;\theta)$ and, as a result, it rests on an assumed form of the p.d.f. f and may not be very robust against possible deviations from such an assumption. Sen (1981c) has considered some further generalizations of these PCS tests for the case where Δ is a multidimensional parameter. Some of the technicalities involved in these progressively censored quantile processes are summarized in the form of exercises (see Problems 11.7.1 through 11.7.8).

The Cox (1972) regression model or the so-called *proportional hazard model* is quite popular in the area of survival analysis (with covariates). This is a quasi-parametric model where the regression of survival time is characterized in a parametric setup, but the *hazard function* is of arbitrary form. We consider the model where n subjects enter into the study at a common entry point; the ith subject (having survival time Y_i and a set of concomitant variates $\mathbf{Z}_i = (Z_{i1}, \ldots, Z_{iq})'$ for some $q \geqslant 1$) has the hazard rate (given $\mathbf{Z}_i = \mathbf{z}_i$)

$$h_i(t) = h_0(t)\exp(\boldsymbol{\beta}'\mathbf{z}_i), \qquad i = 1, \ldots, n, \qquad (t \geqslant 0), \tag{11.7.8}$$

where $h_0(t)$, the hazard rate for $\mathbf{z}_i = \mathbf{0}$, is an unknown, arbitrary nonnegative function and $\boldsymbol{\beta} = (\beta_1, \ldots, \beta_q)'$ parameterizes the regression of the survival time on the covariates. Note that the hazard rate being the logarithmic derivative of the survival function [i.e., 1 − (the d.f.)] satisfies the condition that $\int_0^\infty h_0(t)\,dt = \infty$. We assume that $h_0(t)$ is continuous a.e., so that ties among the Y_i can be neglected with probability 1. To incorporate possible withdrawals of subjects from the scheme, we conceive of a set of r.v. (withdrawal times) $W_1, \ldots, W_n$ which are i.i.d. and independent of the Y_i; we denote by $G(w)$ the d.f. of W and let $Y_i^0 = Y_i \wedge W_i$ and $\delta_i = 0$ or 1 according as Y_i^0 is equal to W_i or Y_i, $i = 1, \ldots, n$. Then, our observable r.v.'s are $(\mathbf{Z}_i, Y_i^0, \delta_i)$, $i = 1, \ldots, n$. If we denote by $g_0(t) = -(\partial/\partial t)\log[1 - G(t)]$, $t \geqslant 0$, then, the log-likelihood function of all the observations is given by

$$\sum_{i=1}^{n} \Bigg\{ \log\left[g_0(Y_{ni}^0) + h_0(Y_{ni}^0)\exp\{ \boldsymbol{\beta}'\mathbf{Z}_{S_i^0}\} \right] - \log\Bigg(\sum_{j=i}^{n} \left[g_0(Y_{ni}^0) + h_0(Y_{ni}^0)\exp\{ \boldsymbol{\beta}'\mathbf{Z}_{S_j^0}\} \right]\Bigg)\Bigg\}, \tag{11.7.9}$$

where the Y_{ni}^0 are the ordered r.v.'s corresponding to the Y_i^0 and the S_i^0 are their antiranks. This likelihood function depends on h_0 and g_0 (as well as the Y_{ni}^0) and hence only a parametric form of analysis is suitable. To be able to use some nonparametric form, as in Cox (1972), we let $T = \{t_1, \ldots, t_m\}$ be the set of ordered failures among the Y_i^0 for which $\delta_i = 1$. Then, at time $t_i - 0$, there is a *risk-set* $\mathcal{R}_i$ of r_i subjects which have neither failed nor dropped out by that time, for $i = 1, \ldots, m$ ($\mathcal{R}_m \subset \ldots \subset \mathcal{R}_1$). Considering the risk set $\mathcal{R}_j$ and the conditional probability of a failure at time $t_j (1 \leqslant j \leqslant m)$, we obtain the following *partial likelihood function*

$$\log L_m^* = \sum_{j=1}^{m} \left\{ \boldsymbol{\beta}' \mathbf{Z}_{S_j^*} - \log\left(\sum_{i \in \mathcal{R}_j} \exp\{ \boldsymbol{\beta}' \mathbf{Z}_i \} \right) \right\}, \qquad (11.7.10)$$

where the S_i^* constitute a subvector of the antiranks $(S_1, \ldots, S_n)$. This does not depend on the h_0 and g_0 and therefore some nonparametric form of analysis may be applied to it.

Suppose now that we desire to test for

$$H_0 : \boldsymbol{\beta} = \mathbf{0} \qquad \text{against} \quad H_1 : \boldsymbol{\beta} \neq \mathbf{0}. \qquad (11.7.11)$$

Cox (1972) considered the test based on the statistic

$$\mathcal{L}_{nm}^* = \mathbf{U}^{*\prime}_{nm} \mathbf{J}^{*-}_{nm} \mathbf{U}^*_{nm} \qquad (11.7.12)$$

where

$$\mathbf{U}_{nm}^* = \left(\frac{\partial}{\partial \boldsymbol{\beta}} \right) \log L_{nm}^* \bigg|_{\boldsymbol{\beta} = \mathbf{0}}, \qquad \mathbf{J}_{nm}^* = -\left(\frac{\partial^2}{\partial \boldsymbol{\beta} \partial \boldsymbol{\beta}'} \right) \log L_{nm}^* \bigg|_{\boldsymbol{\beta} = \mathbf{0}}, \qquad (11.7.13)$$

and $\mathbf{A}^-$ stands for the generalized inverse of $\mathbf{A}$. In the setup of the earlier sections, we consider the following PCS tests, outlined in Sen (1981a).

Suppose that for the subset $\{t_1, \ldots, t_k\}$, we consider the likelihood $\log L_{nk}^*$, defined as in (11.17.10) (with m being replaced by k) and also in (11.7.12) and (11.7.13), we replace m by k and define in an analogous way the statistic $\mathcal{L}_{nk}^*$, for $k = 1, \ldots, m$. Then, as in the case of PCS rank ANOCOVA tests in Section 11.6 [see the discussion following (11.6.43)], we conceive of an initial starting number n_0 and review the process at each failure t_j, $j \geqslant n_0$: the experimentation is curtailed at time t_j if $\mathcal{L}_{nj}^*$ exceeds some prefixed number $l_n^{(\alpha)}$, for some $j (n_0 \leqslant j \leqslant m)$, and if no such j exists the experimentation is curtailed at the mth failure t_m along with the acceptance of H_0. Here also we need to find out the critical value $l_n^{(\alpha)}$ such that the PCS test has size α $(0 < \alpha < 1)$. Basically, we intend to show that even parallel to (11.6.64) and (11.6.66), $l_n^{(\alpha)}$ can be approximated by the critical values of the exit time for a suitable Bessel process. As such, we avoid the details of these procedures, and, in the rest of this section, consider only some relevant invariance principles for the partial sequence of partial likelihood functions $\{\mathcal{L}_{nj}^*\}$.

Note that by (11.7.10) and (11.7.13), for every $k: 1 \leqslant k \leqslant m$,

$$\mathbf{U}_{nk}^* = \sum_{j=1}^{k} \left\{ \mathbf{Z}_{S_j^*} - r_j^{-1} \sum_{i \in \mathcal{R}_j} \mathbf{Z}_i \right\}, \tag{11.7.14}$$

$$\mathbf{J}_{nk}^* = \sum_{j=1}^{k} (r_j - 1) r_j^{-1} \mathbf{V}_{nj}, \tag{11.7.15}$$

where

$$\mathbf{V}_{nj} = (r_j - 1)^{-1} \sum_{i \in \mathcal{R}_j} (\mathbf{Z}_i - \overline{\mathbf{Z}}_j^*)(\mathbf{Z}_i - \overline{\mathbf{Z}}_j^*)' \tag{11.7.16}$$

and

$$\overline{\mathbf{Z}}_j^* = r_j^{-1} \sum_{i \in \mathcal{R}_j} \mathbf{Z}_i, \qquad \text{for} \quad j = 1, \ldots, m. \tag{11.7.17}$$

Let $\mathcal{B}_{nk}^*$ be the sigma field generated by the risk-set $\mathcal{R}_k$, $k = 1, \ldots, m$. Then, under H_0, S_j^* has the (discrete) uniform distribution (given $\mathcal{R}_j$) over the set of r_j realizations in $\mathcal{R}_j$, so that for each j $(= 1, \ldots, m)$

$$E\left\{ \mathbf{Z}_{S_j^*} - r_j^{-1} \sum_{i \in \mathcal{R}_j} \mathbf{Z}_i \,|\, \mathcal{B}_{nj}^* \right\} = \mathbf{0} \qquad \text{a.e.,} \tag{11.7.18}$$

$$E\left\{ \left[\mathbf{Z}_{S_j^*} - r_j^{-1} \sum_{i \in \mathcal{R}_j} \mathbf{Z}_i \right] \left[\mathbf{Z}_{S_j^*} - r_j^{-1} \sum_{i \in \mathcal{R}_j} \mathbf{Z}_i \right]' \,|\, \mathcal{B}_{nj}^* \right\}$$
$$= r_j^{-1} \sum_{i \in \mathcal{R}_j} \left[\mathbf{Z}_i - \overline{\mathbf{Z}}_j^* \right] \left[\mathbf{Z}_i - \overline{\mathbf{Z}}_j^* \right]' = r_j^{-1}(r_j - 1)\mathbf{V}_{nj}. \tag{11.7.19}$$

Now, (11.7.18) ensures that under H_0, for every n,

$$\{\mathbf{U}_{nk}^*, \mathcal{B}_{nk}^*;\ 1 \leqslant k \leqslant m\} \qquad \text{is a (zero mean) martingale.} \tag{11.7.20}$$

Also, the Z_i are i.i.d.r.v. and under H_0, their independence or i.d. nature is not affected by the partitioning $\mathcal{R}_m \subset \cdots \subset \mathcal{R}_1$. Further, by (7.2.16), the $\mathbf{V}_{nj}$ are (matrix-valued) U-statistics of degree 2 and therefore using Theorem 3.2.1 along with (11.7.16) and the above, we conclude that under H_0,

$$\{\mathbf{V}_{nk}, \mathcal{B}_{nk}^*;\ 1 \leqslant k \leqslant m\} \qquad \text{is a martingale.} \tag{11.7.21}$$

Let us now assume that

$$\mathbf{\Gamma} = E[\mathbf{Z}_i - E\mathbf{Z}_i][\mathbf{Z}_i - E\mathbf{Z}_i]' \qquad \text{exists and is p.d. finite.} \tag{11.7.22}$$

Thus, if $\|\mathbf{A}\|$ stands for the maximum of the elements of $\mathbf{A}$, then, by (11.7.21) and (11.7.22),

$$\{\|\mathbf{V}_{nk} - \mathbf{\Gamma}\|, \mathcal{B}_{nk}^*;\ 1 \leqslant k \leqslant m\} \qquad \text{is a nonnegative submartingale.} \tag{11.7.23}$$

By (11.7.23), the Kolmogorov inequality for submartingales and Theorem 3.2.1, we arrive at the following (the proof is left as an exercise; see Problem 11.7.9):

$$\max_{1 \leqslant k \leqslant m} \|m^{-1}(\mathbf{J}_{nk}^* - k\boldsymbol{\Gamma})\| \xrightarrow{p} 0 \qquad \text{as} \quad m \to \infty, \tag{11.7.24}$$

where we note that by definition, as $n \to \infty$,

$$\frac{m}{n} \underset{\text{a.s.}}{\to} \pi = \int_0^\infty F_0(x)\, dG(x), \tag{11.7.25}$$

F_0 is the d.f. of the Y_i under H_0 and we assume that F_0 and G are overlapping, so that $0 < \pi < 1$. Further, we note that by (11.7.14),

$$\max_{1 \leqslant k \leqslant m} |\mathbf{U}_{nk} - \mathbf{U}_{nk-1}| \leqslant 2 \max_{1 \leqslant i \leqslant n} |\mathbf{Z}_i| = 2\tilde{Z}_n, \qquad \text{say}, \tag{11.7.26}$$

where, using the fact that the $|\mathbf{Z}_i|$ are i.i.d.r.v. with a finite second moment, we conclude that for every $\epsilon > 0$,

$$P\left\{ \max_{1 \leqslant i \leqslant n} |\mathbf{Z}_i| \geqslant \epsilon\sqrt{n} \right\} \to 0 \qquad \text{as} \quad n \to \infty. \tag{11.7.27}$$

Further, as in (11.7.18) and (11.7.19), using the uniform permutation probability measure, we obtain by (11.7.24) and (11.7.27),

$$\begin{aligned}
&\left\| m^{-1} \sum_{i=1}^{m} E\left\{ (\mathbf{U}_{ni} - \mathbf{U}_{ni-1})(\mathbf{U}_{mi} - \mathbf{U}_{ni-1})' I\left(|\mathbf{U}_{ni} - \mathbf{U}_{ni-1}| > \epsilon\sqrt{n} \right) | \mathcal{B}_{ni-1}^* \right\} \right\| \\
&= \left\| m^{-1} \sum_{j=1}^{m} r_j^{-1} \sum_{i \in \mathcal{R}_j} (\mathbf{Z}_i - \bar{\mathbf{Z}}_j^*)(\mathbf{Z}_i - \bar{\mathbf{Z}}_j^*)' I\left(|\mathbf{Z}_i - \bar{\mathbf{Z}}_j^*| > \epsilon\sqrt{n} \right) \right\| \\
&\leqslant \left[\operatorname{Tr}\left(m^{-1} \mathbf{J}_{nm}^* \right) \right] I\left(\tilde{Z}_n > \sqrt{n}\,\epsilon/2 \right) \to 0.
\end{aligned} \tag{11.7.28}$$

Finally, by (11.7.15) and (11.7.19), for every $k : 1 \leqslant k \leqslant m$,

$$\sum_{j \leqslant k} E\left\{ (\mathbf{U}_{nj} - \mathbf{U}_{nj-1})(\mathbf{U}_{nj} - \mathbf{U}_{nj-1})' \mid \mathcal{B}_{nj}^* \right\} = \mathbf{J}_{nk}^*. \tag{11.7.29}$$

Therefore, if we define a stochastic process $\mathbf{W}_n = \{\mathbf{W}_n(t),\ t \in [0, 1]\}$ by letting

$$\begin{aligned}
&\mathbf{W}_n\left(\frac{k}{m} \right) = \mathbf{J}_{nm}^{*\,-1/2} \mathbf{U}_{nk}^*, \qquad k = 1, \ldots, m; \qquad \mathbf{W}_n(0) = \mathbf{0}; \\
&\mathbf{W}_n(t) = \mathbf{W}_n\left(\frac{k}{m} \right), \qquad \text{for} \quad \frac{k}{m} \leqslant t < \frac{k+1}{m}, \qquad k = 0, \ldots, m,
\end{aligned} \tag{11.7.30}$$

then, by (a direct multivariate extension of) Theorem 2.4.7, (11.7.24), (11.7.25), and (11.7.27) through (11.7.29), we conclude that under (11.7.22) and H_0,

$$\mathbf{W}_n \xrightarrow{\mathfrak{D}} \mathbf{W} = \{\mathbf{W}(t), t \in [0, 1]\}, \tag{11.7.31}$$

where $B = \{B(t) = \{[\mathbf{W}(t)]'[\mathbf{W}(t)]\}^{1/2}, t \in [0, 1]\}$ is a Bessel process. Since, by (11.7.24), $\mathbf{J}^*_{nk} - m^{-1}k\mathbf{J}^*_{nm} \xrightarrow{P} \mathbf{0}$, for every $k : 1 \leqslant k \leqslant m$, it follows from (11.7.31) that on letting $\eta = m^{-1}n_0$, under (11.7.22) and H_0, for every $\eta > 0$,

$$\max_{n_0 \leqslant k \leqslant m} \mathfrak{L}^*_{nk} \xrightarrow{\mathfrak{D}} \sup\{t^{-1}B^2(t) : \eta \leqslant t \leqslant 1\}. \tag{11.7.32}$$

This proves the desired weak convergence result.

We may also extend the invariance principle in (11.7.31) to the case where H_0 may not hold. We consider a sequence $\{K_n\}$ of local alternatives, where

$$K_n : (11.7.8) \text{ holds for } \qquad \boldsymbol{\beta} = \boldsymbol{\beta}_{(n)} = n^{-1/2}\boldsymbol{\lambda}, \tag{11.7.33}$$

for some $\boldsymbol{\lambda} \in R^p$. If we consider the log-likelihood function (for K_n versus H_0), then it is possible to show that (4.3.26) holds with $\sigma^2 = \pi E(\boldsymbol{\lambda}'\mathbf{Z}_1\mathbf{Z}_1'\boldsymbol{\lambda})$ (see Problem 11.7.10 in this respect), and this ensures the contiguity of the probability measure under $\{K_n\}$ to that under H_0. By virtue of this contiguity, the tightness of $\{\mathbf{W}_n\}$, under H_0, proved in the context of (11.7.31), remains intact under $\{K_n\}$ as well. Also, the convergence of the finite-dimensional distributions of $\{\mathbf{W}_n\}$, under $\{K_n\}$, can be established by using (4.3.27) and verifying the asymptotic joint (multi-) normality of the partial likelihood in (11.7.10) and the total likelihood; we leave the details of this verification as an exercise (see Problem 11.7.11). Thus if we define by $1 - \Psi(y) = [1 - G(y)][1 - F_0(y)]$, $y \in R$, π as in (11.7.25), and

$$\pi(t) = \int_0^{\Psi^{-1}(t)} h_0(u)\{h_0(u) + g_0(u)\}^{-1}\, d\Psi(u), \tag{11.7.34}$$

then, on letting $t^* = \inf\{u : \pi(u) \geqslant t\pi(1)\}$, $t \in [0, 1]$, and

$$\zeta^* = \{\zeta^*(t) = \pi^{-1/2}\pi(t^*)\boldsymbol{\Gamma}^{1/2}\boldsymbol{\lambda} = t\pi^{1/2}\boldsymbol{\Gamma}^{1/2}\boldsymbol{\lambda}, \qquad t \in [0, 1]\}, \tag{11.7.35}$$

we obtain from the above that under $\{K_n\}$ and the assumed regularity conditions,

$$\mathbf{W}_n \xrightarrow{\mathfrak{D}} \mathbf{W} + \zeta^*. \tag{11.7.36}$$

The last result provides the asymptotic power of the PCS test based on $\max\{\mathfrak{L}^*_{nk} : n_0 \leqslant k \leqslant m\}$ in terms of the boundary crossing probability of a drifted Bessel process. It also provides expression for the asymptotic efficiency of the terminal procedure with respect to the same when there is no withdrawal; we pose this as an exercise (see Problem 11.7.13). For the asymptotic efficiency of the PCS testing procedure with respect to the terminal procedure (based on $\mathfrak{L}^*_{nm}$ alone), as in Section 11.5, no simple expression is available and numerical investigations are needed to supplement the theory.

EXERCISES

11.2.1 Let $T_{n,k} = (-\partial/\partial\theta)\log p_\theta(\mathbf{Z}_n^{(k)})$, where $p_\theta(\mathbf{Z}_n^{(k)})$ is defined as in (11.2.1). Show that if $f_\theta(x) = \theta^{-1}e^{-x/\theta}$ for $x \in R^+$, then (i) for each n, $T_{n,k}$ is a difference of two functions monotonic in k and (ii) $T_{n,r}$ has a simple distribution related to the chi square d.f. with $2r$ degrees of freedom. Hence consider some PCS tests based on the partial sequence $\{T_{n,k}, k \leqslant r\}$ where $r = [np] + 1$ for some $p \in (0, 1]$ (Epstein and Sobel, 1954, 1955).

11.2.2. Show (by means of a counter example) that if $f_\theta(x)$ is not an exponential p.d.f., the monotonicity property of $T_{n,k}$ may not hold.

11.2.3. Show that $T_{n,k}$ is a linear combination of functions of order statistics which conforms to the pattern studied in Section 7.4. Hence, or otherwise, study invariance principles for $\{T_{n,k}, k \leqslant r\}$ (Gardiner and Sen, 1978).

11.2.4. Show that $\{T_{n,k}, k \leqslant n\}$ is a martingale (for each n) when θ holds. Hence, use Theorem 2.4.7 to derive an invariance principle for the partial sequence $\{T_{n,k}, k \leqslant r\}$ (Sen, 1976c).

11.3.1. Consider the case of the two-sample median statistic and show that here $A^2_{n,r} = A^2_n$ for every $r \geqslant [\frac{1}{2}(n+1)]$. Thus, in this case, if a PCS testing procedure is employed, it suffices to stop (at the latest) when the combined sample median failure has occurred. Comment on the average cost of sampling in using this test and others based on the Wilcoxon scores or the Savage (log-rank) scores.

11.4.1. Using (11.4.7) and (11.4.8), show that (11.4.21) holds.

11.4.2. Verify (11.4.22). For the first convergence result use (11.4.21) along with the mean square equivalence of $L_{n,r}$ and $\tilde{L}_{n,p}$ (under H_0); for the second result use the projection result in (4.2.23).

11.4.3. In (11.4.32), consider scale-alternatives $F_{n,i}(x) = F(xe^{-\theta d_{ni}})$, $i \geqslant 1$, and extend (11.4.39) along with the expression for the drift μ.

11.4.4. In (11.4.33), consider the joint regression-scale model: $F_{n,i}(x) = F(e^{-\theta_1 d_{ni}}(x - \theta_2 c_{ni}))$, $i \geqslant 1$, where θ_1, θ_2 are real parameters and the

c_{ni}, d_{ni} satisfy the regularity conditions of Theorem 11.4.4 and Problem 11.4.3. Extend Theorem 11.4.4 to such a general model and provide expression for the corresponding drift function.

11.5.1. Under (11.5.13) through (11.5.18), show that as $\theta \to 0, H_\theta(x) = F(x) + \theta \bar{d} f(x) + o(\theta)$ and $G_\theta(x) = \theta \rho_2^* f(x) + o(\theta)$. Integrating (11.5.23) by parts, show that (11.5.32) holds.

11.5.2. **(a)** Show that if $r = [np] + 1$ for some $p \in (0, 1)$ and if the X_i are i.i.d.r.v.'s with an absolutely continuous d.f. F (ensuring the uniqueness of the $\xi_{n,k}$) for which $\int_{-\infty}^{\infty} |x|^a dF(x) < \infty$, for some $a > 0$, then (11.5.50) holds.

(b) If the X_i are independent with d.f.'s $F_1, \ldots, F_n$, differing only locally from a common d.f. F satisfying the conditions in part (a), then (11.5.50) holds.

Hints. (i) Show that the density of the order statistics $X_{n,k}$ converges exponentially to 0 as $X_{n,k}$ departs from $\xi_{n,k}$ (c.f. Sen, 1959), and verify (11.5.50) by integrating by parts. (ii) When the X_i are non i.d., use the dominance theorem for the expected $X_{n,k}$ in terms of the average d.f. $\bar{F}_n = n^{-1} \sum_{i=1}^n F_i$ [c.f. Sen, 1970c and use the result in (i)].

11.5.3. Let $k_n = [n^{1-\eta}]$ where $\eta \in (0, 1)$. Show that if the underlying d.f. is simple exponential, then for $r^* = n - k_n$, $\lim_{n \to \infty} E(X_{n,n} - X_{n,r^*})^2 / \xi_{n,n}$ can be made aribtrarily small by choosing η small. The same conclusion holds for the case of normal and gamma d.f.'s. [In either case, use the inequality $(X_{n,n} - X_{n,r^*})^2 \leqslant X_{n,n}^2 - X_{n,r^*}^2$ and use the asymptotic expressions for $EX_{n,n}^2$ and EX_{n,r^*}^2.]

11.5.4. Let $r(x) = f(x)/[1 - F(x)]$ be the hazard function and let $c(x) = (d/dx)(1/r(x))$. If $r(x)$ is an increasing function of x (in the right-hand tail) and $c(x) \to 0$ as $x \to \infty$, then, F is termed an exponential type d.f. of the convex type, while, if $c(x) \to 0$ but $r(x)$ is nondecreasing, it is termed a concave exponential type. Finally, if $c(x) \to a > 0$ as $x \to \infty$, F is termed a Cauchy-type d.f. Show that for the convex exponential type d.f., (11.5.50) holds in the case of i.i.d.r.v.'s. However, this may not be the case in the case of concave exponential type d.f. Further, for the Cauchy type of d.f., $EX_{n,n}^k (k = 1, 2)$ may not exist unless $a < k^{-1}$.

11.6.1. Let $X^0 = \{X^0(t),\ t \in [0, 1]\}$ be a standard Brownian Bridge, and $X = \{X(t),\ t \in [0, \infty]\}$ be a standard Wiener process on R^+. Then, using (2.7.7), show that for every real a and $b : 0 < b \leqslant 1$,

$$P\left\{ \sup_{0 \leqslant t \leqslant b} X^0(t) < a \right\} = P\left\{ \sup_{0 \leqslant t \leqslant b/(1-b)} (1 + t)^{-1} X(t) < a \right\}$$

$$= P\left\{ X(t) < a + at, \quad \forall 0 \leqslant t \leqslant \frac{b}{1 - b} \right\}.$$

Next use (2.7.17) to provide an expression for the above probability (Schey, 1977).

11.6.2. In the setup of the preceding problem, show that

$$P\left\{\sup_{0\leqslant t\leqslant b}|X^0(t)| < a\right\} = P\{-a - at < X(t) < a + at,$$
$$\forall 0 \leqslant t \leqslant b/(1-b)\}$$
$$= 1 - P\{X(t) \geqslant a + at \quad \text{for a} \quad t(\leqslant b/(1-b))$$
$$\text{smaller than any } t(\leqslant b/(1-b)) \quad \text{for which} \quad X(t) \leqslant -a - at\}$$
$$- P\{X(t) \leqslant -a - at \quad \text{for a} \quad t(\leqslant b/(1-b))$$
$$\text{smaller than any } t(\leqslant b/(1-b)) \quad \text{for which} \quad X(t) \geqslant a + at\}$$
$$= 1 - 2P\{X(t) \geqslant a + at \quad \text{for a} \quad t(\leqslant b/(1-b))$$
$$\text{smaller than any } t(\leqslant b/(1-b)) \quad \text{for which} \quad X(t) \leqslant -a - at\}.$$

Use (2.7.17) to provide an expression for the above probability. (Numerical values of the critical point a for specific values of b are tabulated in Koziol and Byar, 1975.)

11.6.3. Show that if $q(t) = (1 - t)^{-1}w(t)$, $0 < t \leqslant 1$ satisfies the regularity conditions mentioned before (2.4.25), then (2.4.29) holds and $H_n(\cdot)/w(S_n(\cdot))$ converges weakly to a Gaussian function.

11.6.4. In the preceding problem, in particular let $w(t) = 1 - t$, so that $q(t) = 1$ $0 < t \leqslant 1$ and verify that $H_n(\cdot)$ converges weakly to a Brownian Bridge. [A more detailed proof (without using the martingale property) is due to Hájek (1963) and contained in Hájek and Šidák (1967).]

11.6.5. Consider a triangular array $\{X_{ni}, i \leqslant n;\ n \geqslant 1\}$ of row-wise independent r.v. where for each n (consider the hypothesis H_n)

$$F_{ni}(x) = P\{X_{ni} \leqslant x\} = F(x - \beta_0 - \beta_1 c_{ni}^*), \qquad i = 1, \ldots, n, \quad x \in R,$$

the c_{ni}^* are defined by (11.2.11) and it is assumed that $\max_{1\leqslant i\leqslant n}|c_{ni}^*| \to 0$ as $n \to \infty$. Further, assume that F has an absolutely continuous p.d.f. f with a finite Fisher information $I(f)$. Let $\xi = \{\xi(t) = \beta_1 f(F^{-1}(t)),\ 0 \leqslant t \leqslant 1\}$. Show that under the above conditions, $H_n(\cdot)/w(S_n(\cdot))$ converges weakly to $\{W^0(\cdot) + \xi(\cdot)\}/w(\cdot)$, when $w(\cdot)$ satisfies the same regularity conditions as in the null hypothesis ($\beta_1 = 0$) case (Sinha and Sen, 1979a).

Hints. Show first that the contiguity of the probability measures under $\{H_n\}$ with respect to that under the null hypothesis is implied by the regularity conditions. Secondly, invoke this contiguity and the tightness of the process under the null hypothesis to yield the parallel result under

$\{H_n\}$. Finally, establish the convergence of the f.d.d.'s by the classical central limit theorem.

11.6.6. For the model in the preceding problem, show that the asymptotic power function of the test based on K_n^+ (or K_n) under $\{H_n\}$ is given by $P\{W^0(t) + \xi(t) > D_\alpha^+$, for some $t \in [0, 1]\}$ (or $P\{|W^0(t) + \xi(t)| > D_\alpha$, for some $t \in [0, 1]\}$), where D_α^+ and D_α are defied after (11.4.30). Extend these results to the case of the tests based on the weighted statistics in (11.6.5).

11.6.7. Let $J_j = (a_{j-1}, a_j]$ and $j = 1, \ldots, k+1$ (where $a_0 = -\infty < a_1 < \cdots < a_k < a_{k+1} = +\infty$) be a set of ordered time intervals and let $J_r^* = \cup_{s \geqslant r} J_s$ for $r = 1, \ldots, k+1$. For n independent r.v.'s, $X_1, \ldots, X_n$ following the model (11.6.1), consider the observable r.v. $X_1^*, \ldots, X_n^*$ where $X_i^* = \sum_{j=1}^{k+1} J_j Z_{ij}$ and $Z_{ij} = 1$ or 0 according as $X_i \in J_j$ or not ($j = 1, \ldots, k+1$), for $i = 1, \ldots, n$. Then, in a PCS, at the time point a_l, the observable r.v.'s are $X_{i,l}^* = \sum_{j \leqslant l} J_j Z_{ij} + J_{l+1}^*, Z_{l+1}^*$; $Z_{l+1}^* = \sum_{j>l} Z_{ij}$, $l = 1, \ldots, k$, $i = 1, \ldots, n$. Let then $n_j = (\sum_{i=1}^n Z_{ij})$, $n_j^* = \sum_{s \geqslant j} n_s$, $j = 1, \ldots, k+1$, $F_{n,0} = 0$, $F_{n,j} = n^{-1}(\sum_{s \leqslant j} n_s)$, $j = 1, \ldots, k$, and $F_{n,k+1} = 1$. For some nondecreasing and square integrable score function ϕ, defined as in Section 11.3, set $\hat{\Delta}_{n,j} = (n/n_j)\int_{F_{n,j-1}}^{F_{n,j}} \phi(u)du$, if $n_j > 0$ and equal to $\phi(F_{n,j})$, if $n_j = 0$, for $j = 1, \ldots, k+1$. Further, for every l, let $\hat{\Delta}_{n,l+1}^* = (n/n_{l+1}^*) \cdot \int_{F_{n,l}}^1 \phi(u)du$, if $n_{l+1}^* > 0$ and it is equal to 0, otherwise, for $l = 1, \ldots, k$. Then, at time point a_r, a rank statistic based on these grouped observations may be set as $L_{n,r} = \sum_{i=1}^n (c_i - \bar{c}_n)[\sum_{j=1}^r Z_{ij}\hat{\Delta}_{n,j} + Z_{i,r+1}^* \hat{\Delta}_{n,r+1}^*]$, $r = 1, \ldots, k$. A PCS test for the null hypothesis $H_0: \beta = 0$ [with respect to (11.6.1)] is then based on the sequence $\{L_{n,r}, r = 0, 1, \ldots, k\}$ where $L_{n,0} = 0$. The procedure is similar to that in (11.3.1) and sketched after (11.3.3); however, the possible termination points are the a_j, $j = 1, \ldots, k$ (Majumdar and Sen, 1977).

11.6.8. For the problem 11.6.7, define $B_{n,r}^2 = n^{-1}[\sum_{j=1}^r \hat{\Delta}_{n,j}^2 n_j + n_{r+1}^* \hat{\Delta}_{n,r+1}^{*}]$, $r = 0, \ldots, k$ (where $B_{n,0}^2 = 0$, by definition) and let $B_{n,k+1}^2 = B_n^2$, $t_r = B_{n,r}^2 / B_n^2$, $r = 1, \ldots, k$. Show that under H_0, $(L_{n,r}/C_n B_{n,r}$, $r = 1, \ldots, k)$ has asymptotically a multinormal distribution with null mean vector and dispersion matrix with the elements $((t_r \wedge t_{r'}))$. Hence show that for $\max_{r \leqslant k} L_{n,r}/C_n B_{n,r}$ and $\max_{k \leqslant r} |L_{n,r}|/C_n B_{n,r}$ the right-hand sides of (11.4.29) and (11.4.30) provide upper bounds for the asymptotic tail probabilities (Majumdar and Sen, 1977).

11.6.9. Consider the sequence $\{H_n\}$ of local alternatives in the Problem 11.6.5 and under this sequence of alternatives, for the tests based on the statistics in the preceding problem, express the asymptotic power function in terms of the boundary crossing probabilities of a drifted Wiener process (Majumdar and Sen, 1977).

11.6.10 Define $Y_n(\cdot)$ as in the discussion following (11.6.19) and show that under (11.6.13), (11.6.14), and the hypothesis of Theorem 11.4.2, Y_n weakly converges to B, where B is a p-parameter Bessel process with no drift. (Majumdar and Sen, 1978a).

Hints. Note that $|Y_n(t) - Y_n(s)| \leqslant \sqrt{p}\,\{\max_{1 \leqslant j \leqslant p} |W_{nj}^{(r)}(t) - W_{nj}^{(r)}(s)|\}$, for every $0 \leqslant s < t \leqslant 1$, where for each $W_{nj}^{(r)}$, the tightness of the process is insured by Theorem 11.4.2. Hence the tightness of Y_n is in order. For the convergence of the f.d.d.'s use the Cramér-Wald theorem along with the martingale property of the sequence $\{\mathbf{L}_{n,k},\ 0 \leqslant k \leqslant n\}$.

11.6.11. Define $\mathbf{c}_{ni}^* = \mathbf{C}_n^{-1/2}(\mathbf{c}_i - \bar{\mathbf{c}}_n)$, $i = 1, \ldots, n$; $n \geqslant 1$, where the $\mathbf{c}_i, \bar{\mathbf{c}}_n$, and $\mathbf{C}_n$ are defined as in Section 11.6 and consider a sequence $\{H_n\}$ of alternative hypotheses, where under H_n, (11.6.9) holds with the $\mathbf{c}_i$ being replaced by $\mathbf{c}_{ni}^*$. Define $\mathbf{W}_n^{(r)}$ as in (11.6.19) and show that under $\{H_n\}$ and the regularity conditions of Theorem 11.4.4, $\mathbf{W}_n^{(r)}$ converges weakly to $\mathbf{W} + \boldsymbol{\mu}$, where $\boldsymbol{\mu} = \{\boldsymbol{\mu}(t) = A_\psi \rho_1(s(t), p)\boldsymbol{\beta},\ t \in I\}$ and $A_\psi, \rho_1(s,t)$ and $s(t)$ are all defined as in Section 11.4 (Majumdar and Sen, 1978a).

Hint. Use the contiguity of the probability measures under $\{H_n\}$ with respect to that under H_0 (to be proved under the assumed regularity conditions) to prove the tightness of $\mathbf{W}_n^{(r)}$ under $\{H_n\}$. For the convergence of the f.d.d.'s, proceed as in the proof of Theorem 11.4.4.

11.6.12. Use (11.6.20) and the weak convergence result in Problem 11.6.11 to show that under $\{H_n\}$, $K_{n,r}$ converges in law to $\sup\{\|\mathbf{W}(t) + \boldsymbol{\mu}(t)\| : t \in I\}$, where $\|\mathbf{a}\|$ stands for the Euclidean norm (Majumdar and Sen, 1978a).

11.6.13. Use (11.6.20) and the weak convergence result in Problem 11.6.11 to show that under $\{H_n\}$, $M_{n,r}$ weakly converges to $\int_0^1 \|\mathbf{W}(t) + \boldsymbol{\mu}(t)\|^2\, dt$. Hence express the asymptotic power functions of the two tests in (11.6.16) and (11.6.17) in terms of the boundary crossing probabilities for the drifted Bessel process $\|\mathbf{W}(t) + \boldsymbol{\mu}(t)\|$ (Majumdar and Sen, 1978a).

11.6.14. Define the tied-down Bessel process B^0 as in (2.7.26). Show that under H_0: $\boldsymbol{\beta} = \mathbf{0}$ [for the model (11.6.9)], $\|\mathbf{H}_n(\cdot)\|$ [with the Euclidean norm $\|\cdot\|$] in (11.6.21) converges weakly to B^0. Hence, a PCS testing procedure can be considered and this runs parallel to the case of $p = 1$ treated in Section 11.6 (with the critical values of the Brownian Bridge being replaced by those of the p-parameter Bessel process). Further, under $\{H_n\}$ (see Problem 11.6.11), $\|\mathbf{H}_n(\cdot)\|$ converges weakly to $\|\mathbf{W}^0(t) + \boldsymbol{\xi}(\cdot)\|$, where $\boldsymbol{\xi} = \{\boldsymbol{\xi}(t) = \boldsymbol{\beta} f(F^{-1}(t)),\ t \in I\}$ and $B^0 = \|\mathbf{W}^0(\cdot)\|$ (Sinha and Sen, 1979b).

Hint. Same as in Problem 11.6.5.

11.6.15. Define the $L_{n,r}$ as in (11.2.16), C_n^2 as in (11.2.11) and the $A_{n,r}^2$ as in (11.2.20). Consider a two-dimensional time-parameter stochastic process $W_n = \{W_n(\mathbf{t}),\ \mathbf{t} \in E^2\}$ by letting $W_n(\mathbf{t}) = A_n^{-1} C_n^{-1} L_{n(t_1), r(t_1, t_2)}$, $\mathbf{t} \in E^2$, where $n(t_1) = \max\{k : C_k^2 \leqslant t_1 C_n^2\}$ and $r(t_1, t_2) = \max\{r : A_{n(t_1),r}^2 \leqslant t_2 A_{n(t_1),n(t_1)}^2\}$, $\mathbf{t} \in E^2$. Show that under the model (11.2.9) and some regularity conditions, W_n converges weakly to a standard Brownian sheet W on the unit square E^2. (The proof of the convergence of the f.d.d.'s of W_n to those of W follows arguments similar to the ones in Theorem 11.4.2. However, to prove the tightness of this two-dimensional time-parameter process, we need to employ some of the inequalities in Section 2.3; this is somewhat more involved and is beyond the scope of the treatment of the current chapter. We refer the reader to Sen, 1976a, for some of these details.)

11.6.16. In a staggered entry plan, where the units enter into the scheme at possibly different points of time, at any time point with respect to the entrants prior to that point, one can construct a sequence of censored linear rank statistics corresponding to the array of the exposure times of these units. Thus, by varying the observation time over the span of the experimentation, one obtains a two-dimensional array of the $L_{k,q}$, defined as in the preceding problem, where k ranges over 1 to n and q over 1 through k. The maximum of these $L_{k,q}$ (standardized by $A_n^{-1} C_n^{-1}$) may then be taken as the PCS test statistic. Use the weak convergence result in Problem 11.6.15 to show that the asymptotic tail probability of the test statistic is dominated by that of the supremum of the standard Brownian sheet and incorporate this result in the formulation of a PCS test for the staggering entry problem. (Majumdar and Sen, 1978b).

11.6.17. In the setup of the preceding problem, suppose that the units are subject to random withdrawal, with withdrawal time distributed independently of the failure times. (Thus the observable random variables are the minimum of the actual failure times and the withdrawal times.) Show that the results of the preceding problem remain valid for this random withdrawal model as well.

11.6.18. Under (11.4.32) through (11.4.34), granted the contiguity of the probability measures under $\{H_n\}$ with respect to that under H_0, extend the weak convergence result in Problem 11.6.15 to the nonnull case and incorporate the same in the study of the asymptotic power function of the PCS tests in Problems 11.6.16 and 11.6.17. Show that the random withdrawal model incorporates some loss of power in this context. (Majumdar and Sen, 1978b).

11.6.19. Parallel to (11.6.3), we define $H_{n,k}(x) = C_n^{-1} \sum_{i=1}^{k} (c_i - \bar{c}_k) c(x - X_i)$, $x \in R$ and $k \leqslant n$[$H_{n,0}(x) = 0$, $\forall x \in R$, conventionally.] Consider then a two-dimensional time-parameter stochastic process $W_n = \{W_n(t,x) = H_{n,n(t)}(x), 0 \leqslant t \leqslant 1, x \in R\}$. Let $W_n^* = \{W_n^*(t,s) = W_n(t, F^{-1}(s)), (t,s) \in E^2\}$ and let $W^* = \{W^*(t,s), (t,s) \in E^2\}$ be a Kiefer process, defined by (2.3.49). Show that under the Noether condition (11.4.5) and H_0 in (11.2.9), W_n^* converges weakly to W^*. Incorporate this result in the formulation of a PCS testing procedure for the staggering entry plan based on the weighted empirical process $H_{n,k}(x)$, $x \in R$, $k \leqslant n$. Study the asymptotic power properties of the PCS test for contiguous alternatives in (11.4.32) through (11.4.34) (Sinha and Sen, 1981).

11.6.20. For every $s(\geqslant 1)$ and $k_r, r = 1, \ldots, s$, such that $n^{-1}k_r \to t_r \in [0, 1]$, $r = 1, \ldots, s$, establish the asymptotic (joint-) normality of $\mathbf{C}_n^{-1/2}\mathbf{L}_{n,k_r}$, $r = 1, \ldots, s$, where the $\mathbf{L}_{n,k}$ are defined by (11.6.33). [Use the projection result in (4.2.23) for each coordinate of $\mathbf{L}_{n,k}$, as in (11.4.22), and the Cramér-Wold theorem.]

11.6.21. Define ${}_{\epsilon}\mathbf{W}_{n,r}$ as in (11.6.61) and establish its tightness (Sen, 1979e).

Hint. For the tightness part, show first that by virtue of (11.6.58), looking at (11.6.61), it suffices to show that $\max\{\|\mathbf{C}_n^{-1/2}\mathbf{L}_{n,k}\| : k \leqslant n\} = O_p(1)$ and each of the marginal processes in $\{\mathbf{C}_n^{-1/2}\mathbf{L}_{n,k}, k \leqslant n\}$ is tight. The martingale property in (11.6.59) along with the Kolmogorov inequality yield the former result, while, for the latter one, we may repeat the proof of the tightness part of Theorem 11.4.2.

11.6.22. Consider a sequence $\{H_n^*\}$ of alternative hypotheses, where under H_n^*, (11.6.28) holds with $(\mathbf{c}_i - \bar{\mathbf{c}}_n)$ being replaced by $\mathbf{C}_n^{-1/2}(\mathbf{c}_i - \bar{\mathbf{c}}_n)$, $i = 1, \ldots, n$. Also assume that the $(p+1)$-variate d.f. F_i^* are all absolutely continuous with absolutely continuous density functions. (Note that the p-variate marginal p.d.f.'s, corresponding to the p covariates, do not differ from each other.) Establish the contiguity of the probability measures under $\{H_n^*\}$ with respect to that under H_0, and hence, or otherwise, extend the weak convergence result in (11.6.64) to that under $\{H_n^*\}$ where a drifted Bessel process arises in the limiting case.

11.6.23. Use the weak convergence result in the preceding problem to provide an expression for the asymptotic power function of the PCS analysis of covariance test (Sen, 1979e, 1981d).

11.7.1. Let $\{d_{ni}, 1 \leqslant i \leqslant n;\ n \geqslant 1\}$ be a triangular array of real numbers satisfying $\sum_{i=1}^{n} d_{ni} = 0$ and $\sum_{i=1}^{n} d_{ni}^2 = 1$. Also, let $q = \{q(t), 0 < t < 1\}$ be a continuous, nonnegative, U-shaped, and square integrable function inside $(0,1)$. Finally, let $\mathbf{Q} = (Q_1, \ldots, Q_n)$ take on each permutation of

$(1, \ldots, n)$ with the common probability $(n!)^{-1}$. Then,

$$P\left\{\max_{1 \leqslant k \leqslant n-1} q(k/n)\left|\sum_{i=1}^{k} d_{nQ_i}\right| \geqslant 1\right\} \leqslant \int_0^1 q^2(t)\,dt.$$

(Sen, 1979b).

Hint. Show that $\{(n-k)^{-1}\sum_{i=1}^{k} d_{nQ_i},\ 0 \leqslant k \leqslant n-1\}$ is a martingale sequence and then use the Chow extension of the Hájek-Rényi inequality in (2.2.3).

11.7.2. If the X_i are i.i.d.r.v.'s with $\int |g(x)|dF(x) < \infty$, then, for $r = [np] + 1$, for some $p \in (0, 1)$, as $n \to \infty$, $\max\{|\overline{G}(X_{n,i}) + \overline{G}_n(X_{n,i})| : 1 \leqslant i \leqslant r\} \to 0$ a.s., where $\overline{G}(x)$ and $\overline{G}_n(x)$ are defined by (11.7.3) and (11.7.6), respectively, and the $X_{n,i}$ are the order statistics (Sen, 1979b).

Hint. Express $\overline{G}_n(x)$ in terms of the empirical d.f. and use Glivenko-Cantelli theorem type arguments.

11.7.3. Define the T_{nk} and T_{nk}^* as in (11.7.4) and (11.7.7) and use the results of the preceding two problems to show that under $H_0 : \Delta = 0$, for r defined as in Problem 11.7.2, $\max\{|T_{nk} - T_{nk}^*| : 1 \leqslant k \leqslant r\} \to 0$, in probability, as $n \to \infty$ (Sen, 1979b).

11.7.4. Define $\nu_p^2 = \int_{-\infty}^{F^{-1}(p)} g^2(x)\,dF(x) + (1-p)^{-1}\left(\int_{-\infty}^{F^{-1}(p)} g(x)\,dF(x)\right)^2$, for $0 < p < 1$. Show that under (i) $\int g^2(x)\,dF(x) < \infty$ and (ii) $\max\{|c_{ni}^*| : 1 \leqslant i \leqslant n\} \to 0$ as $n \to \infty$, when the X_i are i.i.d.r.v.'s, then $(\max_{1 \leqslant k \leqslant r} T_{nk}^*)/\nu_p$ and $(\max_{1 \leqslant k \leqslant r} |T_{nk}^*|)/\nu_p$ converge in law to $\sup_{0 \leqslant t \leqslant 1} W(t)$ and $\sup_{0 \leqslant t \leqslant 1} |W(t)|$, respectively, where $W = \{W(t),\ t \in [0, 1]\}$ is a standard Wiener process. By virtue of Problem 11.7.3, similar results hold for the T_{nk} (Sen, 1979b).

Hint. Show that under the hypothesis of this problem, $\{T_{nk}^*,\ 0 \leqslant k \leqslant n-1\}$ (or $\{T_{nk},\ 0 \leqslant k \leqslant n-1\}$) is a martingale array for which the weak invariance principle in Section 2.4 holds. This, in turn, provides the desired results.

11.7.5. Extend the results in the preceding problem to the case of contiguous alternatives treated in Problem 11.6.5. (Sen, 1979b).

11.7.6. For an arbitrary and square integrable $g = \{g(x),\ x \in R\}$, not necessarily the one in (11.7.3), the results in the preceding problems work out well so long as $r = [np] + 1$ for some $p < 1$. However, the convergence of $\overline{G}_n(x)$ to $\overline{G}(x)$ may not hold properly when $x \to \infty$. As such, the result in Problem 11.7.3 may not hold when $r = n$. But whenever $\{T_{nk},\ 0 \leqslant k \leqslant n\}$ is a martingale array, for some arbitrarily small $\eta(> 0)$, $\max\{|T_{nn} - T_{nk}| : n(1-\eta) \leqslant k \leqslant n\}/[\operatorname{Var}(T_{nn})]^{1/2}$ can be made arbitrarily small when

$\lim_{p \to 1} \nu_p = \nu_1$ exists. Hence the results continue to hold for $r = n$. Show that the latter condition holds whenever $\int g(x)dF(x) = 0$ and $\int g^2(x)dF(x) < \infty$, where the integrals are over the entire real line.

11.7.7. As a multiparameter extension of the model (11.7.1), let $f_i(x) = f(x; \boldsymbol{\Delta}(\mathbf{c}_i - \bar{\mathbf{c}}_n))$, $i = 1, \ldots, n$, $x \in R$, where $\boldsymbol{\Delta}$ is an $m \times q$ matrix of unknown parameters and the $\mathbf{c}_i$ are specified q-vectors, $\bar{\mathbf{c}}_n = n^{-1}\sum_{i=1}^{n} \mathbf{c}_i$. For $\theta \in R^q$, let $\mathbf{g}(x) = -(\partial/\partial\boldsymbol{\theta})\log f(x; \boldsymbol{\theta})|_{\boldsymbol{\theta}=\mathbf{0}}$ and define $\overline{\mathbf{G}}(x)$ and $\overline{\mathbf{G}}_n(x)$ as in (11.7.3) and (11.7.6). Finally, let $\mathbf{T}_{nk}^* = \sum_{i=1}^{k} \mathbf{g}(X_{n,i})[\mathbf{c}_{S_{ni}} - \bar{\mathbf{c}}_n) + (n - k)^{-1}\sum_{s=1}^{k}(\mathbf{c}_{S_{ns}} - \bar{\mathbf{c}}_n)]'$ for $k = 1, \ldots, n-1$; $T_{n0}^* = 0$ and $T_{nn}^* = T_{nn-1}^*$. Under $H_0 : \boldsymbol{\Delta} = \mathbf{0}$, through a martingale property of the $\mathbf{T}_{nk}^*$, study invariance principles relating to them. Also, let $\mathbf{V}_{nk} = n^{-1}\{\sum_{i=1}^{k}[\mathbf{g}(X_{n,i})][\mathbf{g}(X_{n,i})]' + (n-k)^{-1}[\sum_{i=1}^{k}\mathbf{g}(X_{n,i})][\sum_{i=1}^{k}\mathbf{g}(X_{n,i})]'$, for $k = 1, \ldots, n-1$. We roll out the T_{nk}^* into mq-vectors, write these as $\mathbf{T}_{nk}^{**}$ and define $\mathcal{L}_{nk} = \mathbf{T}_{nk}^{**\prime}(\mathbf{V}_{nk} \otimes \mathbf{c}_n)^{-1}\mathbf{T}_{nk}^{**}$ for $k = 1, \ldots, n$, where $\mathbf{V}_{nn}$ is taken as equal to $\mathbf{V}_{nn-1}$. Then, show that under H_0, $\max\{\mathcal{L}_{nk} : n\epsilon \leqslant k \leqslant n\}$ converges in law to $\mathcal{L}^* = \sup\{t^{-1}B(t) : \epsilon \leqslant t \leqslant 1\}$, where $0 < \epsilon < 1$ and $B = \{B(t), t \in [0, 1]\}$ is an mq-variate Bessel process. Extend this result to the case of contiguous alternatives treated in Problem 11.6.11 (Sen, 1981c).

11.7.8. Use the results in the preceding problem for the simultaneous testing of location and scale in the two-sample case (where $m = 2$ and $q = 1$) under PCS.

11.7.9. Use (11.7.15), (11.7.23), the Kolmogorov inequality, and the U-statistic property of $\mathbf{V}_{nk} - \boldsymbol{\Gamma}$ and verify (11.7.24) (Sen, 1981a).

11.7.10. For testing H_0: $\boldsymbol{\beta} = \mathbf{0}$ versus K_n in (11.7.33), verify that the log-likelihood statistic satisfies (4.3.26) with $\sigma^2 = \pi E(\boldsymbol{\lambda}'\mathbf{Z}_1\mathbf{Z}_1'\boldsymbol{\lambda})$ and hence the sequence of probability measures under $\{K_n\}$ is contiguous to that under H_0 (Sen, 1981a).

11.7.11. Verify that the likelihood function in Problem 11.7.10 and the partial likelihood function in (11.7.10) have jointly (under H_0) asymptotically a bivariate normal distribution. Use (4.3.27) to verify the asymptotic normality of the log-partial likelihood under K_n in (11.7.33) (Sen, 1981a).

11.7.12. Use (11.7.31) and the results in the preceding two problems to verify that (11.7.36) holds (Sen, 19781a).

11.7.13. Use (11.7.35) and (11.7.36) to compute the Pitman A.R.E. of the terminal tests based on $\mathcal{L}_{nm}^*$ with respect to the total likelihood-ratio statistic $\mathcal{L}_{nn}$ (Sen, 1981a).

Appendix

A.1. INTRODUCTION

For a smooth reading of the book, proofs of some basic theorems were omitted from the text. These proofs are presented here. First, we consider the Skorokhod-Strassen embedding of Wiener processes for martingales in Section A.2. This is then employed in Section A.3 in the proof of Theorems 2.4.1, 2.4.2, 2.4.4, 2.4.7 and Corollary 2.4.4.1. The last two sections deal with some other results on rank statistics and some statistical tables.

A.2. PROOF OF THEOREM 2.5.1

First we consider the Strassen extension of the basic Skorokhod representation for martingales.

THEOREM A.2.1. *For a martingale* $\{X_n, \mathfrak{F}_n;\ n \geqslant 1\}$ *satisfying* (2.5.1), *there exists a probability space* $(\Omega^*, \mathfrak{B}^*, P^*)$ *with a standard Wiener process* $W = \{W(t), t \in R^+\}$ *and a sequence* $\{T_n;\ n \geqslant 1\}$ *of nonnegative r.v.'s, such that the sequence* $\{X_n\}$ *has the same distribution as* $\{W(\sum_{k \leqslant n} T_k)\}$.

Proof. For sums of i.i.d.r.v.'s, proofs of this basic theorem (with further properties of the T_n) are given in the books of Skorokhod (1965), Breiman (1968), and Freedman (1971), as well as others. We closely follow their treatment and provide the necessary modifications. On $(\Omega^*, \mathfrak{B}^*, P^*)$, consider a sequence of random vectors $\{(L_n, L_n^*);\ n \geqslant 1\}$ such that $L_n \leqslant 0 \leqslant L_n^*$, $\forall n \geqslant 1$, and define

$$T_n = \inf\{t : W(T_0 + \cdots + T_{n-1} + t) - W(T_0 + \cdots + T_{n-1}) \notin (L_n, L_n^*)\}, \qquad \forall n \geqslant 1 \quad \text{(A.2.1)}$$

where $T_0 = 0$. Note that $[T_1 > t] \equiv [W(s) \in (L_1, L_1^*),\ \forall 0 \leqslant s \leqslant t]$, so that

T_1 is properly defined and nonnegative. Also, $\{W(T_1+s)-W(T_1), s \geq 0\}$ is independent of $\{W(\tau), \tau \leq T_1\}$ or $\{W(\tau), \tau \leq T_1; L_1, L_1^*\}$ [as T_1 is $\tilde{\mathfrak{F}}_1$-measurable where $\tilde{\mathfrak{F}}_1 = \mathfrak{F}(L_1, L_1^*, W(s), s \leq T_1)$] and further, $\{W(T_1 + s) - W(T_1), s \geq 0\} \stackrel{\mathfrak{D}}{=} \{W(s), s \geq 0\}$. If, $\tilde{\mathfrak{F}}_k = \mathfrak{F}(L_i, L_i^*, i \leq k, W(\tau), \tau \leq T_1 + \cdots + T_k)$, $k \geq 1$, then, by induction, it follows that T_k is $\tilde{\mathfrak{F}}_k$-measurable, $\{W(T_1 + \cdots + T_k + s) - W(T_1 + \cdots + T_k), s \geq 0\}$ is independent of $\{W(\tau), \tau \leq T_1 + \cdots + T_k; L_i, L_i^*, i \leq k\}$ and $\{W(T_1 + \cdots + T_k + s) - W(T_1 + \cdots + T_k), s \geq 0\} \stackrel{\mathfrak{D}}{=} \{W(s), s \geq 0\}$, $\forall k \geq 1$. Thus the real problem is to choose the sequence $\{L_n, L_n^*; n \geq 1\}$ such that $\{W(T_1 + \cdots + T_n)\}$ has the same distribution as $\{X_n\}$.

Now, $\{W(t), t \in R^+\}$ and $\{W^2(t) - t, t \in R^+\}$ are both martingales, so that by a version of (2.2.12) (for the continuous time-parameter case), for every $k \geq 1$,

$$E\{W(T_1 + \cdots + T_k) - W(T_1 + \cdots + T_{k-1}) \mid \tilde{\mathfrak{F}}_{k-1}\} = 0 \quad \text{a.s.,} \tag{A.2.2}$$

$$\begin{aligned} E\{[W(T_1 + \cdots + T_k) - W(T_1 + \cdots + T_{k-1})]^2 \mid \tilde{\mathfrak{F}}_{k-1}\} \\ = E\{T_k \mid \tilde{\mathfrak{F}}_{k-1}\} \quad \text{a.s.} \end{aligned} \tag{A.2.3}$$

Define $\{Y_n\}$ as in before (2.5.1) [so that $E(Y_n \mid \mathfrak{F}_{n-1}) = 0$ a.s., for every $n \geq 1$]. Then we need to show that for every $n \geq 1$, the conditional distribution of Y_n, given $\mathfrak{F}_{n-1}$, must agree with the conditional distribution of $W(T_1 + \cdots + T_n) - W(T_1 + \cdots + T_{n-1})$, given $\tilde{\mathfrak{F}}_{n-1}$. Note that by (A.2.2), (A.2.3) and the above, we claim that we must have every $n \geq 1$,

$$E\{Y_n^2 \mid \mathfrak{F}_{n-1}\} = E\{T_n \mid \tilde{\mathfrak{F}}_{n-1}\} \quad \text{a.s.,} \tag{A.2.4}$$

so that by (2.5.2) and (A.2.4),

$$\tilde{V}_n = \sum_{k=1}^{n} E\{T_k \mid \tilde{\mathfrak{F}}_{k-1}\} = V_n \quad \text{a.s.,} \quad \forall n \geq 1. \tag{A.2.5}$$

We complete the proof of Theorem A.2.1 by using the following

Lemma A.2.2. For a martingale-difference sequence $\{Y_n\}$, satisfying (2.5.1), for every $n \geq 1$, there are random boundaries $L_n \leq 0 \leq L_n^*$ such that the conditional distribution of Y_n given $\mathfrak{F}_{n-1}$ agrees with the conditional distribution of $W(T_1 + \cdots + T_n) - W(T_1 + \cdots + T_{n-1})$ given $\tilde{\mathfrak{F}}_{n-1}$ (a.s.).

The proof of this lemma runs parallel to the proof of Proposition 13.7 of Breiman (1968, pp. 277–278, dealing with i.i.d.r.v.'s), and hence is omitted.

We return now to the proof of Theorem 2.5.1. Corresponding to the sequence $\{Y_n\}$, we introduce the r.v.'s

$$Y_n^* = \tilde{Y}_n - E(\tilde{Y}_n | \mathfrak{F}_{n-1}), \qquad n \geqslant 1, \tag{A.2.6}$$

with

$$\tilde{Y}_n = \begin{cases} Y_n, & \text{if } Y_n^2 \leqslant f(V_n), \\ (\operatorname{sgn} Y_n)\left[2f^{1/2}(V_n) - \dfrac{f(V_n)}{|Y_n|}\right], & \text{if } Y_n^2 > f(V_n);\ n \geqslant 1, \end{cases} \tag{A.2.7}$$

where V_n and $f(V_n)$ are defined after (2.5.1). Then, $\{Y_n^*\}$ is also a martingale and the following holds: for every $n \geqslant 1$,

$$|\tilde{Y}_n| \leqslant 2f^{1/2}(V_n) \qquad \text{and} \quad |Y_n^*| \leqslant 4f^{1/2}(V_n) \qquad \text{a.s.,} \tag{A.2.8}$$

$$\begin{aligned} E\left[\tilde{Y}_n^4 | \mathfrak{F}_{n-1}\right] &\leqslant 44f(V_n)E\left[\tilde{Y}_n^2|\mathfrak{F}_{n-1}\right] \leqslant 44f(V_n)E(Y_n^2|\mathfrak{F}_{n-1}) \\ &= 44(V_n - V_{n-1})f(V_n); \end{aligned} \tag{A.2.9}$$

$$E(\tilde{Y}_n|\mathfrak{F}_{n-1})^2 \leqslant 4\int_{x^2>f(V_n)} x^2\, dP\{Y_n \leqslant x|\mathfrak{F}_{n-1}\}; \tag{A.2.10}$$

$$|E(Y_n^2|\mathfrak{F}_{n-1}) - E(Y_n^{*2}|\mathfrak{F}_{n-1})| \leqslant 6\int_{x^2>f(V_n)} x^2\, dP\{Y_n \leqslant x|\mathfrak{F}_{n-1}\}. \tag{A.2.11}$$

Thus, writing $V_n^* = \sum_{k=1}^n E(Y_k^{*2}|\mathfrak{F}_{k-1})$, $n \geqslant 1$, we have from the above

$$\begin{aligned} |V_n - V_n^*| &= |\tilde{V}_n - V_n^*| \\ &\leqslant 6\sum_{k=1}^n \int_{x^2>f(V_k)} x^2\, dP\{Y_k \leqslant x|\mathfrak{F}_{k-1}\} \\ &\leqslant 6f(V_n)\left(\sum_{k\geqslant 1}\left[f(V_k)\right]^{-1}\int_{x^2>f(V_k)} x^2\, dP\{Y_k \leqslant x|\mathfrak{F}_{k-1}\}\right) \end{aligned} \tag{A.2.12}$$

where, by (2.5.7), the last term on the rhs of (A.2.12) is finite a.s. Let then

$$g(t) = \left[tf(t)\right]^{1/2}\log(t \vee 2) \qquad \text{and} \quad h(t) = \left[tf(t)\right]^{1/4}\log(t \vee 2), \qquad t > 0, \tag{A.2.13}$$

where $f(t)$ is defined by (2.5.6). Then, by (A.2.12) and (A.2.13), we have

$$|V_n - V_n^*| = o(g(V_n)) \qquad \text{a.s.,} \quad \text{as} \quad n \to \infty. \tag{A.2.14}$$

Further, by (2.5.7), (A.2.7) and (A.2.10),

$$\begin{aligned} &\sum_{n\geqslant 1}\left[h(V_n)\right]^{-2}E\left\{(Y_n - Y_n^*)^2|\mathfrak{F}_{n-1}\right\} \\ &\qquad \leqslant 6\sum_{n\geqslant 1}\left[h(V_n)\right]^{-2}\int_{x^2>f(V_n)} x^2\, dP\{Y_n \leqslant x\,|\,\mathfrak{F}_{n-1}\} < \infty \qquad \text{a.s.} \end{aligned} \tag{A.2.15}$$

Since $\{Y_n - Y_n^*\}$ is also a martingale sequence, by (A.2.15) and the Chow (1965, Theorem 5) law of large numbers for martingales, we have

$$[h(V_n)]^{-1}\Big\{\sum_{k \leqslant n}(Y_k - Y_k^*)\Big\} \to 0 \quad \text{a.s., on } \{V_n \to \infty\}. \quad \text{(A.2.16)}$$

Hence, to prove (2.5.8), it suffices to consider the martingale sequence $\{Y_n^*, \mathfrak{F}_n;\ n \geqslant 1\}$. For this sequence, we adopt Theorem A.1.1 and construct the Wiener process W along with the stopping times $\{T_i\}$, so that (A.2.4) and (A.2.5) hold with the Y_k being replaced by Y_k^*. Further, using the strong Markov property of the Wiener process along with Lemma A.2.2 and (A.2.9), we have $E(T_n^2|\mathfrak{F}_{n-1}^*) \leqslant K_2 E(Y_n^{*4}|\mathfrak{F}_{n-1}) \leqslant K_2^* f(V_n)[V_n - V_{n-1}]$ a.s., for every $n \geqslant 1$, where K_2 and K_2^* are positive constants. Let then $T_n^* = [T_n - E(T_n | \mathfrak{F}_{n-1}^*)]/g(V_n)$, $n \geqslant 1$ (which is also a martingale sequence). Observe that

$$\begin{aligned}\sum_{n \geqslant 1} E(T_n^{*2}|\mathfrak{F}_{n-1}^*) &\leqslant \sum_{n \geqslant 1}[g(V_n)]^{-2} E(T_n^2|\mathfrak{F}_{n-1}^*) \\ &\leqslant K_2^* \sum_{n \geqslant 1}\left[\frac{f(V_n)}{g^2(V_n)}\right](V_n - V_{n-1}) \\ &\leqslant K^* \sum_{n \geqslant 1}\left[(V_n - V_{n-1})/V_n(\log(V_n \vee 2))^2\right] < \infty \quad \text{a.s.,}\end{aligned} \quad \text{(A.2.17)}$$

where $K^* < \infty$. Thus, as in (A.2.15) and (A.2.16), we have

$$[g(V_n)]^{-1}\Big\{\sum_{k \leqslant n}(T_k - E(T_k | \mathfrak{F}_{k-1}^*))\Big\} \to 0 \quad \text{a.s. on } \{V_n \to \infty\}, \quad \text{(A.2.18)}$$

which implies that

$$\Big[\sum_{k \leqslant n} T_k - V_n\Big] = o(g(V_n)) \quad \text{a.s. on } \{V_n \to \infty\}. \quad \text{(A.2.19)}$$

The proof of (2.5.8) is then completed by using Theorem A.2.1, (A.2.19) and the following

Lemma A.2.3. For g and h defined by (A.2.13) and any sequence $\{a_n\}$ of positive numbers with $a_n \uparrow \infty$ as $n \uparrow \infty$, $\forall \epsilon > 0$,

$$P\Big\{\sup_{t \geqslant a_n}\big[\sup\{|W(t) - W(s)|/h(s) : t \leqslant s \leqslant t + o(g(t))\}\big] > \epsilon\Big\} \to 0 \quad \text{as } n \uparrow \infty. \quad \text{(A.2.20)}$$

Proof. Since for any real $a < b$,

$$P\left\{\sup_{a \leqslant t \leqslant b} \frac{|W(t) - W(a)|}{h(t)} > \epsilon\right\}$$

$$\leqslant 4P\{W(b) - W(a) > \epsilon h(a)\}$$

$$= 4P\left\{W(1) > \frac{\epsilon h(a)}{(b-a)^{1/2}}\right\}$$

$$\leqslant [4/\epsilon h(a)]\left[\frac{b-a}{2\pi}\right]^{1/2} \cdot \exp\left\{-\frac{\epsilon^2 h^2(a)}{2(b-a)}\right\}, \qquad \text{(A.2.21)}$$

and by (A.2.13), $h^2(t)/g(t) = \log(t \vee 2)$, the proof of (A.2.20) follows by the usual truncation of the range (a_n, ∞) into a countable number of grid points, using (A.2.21) for each such subinterval and showing that the sum of the rhs of (A.2.21) over these countable number of grid points converges to 0 as $n \to \infty$. Q.E.D.

We conclude this section with the remark that in Theorem A.2.1 and Lemma A.2.2, if the Y_k are independent, then, by construction, the T_k are also so. If further, the Y_k are i.i.d.r.v.'s, then the T_k are also so. In the case of independent Y_i, both (A.2.4) and (A.2.5) relate to nonstochastic quantities.

A.3. PROOFS OF THEOREMS 2.4.1, 2.4.2, 2.4.7 AND COROLLARY 2.4.4.1

Let $S^*_{n,k} = S_{n,k} - \mu_{n,k}, k \geqslant 1$ and $S^*_{n,0} = 0$, where $S_{n,k}$ and $\mu_{n,k}$ are defined by (2.4.1) and (2.4.2). Further, in (2.4.7), we replace the $S_{n,k}$ by $S^*_{n,k}$ and denote the resulting process by W^*_n. Then, by (2.4.8), $\sup_{0 \leqslant t \leqslant 1}|W_n(t) - W^*_n(t)| = \max_{k \leqslant n}|\sum_{r=1}^{k} \mu_{n,r}| \leqslant \sum_{k=1}^{n}|\mu_{n,k}| \xrightarrow{p} 0$ as $n \to \infty$. On the other hand, the $S^*_{n,k}$ form a triangular array of martingales where Theorem 2.4.7 can be applied. Hence, from the above we conclude that whenever Theorem 2.4.7 and (2.4.8) hold, Theorem 2.4.2 also holds. Note that $|S_{n,k_n} - S^*_{n,k_n}| = |\sum_{k=1}^{k_n} \mu_{n,k}| \xrightarrow{p} 0$, by (2.4.3), and hence the asymptotic normality of S_{n,k_n} (in Theorem 2.4.7) ensures (2.4.5). The only difference in this indirect proof of Theorem 2.4.1 is that the second condition in (2.4.3) is slightly less restrictive than its counterpart in (2.4.8); this apparent discrepancy can be immediately resolved if one goes through the proof of Theorem 2.4.7 (given below) and note that just for the asymptotic normality of $W_n(1)$, (2.4.3) suffices. Hence we present only the proofs of Corollary 2.4.4.1 and Theorem 2.4.7.

Proof of Corollary 2.4.4.1. By definition in (2.4.13), for every $\epsilon > 0$ and $\delta > 0$,

$$P\left[\sup\{|W_n(t) - W_n(s)| : |t - s| < \delta\} > \epsilon\right] \leqslant \sum_{k\delta < 1} P\left[\sup\{|W_n(t) - W_n(k\delta)| : k\delta < t \leqslant (k+1)\delta\} > \epsilon/4\right] \tag{A.3.1}$$

and

$$\sup\{|W_n(t) - W_n(k\delta)| : k\delta < t \leqslant (k+1)\delta\} \leqslant 2\max\left\{|s_n^{-1}(S_r - S_{q_k})| : q_k \leqslant r - 1 \leqslant q_{k+1}\right\}, \tag{A.3.2}$$

where $q_k = \max\{r : s_r^2 \leqslant k\delta s_n^2\}$, for $k \geqslant 0$. Since $\{(S_r - S_{q_k});\ r \geqslant q_k\}$ is a martingale, by using (2.2.9) and (2.2.10) (with $r = s = 2$), the rhs of (A.3.1) is bounded from above by

$$\sum_{k\delta < 1} (16/\epsilon)\left[E\{W_n((k+1)\delta) - W_n(k\delta)\}^2 \cdot P\{|W_n((k+1)\delta) - W_n(k\delta)| > \epsilon/8\}\right]^{1/2}, \tag{A.3.3}$$

where $E[W_n((k+1)\delta) - W_n(k\delta)]^2$ converges to δ and by the convergence of the f.d.d.'s of W_n to those of W, $P\{|W_n((k+1)\delta) - W_n(k\delta)| > \epsilon/8\} \to P\{|W((k+1)\delta) - W(k\delta)| > \epsilon/8\} = 2[1 - \Phi(\epsilon/8\delta^{1/2})]$, where Φ is the standard normal d.f. and $1 - \Phi(a)$ converges exponentially to 0 as $a \to \infty$. Thus, for every $\epsilon > 0$ and $\eta > 0$, there exists a $\delta > 0$, such that $2\delta[1 - \Phi(\epsilon/8\delta^{1/2})] \leqslant (\epsilon\eta)^2/32$, and hence, for n sufficiently large, (A.3.3) is smaller than η. This ensures the tightness of $\{W_n\}$. Q.E.D.

Proof of theorem 2.4.7. Let us define $Z_{n,k}$ and $\mathcal{F}_{n,k}$ as in the text preceding (2.4.19) and let

$$Z_{n,k}^* = Z_{n,k} I(|Z_{n,k}| < \epsilon) - E\left[Z_{n,k} I(|Z_{n,k}| < \epsilon) | \mathcal{F}_{n,k-1}^*\right], \qquad k \geqslant 1; \tag{A.3.4}$$

$$S_{n,k}^* = \sum_{r=1}^{k} Z_{n,r}^*, \qquad k \geqslant 1 \quad \text{and } S_{n,0}^* = Z_{n,0}^* = 0, \tag{A.3.5}$$

where $\mathcal{F}_{n,k}^*$ is the σ-field generated by $Z_{n,r}^*$, $r \leqslant k$, for $k \geqslant 0$. Further, in (2.4.19), replacing the $S_{n,k}$ by $S_{n,k}^*$, we denote the resulting process by W_n^*.

Also, we assume the equivalence of (A), (B), (C), (D), and (E) in (2.4.20) through (2.4.24); the proof follows by some standard steps (see Scott, 1973) and is omitted. Then,

$$\sup\{|W_n(t) - W_n^*(t)| : 0 \leqslant t \leqslant 1\}$$
$$\leqslant \sum_{k=1}^{n} |Z_{n,k}| I(|Z_{n,k}| > \epsilon) + \sum_{k=1}^{n} |E[Z_{n,k} I(|Z_{n,k}| < \epsilon) | \mathfrak{F}_{n,k-1}^*]|. \tag{A.3.6}$$

Now, $E(Z_{n,k}|\mathfrak{F}_{n,k-1}^*) = 0$ a.s. and $\mathfrak{F}_{n,k}^* \subset \mathfrak{F}_{n,k}$ for every k. Hence, $\sum_{k=1}^n |E[Z_{n,k} \cdot I(|Z_{n,k}| < \epsilon)|\mathfrak{F}_{n,k-1}^*]| = \sum_{k=1}^n E[Z_{n,k} I(|Z_{n,k}| > \epsilon)$ $|\mathfrak{F}_{n,k-1}^*]| \leqslant \epsilon^{-1} \sum_{k=1}^n E[Z_{n,k}^2 \cdot I(|Z_{n,k}| > \epsilon)|\mathfrak{F}_{n,k-1}] \overset{p}{\to} 0$, by (2.4.21). Also, $P\{|\sum_{k=1}^n Z_{n,k} I(|Z_{n,k}| > \epsilon)| \neq 0\} \leqslant \sum_{k=1}^n P\{|Z_{n,k}| > \epsilon\} \leqslant \epsilon^{-2} \sum_{k=1}^n$ $E[Z_{n,k}^2 I(|Z_{n,k}| > \epsilon)] \to 0$. Therefore, by (A.2.6), W_n and W_n^* are convergent equivalent and it suffices to show that W_n^* weakly converges to W. At this stage, we make use of the Skorokhod-Strassen representation of martingales considered in the earlier section. By Theorem A.2.1, there exist nonnegative r.v.'s $T_{n,k}, k \geqslant 1$ (and $T_{n,0} = 0$) and a Wiener process $W = \{W(t);\ t \in R^+\}$, such that

$$\left\{ W\left(\sum_{j=0}^{k} T_{n,j} \right), k = 1, \ldots, n \right\} \overset{\mathfrak{D}}{=} \{S_{n,k}^*, k = 1, \ldots, n\}. \tag{A.3.7}$$

Thus, to establish the convergence of the f.d.d.'s of W_n^* to those of W, it suffices to show that on defining $k_n(t)$ as in (2.4.19), for every $t \in [0, 1]$,

$$T_{n,0} + \cdots + T_{n,k_n(t)} \overset{p}{\to} t \qquad \text{as} \quad n \to \infty. \tag{A.3.8}$$

Let $\bar{S}_{n,k} = W(\sum_{j=0}^k T_{n,j})$, $k \geqslant 0$, $\bar{Z}_{n,k} = \bar{S}_{n,k} - \bar{S}_{n,k-1}$ for $k \geqslant 1$, $\bar{\mathfrak{F}}_{n,k}$ be the σ-field generated by $\bar{Z}_{n,r}$, $r \leqslant k$ and let $\mathfrak{F}_{n,k}$ be the σ-field generated by $\bar{Z}_{n,r}, r \leqslant k$ and $W(t)$, $t \leqslant \sum_{j=0}^k T_{n,j}$, for $k \geqslant 0$. Then, by Theorem A.2.1, $E(T_{n,k}|\check{\mathfrak{F}}_{n,k-1}) = E(\bar{Z}_{n,k}^2|\mathfrak{F}_{n,k-1})$ for every $n \geqslant k \geqslant 1$. Hence, for every $\delta > 0, t \in [0, 1]$, by (A.3.7),

$$P\left\{ \left| \sum_{i=1}^{k_n(t)} E[t_{n,i}|\hat{\mathfrak{F}}_{n,i-1}] - t \right| > 2\delta \right\} = P\left\{ \left| \sum_{i=1}^{k_n(t)} E[\bar{Z}_{n,i}^2|\mathfrak{F}_{n,i-1}] - t \right| > 2\delta \right\}$$
$$\leqslant \left\{ \left| \sum_{i=1}^{k_n(t)} E[Z_{n,i}^{*2}|\mathfrak{F}_{n,i-1}^*] - Z_{n,i}^{*2} \right| > \delta \right\}$$
$$+ P\left\{ \left| \sum_{i=1}^{k_n(t)} Z_{n,i}^{*2} - t \right| > \delta \right\}. \tag{A.3.9}$$

By (A.3.4), (2.4.20), and (2.4.21), the second term on the rhs of (A.3.9) converges to 0 as $n \to \infty$, while the first term is bounded from above by $\delta^{-2}\sum_{k=1}^{n} E\{[Z_{n,i}^{*2} - E(Z_{n,i}^{*2} \mid \mathfrak{F}_{n,i-1}^{*})]^2\} \leqslant \delta^{-2}\sum_{i=1}^{n} E[Z_{n,i}^{*4}] \leqslant (2\epsilon/\delta)^2 \sum_{i=1}^{n} E[Z_{n,i}^{*2}]$, and hence, by (2.4.21), it can be made arbitrarily small by choosing (ϵ/δ) small. Thus

$$\sum_{i=1}^{k_n(t)} E(T_{n,i} \mid \check{\mathfrak{F}}_{n,i-1}) \xrightarrow{p} t \quad \text{as} \quad n \to \infty. \tag{A.3.10}$$

Also,

$$\begin{aligned} P&\left\{\left|\sum_{i=1}^{k_n(t)} \left[T_{n,i} - E(T_{n,i} \mid \check{\mathfrak{F}}_{n,i-1})\right]\right| > \delta\right\} \\ &\leqslant \delta^{-2} \sum_{i=1}^{n} E\left[T_{n,i} - E(T_{n,i} \mid \check{\mathfrak{F}}_{n,i-1})\right]^2 \\ &\leqslant \delta^{-2} L_2 \sum_{k=1}^{n} E(\bar{Z}_{n,k}^{4}) = \delta^{-2} L_2 \sum_{k=1}^{n} E(Z_{n,k}^{*4}) \\ &\leqslant L_2(\epsilon/\delta)^2 \sum_{k=1}^{n} E(Z_{n,k}^{*2}) \end{aligned} \tag{A.3.11}$$

and this can be made arbitrarily small by choosing (ϵ/δ) small. Hence (A.3.8) follows from (A.3.10) and (A.3.11). This completes the proof of the convergence of f.d.d.'s of W_n^* (and hence, of W_n) to those of W. The proof of the tightness of W_n or W_n^* then follows directly by adapting the proof of Corollary 2.4.4.1.

Q.E.D.

A.4. SOME A.S. LINEARITY THEOREMS ON NONPARAMETRIC STATISTICS

As has been indicated in the concluding lines of Section 4.5, we expand here (4.5.7) to the case where the domain of b is extended to $|b| \leqslant C(\log n)^k$ for some $k \geqslant 1$ and the "in probability" statement is replaced by suitable rates of convergence insuring a.s. convergence. In this context, we assume that

$$\max_{1 \leqslant i \leqslant N} |c_{Ni}^{*}| = O(N^{-1/2}) \text{ and } \liminf_{N} N^{-1} C_N^2 \geqslant C_0 > 0. \tag{A.4.1}$$

Further, ϕ is absolutely continuous, nondecreasing, and

$$\left|\left(\frac{d}{du}\right)\phi(u)\right| \leqslant K\left[u(1-u)\right]^{-(3/2)+\delta}, \qquad \forall u \in (0,1), \tag{A.4.2}$$

where $\delta > \frac{1}{4}$. Finally, we assume that F belongs to the class $\mathfrak{F}_0$ of all absolutely continuous d.f.'s for which the p.d.f. f and its first derivative f'

are bounded a.e. and

$$\lim_{x \to \pm\infty} f(x)\phi^{(1)}(F(x)) \qquad \text{are finite.} \tag{A.4.3}$$

Let then

$$J_N^* = \{ b : |b| \leqslant C(\log N)^k \}, \qquad \text{for some finite and positive } (C, k). \tag{A.4.4}$$

Then we have the following theorem which is an improved version of an one due to Ghosh and Sen (1972).

THEOREM A.4.1. *Under* (A.4.1) *through* (A.4.3), *for every* $\delta(<\frac{1}{2})$ *and* $\epsilon(>0)$, *there exist a sample size* $N_{\delta\epsilon}$ *and a positive* K_δ $(<\infty)$, *such that*

$$P\left\{ \sup_{b \in J_N^*} |\tilde{L}_N(b) - \tilde{L}_N(0) + bAI^{1/2}(f)\rho_1| > \epsilon \right\} \leqslant K_\delta N^{-s}, \qquad \forall N \geqslant N_{s\epsilon}, \tag{A.4.5}$$

where $\tilde{L}_N(b)$, A, $I(f)$, *and* ρ_1 *are all defined as in Section* 4.5 *and* $s < 2\delta/(1-2\delta)$.

Proof. We work here with the scores $a_N(i)$ defined as $\phi(i/(N+1))$, $i = 1, \ldots, N$ and let

$$\begin{aligned} F_N(x; b) &= (N+1)^{-1} \sum_{i=1}^{N} c(x + bc_{Ni}^* - X_i), \\ S_N^*(x; b) &= \sum_{i=1}^{N} c_{Ni}^* c(x + bc_{Ni}^* - X_i), \end{aligned} \tag{A.4.6}$$

for x and b belonging to the real line R, so that

$$\tilde{L}_N(b) = \int_{-\infty}^{\infty} \phi(F_N(x; b))\, dS_N^*(x; b), \qquad \forall b \in R. \tag{A.4.7}$$

Then, by virtue of (4.5.10), where we replace C by $C(\log N)^k$ and K_ϵ by $K_{N,\epsilon} = (\log N)^k K_\epsilon$, we need to show that for every $b \in J_N^*$ and $s > 0$, there exists an $s' > s$, a finite positive K_s and an $N_{s\epsilon}$, such that

$$P\{ |\tilde{L}_N(b) - \tilde{L}_N(0) + bAI^{1/2}(f)\rho_1| > \epsilon \} \leqslant K_s N^{-s'}, \qquad \forall N > N_{s\epsilon}. \tag{A.4.8}$$

Let us now choose two sequences $\{a_N\}$ and $\{b_N\}$ of real numbers such that $F(a_N) = 1 - F(b_N) = N^{-(1-\delta)}$, where δ is defined as in (A.4.2). Then, by (A.4.1),

$$\begin{aligned} \left| \int_{-\infty}^{a_N} \phi(F_N(x;0))\, dS_N^*(x;0) \right| &\leqslant CN^{1/2} \int_{-\infty}^{a_N} |\phi(F_N(x;0))|\, dF_N(x;0) \\ &\leqslant C^* N^{1/2} \left[F_N(a_N; 0) \right]^{(1/2)+\delta}, \end{aligned}$$

where $C^*(<\infty)$. Since $\delta \in (\frac{1}{4}, \frac{1}{2})$, $(1-\delta)(\frac{1}{2}+\delta) > \frac{1}{2}$, by (2.6.29) and the above, the integral converges to 0 with a probability $\geq 1 - 2N^{-s'}$. A similar case holds for the upper tail (b_N, ∞) and for the case of $F_N(x;b)$ and $S_N^*(x;b)$. On the other hand, we note that by virtue of the above result, for every $\epsilon > 0$, there exists a $c < \infty$, such that as $N \to \infty$,

$$P\left\{ |\tilde{L}_N(b) - \tilde{L}_N(0) + bAI^{1/2}(f)\rho_1 - \sum_{j=1}^{4} I_{Nj}(b)| > \epsilon \right\} \leq cN^{-s'}, \tag{A.4.9}$$

where

$$I_{N1}(b) = \int_{a_N}^{b_N} [\phi(F_N(x;b)) - \phi(F_N(x;0))] dS_N^*(x;b), \tag{A.4.10}$$

$$I_{N2}(b) = \int_{a_N}^{b_N} \phi(F_N(x;0)) d[S_N^*(x;b) - S_N^*(x;0) - bf(x)], \tag{A.4.11}$$

$$I_{N3}(b) = b\int_{a_N}^{b_N} [\phi(F_N(x;0)) - \phi(F(x))] \left[\frac{f'(x)}{f(x)} \right] df(x), \tag{A.4.12}$$

$$I_{N4}(b) = b\left(\int_{-\infty}^{a_N} + \int_{b_N}^{\infty} \right) \phi(F(x)) \left[\frac{f'(x)}{f(x)} \right] df(x). \tag{A.4.13}$$

Now $I_{N4}(b)$ is nonstochastic and by the Schwarz inequality and the definitions of $\{a_N\}$ and $\{b_N\}$, it converges to 0 as $N \to \infty$, uniformly in $b \in J_N^*$. Similarly,

$$I_{N3}^2(b) \leq b^2 I(f) \int_{a_N}^{b_N} [\phi(F_N(x;0)) - \phi(F(x))]^2 dF(x), \tag{A.4.14}$$

where, by (2.6.29) and (A.4.2), for every $b \in J_N^*$, the rhs of (A.4.14) is $O((\log N)^{2k} N^{-1/2})(\to 0$ as $N \to \infty)$ with a probability $\geq 1 - 2N^{-s'}$. Hence, it suffices to show that both $I_{N1}(b)$ and $I_{N2}(b)$ can be bounded by $\epsilon/4$, uniformly in $b \in J_N^*$, with a probability $\geq 1 - 2N^{-s'}$, for every $N \geq N_\epsilon$.

Note that by the monotonicity of $\phi(u)$, $0 < u < 1$ and (A.4.2), the total variation of $\phi(F_N(x;0))$ over $[a_N, b_N]$ is $O(N^{(1-\delta)((1/2)-\delta)})$, where, for $\delta > \frac{1}{4}$, $(1-\delta)(\frac{1}{2}-\delta) < \frac{3}{16}(<\frac{1}{4})$. Thus, for the treatment of $I_{N2}(b)$, integrating it by parts, it suffices to show that for every $s' > 0$, there exist positive constants $K_{s'}^{(1)}$, $K_{s'}^{(2)}$ and a sample size $N_{s'}$, such that for every $N \geq N_{s'}$,

$$P\left\{ \sup_{b \in J_N^*} \sup_{a_N \leq x \leq b_N} |S_N^*(x;b) - S_N^*(x;0) - bf(x)| \geq K_{s'}^{(1)} N^{-1/4} (\log N)^{k+1} \right\}$$
$$\leq K_{s'}^{(2)} N^{-s'}. \tag{A.4.15}$$

Similarly, by (A.4.1), $|S_N^*(x;b)| \leq O(N^{1/2}) F_N(x;b)$ for all real x and

$\{F_N(x;b)[1-F_N(x;b)]\}^{-(1/2)+\epsilon} \leqslant O(N^{(1-\delta)((1/2)-\epsilon)})$, $\epsilon>0$, for every $x\epsilon[a_N, b_N]$ and $b\in J_N^*$, with a probability $\geqslant 1-2N^{-s'}$. Hence, for the treatment of $I_{N1}(b)$, it suffices to show that for every $s'>0$, there exist positive constants $K_{s'}^{(1)}$, $K_{s'}^{(2)}$, and a sample size $N_{s'}$, such that for every $N \geqslant N_{s'}$,

$$P\left\{\sup_{b\in J_N^*}\sup_{a_N\leqslant x\leqslant b_N} N^{1/2}|F_N(x;b)-F_N(x;0)| > K_{s'}^{(1)}N^{-1/4}(\log N)^{k+1}\right\} \leqslant K_{s'}^{(2)}N^{-s'}. \tag{A.4.16}$$

[Note that in the above we need to use (2.6.29) with an ϵ just smaller than δ.] Since by (A.4.1), the c_{Ni}^* are $O(N^{-1/2})$ and $\sum_{i=1}^{N}c_{Ni}^*=0$, the proof of (A.4.16) follows precisely along the line of the proofs of Theorems 7.3.1, 7.3.2, and (7.3.9); the only modification is to write $F_N(x;b)$ as the sum of two components (consisting of the negative and the nonnegative c_{Ni}^*) of which one is nonincreasing and the other is nondecreasing in b, while both are nondecreasing in x. Hence, the details are omitted. For the proof of (A.4.15), we also express $S_N^*(x;b)-S_N^*(x;0)-bf(x)$ as the sum of two components (with the c_{Ni}^* negative and nonnegative, respectively) which enables us to replace the suprema by maxima over a certain number of b and x values, and then, as in the proof of Theorem 7.3.1, use the Bernstein inequality to obtain a power rate for each of these terms. Again, the details are omitted. Q.E.D.

If in (A.4.5), we need a more refined order of the remainder term, that is, we replace ϵ by a term that converges to 0 as $N\to\infty$ (in a certain manner), then we need more stringent regularity conditions on the score function ϕ. Such a theorem has been worked out by Ghosh and Sen (1972).

Let us consider next the case of signed rank statistics $S_N(b)$, and as in (10.3.38), we assume that for some generic constant $K(<\infty)$,

$$\left|\left(\frac{\partial^r}{\partial u^r}\right)\phi(u)\right| \leqslant K(1-u)^{-\delta-r},\ r=0,1,2,\quad u\in(0,1), \tag{A.4.17}$$

where $\delta<\frac{1}{4}$. Define J_N^* as in (A.4.4).

THEOREM A.4.2. *Under* (A.4.17), (A.4.4), *for every* $\delta(<\frac{1}{4})$ *and* $\epsilon(>0)$, *there exist a sample size* $N_{\delta\epsilon}$ *and a positive constant* K_δ $(<\infty)$, *such that,*

$$P\left\{\sup_{b\in J_N^*} N^{-1/2}|S_N(b)-S_N(0)+nbB^*(F)|>\epsilon\right\} \leqslant KN^{-s},\qquad N\geqslant N_{\delta\epsilon}, \tag{A.4.18}$$

where $s<(1-2\delta)/2\delta$ *and* $B^*(F)$ *is defined by* (10.3.41).

Proof. We use the same decomposition as in (5.5.6) and (5.5.7), but then we follow the line of the proof of Theorem A.4.1. The role of $S_N^*(x;b)$

is played here by $H_{N,b}(x)$, defined after (5.5.5). Hence the details are omitted. Q.E.D.

In (A.4.17), if we want a more refined order for the remainder term (i.e., we replace ϵ by a term explicitly converging to 0 as $N \to \infty$), then we need to impose stronger regularity conditions on the score function and the d.f. F. Sen and Ghosh (1971) studied such a result under quite stringent regularity conditions, while Sen (1980c) obtained an intermediate result under slightly more restrictive conditions. We conclude this section with the remark that both (A.4.3) and (A.4.17) hold for a broad class of score functions, including the Wilcoxon scores ($\phi(u) = u$), normal scores where δ can be taken as equal to $\frac{1}{2}$ in (A.4.3) or 0 in (A.4.17) and the exponential scores where $\phi(u)$ behaves as $-\log(1-u)$ and hence (A.4.3) holds.

A.5. Some Selected Tables

We define the Bessel and tied-down Bessel processes as in (2.3.50) and (2.7.26), respectively, and let the lhs of (2.7.24) and (2.7.27) be denoted by $1 - \Psi_k(x)$ and $\Psi_q^0(x)$, respectively. Further, for every $k \geqslant 1$ and $0 < \alpha < 1$, define $\zeta_{k,\alpha}$ and $\zeta_{k,\alpha}^0$ by letting

$$\Psi_k(\zeta_{k,\alpha}) = 1 - \alpha \qquad \text{and} \quad \Psi_k^0(\zeta_{k,\alpha}^0) = 1 - \alpha.$$

Then Table 1, adapted with permission from Kiefer (1959) and DeLong (1980), presents the critical levels $\zeta_{k,\alpha}$ and $\zeta_{k,\alpha}^0$ for some typical k and α.

TABLE 1

	$\alpha = 0.01$		$\alpha = 0.05$		$\alpha = 0.10$	
k	$\zeta_{k,\alpha}$	$\zeta_{k,\alpha}^0$	$\zeta_{k,\alpha}$	$\zeta_{k,\alpha}^0$	$\zeta_{k,\alpha}$	$\zeta_{k,\alpha}^0$
1	2.807	1.628	2.241	1.358	1.960	1.224
2	3.242	1.843	2.695	1.584	2.419	1.454
3	3.562	2.001	3.023	1.748	2.750	1.617
4	3.827	2.133	3.294	1.882	3.023	1.756
5	4.059	2.248	3.530	2.000	3.260	1.875

Consider next the standardized Bessel and tied-down Bessel processes, for which the boundary crossing probabilities were considered in (2.7.29), (2.7.30), and (2.7.31). Let $\Psi_{k,T}^*(c)$ stand for (2.7.31) when B is a k parameter Bessel process and let $\zeta_{k,T,\alpha}^*$ be defined by

$$\Psi_{k,T}^*(\zeta_{k,T,\alpha}^*) = 1 - \alpha.$$

Then, the following table, adapted with permission from DeLong (1981), presents the critical levels $\zeta^*_{k,T,\alpha}$ for some typical values of k, T, and α. For $k = 1$, DeLong has also considered the critical values for the one-sided case, that is,

$$P\left\{ \sup_{1 \leqslant t \leqslant T} t^{-1/2} W(t) \leqslant \zeta^{**}_{T,\alpha} \right\} = 1 - \alpha,$$

and these are presented along with.

			$\zeta^*_{k,T,\alpha}$			
T	α	$\zeta^{**}_{T,\alpha}$	$k = 1$	$k = 2$	$k = 3$	$k = 4$
	.01	2.88	3.12	3.57	3.90	4.18
2	.05	2.22	2.52	3.01	3.35	3.64
	.10	1.87	2.22	2.72	3.07	3.36
	.01	3.08	3.32	3.76	4.09	4.36
5	.05	2.45	2.74	3.22	3.56	3.84
	.10	2.12	2.45	2.94	3.29	3.58
	.01	3.18	3.41	3.85	4.18	4.45
10	.05	2.57	2.85	3.32	3.66	3.94
	.10	2.24	2.57	3.05	3.40	3.69
	.01	3.26	3.48	3.92	4.25	4.52
20	.05	2.65	2.93	3.40	3.74	4.02
	.10	2.33	2.65	3.14	3.48	3.77
	.01	3.33	3.55	3.99	4.32	4.58
50	.05	2.74	3.01	3.48	3.82	4.10
	.10	2.43	2.74	3.22	3.57	3.85
	.01	3.38	3.60	4.04	4.36	4.62
100	.05	2.80	3.07	3.53	3.87	4.14
	.10	2.49	2.80	3.28	3.62	3.90

For various other distributions arising typically in statistical inference problems (such as the normal, student t, variance-ratio, chi-square, binomial, Poisson, and other distributions), the critical values have been tabulated in various statistical tables (e.g., *Biometrika Tables for Statisticians*, E. S. Pearson and H. O. Hartley, Eds. Cambridge University Press, New York, 1966). These tables are not reproduced here.

Bibliography

This bibliography is by no means exhaustive, although it does include those publications noted in the text. For journals frequently referred to, the following abbreviations are used; for other journals and publications, standard abbreviations are adapted.

AISM	*Annals of the Institute of Statistical Mathematics*
AMS	*Annals of Mathematical Statistics*
AP	*Annals of Probability*
AS	*Annals of Statistics*
BC	*Biometrics*
BK	*Biometrika*
CSAB	*Calcutta Statistical Association Bulletin*
CS,A	*Communications in Statistics: Theory & Methods, Series A*
JASA	*Journal or the American Statistical Association*
JAP	*Journal of Applied Probability*
JMA	*Journal of Multivariate Analysis*
JRSS,B	*Journal of the Royal Statistical Society, Series B*
PCPS	*Proceedings of the Cambridge Philosophical Society*
S,A	*Sankhya, Series A (S,B: Series B)*
TC	*Technometrics*
TVP	*Teoria Veroyatnostey i ee Primenyia*
ZWVG	*Zeitschrift für Wahrscheinlichkeitstheorie verwandte Gebiete.*

Abdalimov, B. and Malevic, T. L. (1970). Sharpening the limit theorem for *U*-statistics. *Izv. Acad. Nauk Uzbek, SSR* **14**, 6–12.

Adichie, J. N. (1967). Estimates of regression parameters based on rank tests. *AMS* **38**, 894–904.

Alling, D. W. (1963). Early decision in the Wilcoxon two-sample test. *JASA* **58**, 713–720.

Anderson, T. W. (1960). A modification of the sequential probability ratio test to reduce the sample size. *AMS* **31**, 165–197.

Anderson, T. W. and Darling, D. A. (1952). Asymptotic theory of certain goodness of fit criteria based on stochastic processes. *AMS* **23**, 193–212.

Anscombe, F. J. (1952). Large sample theory of sequential estimation. *PCPS* **48**, 600–607.

Antille, A. (1972). Linearité asymptotique d'une statistique de rang. *ZWVG* **24**, 309–324.

Antille, A. (1979). On the invariance principle for signed rank statistics. *ZWVG* **47**, 315–324.

Armitage, P. (1975). *Sequential Medical Trials*, 2nd ed. Oxford: Blackwell.

Ash, R. B. (1972). *Real Analysis and Probability*. New York: Academic Press.

Bahadur, R. R. (1966). A note on quantiles in large samples. *AMS* **37**, 557–580.

Bahadur, R. R. (1971). *Some Limit Theorems in Statistics*. SIAM Publications in Applied Mathematics, Philadelphia.

Bahadur, R. R. and Savage, L. J. (1956). The nonexistence of certain statistical procedures in nonparametric inference. *AMS* **37**, 1115–1122.

Barr, D. R. and Davidson, T. G. (1973). A Kolmogorov-Smirnov test for censored samples. *TC* **15**, 739–757.

Bartlett, M. S. (1946). The large sample theory of sequential tests. *PCPS* **42**, 239–244.

Berk, R. H. (1966). Limiting behavior of posterior distributions when the model is incorrect. *AMS* **37**, 51–58.

Behnen, K. (1971). Asymptotic optimality and ARE of certain rank-order tests under contiguity. *AMS* **42**, 325–329.

Behnen, K. (1972). A characterization of certain rank-order tests with bounds for the asymptotic relative efficiency. *AMS* **43**, 1122–1135.

Behnen, K. and Neuhaus, G. (1975). A central limit theorem under contiguous alternatives. *AS* **3**, 1349–1354.

Bhattacharyya, B. B. and Sen, P. K. (1977). Weak convergence of the Rao-Blackwell estimator of a distribution function. *AP* **5**, 500–510.

Bhattacharya, P. K. (1974). Convergence of sample paths of normalized sums of induced order statistics. *AS* **2**, 1034–1039.

Bhattacharya, P. K. (1976). An invariance principle in regression analysis. *AS* **4**, 621–624.

Bhattacharya, R. N. and Rao, R. R. (1976). *Normal Approximation and Asymptotic Expansions*. New York: Wiley.

Bhuchongkul, S. (1964). A class of nonparametric tests for independence in bivariate populations. *AMS* **35**, 138–149.

Bickel, P. J. (1967). Some contributions to the theory of order statistics. *Proc. Fifth Berkeley Symp. Math. Statist. Prob.* (Ed. L. LeCam et al.). Los Angeles: University of California Press, Vol. 1, 575–591.

Bickel, P. J. and Wichura, M. J. (1971). Convergence criteria for multi-parameter stochastic processes and some applications. *AMS* **42**, 1656–1670.

Billingsley, P. (1968). *Convergence of Probability Measures*. New York: Wiley.

Billingsley, P. (1971). *Weak Convergence of Measures: Applications in Probability*. SIAM Publication in Applied Mathematics, Philadelphia.

Birnbaum, Z. W. and Marshall, A. W. (1961). Some multivariate Chebyshev inequalities with extensions to continuous parameter processes. *AMS* **32**, 687–703.

Blackman, J. (1955). On the approximation of a distribution function by an empiric distribution. *AMS* **26**, 256–267.

Blom, G. (1962). Nearly best linear estimates of location and scale parameters. In *Contribution to Order Statistics* (Ed. A. E. Sarhan and B. G. Greenberg). New York: Wiley, pp. 34–46.

Bönner, N. (1976). *Sequential Rank Tests for Independence*. Ph.D. dissertation, University of Freiburg, Breisgau, West Germany.

Bönner, N. and Kirschner, H. P. (1977). Note on conditions for weak convergence of von Mises' differentiable statistical functions. *AS* **5**, 405–407.

Bönner, N., Muller-Funk, U., and Witting, H. (1980). A Chernoff-Savage theorem for correlation rank statistics with applications to sequential tests. In *Asymptotic Theory of Statistical Tests and Estimation* (Ed. I. M. Chakravarti). New York: Academic Press, pp. 85–125.

Boos, D. D. (1979). A differential for L-statistics. *AS* **7**, 955–959.

Boos, D. D and Serfling, R. J. (1980). A note on differentials and the CLT and LIL for statistical functions, with applications to M-estimates. *AS* **8**, 618–624.

Bradley, R. A., Martin, D. C., and Wilcoxon, F. (1965). Sequential rank tests, I. Monte Carlo studies of the two-sample procedure. *TC* **7**, 463–483.

Bradley, R. A., Martin, D. C., and Wilcoxon, F. (1966). Sequential rank tests, II. A modified two-sample procedure. *TC* **8**, 615–623.

Braun, H. (1976). Weak convergence of sequential rank statistics. *AS* **4**, 554–575.

Breimen, L. (1968). *Probability*. Reading, Mass.: Addison-Wesley.

Breslow, N. (1969). On large sample sequential analysis with applications to survivorship data. *JAP* **6**, 261–274.

Breslow, N. (1970). A generalized Kruskal-Wallis test for comparing K samples subject to unequal pattern of censorship. *BK* **57**, 579–594.

Brillinger, D. R. (1969). An asymptotic representation of the sample distribution function. *Bull. Amer. Math. Soc.* **75**, 545–547.

Brown, B. M. (1971). Martingale central limit theorems. *AMS* **42**, 59–66.

Butler, C. C. (1969). A test for symmetry using the sample distribution function. *AMS*, **40**, 2209–2210.

Carroll, R. J. (1977a). On the uniformity of sequential procedures. *AS* **5**, 1039–1046.

Carroll, R. J. (1977b). On the asymptotic normality of stopping times based on robust estimators. *S, A* **39**, 355–377.

Carroll, R. J. (1978). On almost sure expansions for M-estimates. *AS* **6**, 314–318.

Chatterjee, S. K. (1962). Sequential inference procedures of Stein's type for a class of multivariate regression problems. *AMS* **33**, 1039–1064.

Chatterjee, S. K. and Sen, P. K. (1973a). Nonparametric testing under progressive censoring. *CSAB* **22**, 13–50.

Chatterjee, S. K. and Sen, P. K. (1973b). On Kolmogorov-Smirnov type tests for symmetry. *AISM* **25**, 288–300.

Chernoff, H. and Savage, I. R. (1958). Asymptotic normality and efficiency of certain nonparametric test statistics. *AMS* **29**, 972–994.

Chernoff, H., Gastwirth, J. L., and Johns, M. V. (1967). Asymptotic distribution of linear combinations of functions of order statistics with applications to estimation. *AMS* **38**, 52–72.

Chibisov, D. M. (1964). Some theorems on the limiting behavior of empirical distribution functions. *Selected Transl. Math. Statist. Prob.* **6**, 147–156.

Chow, Y. S. (1960). A martingale inequality and the law of large numbers. *Proc. Amer. Math. Soc.* **11**, 107–110.

Chow, Y. S. (1965). Local convergence of martingales and the law of large numbers. *AMS* **36**, 552–558.

Chow, Y. S. (1967a). On the expected value of a stopped submartingale. *AMS* **38**, 608–609.

Chow, Y. S. (1967b). On the strong law of large numbers for martingales. *AMS*, **38**, 610.

Chow, Y. S., Hsiung, C. A., and Lai, T. L. (1979). Extended renewal theory and moment convergence in Anscombe's theorem. *AP* **7**, 304–318.

Chow, Y. S. and Robbins, H. (1965). On the asymptotic theory of fixed-width sequential confidence intervals for the mean. *AMS* **36**, 457–462.

Chow, Y. S., Robbins, H., and Siegmund, D. (1971). *Great Expectations: The Theory of Optimal Stopping*. Boston: Houghton Mifflin.

Chow, Y. S., Robbins, H., and Teicher, H. (1965). Moments of randomly stopped sums, *AMS* **36**, 787–799.

Chow, Y. S. and Yu, K. F. (1981). The performance of a sequential procedure for the estimation of the mean. *AS* **9**, 184–188.

Chung, K. L. (1974). *A Course in Probability Theory*, 2nd ed. New York: Academic Press.

Cox, D. R. (1963). Large sample sequential tests of composite hypotheses. *S,A* **25**, 5–12.

Cox, D. R. (1972). Regression models and life tables. *JRSS, B* **34**, 187–220.

Cox, D. R. (1975). Partial likelihoods. *BK* **62**, 269–276.

Csáki, E. (1977). The law of iterated logarithm for normalized empirical distribution function. *ZWVG* **38**, 147–167.

Csörgö, M. and Révész, P. (1975a, b). A new method to prove Strassen type laws of invariance principles, I, II. *ZWVG* **31**, 251–259, 261–269.

Csörgö, M. and Révész, P. (1975c). Some notes on the empirical distribution function and the quantile process. *Coll. Math. Soc. Janos Bolyai* **11**, 53–71.

Csörgö, M. and Révész, P. (1978). Strong approximations of the quantile process. *AS* **6**, 882–894.

Csörgö, M. and Révész, P. (1980). *Strong Approximations in Probability and Statistics*. New York: Academic Press.

Daniels, H. A. (1945). The statistical theory of the strength of bundles of threads. *Proc. Roy. Soc. Ser. A* **183**, 405–435.

Dantzig, G. B. (1940). On the non-existence of tests of "Student's hypothesis" having power function independent of σ. *AMS* **11**, 186–192.

Darling, D. A. (1957). The Kolmogorov-Smirnov, Cramér-von Mises tests. *AMS* **28**, 828–838.

Darling, D. A. and Robbins, H. (1967a). Iterated logarithm inequalities. *Proc. Nat. Acad. Sci. USA* **57**, 1188–1192.

Darling, D. A. and Robbins, H. (1967b). Inequalities for the sequence of sample means. *Proc. Nat. Acad. Sci. USA* **57**, 1577–1580.

Darling, D. A. and Robbins, H. (1967c). Confidence sequences for mean, variances and median. *Proc. Nat. Acad. Sci. USA* **58**, 66–68.

Darling, D. A. and Robbins, H. (1967d). Finding the size of a finite population *AMS* **38**, 1392–1398.

Darling, D. A. and Robbins, H. (1968a). Some further remarks on inequalities for sample sums. *Proc. Nat. Acad. Sci. USA* **60**, 1175–1182.

Darling, D. A. and Robbins, H. (1968b). Some nonparametric sequential tests with power 1. *Proc. Nat. Acad. Sci. USA* **61**, 805–809.

David, H. A. (1980). *Order Statistics*, 2nd ed. New York: Wiley.

Davis, C. E. (1978). A two-sample Wilcoxon test for progressively censored data. *CS, A7*, 389–398.

De Long, D. (1980). Some asymptotic properties of a progressively censored nonparametric test for multiple regression. *JMA* **10**, 363–370.

De Long, D. (1981). Crossing probabilities for a square root boundary by a Bessel process. *CS, A* **10**, 2197–2213.

De Long, E. R. and Sen, P. K. (1981). Estimation of $P\{X < Y\}$ based on progressively truncated versions of the Wilcoxon-Mann-Whitney statistics. *CS, A* **10**, 963–981.

Donsker, M. D. (1951). An invariance principle for certain probability limit theorems. *Mem. Amer. Math. Soc.* **6**, 1–12.

Doob, J. L. (1949). Heuristic approach to the Kolmogorov-Smirnov theorems. *AMS* **20**, 393–403.

Doob, J. L. (1967). *Stochastic Processes*, 2nd ed. New York: Wiley.

Drogin, R. (1972). An invariance principle for martingales. *AMS* **43**, 602–620.

Dufour, R. and Maag, J. R. (1978). Distribution results for modified Kolmogorov-Smirnov statistics for truncated or censored samples. *TC* **20**, 29–32.

Dutta, K. and Sen, P. K. (1971). On the Bahadur representation of sample quantiles in some stationary multivariate autoregressive processes. *JMA* **1**, 167–185.

Durbin, J. (1973). *Distribution Theory for Tests Based on the Sample Distribution Function*, SIAM Publications in Applied Mathematics, Philadelphia.

Dvoretzky, A. (1972). Central limit theorems for dependent random variables. *Proc. Sixth Berkeley Symp. Math. Statist. Prob.* (Ed. L. LeCam et al.). Los Angeles: University of California Press, Vol 2, pp. 513–555.

Dvoretzky, A., Kiefer, J., and Wolfowitz, J. (1953). Sequential decision problems for processes with continuous time parameter. Testing hypotheses. *AMS* **24**, 254–264.

Dvoretzky, A., Kiefer, J., and Wolfowitz, J. (1956). Asymptotic minimax character of the sample distribution function and the classical multinomial estimator. *AMS* **27**, 642–669.

Efron, B. (1967). The two-sample problem with censored data. *Proc. Fifth Berkeley Symp. Math. Statist. Prob.* (Ed. L. LeCam et al.). Los Angeles: University of California Press, Vol 4, pp. 831–854.

Efron, B. (1977). The efficiency of Cox's likelihood function for censored data. *JASA* **72**, 557–565.

Eicker, F. (1970). A new proof of the Bahadur-Kiefer representation of sample quantiles. In *Nonparametric Techniques in Statistical Inference* (Ed. M. L. Puri), New York: Cambridge University Press, pp. 321–342.

Epstein, B. and Sobel, M. (1954). Some theorems relevant to life testing from an exponential distribution. *AMS* **25**, 373–381.

Epstein, B. and Sobel, M. (1955). Sequential life testing in the exponential case. *AMS* **26**, 82–93.

Farrell, R. H. (1966). Bounded length confidence intervals for the p-point of a distribution function, III. *AMS* **37**, 586–592.

Fillippova, A. A. (1961). Mises' theorem on the asymptotic behavior of functionals of empirical distribution functions and its statistical applications. *TVP* **7**, 24–57.

Fraser, D. A. S. (1957). *Nonparametric Methods in Statistics*. New York: Wiley.

Freedman, D. (1971). *Brownian Motion and Diffusion*. San Francisco: Holden-Day.

Gaenssler, P. and Stute, W. (1979). Empirical processes: a survey of results for independent and identically distributed random variables. *AP* **7**, 193–243.

Gardiner, J. C. and Sen, P. K. (1978). Asymptotic normality of a class of time-sequential statistics and applications. *CS, A* **7**, 373–388.

Gardiner, J. C. and Sen, P. K. (1979). Asymptotic normality of a variance estimator of a linear combination of a function of order statistics. *ZWVG* **50**, 205–221.

Geertsema, J. C. (1970). Sequential confidence intervals based on rank tests. *AMS* **41**, 1016–1026.

Gehan, E. A. (1965). A generalized two-sample Wilcoxon test for doubly censored data. *BK* **52**, 203–224.

Ghosh, B. K. (1970). *Sequential Tests of Statistical Hypotheses*. Reading, Mass.: Addison-Wesley.

Ghosh, J. K. (1971). A new proof of the Bahadur representation of quantiles and an application. *AMS* **42**, 1957–1961.

Ghosh, M. (1972). On the representation of linear functions of order statistics. *S, A* **34**, 349–356.

Ghosh, M. and Mukhopadhyay, N. (1979). Sequential point estimation of the mean when the distribution is unspecified. *CS, A* **8**, 637–652.

Ghosh, M. and Mukhopadhyay, N. (1981). Consistency and asymptotic efficiency of two-stage and sequential estimation procedures. *S, A* **43**, (in press).

Ghosh, M. and Sen, P. K. (1970). On the almost sure convergence of von Mises' differentiable statistical functions. *CSAB* **19**, 41–44.

Ghosh, M. and Sen, P. K. (1971a). On a class of rank order tests for regression with partially informed stochastic predictors. *AMS* **42**, 650–661.

Ghosh, M. and Sen, P. K. (1971b). Sequential confidence intervals for the regression coefficient based on Kendall's tau. *CSAB* **20**, 23–36.

Ghosh, M. and Sen, P. K. (1972). On bounded confidence intervals for the regression coefficient based on a class of rank statistics. *S, A* **34**, 33–52.

Ghosh, M. and Sen, P. K. (1973). On some sequential simultaneous confidence intervals procedures. *AISM* **25**, 123–134.

Ghosh, M. and Sen, P. K. (1976). Asymptotic theory of sequential tests based on linear functions of order statistics. In *Essays in Prob. & Statist: Ogawa Vol.* (Ed. S. Ikeda et al.). Tokyo: Tsuso, pp. 480–496.

Ghosh, M. and Sen, P. K. (1977). Sequential rank tests for regression. *S, A* **39**, 45–62.

Ghosh, M., Sinha, B. K., and Mukhopadhyay, N. (1976). Multivariate sequential point estimation. *JMA* **6**, 281–294.

Goodman, L. A. (1953). Sequential sampling tagging for population size problems. *AMS* **24**, 56–69.

Grams, W. F. and Serfling, R. J. (1973). Convergence rates for *U*-statistics and related statistics. *AS* **1**, 153–160.

Hájek, J. (1961). Some extensions of the Wald-Wolfowitz-Noether theorem. *AMS*, **32**, 506–523.

Hájek, J. (1962). Asymptotically most powerful rank order tests. *AMS* **33**, 1124–1147.

Hájek, J. (1963). Extensions of the Kolmogorov-Smirnov tests to regression alternatives. *Proc. Bernoulli-Bayes-Laplace Seminar* (Ed. L. LeCam), Los Angeles: University of California Press, pp. 45–60.

Hájek, J. (1968). Asymptotic normality of simple linear rank statistics under alternatives. *AMS* **39**, 325–346.

Hájek, J. (1974). Asymptotic sufficiency of the vector of ranks in the Bahadur sense. *AS* **2**, 75–83.

Hájek, J. and Šidák, Z. (1967). *Theory of Rank Tests*. New York: Academic Press.

Hall, P. (1977). Martingale invariance principles. *AP* **5**, 875–887.

Hall, P. (1979). On the invariance principle for *U*-statistics. *Stoch. Proc. Appl.* **9**, 163–174.

Hall, W. J. (1969). Embedding submartingales in Wiener processes with drift, with applications to sequential analysis. *JAP* **6**, 612–623.

Hall, W. J. (1970). On Wald's equation in continuous time. *JAP* **7**, 59–68.

Hall, W. J. (1974). Two asymptotically efficient sequential *t*-tests. *Proc. Prague Conf. Asympt. Methods*. (Ed. J. Hájek). Prague: Academia, pp. 89–108.

Hall, W. J. and Loynes, R. M. (1977). Weak convergence of processes related to likelihood ratios. *AS* **5**, 330–341.

Halmos, P. R. (1946). The theory of unbiased estimation. *AMS* **17**, 34–43.

Halperin, M. (1960). Extensions of the Wilcoxon-Mann-Whitney test to samples censored at the same fixed point. *JASA* **55**, 125–138.

Halperin, M. and Ware, J. (1974). Early decision in a censored Wilcoxon two-sample test for accumulating survival data. *JASA* **69**, 414–422.

Hodges, J. L., Jr. and Lehmann, E. L. (1963). Estimates of location based on rank tests. *AMS* **34**, 598–611.

Hoeffding, W. (1948a). A class of statistics with asymptotically normal distribution. *AMS* **19**, 293–325.

Hoeffding, W. (1948b). A nonparametric test for independence. *AMS* **19**, 546–557.

Hoeffding, W. (1963). Probability inequalities for sums of bounded random variables. *JASA* **58**, 13–30.

Huber, P. J. (1964). Robust estimation of a location parameter. *AMS* **35**, 73–101.

Huber, P. J. (1969). Theorie de l'inference statistique robuste. *Seminar de Math. Superieures*. University of Montreal.

Huber, P. J. (1973). Robust regression: asymptotics, conjectures and Monte Carlo. *AS* **1**, 799–821.

Huber, P. J. (1977). Robust methods of estimation of regression coefficients. *Oper. Statist. Ser. Statist.* **8**, 41–53.

Huber, P. J. (1981). Robust Statistics. New York: Wiley.

Hušková, M. (1970). Asymptotic distribution of simple linear rank statistics for testing symmetry. *ZWVG* **12**, 308–322.

Hušková, M. (1981). On bounded length sequential confidence interval for parameter in regression model based on ranks. *Coll Nonparametric Infer., Janos Bolyai Math. Soc.* (to appear).

Jaeckel, L. A. (1971). Robust estimation of location: symmetry and asymmetric contamination. *AMS* **42**, 1020–1034.

Jung, J. (1962). Approximation to the best linear estimates. In *Contributions to Order Statistics*. (Ed. A. E. Sarhan and B. G. Greenberg). New York: Wiley, pp. 28–33.

Jurečková, J. (1969). Asymptotic linearity of a rank statistic in regression parameter. *AMS* **40**, 1889–1900.

Jurečková, J. (1971a). Nonparametric estimates of regression coefficients. *AMS* **42**, 1328–1338.

Jurečková, J. (1971b). Asymptotic independence of rank test statistic for testing symmetry on regression. *S, A* **33**, 1–18.

Jurečková, J. (1973). Central limit theorem for Wilcoxon rank statistic process. *AS* **1**, 1046–1060.

Jurečková, J. (1977). Asymptotic relations of M-estimates and R-estimates in linear regression models. *AS* **5**, 664–672.

Jurečková, J. (1980). Asymptotic representation of M-estimators of location. *Oper. Statist. Ser. Statist.* **11**, 61–73.

Jurečková, J. and Sen, P. K. (1981a). Invariance principles for some stochastic processes related to M-estimators and their role in sequential statistical inference. *S, A* **43**, (In press.)

Jurečková, J. and Sen, P. K. (1981b). Sequential procedures based on M-estimators with discontinuous score functions. *J. Statist. Plan. Infer.* **5**, in press.

Jurečková, J. and Sen, P. K. (1981c). M-estimators and L-estimators of location: uniform integrability and asymptotically risk-efficient sequential versions. (to appear).

Kac, M., Kiefer, J., and Wolfowitz, J. (1955). On tests of normality and other tests of goodness of fit based on distance methods. *AMS* **26**, 189–211.

Kiefer, J. (1959). K-sample analogue of the Kolmogorov-Smirnov and Cramér-von Mises tests. *AMS* **30**, 420–447.

Kiefer, J. (1961). On large deviations of the empiric D. F. of vector chance variables and a law of iterated logarithm. *Pacific J. Math.* **11**, 649–660.

Kiefer, J. (1967). On Bahadur's representation of sample quantiles. *AMS* **38**, 1323–1342.

Kiefer, J. (1970). Deviations between the sample quantile process and the sample D. F. In *Nonparametric Techniques in Statistical Inference* (Ed. M. L. Puri). New York: Cambridge University Press, pp. 299–319.

Kiefer, J. (1972a). Iterated logarithm analogues for sample quantiles when $p_n \downarrow 0$. *Proc. Sixth Berkeley Symp. Math. Statist. Prob.* (Ed. L. LeCam et al.). Los Angeles: University of California Press. Vol 1, 227–244.

Kiefer, J. (1972b). Skorokhod embedding of multivariate rv's and the sample DF. *ZWVG* **24**, 1–35.

Kingman, J. F. C. (1969). An ergodic theorem. *Bull. London Math. Soc.* **1**, 339–340.

Komlos, J., Major, P., and Tusnady, G. (1975). An approximation of partial sums of independent r.v.'s and the sample DF. *ZWVG* **32**, 111–131.

Koul, H. L. (1969). Asymptotic behavior of Wilcoxon type confidence regions in multiple regression. *AMS* **40**, 1950–1979.

Koul, H. L. (1970). Some convergence theorems for ranks and weighted empirical cumulatives. *AMS* **41**, 1768–1773.

Kóziol, J. A. and Byar, D. P. (1975). Percentage points of the asymptotic distributions of one- and two-sample K-S statistics for truncated or censored data. *TC* **17**, 507–510.

Kóziol, J. A. and Petkau, A. J. (1978). Sequential testing of equality of two survival distributions using the modified Savage statistics. *BK* **65**, 615–623.

Laha, R. G. and Rohatgi, V. K. (1979). *Probability Theory*. New York: Wiley.

Lai, T. L. (1975a). On Chernoff-Savage statistics and sequential rank tests. *AS* **3**, 825–845.

Lai, T. L. (1975b). A note on first exit time with applications to sequential analysis. *AS* **3**, 825–845.

Lai, T. L. (1978). Pitman efficiencies of sequential tests and uniform limit theorems in nonparametric statistics. *AS* **6**, 1027–1047.

Lai, T. L. and Siegmund, D. (1977). A nonlinear renewal theory with applications to sequential analysis. *AS* **5**, 946–954.

Lai, T. L. and Siegmund, D. (1979). A nonlinear renewal theory with applications to sequential analysis, II *AS* **7**, 60–76.

LeCam, L. (1957). Convergence in distribution of stochastic processes. *Univ. Calif. Publ. Statist.* **2**, 207–236.

LeCam, L. (1960). Locally asymptotically normal families of distributions. *Univ. Calif. Publ. Statist.* **3**, 37–98.

Lindvall, T. (1973). Weak convergence of probability measures and random functions in the function space $D[0, \infty)$. *JAP* **10**, 109–121.

Loeve, M. (1963). *Probability Theory*, 3rd. ed. Princeton: Van Nostrand.

Lombard, F. (1979). Sequential procedures based on Kendall's tau statistics. *South Afrikan. J. Statist.* **11**, 79–87.

Loynes, R. M. (1970). An invariance principle for reverse martingales. *Proc. Amer. Math. Soc.* **25**, 56–64.

Loynes, R. M. (1978). On the weak convergence of U-statistics processes and of the empirical processes. *PCPS(Math.)* **83**, 269–272.

Majumdar, H. (1977). Rank order tests for multiple regression for grouped data under progressive censoring. *CSAB* **26**, 1–16.

Majumdar, H. and Sen, P. K. (1977). Rank order tests for grouped data under progressive censoring. *CS, A* **6**, 507–524.

Majumdar, H. and Sen, P. K. (1978a). Nonparametric tests for multiple regression under progressive censoring. *JMA* **8**, 73–95.

Majumdar, H. and Sen, P. K. (1978b). Nonparametric testing for simple regression under progressive censoring with staggering entry and random withdrawal. *CS, A* **7**, 349–371.

Majumdar, H. and Sen, P. K. (1978c). Invariance principles for jackknifing U-statistics for finite population sampling and some applications. *CS, A* **7**, 1007–1025.

McLeish, D. (1974). Dependent central limit theorems and invariance principles. *AP* **2**, 620–628.

Meyer, P. A. (1966). *Probability and Potentials*. Blaisdell: Waltham.

Miller, R. G., Jr. (1970). A sequential rank test. *JASA* **65**, 1554–1561.

Miller, R. G., Jr. (1972). Sequential rank tests: one sample case. *Proc. Sixth Berkeley Symp. Math. Statist. Prob.* (Ed. L. LeCam et al.). Los Angeles: University of California Press, Vol. 1, pp. 97–108.

Miller, R. G., Jr. and Sen, P. K. (1972). Weak convergence of U-statistics and von Mises' differentiable statistical functions. *AMS* **43**, 31–41.

Moore, D. S. (1968). An elementary proof of asymptotic normality of linear functions of order statistics. *AMS* **39**, 263–265.

Mukhopadhyay, N. (1980). Consistent and asymptotically efficient two-stage procedure to construct fixed-width confidence intervals for the mean. *Metrika* **27**, 281–284.

Müller, D. W. (1968). Verteilungs Invarianzprinzipien für das Gesetz der grossen Zahlen. *ZWVG* **10**, 173–192.

Müller, D. W. (1970). Nonstandard proofs of invariance principles in probability theory. In *Applications of Model Theory to Algebra, Analysis and Probability*. New York: Holt, Reinehart & Winston, pp. 186–194.

Müller-Funk, U. (1979). Nonparametric sequential tests for symmetry. *ZWVG* **46**, 325–342.

Nandi, H. K. and Sen, P. K. (1963). On the properties of U-statistics when the observations are not independent. Part Two: Unbiased estimation of the parameters of a finite population. *CSAB* **12**, 125–148.

Neuhaus, G. (1971). On weak convergence of stochastic processes with multidimensional time parameter. *AMS* **42**, 1285–1295.

Neuhaus, G. and Sen, P. K. (1977). Weak convergence of tail-sequence processes for sample distributions and averages. *Mitteilungen Math. Seminar Giessen* **123**, 25–35.

O'Reilly, N. E. (1974). On the weak convergence of empirical processes in sup-norm metrics. *AP* **2**, 642–651.

Parthasarathy, K. R. (1967). *Probability Measures on Metric Spaces*. New York: Academic.

Peto, R. and Peto, J. (1973). Asymptotically efficient rank invariant test procedures. *JRSS, B* **35**, 187–207.

Prokhorov, Yu. V. (1956). Convergence of random processes and limit theorems in probability. *TVP* **1**, 157–214.

Puri, M. L. and Sen, P. K. (1971). *Nonparametric Methods in Multivariate Analysis*. New York: Wiley.

Pyke, R. (1970). Asymptotic results for rank statistics. In *Nonparametric Techniques in Statistical Inference*. (Ed. M. L. Puri) New York: Cambridge University Press, pp. 21–37.

Pyke, R. (1975). Multi-dimensional empirical processes: Some comments. In *Statistical Inference and Related Topics*. (Ed. M. L. Puri) New York: Academic Press, pp. 45–58.

Pyke, R. and Shorack, G. R. (1968a). Weak convergence of a two-sample empirical process and a new approach to Chernoff-Savage theorems. *AMS* **39**, 755–771.

Pyke, R. and Shorack, G. R. (1968b). Weak convergence and a Chernoff-Savage theorem for random sample sizes. *AMS* **39**, 1675–1685.

Rao, P. V. and Littell, R. C. (1976). An estimator of relative potency. *CS, A* **5**, 183–189.

Rao, P. V., Schuster, E. F., and Littell, R. C. (1975). Estimation of shift and center of symmetry based on Kolmogorov-Smirnov statistics. *AS* **3**, 862–873.

Reynolds, M. R. (1975). A sequential signed-rank test for symmetry. *AS* **3**, 382–400.

Robbins, H. (1959). Sequential estimation of the mean of a normal population. In *Probability and Statistics*. (H. Cramer Vol.), Uppsala: Almquist & Wicsell, pp. 235–245.

Robbins, H. (1970). Statistical methods related to the law of the iterated logarithm. *AMS* **41**, 1397–1409.

Robbins, H. and Siegmund, D. (1968). Iterated logarithm inequalities and related statistical procedures. *Math. Decis. Sci.* **2**, 267–279 (American Mathematical Society).

Robbins, H. and Siegmund, D. (1969). Probability distributions related to the law of iterated logarithm. *Proc. Nat. Acad. Sci. USA* **62**, 11–13.

Robbins, H. and Siegmund, D. (1970). Boundary crossing probabilities for the Wiener process and sample sums. *AMS* **41**, 1410–1429.

Robbins, H. and Siegmund, D. (1974). The expected sample size of some tests of power one. *AS* **2**, 515–536.

Robbins, H., Siegmund, D., and Wendel, J. (1968). The limiting distribution of the last time $S_n \geqslant n\epsilon$. *Proc. Nat. Acad. Sci. USA* **61**, 1228–1230.

Rosenblatt, M. (1952). Limit theorems associated with variants of the von Mises' statistics. *AMS* **23**, 617–623.

Roussas, G. (1972). *Contiguity of Probability Measures: Some Applications in Statistics*. New York: Cambridge University Press.

Rubin, H. and Sethuraman, J. (1965). Probabilities of moderate deviations. *S, A* **27**, 326–346.

Ruymgaart, F. H. (1973). *Asymptotic Theory of Rank Tests for Independence*. Math. Centre Amsterdam, Tract No. 43.

Ruymgaart, F. H. (1974). Asymptotic normality of nonparametric tests for independence. *AS* **2**, 892–910.

Ruymgaart, F. H., Shorack, G. R., and van Zwet, W. R. (1972). Asymptotic normality of nonparametric tests for independence. *AMS* **43**, 1122–1135.

Samuel, E. (1968). Sequential maximum likelihood estimation of the size of a population. *AMS* **39**, 1057–1068.

Samuel-Cahn, E. (1974a). Repeated significance test II for hypotheses about the normal distribution. *CS, A* **3**, 711–733.

Samuel-Cahn, E. (1974b). Repeated significance tests I and II. Generalizations. *CS, A* **3**, 735–743.

Sarhan, A. E. and Greenberg, B. G. (Eds.). (1962). *Contributions to Order Statistics*. New York: Wiley.

Sarkadi, K. (1974). On the convergence of the expectation of the sample quantile. *Coll. Math. Soc. Janos Bolyai* **11**, 341–345.

Savage, I. R. and Sethuraman, J. (1966). Stopping time of a rank order sequential probability ratio test based on Lehman alternatives. *AMS* **37**, 1154–1160.

Savage, I. R. and Sethuraman, J. (1972). Asymptotic distribution of the log likelihood ratio based on ranks in the two sample problem. *Proc. Sixth Berkeley Symp. Math. Statist. Prob.* (Ed. L. LeCam et al.). Los Angeles: University of California Press. Vol. 1, 437–458.

Schey, H. M. (1977). The asymptotic distribution of the one-sided Kolmogorov-Smirnov statistic for truncated data. *CS, A* **6**, 1361–1366.

Schuster, E. F. and Narvarte, J. A. (1973). A new nonparametric estimator of the center of a symmetric distribution. *AS* **1**, 1096–1104.

Scott, D. J. (1973). Central limit theorems for martingales and for processes with stationary increments using a Skorokhod representation approach. *Adv. Appl. Prob.* **5**, 119–137.

Sen, P. K. (1959). On the moments of the sample quantiles. *CSAB* **9**, 1–20.

Sen, P. K. (1960). On some convergence properties of U-statistics. *CSAB* **10**, 1–18.

Sen, P. K. (1961). A note on large sample behaviour of sample extreme values from distributions with finite end-points. *CSAB* **10**, 106–115.

Sen, P. K. (1963). On the estimation of relative potency in dilution (-direct) assays by distribution-free methods. *BC* **19**, 532–552.

Sen, P. K. (1964). On some properties of the rank weighted means. *J. Ind. Soc. Agri. Statist.* **16**, 51–61.

Sen, P. K. (1968). Asymptotic normality of sample quantiles for m-dependent processes. *AMS* **39**, 1724–1730.

Sen, P. K. (1970a). On some convergence properties of one-sample rank order statistics. *AMS* **41**, 2140–2143.

Sen, P. K. (1970b). The Hájek-Rènyi inequality for sampling from a finite population. *S, A* **32**, 181–188.

Sen, P. K. (1970c). A note on order statistics from heterogeneous distributions. *AMS* **41**, 2137–2139.

Sen, P. K. (1972a). On the Bahadur representation of sample quantiles for sequences of ϕ-mixing random variables. *JMA* **2**, 77–95.

Sen, P. K. (1972b). Weak convergence and relative compactness of martingale processes with applications to nonparametric statistics. *JMA* **2**, 345–361.

Sen, P. K. (1972c). A Hájek-Rènyi type inequality for generalized *U*-statistics. *CSAB* **21**, 171–179.

Sen, P. K. (1972d). Finite population sampling and weak convergence to a Brownian bridge. *S, A* **34**, 85–90.

Sen, P. K. (1973a). Asymptotic sequential tests for regular functionals of distribution functions. *TVP* **18**, 235–249.

Sen, P. K. (1973b). An asymptotically optimal test for the bundle strength of filaments. *JAP* **10**, 586–596.

Sen, P. K. (1973c). On weak convergence of empirical processes for random number of independent stochastic vectors. *PCPS* **73**, 135–140.

Sen, P. K. (1973d). On fixed-size confidence bands for the bundle strength of filaments. *AS* **1**, 526–537.

Sen, P. K. (1973e). An almost sure invariance principle for multivariate Kolmogorov-Smirnov statistics. *AP* **1**, 488–496.

Sen, P. K. (1974a). The invariance principle for one sample rank order statistics. *AS* **2**, 49–62.

Sen, P. K. (1974b). Almost sure behaviour of *U*-statistics and von Mises' differentiable statistical functions. *AS* **2**, 387–395.

Sen, P. K. (1974c). Weak convergence of generalized *U*-statistics. *AP* **2**, 90–102.

Sen, P. K. (1974d). On L_p-convergence of *U*-statistics. *AISM* **26**, 55–60.

Sen, P. K. (1975). Rank statistics, martingales and limit theorems. In *Statistical Inference and Related Topics* (Ed. M. L. Puri). New York: Academic Press, pp. 129–158.

Sen, P. K. (1976a). A two-dimensional functional permutational central limit theorem for linear rank statistics. *AP* **4**, 13–26.

Sen, P. K. (1976b). Asymptotically optimal rank order tests for progressive censoring. *CSAB* **25**, 65–78.

Sen, P. K. (1976c). Weak convergence of a tail sequence of martingales. *S, A* **38**, 190–193.

Sen, P. K. (1976d). An almost sure invariance principle for the extrema of certain sample functions. *AP* **4**, 81–88.

Sen, P. K. (1976e). A note on invariance principles for induced order statistics. *AP* **4**, 476–479.

Sen, P. K. (1976f). Weak convergence of progressively censored likelihood ratio statistics and its role in asymptotic theory of life testing. *AS* **4**, 1247–1257.

Sen, P. K. (1977a). Some invariance principles relating to Jackknifing and their role in sequential analysis. *AS* **5**, 315–329.

Sen, P. K. (1977b). Almost sure convergence of generalized *U*-statistics. *AP* **5**, 287–290.

Sen, P. K. (1977c). On Wiener process embedding for linear combinations of order statistics. *S, A* **39**, 138–143.

Sen, P. K. (1977d). Tied-down Wiener process approximations for aligned rank order statistics and some applications. *AS* **5**, 1107–1123.

Sen, P. K. (1978a). Nonparametric repeated significance tests. In *Developments in Statistics* (Ed. P. R. Krishnaiah). New York: Academic Press, Vol. 1, pp. 227–264.

Sen, P. K. (1978b). An invariance principle for linear combinations of order statistics. *ZWVG* **42**, 327–340.

Sen, P. K. (1978c). Invariance principles for rank discounted partial sums and averages. *ZWVG* **42**, 341–352.

Sen, P. K. (1978d). Invariance principles for linear rank statistics revisited. *S, A*, **40**, 215–236.

Sen, P. K. (1979a). Invariance principles for the coupon collector's problem: a martingale approach. *AS* **7**, 372–380.

Sen, P. K. (1979b). Weak convergence of some quantile processes arising in progressively censored tests. *AS* **7**, 414–431.

Sen, P. K. (1979c). Nonparametric repeated significance tests for some analysis of covariance models. *CS, A* **8**, 819–841.

Sen, P. K. (1979d). Nonparametric tests for bivariate interchangeability under competing risks. In *Contributions to Statistics: J. Hájek Mem. Vol.* (Ed. J. Jurečková). Prague: Academia, pp. 211–228.

Sen, P. K. (1979e). Rank analysis of covariance under progressive censoring. *S, A* **41**, 147–169.

Sen, P. K. (1980a). On time-sequential point estimation of the mean of an exponential distribution. *CS, A* **9**, 27–38.

Sen, P. K. (1980b). On almost sure linearity theorems for signed rank order statistics. *AS* **8**, 313–321.

Sen, P. K. (1980c). Asymptotic theory of some tests for a possible change in the regression slope occurring at an unknown time point. *ZWVG* **52**, 203–218.

Sen, P. K. (1980d). Nonparametric simultaneous inference for some MANOVA models. In *Handbook of Statistics* (Ed. P. R. Krishnaiah) Amsterdam: North Holland, Vol. 1, pp. 673–702.

Sen, P. K. (1980e). On nonparametric sequential point estimation of location based on general rank order statistics. *S, A* **42**, 223–240.

Sen, P. K. (1981a). The Cox regression model, invariance principles for some induced quantile processes and some repeated significance tests. *AS* **9**, 109–121.

Sen, P. K. (1981b). Weak convergence of an iterated renewal process. *JAP* **18**, 291–296.

Sen, P. K. (1981c). Asymptotic theory of some time-sequential tests based on progressively censored quantile processes. In *Statistics and Probability: Essays in honor of C. R. Rao* (Ed. G. Kallianpur, et al.). Amsterdam: North Holland, (in press).

Sen, P. K. (1981d). Rank analysis of covariance under progressive censoring, II. *Proc. Internat. Confer. Statist. Infer. Rel. Topics* (Ed. A. K. M. E. Saleh et al.). Amsterdam: North Holland, (in press).

Sen, P. K. (1981e). Some invariance principles for mixed rank statistics and induced order statistics and some applications. *CS, A* **10**, 1691–1718.

Sen, P. K. (1981f). On asymptotic normality in sequential sampling tagging. (to appear).

Sen, P. K. (1981g). The *UI*-principle and LMP rank tests. *Coll. Nonparametric Infer., Janos Bolyai Math. Soc.* (to appear).

Sen, P. K. (1981h). On invariance principles for LMP conditional test statistics, *CSAB* **30**, in press.

Sen, P. K. (1982a). A renewal theorem for an urn model. *AP* **10**, in press.

Sen, P. K. (1982b). Invariance principles for recursive residuals. *AS* **10**, in press.

Sen, P. K. and Bhattacharyya, B. B. (1976). Asymptotic normality of the extrema of certain sample functions. *ZWVG* **34**, 113–118.

Sen, P. K., Bhattacharyya, B. B., and Suh, M. W. (1973). Limiting behavior of the extrema of certain sample functions. *AS* **1**, 297–311.

Sen, P. K. and Ghosh, M. (1971). On bounded length sequential confidence intervals based on one-sample rank order statistics. *AMS* **42**, 189–203.

Sen, P. K. and Ghosh, M. (1972). On strong convergence of regression rank statistics. *S, A* **34**, 335–348.

Sen, P. K. and Ghosh, M. (1973a). A Chernoff-Savage representation of rank order statistics for stationary ϕ-mixing processes. *S, A* **35**, 153–172.

Sen, P. K. and Ghosh, M. (1973b). A law of iterated logarithm for one sample rank order statistics and some applications. *AS* **1**, 568–576.

Sen, P. K. and Ghosh, M. (1973c). Asymptotic properties of some sequential nonparametric estimators in some multivariate linear models. In *Multivariate Analysis-III* (Ed. P. R. Krishnaiah). New York: Academic, pp. 299–316.

Sen, P. K. and Ghosh, M. (1974a). On sequential rank tests for location. *AS* **2**, 540–552.

Sen, P. K. and Ghosh, M. (1974b). Some invariance principles for rank statistics for testing independence. *ZWVG* **29**, 93–108.

Sen, P. K. and Ghosh, M. (1980). On the Pitman-efficiency of sequential tests. *CSAB* **29**, 65–72.

Sen, P. K. and Ghosh, M. (1981). Sequential point estimation of estimable parameters based on U-statistics. *S, A* **43**, in press.

Sen, P. K. and Krishnaiah, P. R. (1974). On a class of simultaneous rank order tests in MANOCOVA. *AISM* **26**, 135–145.

Sen, P. K. and Puri, M. L. (1970). Asymptotic theory of likelihood ratio and rank order tests in some multivariate linear models. *AMS* **41**, 87–100.

Sen, P. K. and Tsong, Y. (1980). On functional central limit theorems for certain continuous time-parameter stochastic processes. *JMA* **10**, 371–378.

Sen, P. K. and Tsong, Y. (1981). An invariance principle for progressively truncated likelihood ratio statistics. *Metrika* **28** (in press).

Serfling, R. J. (1974). Probability inequalities for the sum in sampling without replacement. *AS* **2**, 39–48.

Serfling, R. J. (1980). *Approximation Theorems of Mathematical Statistics*. New York: Wiley.

Setheraman, J. (1970). Stopping time of a rank order sequential probability ratio test based on Lehmann alternatives, II. *AMS* **41**, 1322–1333.

Shorack, G. R. (1969). Asymptotic normality of linear combinations of order statistics. *AMS* **40**, 2041–2050.

Shorack, G. R. (1972a). Functions of order statistics. *AMS* **43**, 412–427.

Shorack, G. R. (1972b). Convergence of quantile and spacing processes with applications. *AMS* **43**, 1400–1411.

Shorack, G. R. and Wellner, J. A. (1978). Linear bounds on the empirical distribution function, *AP* **6**, 349–353.

Sinha, A. N. and Sen, P. K. (1979a). Progressively censored tests for clinical experiments and life testing problems based on weighted empirical distributions. *CS, A* **8**, 819–841.

Sinha, A. N. and Sen, P. K. (1979b). Progressively censored tests for multiple regression based on weighted empirical distributions. *CSAB* **28**, 57–82.

Sinha, A. N. and Sen, P. K. (1981). Tests based on empirical processes for progressive censoring schemes with staggering entry and random withdrawal. (to appear).

Skorokhod, A. V. (1956). Limit theorems for stochastic processes. *TVP* **1**, 261–290.

So, Y. C. and Sen, P. K. (1981). Repeated significance tests based on likelihood ratio statistics. *CS, A* **10**, 2149–2176.

Sproule, R. N. (1969). A sequential fixed width confidence interval for the mean of a *U*-statistic. Ph.D. dissertation, University of North Carolina, Chapel Hill.

Sproule, R. N. (1974). Asymptotic properties of *U*-statistics. *Trans. Amer. Math. Soc.* **199**, 55–64.

Starr, N. (1966). On the asymptotic efficiency of a sequential point estimation. *AMS* **37**, 1173–1185.

Starr, N. and Woodroofe, M. (1969). Remarks on sequential point estimation. *Proc. Nat. Acad. Sci. USA* **63**, 285–288.

Starr, N. and Woodroofe, M. (1972). Further remarks on sequential point estimation: the exponential case. *AMS* **43**, 1147–1154.

Stein, C. (1945). A two-sample test for a linear hypothesis whose power function is independent of σ. *AMS* **16**, 243–258.

Steyn, H. S. and Geertsema, J. C. (1974). Nonparametric confidence sequence for the centre of a symmetric distribution. *South Afrikan J. Statist.* **8**, 25–34.

Stigler, S. M. (1969). Linear functions of order statistics. *AMS* **40**, 770–784.

Stigler, S. M. (1974). Linear functions of order statistics with smooth weight functions. *AS* **2**, 676–693.

Stone, C. (1961). Limit theorems for birth and death processes and diffusion processes. Ph.D. dissertation, Stanford University.

Stone, C. (1963). Weak convergence of stochastic processes defined on a semi-finite time interval. *Proc. Amer. Math. Soc.* **14**, 694–696.

Stout, W. F. (1974). *Almost Sure Convergence*. New York: Academic.

Straf, M. L. (1972). Weak convergence of stochastic processes with several parameters. *Proc. Sixth Berkeley Symp. Math. Stat. Prob.* (Ed. L. LeCam et al.). Los Angeles: University of California Press, Vol. 2, 187–222.

Strassen, V. (1964). An invariance principle for the law of iterated logarithm. *ZWVG* **3**, 211–226.

Strassen, V. (1967). Almost sure behavior of sums of independent random variables and martingales. *Proc. Fifth Berkeley Symp. Math. Stat. Prob.* (Ed. L. LeCam, et al.). Los Angeles: University of California Press, Vol. 2, pp. 315–343.

Suh, M. W., Bhattacharyya, B. B., and Grandage, A. H. E. (1970). On the distribution and moments of the strength of a bundle of filaments. *JAP* **7**, 712–720.

Tucker, H. G. (1967). *A Graduate Course in Probability*. New York: Academic Press.

Van Eeden, C. (1972). An analogue for signed rank statistics of Jurečková's asymptotic linearity theorems for rank statistics. *AMS* **43**, 791–802.

Von Bhar, B. (1965). On the convergence of moments in the central limit theorem. *AMS* **36**, 808–818.

Von Mises, R. (1947). On the asymptotic distribution of differentiable statistical functions. *AMS* **18**, 309–348.

Van Zwet, W. R. (1980). A strong law for linear functions of order statistics. *AP* **8**, 986–990.

Wald, A. (1947). *Sequential Analysis*. New York: Wiley.

Wald, A. and Wolfowitz, J. (1948). Optimal character of the sequential probability ratio test. *AMS* **19**, 326–339.

Weed, H. D. and Bradley, R. A. (1971). Sequential one sample grouped signed rank test for symmetry. *JASA* **66**, 321–326.

Wellner, J. A. (1974). Convergence of the sequential uniform empirical process with bounds for centered Beta r.v.'s and a log-log law. Ph.D. dissertation, University of Washington, Seattle.

Wellner, J. A. (1977a). A martingale inequality for the empirical process. *AP* **5**, 303–308.

Wellner, J. A. (1977b). A law of iterated logarithm for functions of order statistics. *AS* **5**, 481–494.

Wellner, J. A. (1977c). A Glivenko-Cantelli theorem and strong law of large numbers for functions of order statistics. *AS* **5**, 473–480.

Whitt, W. (1970). Weak convergence of probability measures on the function space $C[0, \infty)$. *AMS* **41**, 939–944.

Wilcoxon, F., Rhodes, L. J., and Bradley, R. A. (1963). Two sequential two-sample grouped rank tests with applications to screening experiments. *Biometrics* **19**, 58–84.

Williams, G. W. and Sen, P. K. (1973). Asymptotically optimal sequential estimation of regular functionals of several distributions based on generalized U-statistics. *JMA* **3**, 469–482.

Williams, G. W. and Sen, P. K. (1974). On bounded maximum width sequential confidence ellipsoids based on generalized U-statistics. *JMA* **4**, 453–468.

Woodroofe, M. (1977). Second order approximations for sequential point and interval estimation. *AS* **5**, 984–995.

Yohai, V. J. (1974). Robust estimation in the linear model. *AS* **2**, 562–567.

Yohai, V. J. and Maronna, R. A. (1979). Asymptotic behavior of M-estimators for the linear model. *AS* **7**, 258–268.

Author Index

Abdalimov, B., 398
Adichie, J. N., 119, 399
Alling, D. W., 399
Anderson, T. W., 43, 399
Anscombe, F. J., 299
Antille, A., 301, 331, 399
Armitage, P., 399
Ash, R. B., 13, 399

Bahadur, R. R., 5, 138, 170, 171, 172, 350, 353, 399
Barr, D. R., 399
Bartlett, M. S., 253, 399
Behnen, K., 99, 157, 399
Berk, R. H., 2, 50, 399
Bernstein, S. N., 173, 319
Bhattacharya, P. K., 201, 399
Bhattacharya, R. N., 6, 399
Bhattacharyya, B. B., 3, 4, 6, 221, 229, 399, 408
Bhuchongkul, S., 163, 399
Bickel, P. J., 12, 18, 20, 22, 38, 399
Billingsley, P., 2, 12, 21, 22, 23, 24, 60, 95, 399
Birnbaum, Z. W., 15, 16, 32, 185, 399
Blackman, J., 224, 225, 400
Blom, G., 400
Bönner, N., 163, 165, 272, 400
Boos, D. D., 400
Bradley, R. A., 400, 412, 413
Braun, H., 3, 39, 41, 105, 108, 400
Breiman, L., 385, 386, 400
Breslow, N., 400
Brillinger, D. R., 400
Brown, B. M., 4, 12, 14, 29, 30, 31, 95, 400
Butler, C. C., 225, 400
Byar, D. P., 357, 378, 406

Carroll, R. J., 400
Chatterjee, S. K., 3, 7, 225, 342, 400
Chernoff, H., 1, 97, 114, 163, 198, 200, 400
Chibisov, D. M., 25, 39, 400
Chow, Y. S., 13, 14, 280, 282, 303, 310, 401
Chung, K. L., 401
Cox, D. R., 3, 7, 253, 342, 371, 401
Csáki, E., 40, 401
Csorgo, M., 35, 40, 198, 401

Daniels, H. A., 217, 401
Dantzig, G. B., 280, 282, 401
Darling, D. A., 6, 233, 234, 245, 294, 399, 401
David, H. A., 170, 402
Davidson, T. G., 399
Davis, C. E., 342, 402
DeLong, D., 45, 46, 47, 368, 396, 397, 402
DeLong, E. R., 402
Donsker, M. D., 4, 32, 402
Doob, J. L., 13, 14, 39, 41, 42, 74, 402
Drogin, R., 12, 402
Dufour, R., 402
Durbin, J., 402
Dutta, K., 176, 402
Dvoretzky, A., 27, 39, 402

Efron, B., 402
Eicker, F., 176, 402
Epstein, B., 376, 402

Farrell, R. H., 402
Fillippova, A. A., 402
Fraser, D. A. S., 402
Freedman, D., 385, 402

Gaenssler, P., 40, 402
Gardiner, J. C., 3, 7, 197, 200, 339, 342, 403
Gastwirth, J. L., 198, 200, 400
Geertsema, J. C., 403, 412
Gehan, E. A., 403
Ghosh, B. K., 253, 403
Ghosh, J. K., 176, 403
Ghosh, M., 2, 3, 5, 6, 40, 56, 90, 93, 110, 114, 120, 129, 132, 140, 148, 151, 155, 160, 198, 272, 275, 279, 306, 310, 329, 331, 332, 393, 395, 396, 403, 410, 411
Goodman, L. A., 403
Grams, W. F., 56, 403
Greenberg, B. G., 170, 199, 408

Hájek, J., 1, 2, 7, 13, 86, 90, 97, 98, 99, 120, 123, 148, 198, 356, 403, 404
Hall, P., 12, 404
Hall, W. J., 265, 404
Halmos, P. R., 50, 404
Halperin, M., 342, 404
Hartley, H. O., 397
Hodges, J. L. Jr., 404
Hoeffding, W., 4, 48, 50, 52, 84, 96, 109, 124, 404
Hsiung, C. A., 401
Huber, P. J., 3, 5, 6, 203, 205, 404
Husková, M., 306, 404

Jaekel, L. A., 404
Johns, M. V., 198, 202, 404
Jung, J., 199, 404
Jurecková, J., 3, 5, 6, 117, 124, 142, 211, 216 216,227, 228, 301, 306, 317, 331, 332, 404, 405

Kac, M., 405
Kiefer, J., 5, 6, 26, 27, 39, 46, 138, 176, 396, 402, 405
Kingman, J. F. C., 2, 50, 406
Kirschner, H. P., 400
Kolmogorov, A., 13, 23, 225, 356
Komlos, J., 35, 38, 198, 405
Koul, H. L., 406
Koziol, J. A., 357, 378, 405
Krishnaiah, P. R., 411

Laha, R. G., 406
Lai, T. L., 3, 4, 6, 114, 265, 328, 401, 405, 406
LeCam, L., 17, 98, 99, 406
Lehamnn, E. L., 404
Lindvall, T., 23, 406
Littell, R. C., 225, 226, 407
Loève, M., 13, 407
Loynes, R. M., 4, 12, 30, 48, 265, 404, 406

Maag, J. R., 402
McLeish, D. L., 4, 12, 28, 406
Major, P., 35, 38, 406
Majumdar, H., 7, 278, 342, 359, 360, 363, 368, 380, 381, 406
Malevic, T. L., 398
Maronna, R. A., 413
Marshall, A. W., 15, 16, 32, 185, 399
Martin, D. C., 400
Meyer, P. A., 13, 198, 406
Miller, R. G., Jr., 4, 48, 59, 406
Moore, D., 200, 406
Mukhopadhyay, N., 310, 328, 329, 332, 403, 406
Müller, D. W., 25, 406
Müller-Funk, U., 140, 163, 165, 272, 400, 407

Nandi, H. K., 66, 407
Narvarte, J. A., 225, 408
Neuhaus, G., 12, 18, 20, 38, 99, 399, 407
Noether, G., 69, 93

O'Reilly, N. E., 25, 39, 40, 407

Parthasarathy, K. R., 12, 42, 407
Pearson, E. S., 397
Petkau, A. J., 406
Peto, J., 407
Peto, R., 407
Pitman, E. J. G., 252, 303
Prokhorov, Yu. V., 17, 407
Puri, M. L., 1, 50, 96, 97, 123, 407, 411
Pyke, R., 2, 5, 39, 41, 97, 98, 105, 107, 108, 225, 407

Rao, P. V., 225, 226, 407

Rao, R. R., 6, 399
Rényi, A., 13
Révész, P., 35, 40, 198, 401
Reynolds, M. R., 407
Rhodes, L. J., 413
Robbins, H., 6, 14, 45, 233, 234, 235, 276, 280, 281, 282, 294, 303, 310, 328, 401, 407
Rohatgi, V. K., 406
Rosenblatt, M., 407
Roussas, G., 98, 407
Rubin, H., 408
Ruymgaart, F. H., 163, 408

Samuel, E., 202, 408
Samuel-Cahn, E., 408
Sarhan, A. E., 170, 199, 408
Sarkadi, K., 408
Sacage, I. R., 1, 97, 114, 163, 400, 408
Savage, L. J., 399
Schey, H., 357, 378, 408
Schuster, E. F., 225, 226, 408
Scott, D. J., 12, 31, 391, 408
Sen, P. K., 1-7, 30, 31, 33, 48-50, 56, 59, 66, 70, 80, 84, 90, 93, 97, 110, 114, 120, 123, 129, 132, 145, 147-151, 155, 160, 176, 178, 186, 192, 200-202, 211, 221, 225, 227-230, 243, 252, 261, 272, 275, 279, 306, 310, 317, 331-334, 339, 342, 356, 359, 360, 363, 368, 371, 378-384, 393, 395, 396, 399, 400, 401, 402, 403, 405, 406, 407, 408-411, 413
Serfling, R. J., 56, 400, 403, 411
Sethuraman, J., 408, 411
Shorack, G. R., 2, 3, 19, 20, 25, 38, 39, 41, 97, 98, 105, 107, 108, 163, 186, 407, 408, 411
Sidák, Z., 1, 90, 98, 99, 123, 356, 404
Siegmund, D., 45, 234, 276, 328, 361, 401, 406, 407
Sinha, A. N., 7, 342, 356, 378, 380, 383, 411
Skorokhod, A. V., 12, 32, 33, 56, 64, 109, 132, 187, 305, 411
Smirnov, 225, 256
So, Y. C., 252, 411
Sobel, M., 376, 402
Sproule, R. N., 49, 80, 411
Starr, N., 280, 282, 310, 412
Stein, C., 280, 309, 310, 329, 412
Steyn, H. S., 412
Stigler, S. M., 198, 200, 412
Stone, C., 23, 412
Stout, W. F., 13, 412
Straft, M. L., 12, 18, 20, 412
Strassen, V., 2, 5, 12, 34, 64, 109, 132, 187, 276, 385, 412
Stute, W., 40, 402
Suh, M. W., 3, 227, 410, 412

Teicher, H., 14, 401
Tsong, Y., 3, 7, 33, 411
Tucker, H., 13, 412
Tusnady, G., 35, 38, 198, 406
Van der Wearden, B. L., 123
Van Eeden, C., 141, 142, 412
van Zwet, W. R., 163, 191, 408, 412
von Bhar, B., 333, 412
von Mises, R., 4, 48, 49, 56, 59, 64, 412

Wald, A., 2, 6, 252, 253, 412
Ware, J., 342, 404
Weed, H. D., 412
Wellner, J., 3, 35, 39-41, 186, 198, 200, 407, 411, 412
Wendel, J., 407
Whitt, W., 23, 412
Wichura, M., 12, 18, 20, 22, 38, 399
Wilcoxon, F., 123, 400, 413
Williams, G. W., 413
Witting, H., 163, 165, 272, 400
Wolfowitz, J., 39, 402, 406, 412
Woodroofe, M., 280, 310, 328, 412, 413

Yohai, V. J., 413
Yu, K. F., 310, 401

Subject Index

Aligned signed rank statistic, 145, 262
weak convergence, 145
Almost sure convergence:
bivariate rank statistics, 167, 168
LCFOS, 192, 193, 195, 197
linear rank statistic, 120
signed rank statistic, 146
U-statistics, 50
variance estimators, 80-82, 195, 214, 224, 272
von Mises' functionals, 51
ANOCOVA, 363-369
Anti-ranks, 340, 364
Asymptotically risk efficient, 6, 310, 312
Asymptotic consistency, 258, 284, 296
Asymptotic efficiency, 285, 297
Asymptotic equivalence of M-, R- and L-estimators, 214-217
Asymptotic linearity of rank statistics:
in location parameter, 5, 141, 142, 395
in regression parameters, 5, 114, 116, 117, 118, 393
Asymptotic normality, 2, 27, 29, 59, 90, 99, 107, 127, 219
of stopping times, 290, 297, 312, 322, 326
Asymptotic relative efficiency, 251, 272, 273, 274, 275, 302, 303, 325, 350-354
Asymptotic risk efficiency, 310, 312, 323, 324
Average sample number (ASN), 2, 6, 236, 267, 271, 285, 289, 296-297, 327

Bahadur efficiency, 250, 352, 353
Bahadur representation of sample quantiles, 5, 138, 172, 173
Bessel process, 27, 44-47, 247, 278, 369, 380, 384, 396, 397
Bivariate independence problem, 150, 151, 241, 265, 362
Boundary crossing probabilities for Wiener processes, 41-47, 259, 275, 276
Bounded width confidence intervals, 6, 282, 292, 295
Brownian bridge, 26, 27, 37, 43
for U-statistics, 66-70
Brownian motion, 26, 34, 41-43, 44-45
Brownian sheet, 26, 381
tied down, 26, 27
Bundle strength of filaments, 3, 5, 6, 204, 217

$C[10,1]^p$ space, 4, 16, 17, 23
$C[R^k]$ space, 23, 24
Capture, mark and recapture method, 201, 202
Censoring:
type I, 336
type II, 336
Confidence coefficient, 281
interval, 280, 281
limits, 281
sequences, 306
Contiguity of probability measures, 5, 98, 99, 101, 158, 347
Coverage probability, 280

$D[0,1]^p$ space, 4, 16, 17, 21, 23, 100
$D[R^k]$ space, 23, 24
Discontinuity of the first kind, 19
Distribution-free, 89, 127, 151-153, 295, 342
d_q metric, 25, 32, 33, 63, 96

Embedding of Wiener process, 2, 4, 33, 64, 109, 132, 159, 175, 187
Empirical distribution, 4, 12, 36-41, 49, 105, 121, 134, 143, 164, 209, 393
Empirical process, 4, 12, 36, 41, 105, 106, 121, 226, 357, 361, 393
Estimatable parameter, 48, 49, 71
  confidence interval, 290, 291
  degree, 49, 71
  domain, 49, 71
  point estimation (sequential), 311
  stationary of order *d*, 52
  tests for, 240, 261
Estimator of location:
  Blackman, 224-226
  L-estimator, 217
  M-estimator, 205, 215
  Rao, Schuster, Littell, 225
  R-estimator, 145, 215
  Schuster-Narvarte, 225
Extrema of sample functions, 4, 5, 6, 217, 218

Finite-dimensional distributions, 21, 24
Finite population sampling, 4, 66
Fisher information, 102, 116, 131, 142, 206

Generalized *U*-statistics, 4, 71
  invariance principles, 73-79
Generalized von Mises' functional, 4, 71
  invariance principles, 73-79
Goodness of fit tests, 242

Hypothesis of independence, 150
  randomness, 89
  sign invariance, 127

Induced order statistics, 201
Inequality:
  Bernstein, 173, 319
  Birnbaum-Marshall, 15, 16
  Brown, 14
  Doob, 13, 14
  Hájek-Rényi-Chow, 13
  Hoeffding, 305, 320
  Hölder, 14, 192, 318, 324
  Jensen, 52
  Kolmogorov, 13
Invariance principles strong, 2, 3, 4, 5, 6, 9, 33, 34
  bivariate rank statistics, 160, 164
  empirical process, 38
  extrema of sample functions, 221-223
  LCFOS, 175, 188
  linear rank statistics, 109-113
  martingales, 33-34
  M-estimators, 211-214
  signed rank statistics, 132
  U-statistics and von Mises' functionals, 64-66
Invariance principles weak, 2, 3, 4, 5, 6, 9, 27-33
  bivariate rank statistics, 155-159
  empirical process, 36-41, 105-108
  extrema of sample functions, 221-223
  generalized *U*-statistics and von Mises' functionals, 73-79
  LCFOS, 174, 175, 179-181, 183, 186
  linear rank statistics, 93-96, 101-104, 108
  martingales, 28-31
  M-estimators, 207-211
  progressive censoring, 345-348, 368, 375
  signed rank statistics, 129-131
  U-statistics and von Mises' functionals, 56-63

Kiefer process, 26, 38
Kolmogorov condition, 23

Least squares estimators, 203, 204
L-estimators, 203, 216, 217
Life testing, *see* Time sequential tests
Likelihood ratio, 99, 202, 338, 369
Lindeberg condition, 27, 28, 30
Linear combinations of functions of order statistics, 2, 6, 170, 171, 172
  tests, 250
  type I, 171
  type II, 171, 172
Linear rank statistics, 3, 4, 6, 88

Martingale, 3, 4, 11, 13, 28, 30, 31, 90, 127, 152, 154, 344, 357
Maximum likelihood estimator, 203, 204, 253-254
M-estimators, 3, 5, 6, 203, 301, 316, 317
d_q Metric, 25, 32, 33, 63, 96
ρ_q Metric, 39, 40, 41
Minimum risk estimation, 309
Mixed rank statistics, 5, 150, 152, 154, 159, 163, 166, 201, 362
Modulus of continuity, 18, 22

Moments of randomly stopped martingales, 14, 15

Noether condition, 69, 93

One sample location problem, 236-239, 246, 260-262, 271, 299
Operating characteristic (OC), 2, 6, 257, 258, 260

Permutational central limit theorem, 69
Pitman efficiency, 248, 252, 274
Progressive censoring, 3, 7, 336, 342, 358-359, 365, 372
Progressively censored quantiles process, 7, 369, 370
Proportional harard model, 371

Quantiles, 171
  confidence intervals, 298
  process, 7, 369, 370
  tests, 238, 246, 261

Rank discounted partial sums, 201
Regression model (simple and multiple), 90, 242, 247, 248, 278, 303, 347, 350, 360, 379
Regret function, 327
Regular functional, *see* Estimable parameter
Relative compactness, 17, 21
Repeated significance tests, 6, 233, 243
  type A, 244-249
  type B, 244, 248
R-estimation of location, 145, 203, 204
  regression, 119, 203, 204
Reverse martingales, 3, 4, 11, 13, 30, 50, 177
  submartingale, 4, 13
ρ metric, 39, 40, 41
Robust estimation, 3, 6, 203, 216, 217

Score function, 89, 205
Second order efficiency, 329
Separable metric space, 12, 17
Sequential confidence intervals:
  type A, 280
  type B, 295
Sequential interval estimation, 3, 6, 280
Sequential likelihood ratio tests (SLRT), 233, 253-254
Sequential M-test, 263, 266
Sequential point estimation, 3, 6, 280, 309-326
Sequential probability ratio tests (SPRT), 2, 6, 233, 253
Sequential rank order test (SROT), 233, 252, 261-263, 265, 273
Sequential tests, 3, 6, 233
  type A, 244-248
  type B, 244, 249
  type C, 254
  type D, 255
Sequential tests with power one, 6, 233-243
Signed-rank statistic, 5, 6, 126
Sign invariance, 5, 127
Skorokhod J_1 topology, 19, 20, 57
  metric, 19
Skorokhod-Strassen embedding, 2, 4, 5, 6, 12, 64, 65, 109, 110, 132, 133, 187, 385, 386
Stochastic process, 11, 12, 16
Stopping number (variable and time), 14, 235, 244, 253, 255, 256, 257, 282, 283, 295, 301, 309, 311, 316, 343, 353
  expected, 355-357
Submartingales, 4, 13
  convergence, 14
  inequalities, 14, 16

Termination of sequential tests, 256, 257
Test for population mean, 237, 238, 250, 260, 261
Test with power one, 233, 234, 235-243
Tight down Brownian sheet, 26
Tight down Wiener process, 26
Tightness, 17, 19, 22, 23, 100, 390
Time sequential tests, 7, 233, 336
Trimmed mean, 330
Two-sample problems, 90, 241, 247

Uniform topology, 18
Upcrossing inequality, 13
U-statistics, 2, 3, 6, 50, 182, 311

Variance estimation:
  4, 80-83
  LCFOD, 197
  M-estimator, 210, 214
  R-estimator, 316
  U-statistics and von Mises' functionals, 4, 80-83
  von Mises' functional, 4, 6, 49

Wald-Wolfowitz condition, 69
Weak convergence, 2, 4, 5, 6, 11, 16, 17
Wiener process, 2, 4, 5, 6, 11, 12, 25, 26, 41-44
Winsorized mean, 330